CRACKING THE PHILOSOPHERS' STONE

CRACKING THE PHILOSOPHERS' STONE

Origins, Evolution and Chemistry of Gold-Making

J. Erik LaPort

&

Dr Roger Gabrielsson PhD

Quintessence

CRACKING THE PHILOSOPHERS' STONE
by J. Erik LaPort and Roger Gabrielsson PhD
Copyright © 2009 Quintessence

Paperback Edition (B/W)
ISBN-10: 099061980X
ISBN-13: 978-0-9906198-0-2

All Rights Reserved. No part of this publication may be commercially reproduced, stored in a retrieval system, or transmitted, in any form or in any means – by electronic, mechanical, photocopying, recording or otherwise – without prior written permission from the authors apart from fair-use specified below.

The publisher permits and encourages Fair Use of the material contained in this book for the noncommercial purposes of teaching, research, scholarship, non-profit education, factual or non-fiction publications, criticism or commentary.

First English edition published 2015 by Quintessence

Publication Assistance and Digital Printing in the United States, Australia and/or the UK by Lightning Source®

Initial Edit by Paul Hardacre
www.paulhardacre.com

Final Edit by Adam Mayer, MA
Webster University, Arts and Science

Cover art by Victor Hörnfeldt and Lauren Khu

Cover design by Rachel Turton
www.graphicdesignkohtao.com

Original artwork Copyright © 2010 by Lauren Khu

Layout consultant – Jeff Williams

Digital, Paperback and Hardcover Editions Distributed by Ingram

Leatherbound Collectors' Edition Distributed by Third Order Alchemy
www.third-order-alchemy.com

KEY TO THE COVER ART

Ω	The One / All
☉	Gold / Sol
Q	Quintessence
⊕	Tetrasomia / Elixir
☨	Divine Water
✳	Sal Ammoniac
☉̇	Latten / Red Gum
✣	Eudica / White Gum
♁	Antimony
☿	Molybdochalkon
♀	Copper
▽̶	Earth
△̶	Air
▽	Water
△	Fire
✡	Tetrasomia / Conjunction

The symbol on the front cover encrypts the recipe for the Philosophers' Stone, Aurum Potabile and Alchemical Gold. J. Erik LaPort and Victor Hörnfeldt designed the symbol as a visual representation encapsulating the Art of Alexandrian Judeo-Egyptian *chrysopœia* in a single icon encircled by a Cleopatra-style brazen-serpent / orobouros.

DEDICATION

To my daughter Ayla, and the next generation of philosophers …

To aspirants, who, upon becoming captivated by alchemy's riddle of the Philosophers' Stone, have passionately sought entrance to its mystery …

To adepts who, upon solving the riddle of the Stone, have realized that promises of immortality and unlimited wealth were absolutely true, yet in spirit only …

And to those patient souls who perhaps not fully understanding such a fervent pursuit, nevertheless lent patronage and support to aspirants and adepts throughout their quest …

This book is dedicated to you.

CONTENTS

Acknowledgments i

Foreword iii

Author's Preface v

Chemist's Preface xi

Editor's Note xv

Introducing the Stone xix

Chapter 1 – Sophic Substances 1

Chapter 2 – Moses and the Golden-Calf 73

Chapter 3 – Maria Hebrea's Tincture 117

Chapter 4 – White Stone of Hermes 169

Chapter 5 – Stephanos' *Chrysopœia* 219

Chapter 6 – From *Chrysopœia* to *Al-Kīmyā'* 261

Chapter 7 – Paracelsus and the Alkahest 301

Chapter 8 – Labors of Hercules 359

Chapter 9 – Chemical Wedding 397

Chapter 10 – Powder of Projection 419

Chapter 11 – Aurum Potabile 479

Chapter 12 – Universal Medicine 549

Appendix A – Timeline of Developmental Alchemy 571

Appendix B – Latin Dialogue of Maria and Aros 581

Appendix C – Scorpion Formula for Gold-making 591

Alchemical Imagery Index 601

Afterword 607

ACKNOWLEDGMENTS

To my loving wife Sunisa Kaewyindee (Paw), without whose tremendous patience and understanding, this book would never have been possible. To Victor Hörnfeldt for his patronage and steadfast support throughout all stages of research and writing of this book.

The authors would also like to extend thanks to all bygone and present-day alchemists and alchemy researchers who, despite a tradition of secrecy, have endeavored to bring alchemy and the Philosophers' Stone to the forefront. We have gained inspiration and answers through your efforts, which contributed tremendously to the writing of this book.

The information in this book is derived from hundreds of alchemical and scientific texts and articles that made their way into print dating from several centuries BCE to the present, most of which are in the public domain. We have drawn upon several articles and works published during the 20th and 21st centuries that helped us to either affirm a position therein or expand upon it and have included this information as fair use. The authors felt strongly that rather than relegate contributors to a footnote or endnote, to instead employ in-text attribution that allowed their work to be included as part of the narrative, thus serving as the primary citation method.

Among the libraries, archives and journals consulted whom special thanks is owed are RAMS Digital Library, Ambix, Isis, the Gold Bulletin, Royal Society Publishing, the Alchemy Website, the History of Science and Technology in Islam Website, the Alchemy Journal, Livius.org, Rex Research, Sacred-Texts, the Roger Pearse Blog, the Newton Project, Project Gutenberg Online Literature, Perseus Digital Library, the Internet

Archive and several others too numerous to list. These fine primary sources provided access to very rare materials, which contributed greatly to this treatment.

This book would not have been be possible without the help of immediate sources, in particular Rabbi Joel Bakst for his kind encouragement and support regarding the chapter on Moses, and his generous interest in our ongoing work. Rubellus Petrinus provided insightful conversation and encouragement and introduced us to his editor, Paul Hardacre. Indirect thanks to Stuart Nettleton and Robert Allen Bartlett who both published versions of the archetypal recipe. To Dennis W. Hauk who wrote the forward and generously provided source material for inclusion, thank you. Special thanks to Ayman Gabarin who kindly allowed us to quote the Arabic Account and English translation of Ghalib's *Epistle of Morienus* published by his father, Ahmad Y. al-Hassan. In addition, we have quoted Lee Stavenhagen's translations and articles in the Morienus' Chapter, which added greatly to the elucidation of the subject. Anyone with an interest in this unique episode in the history of alchemy should read his book *A Testament of Alchemy*. Thanks to Paul Hardacre, who was very helpful with the initial editing – he provided insightful commentary, considerations and stopped by the lab with his family to help whip up a batch of *Algarot Powder* and *Philosophical Spirit of Vitriol*. Much appreciation to Lauren Khu for her valuable original artwork (and the hidden alchemical symbolism embedded in each picture). To Adam Mayer whose kind words and advice during the final editing process bolstered resolve to finish this treatise, thank you. My good friend Ian Barnes acted as sounding board and and helped tidy up some of the fine points of technical writing. Finally, to my close friend and co-author, Dr. Roger Gabreilsson, my deepest gratitude for stepping out of the box and putting your professional academic career on the line to humor an old friend's passion and explain the particulars, eternal thanks indeed.

FOREWORD

What you have before you is not a book, but a key to one of the most mysterious subjects in human history – the story of the Philosopher's Stone. You will find that this key morphs like mercury into many different shapes depending on what you are looking for. *Cracking the Philosophers' Stone* is not a fixed book but a flexible compendium that can only be opened with a key like this. J. Erik LaPort describes how to use this key in the following pages, and he urges readers to refashion it to suit their needs.

The story of the Philosopher's Stone began with Alexandrian and Arabian alchemists and soon captured the imagination of the entire world. By the Middle Ages, it was not only the touchstone that transformed base metals into gold but also held the secret to eternal life and spiritual perfection. However, as this book makes abundantly clear, the Philosopher's Stone was never just a psychological scheme or philosophical construct to alchemists. Both Eastern and Western alchemists believed it was a tangible, physical object they could create in their laboratories.

I was initiated into alchemy while I was attending graduate school in mathematics at the University of Vienna. Vienna and nearby Prague were at the center of alchemy in sixteenth century, and the university library had a collection of manuscripts dating back to that era. I could see in the writings and drawings of alchemists a beautiful and coherent view of the world that encompassed both physical and spiritual reality, and I determined to learn more.

I soon found a chemistry professor who led a discussion group on alchemy. Such gatherings of scholars were common at the university, and most departments had their own reserved table at local coffee houses. But the alchemy group met at a beer hall.

It was a raucous group that included not only professors and students but also practicing alchemists. One of these was an Italian man who had moved from Prague to Vienna when Russia invaded Czechoslovakia. He agreed to teach me alchemy, and so began my three-year apprenticeship in the art of transformation.

I soon learned that making stones in alchemy is no great mystery, although it is a long process. The techniques used have not all survived in modern chemistry, but they are the same operations used by ancient alchemists. Basically, there are two kinds of stones: vegetable stones made from plants or herbs, and mineral stones made from minerals, crystals or metals, surviving methods for which I've shared with Erik (see page 56). In general, the archetypal Philosopher's Stone is considered to be a mineral stone made from gold, or at least it requires a seed of gold.

This mercurial book, which has the potential to take on many shapes depending on what is on your mind, is your First Matter. I invite you to transform it into your own personal Philosopher's Stone.

Dennis William Hauck PhD

AUTHOR'S PREFACE

During early 2009, my good friend and mad scientist-genius Dr Roger Gabrielsson PhD arrived at Koh Tao for a holiday visit. While here, I asked for his help in researching a novel gold compound, based on an alchemical precedent. It occurred to me while browsing through 16th and 17th century literature on the development of gold salts, that the legendary Philosophers' Stone of alchemy was perhaps a type of gold-salt. Delving a little deeper into the alchemical literature, it became clearly apparent that many famous writers on alchemy specifically stated that the Philosophers' Stone was nothing more than gold in its finest state, whereas other sources were adamant that the Philosophers' Stone was indeed a salt. One thing was certain; all these alchemical authorities treated the Philosophers' Stone as a genuine compound synthesized in a laboratory. After a cursory study, I concluded that the chemical identity of the Philosophers' Stone remained an unsolved mystery that would potentially yield its secrets. I was immediately captivated and initial curiosity quickly gave way to sheer obsessive and compulsive research into the matter. It was an honorable riddle pursued by great minds throughout history and I immediately regarded the quest as a worthy endeavor.

I knew that I could not do this alone and that is when I harangued Roger into collaborating with me as chemistry consultant and co-author. I am deeply indebted to him for his help and unfailing encouragement. I knew that he would not decline because Roger once explained to me that any question of chemistry was valid as far as he was concerned, and that there was no such thing as an absurd or unimportant chemistry question as long as it was framed intelligently. The result of that collaboration and years of passionate research is this treatise – *Cracking the Philosophers' Stone*.

This book is divided into three sections. The first is a general introduction to the Philosophers' Stone, its origins and historical development in Alexandria, Egypt during the pre-Islamic period. The Stone's composition, applications and cultural value are presented in historical context and explored in detail. Section 2 describes the preparation of the basic alchemical ingredients, overall chemistry and confection of the Philosophers' Stone, along with accounts of reproducibility experiments performed at the Q'era-Tech Research facility. Section 3 examines the traditional applications for the Philosophers' Stone and expands on the notion of the alchemical process as a workbench wisdom tradition.

Any exploration of history and this is especially implicit concerning alchemy, is that a conclusion is primarily a work of interpretation. The purpose of this book is to present just such an interpretation, which ultimately leads to a working hypothesis, or more accurately, a series of hypotheses to explain the phenomenon of the Philosophers' Stone, its origins, evolution, chemistry, valuations and applications. We repeatedly stress plausibility statements throughout the book that are not to be construed as statements of fact, but rather address *plausible, possible* or *most likely* scenarios for the subject at hand. This serves to stimulate future research efforts and to advance current understanding while leaving the subject fully exposed to alternate hypotheses and scenarios. We have attempted to remain unbiased; focusing our concerns throughout on numerous scenarios that we felt made most sense given several factors and substantiating evidence. This was not always easy and often the only evidence available was scanty or circumstantial in which case we plainly state as much in the narrative.

We began the journey of decoding the Philosophers' Stone from the dual premise that 1) the stone was an actual chemical compound, and 2) that it had manifested as various diverse products at different times in history. Research into the matter has confirmed both, yet a dangling question remained. What was the chemical identity of the original (archetypal) Philosophers' Stone and could it be rediscovered and reestablished today? The attempt to answer these questions brought me into contact with some wonderfully intriguing and insightful people on a journey through history, traditions and practices that broadened my understanding of the alchemical process in unpredictable yet very welcome ways, least of which was developing a profound admiration for alchemal traditions.

One might assume that the chemistry would prove the greatest challenge. It was actually the language of alchemy however, that presented an almost insurmountable yet ironically marvelous obstacle to writing this book. A tremendous amount of data and a robust conceptual framework lie just below the surface of every cover-name and cover-term encrypting alchemical substances and processes. I realized that if I were to decode alchemical abstractions, it would be of tremendous value to approach someone fluent in the arcane language of alchemy for assistance. After several attempts at collaboration with authorities in the field, followed by snubs, polite responses to the negative or no reply at all, I realized that I was on my own in this regard. I began to self-study the etymology and history of each cover-term, exposing a fantastic world of imagery, historical context, mythology, and trade-jargon, which upon being understood, revealed that each cover-term was far from haphazard. Substances and processes were encyrpted in many layers of subtle and interconnected meaning. Indeed some terms were truly multi-cultural with their practical use spanning millennia and remain in use today. Learning the language of alchemy was as difficult as learning any other – bits and pieces at first, crude phrases later, before becoming fluent in alchemy's dialects and sophisticated abstractions.

The research process was arduous. Upon decoding a cover-name or terminology, we then embarked upon a detailed study of each corresponding substance or process. We attempted to find the earliest record of each, its discoverer and its various applications, which were often at odds with established scientific literature. If it were a substance, we attempted to glean whether it may have been used in a chemical technology prior to the accepted scientific date of discovery, then searched for a body of evidence to support an earlier date. Each time we felt close to understanding a substance or process fully, we then consulted numerous traditional alchemical texts and compared those accounts to what we observed in the laboratory. A primary consideration was that perfect reproducibility may prove challenging on the basis that adepts were artisans and extremely skilled at their art. It could take one's career to develop the skills possessed by an adept-alchemist. As any chemist knows, it can take innumerable tries before a reaction happens in an expected or optimal manner. Only after ensuring that historical and alchemical texts accorded with our lab results and observations, did we feel confident to present the hypotheses herein.

The entire book project took over five years to complete, thousands of hours of research, entire libraries of books and journal articles, ancient and modern, scoured and scrutinized for helpful information. Library efforts were combined with laboratory frustration and elation before the book reached its final form. I began to understand that an Alexandrian alchemist was more an artisan and philosopher than proto-chemist, yet alchemists would go on to assume many roles throughout alchemical history such as pharmacologist and experimental chemist of later alchemical traditions. It also became clear that much of the existing research into Alexandrian alchemy was based on Pseudo-Democritus and his Persian-Babylonian approach to chemical technologies, reinforced by works such as the Leyden and Stockholm papyri that typified this tradition. The Philosophers' Stone however originated with Maria Hebrea and her Judeo-Egyptian school or alternately a Hermetic Greco-Egyptian school – these being completely dissimilar in many important ways to the Pseudo-Democritan type. The challenge was to elucidate this long-held misunderstanding in regards to Alexandrian alchemy beyond a reasonable doubt.

Today's alchemy enthusiasts, historians and chemists understandably have specific areas of interest or specialization in their respective fields of research. Nevertheless, *Cracking the Philosophers' Stone* promises something new and interesting to everyone. While it is possible to read this treatise from cover to cover, it is not necessary to do so, which is to say that this book is written so that each chapter stands on its own merit as a mini-course on the specific aspect of the Philosophers' Stone addressed. For those new to alchemy, begin with chapter 1, followed by 10-12 prior to delving deeper. For those primarily interested in the Philosophers' Stone from a socio-historical perspective, read chapters 2-7, followed by 10-12 and Appendix A. For the rare aspirants, operative alchemists or qualified chemists primarily interested in operative reproducibility of Alexandrian alchemy's substances and processes, read chapters 1, 3, 4, 6-9 and Appendix C to satisfy curiosity before moving on to the remainder of the Stone's historical context. For those more interested in the speculative aspects of the Philosophers' Stone, its philosophical and psychospiritual connotations, its value as an early cosmological model or as a wisdom tradition, chapters 4, 5 and 12 focus on these intriguing valuations of the Stone. It is my sincerest aspiration that each reader discovers something of lasting value by reading this book and enjoys the journey as much as we did. As a pioneer effort, this work is certain to contain errors, which academic scholars and independent researchers will ultimately correct. To them I would like to express my deepest gratitude in advance for furthering the subject.

J. Erik LaPort

CHEMIST'S PREFACE

Some years ago during a summer vacation on the beautiful island of Koh Tao in Thailand, my good friend J. Erik LaPort told me about a plan he had for writing a book on alchemical gold processes to be called *Cracking the Philosophers' Stone*. The book was to be researched and developed using references to old texts and manuscripts that describe the mystical alchemical processes and, unlike its predecessors, would focus on modern chemistry descriptions in relation to experimental proof. To be honest, just the title itself encouraged me to jump on board as co-author and more importantly as the experimentalist. During the early stages of writing the book, it was more or less a historical manuscript about gold, still with missing historical overlaps, scientific background and experimental truth, but there was also a thread of inspiration going through it. When I asked Erik exactly what he wanted me to do, he gave me Pierre Duchesne's photographic illustrations of the Great Work as an example, "test my hypothesis" he said and "record the color changes." The most amazing thing about the contents of the draft was that it more or less claimed to provide full theory and practical instructions on alchemy and more specifically on how to make the real Philosophers' Stone!

Some years later and after endless hours of discussions with Erik, I am now looking back, remembering what a bold claim that actually was. Throughout the drafting of this book, Erik has never doubted his early hypothesis and has never redirected its path, and for a person with no initial training in chemistry to do that is really impressive! The hours of research, decoding of information and experimental work and discussions that are behind the final version of this book should be enough for the reader to realize that before you is an amazing accomplishment in so many ways. The majority of the work and conclusions herein are solely a result of Erik's determination, logic and exceptional clarity in decoding the history of the Philosophers' Stone. As a result, he has laid bare the secret processes and experimental background of one of the last remaining mysteries in the history of alchemy. Such an interesting, objective and no-nonsense book on alchemy has in my opinion never before been written, which fits together perfectly with the older books on alchemy and also as a historical complement to modern chemistry literature.

From my perspective, while on my own quest to understand alchemy I have often found it hard to see any logical meaning to many of the modern and older representations of alchemical crypticism and the spiritual and allegorical approaches to experimental work. These are often described and reflected on as the highest aspirations of the human soul, that the success of experimental work done by an alchemist depends on his/her understanding that lasting transformations only take place once the work is accomplished on all levels of reality – the mental, the spiritual, and the physical. I really do not see, nor understand, how many in the alchemical community can define chemistry simply as a superficial science that seeks to rearrange atoms and molecules to exhibit different properties of the same dead material, with a simple comparison that when an alchemist performs a laboratory experiment, it is the culmination of careful planning to find the right timing and personal purification to create the sacred space in which the transformation can take place.

From my perspective, the "real" work of transformation takes place in the real world, yet alchemy is not chemistry? Kindly put, I am not a believer in alchemy. In its most basic form, it is not possible to transform lead into gold through a series of chemical additions, subtractions, distillations and cooking operations in a rudimentary laboratory as was practiced in medieval and Renaissance Europe. I have reminded myself during times of doubt, that alchemy had a profound impact on the development of chemistry as a science in the sixteenth century. The fantastic experimental works of Johann Baptist Von Helmont (the father of the term "gas"), Isaac Newton (founded physics and discovered calculus), and Robert Boyle (first father of chemistry), were heavily influenced by their belief in alchemy. They in turn inspired Georg Ernst Stahl, Antoine-Laurent Lavoisier and Jöns Jakob Berzelius, whose tremendous achievements led to the "death" of alchemy and the true birth of modern chemistry.

During the development of this book, I have come to the personal conclusion that alchemy is in fact tightly interwoven with modern chemistry, the variable of time being the only difference. Chemistry or alchemy as we know it, or used to know it to be, has had its understandings shaped by years, even decades, of individual experimental study. Regardless of the time in history, I look at chemistry as a coherent, rational whole, and we must accept the worldview that chemistry gives us based on our present physical reality, by our tests of physical reality, by the questions we frame as we test this real universe experimentally, and by the answers we receive from our experimental tests. Chemistry is, above and beyond all else, an experimental science of its time. We accept the reality of atoms, ions, molecules and all the other structural concepts of chemistry because we have no other rational choice. Our experimental tests of the universe, through the scientific method, lead us to them and only to them.

As a trained chemist, my personal view on chemistry is that it is capable, when suitably presented, of making a strong appeal to one's intelligence and imagination. To me it is the most romantic branch of science and its variegated history stretches back through unnumbered generations of alchemists and chemists, and after working on this book, I must say that the present votaries have a rich and humanistic heritage.

Before closing, I would like to point out that for me, this book is an enthusiastic celebration of molecules and chemical synthesis, allowing you to witness the birth of chemistry and science as we know it. More importantly, it is also a unique tribute to the many scientists who were involved in their study, most of whom are described in its pages. On top of that, there are innumerable historical vignettes that interweave chemistry and alchemy in a very appealing way. Although the emphasis of this work is on alchemy, it contains much that will be of interest to those outside the tradition, indeed to anyone with a fascination for the world of chemistry and molecules.

While the processes for the Philosopher's Stone have been described by others, in my opinion it is a more spagyrical approach to alchemy than a chemical one. However, this book does reveal the alchemical methods for evolving the Philosopher's Stone. Inside you will see the transition of matter through the colors that were so well described by the ancient alchemists. I believe that if you study this book with an open mind you will be greatly rewarded. Listen to the words of J. Erik LaPort who has laid the groundwork for this rediscovery and learn from his unique and unveiled efforts.

But the question remains: "is it true?" I cannot say I know the answer. That is for you to decide. The claims were from the beginning unbelievable, yet the explanations and experimental conclusions are now perfectly logical.

Roger Gabrielsson PhD

EDITOR'S NOTE

As an editor of this book, I am one of its first readers. When Erik LaPort asked me to edit his tome, I initially had very little idea of the weight of what he had actually set out to do with this publication. This is not merely an attempt at giving a chronological account of the development of alchemy. It is rather the work of a practising alchemist (although he asserts that he isn't) who sets forth an incredibly exciting new theory of what the Philosopher's Stone really is, and what it has been through the ages in the Middle East, in Europe, and elsewhere.

This book contains daring new ideas and surprising twists, but none that may be described as myth making. Meticulous details in the language of chemistry support historical and textual research. The scholarship behind this book is massive, but the language is still accessible to the enthusiast as well as the advanced student of alchemy.

Before I read LaPort's text, I was convinced – by Carl Gustav Jung actually, whom he of course references – that alchemy was first and foremost metaphorical and its real relevance was limited to the depths of the human soul. Those ways of reading alchemical texts remain and LaPort emphasizes how rich alchemy's cultural and philosophical tradition really is. He presents an entirely plausible hypothesis that allows for the practical use of alchemy, rooted in tradition but relevant to today's man. I am sure that many a reader will experience the kind of excitement that I felt reading this book.

Adam Mayer

CRACKING THE PHILOSOPHERS' STONE

Introducing the Stone

*All truths are easy to understand once they are discovered;
the point is to discover them.*

Galileo Galilei

Tracing the development of most sciences from their ancient origins – and this is particularly true regarding pharmacy and chemistry – ultimately leads to alchemy. For millennia, the beating heart of alchemy was a legendary substance that was believed to confer wealth, power and immortality to anyone who could unlock the secret of its confection. This substance was called the *Philosophers' Stone*. The recipe to create this substance dates back to the Bronze Age in Egypt, Mesopotamia and the Levant. Because of the numerous sources, there seems to be very little doubt that this substance actually existed – the real mystery lies with the secret of its confection. The quest to rediscover the Philosophers' Stone has spanned numerous cultures over thousands of years until the present day.

The Philosophers' Stone was the source of two interrelated products: the *universal medicine*, believed to confer unnatural longevity to all who ingested it; and the *powder of projection* that transmutes copper, lead or mercury into *alchemical gold*. The promise of immortality and unlimited wealth earned the Philosophers' Stone special status as one of the greatest mysteries to captivate early scientists, physicians, philosophers, nobility, and con artists throughout history.

The quest to reveal the process for confecting the Stone exposed one to many unique dangers of the period, among the worst being self-poisoning, madness, torture, or execution. Despite ever-present perils, the promise of rediscovering the secrets of the Philosophers' Stone served as the proverbial carrot-on-a-string that played a large role in driving the donkey cart of scientific pursuits of ancient and mediæval times into the modern age of advanced science and technology.

The philosophers

The Philosophers' Stone dates back thousands of years to a time in history when science, philosophy, magic, and mystery were synonymous. Philosophers in former times addressed important issues such as the nature of reality, forms of knowledge, social organization, technology, and the human condition. These *lovers of wisdom* attempted to understand and harness the forces of Nature through what might today be described as a multidisciplinary holistic approach.

Alchemy began with the manipulation of minerals and metals for medicinal and industrial purposes. It was an ancient science of matter based upon trial and error. Early alchemists sought to discover a single substance that would enable them to perfect matter. The Philosophers' Stone fit the profile so well that, because of their attempts to reveal its secrets, alchemists developed equipment and discovered new substances and processes very relevant to modern science and chemistry.

The secret of the Philosophers' Stone – believed to be a key that would unlock the mysteries of creation and change – played a central role in the philosophical pursuits of all alchemical traditions. Success at confecting the Stone was a rite of passage and a milestone accomplishment passionately pursued by ancient alchemists and philosophers. Those who attained success became legends, whereas others were led to absolute ruin, madness and, in some cases, death arising from their obsessive pursuit of the Stone.

INTRODUCING THE STONE

The Philosophers' Stone made its appearance in Europe around the 12th century, remaining mainly within alchemical circles. By the time of the Italian Renaissance of the 16th century, the legend of the Philosophers' Stone and its power was so great that it rivaled the Holy Grail and the Ark of the Covenant as an object of quest.

The Stone's potential power was no longer secret, posing a great threat to the Catholic Church and Europe's ruling elite. They addressed this threat by criminalising the practice of alchemy for all but court and church sanctioned alchemists. In the frenzy to gain access to the Stone and its power, even royalty were ensnared by the quest. Europe's powerful elite – such as the Medicis of Florence, King Charles II of Britain, Holy Roman Emperor Rudolf II, and Queen Christina of Sweden, among others – not only supported court alchemists, but also outfitted private alchemical laboratories and actively participated in the search for the Philosophers' Stone.

Although Pope John XXII issued a papal bull against alchemy, the words in his prohibition applied to 'poor' alchemists, meaning false ones. He was reputed to have maintained a personal alchemical laboratory in Avignon and was respected by skilled alchemists as a patron of their art. Countless friars, monks, bishops, and archbishops practiced and published alchemical treatises. Religious reformer Martin Luther, through his support of alchemy, played an important role in inspiring German alchemists to continue alchemical pursuits and advance chemical technologies:

> *The Art of Alchemy is rightly and truly the Philosophy of the Sages of old, with which I am well pleased, not only by reason of its virtue and manifold usefulness, which it hath through Distillation and Sublimation in the metals, herbs, waters, oleities, but also by reason of the noble and beautiful likeness which it hath with the Resurrection of the Dead on the Day of Judgment.*

Great scientific figures such as Roger Bacon, Robert Boyle, Johann Rudolf Glauber, and Sir Isaac Newton pursued the Philosophers' Stone enthusiastically. Even founding father of the United States Benjamin Franklin took a keen interest in alchemy. Things appeared to change after Robert Boyle published *The Sceptical Chymist* and Antoine Lavoisier ushered in the age of modern chemistry.

However, the lore of the Philosophers' Stone was kept alive by accounts of colorful figures such as Giacomo Girolamo Casanova, Count Alessandro di Cagliostro and the enigmatic Comte de St. Germain who consorted with European aristocracy, providing elixirs and organising secret societies in grand fashion. Comte de St. Germain initiated Comte François de Chazal who passed on the Rosicrucian tradition to noted Swedish adventurer and doctor, Sigismund Bacstrom. Bacstrom did much to preserve the Rosicrucian tradition and the secrets of confecting the Philosophers' Stone.

The formula

The earliest account clear enough to identify the materials, methods and application of the Philosophers' Stone is found in the teachings of Alexandrian alchemist and teacher Maria Hebrea (also known as Maria Prophetissa) sometime between the 1st and 3rd centuries of the Common Era (CE). In cryptic fashion, she lists the three ingredients and divulges not only the archetypal method, but also reveals knowledge of an alternative faster process to confect the Stone. In Europe these two methods would come to be known as the *Ars Magna* (or Great Art) and the *Ars Brevis* (or Brief Art) respectively.

Maria Hebrea also clearly explained that the application of the Philosophers' Stone was to change the properties of copper. This important clue argues for a technology originating during the Bronze Age and is important because, at the time of Maria's Alexandria, bronze technology was already considered a remnant of great antiquity that had

INTRODUCING THE STONE

reached an advanced level. Maria also implied that the Philosophers' Stone was a legacy of the race of Abraham, yet disclosed nothing of its origins.

One interpretation from a much earlier source reveals an encrypted recipe in accordance with Maria's methodology. A legendary work called the *Emerald Tablet* of Hermes Trismegistus – known to Alexander the Great centuries prior to Maria – suggests Græco-Egyptian origins for the recipe. The names Enoch (Jewish), Thoth (Egyptian) and Hermes (Greek) were synonymous for the mytho-legendary individual who introduced science and technology to humanity during the development of early civilization.

The Stone's value
In Alexandria, the art of alchemy was described by some as *chrysopœia*, meaning *gold making*. This art of *aurifaction* involved changing copper to a unique type of alchemical gold. This was in direct contrast with established gilding technology of the period, which involved surface treating bronze that resulted in what might be described as *aurifiction* meaning *falsified gold*. Alchemists believed that they were harnessing the powers of creating and transmuting inner properties, rather than merely falsifying an outer appearance.

Philosophers and alchemists made keen observations of the natural world, and witnessed transformation and change as a daily phenomenon. In the animal kingdom, eggs transformed into hatchlings and caterpillars morphed into butterflies. From seeds, sprouts emerged and flowers became fruit in the plant kingdom. The philosophers and alchemists realized that man could participate in the process of transformation via practical means such as creating butter and cheese from milk, bread from grains, or wine from grapes and vinegar from wine. Long before living cells or chemicals and their reactions were identified, alchemists witnessed the dynamic processes of transform-ation and change in Nature and termed it *transmutation*.

The mineral kingdom likewise showed signs of growth and regeneration. Seawater crystallized into salt and cinnabar ore 'magically' produced mercury and sulfur. Gold was a byproduct of stibnite mining. The observation that sulphide ores, such as cinnabar and stibnite, commonly occur in the presence of gold caused alchemists to determine that metal ores and base metals were in an organic geological process of maturing into gold.

From the perspective of an alchemist, the idea of transmuting a base metal into alchemical gold was merely the act of assisting the *quickening* of a natural process. If transmutation is understood to mean *a change in the properties of matter*, it engenders a better understanding of the alchemists' mindset. The Stone's ability to affect transmutation in base metals, resulting in alchemical gold, was a very real primary power in the minds of those in search of its secrets.

The Stone's dual value is best described by 13th century Franciscan friar, philosopher and early European advocate of the scientific method, Roger Bacon. Bacon explained that alchemy:

> ... *is the science which teaches how to make and generate a certain medicine, called elixir, which when projected onto metals or imperfect bodies perfects them completely at the moment of projection.*

The second power inherent in the Philosophers' Stone was that it could extend human life to its fullest potential. This was mistakenly interpreted by some to mean that one in possession of the Philosophers' Stone had access to potential immortality. Bacon addresses this property as follows:

That medicine which will remove all impurities and incorruptibilities from the lesser metals will also, in the opinion of the wise, take off so much of the incorruptibility of the body that human life may be prolonged for many centuries.

Similar ideas occur in Chinese, Indian, Persian, and Islamic alchemical traditions long before arriving in Europe. The idea that life might be prolonged by centuries, or possibly to the extent of immortality, was appealing to many who encountered descriptions such as these.

The ability to create gold would provide the alchemist or his patron with unparalleled power through manipulation of local and worldwide economies based upon a gold standard. It was viewed by the benevolent as a weapon with the potential to topple unjust power structures. By devaluing gold, some alchemists hoped to usher in an age of enlightenment, as the Harvard educated American physician and alchemist George Starkey, writing under the pseudonym Eirenæus Philalethes, alludes to in the following:

I hope that in a few years gold (not as given by God, but as abused by man) will be so common that those who are now so mad after it shall contemptuously spurn aside this bulwark of Antichrist. Then will the day of our deliverance be at hand ... for then gold, the great idol of mankind, would lose its value, and we should prize it only for its scientific teaching.

The incredible power attributed to the Philosophers' Stone, and the secret of its manufacture, obsessed many of the greatest minds throughout the history of science and chemistry. The allure of a single substance – created and held in the palm of one's hand – which embodied potentially unlimited wealth, power and immortality ... this proved such a difficult temptation to resist that it kept many seekers of the Philosophers' Stone spellbound.

Transmission

During the 12th century, alchemy and the Philosophers' Stone were making a three-pronged arrival into Europe. Muslims valued science and technology and, according to their writings, held alchemy and the Philosophers' Stone in high regard. Alchemical texts in Arabic entered Europe via Muslim controlled territories in al-Andalus and through Italian and Sicilian trading contact with North Africa and the Middle East. Crusaders returning from the East also imported alchemical works from Constantinople and the Holy Land. It is primarily through Arabic translations of original Greek, Egyptian and Hebrew alchemical texts, along with a corpus found at Constantinople, that European intellectuals were first introduced to alchemy and the Philosophers' Stone.

Knowledge of the Philosophers' Stone remained within tight circles, as a mystery-craft tradition, for approximately 300 years after arriving in Europe. Tales of its power and eyewitness accounts increased its status in Europe to the point where the quest for the Philosophers' Stone was on par with great mysteries like the Holy Grail and the Ark of the Covenant. The Stone's secrecy, yet alleged achievability, added to its allure.

Much like masonry, glass making, metallurgy and art, alchemy was considered a mystery-craft whereby one learned through long apprenticeship with a master artisan. The trade was passed on primarily through oral teachings and physical work. Any written instructions or records were encoded in images or symbolism intelligible only to those initiated into the particular tradition being studied. Throughout alchemy's history, and in each culture that practiced it, transmission of the art followed this manner. This approach allowed for the communication of discoveries and new techniques, and a general sharing of information between brothers-in-the-craft without revealing any secrets to outsiders. Doctor, Rosicrucian and alchemical adept, Sigismund Bacstrom explains that:

> ... many wrote on this art, more for the purpose of showing to others who possessed the secret, that the author also knew it, than from any desire to communicate knowledge to the ignorant – and what is worse, many wrote and published books who knew nothing whatever of the subject.

This fact is what makes modern research into the Philosophers' Stone so challenging. The elaborate alchemical jargon – which has so successfully encrypted the secrets of the Philosophers' Stone – forms a complex matrix of symbols, cover-names, allegory, and emblematic art. For many alchemists, the grand test or rite of passage was to decode the encryption and rediscover the secret substances and processes necessary to confect the Stone. Often only a fraction of the recipe or a few of the easier procedures would be uncovered, revealing just enough to keep the alchemist encouraged.

From the time of the scientific revolution up until fairly recently, alchemical jargon had been derided as charlatanism, superstition and of no relevance to modern scholarship. Glauber's biographer, K.F. Gugel, attempted to impress upon the reader that alchemical imagery was actually the language of science for that era, comparable to the chemical equations of modern chemistry, by declaring that:

> *The convoluted symbolic language [of alchemy] was just as comprehensible to the chemists of his day as modern formulae are to us now.*

Alchemists needed to be fluent in Latin and possibly many other languages in order to read and compare various source works. John T. Young explains that the scientific language of alchemy required extraordinary talent and insight on the part of the alchemist in order to interpret alchemical texts and create symbolic imagery to apply to new discoveries:

> *Hermetic writing was certainly not incomprehensible to the initiated, but its elucidation depended on a combination of skills far more diverse than would be expected of a twentieth-century research specialist in any field. It demanded great practical experience, extensive familiarity with a vast range of rare literature, and in many cases access to a particular key obtainable only through personal contact with the author or his friends [although the key was sometimes discovered independently]. It also demanded highly advanced reading skills of a type regarded nowadays as far more the province of the literary scholar. Symbol, metaphor and often very heavily veiled allusion, not to mention deliberate red herrings and self-proclaimed self-contradictions, were the stock-in-trade of these authors.*

Most important in the quote above is the mention of a master *Key* – usually obtained by an initiate through direct transmission – which allowed the would-be adept to unlock the secret to confecting the Philosophers' Stone. It was this all-important *Key* that was the most guarded and greatest secret of alchemy.

The uninitiated generally experienced an overwhelming sense of impenetrability. Accomplished alchemists, as part of their tradecraft, would attempt to sidetrack or even completely discourage seekers by contradicting themselves in their writings, unnecessarily repeating or presenting procedures out of order, or including imagery that had nothing to do with the actual recipe. Starkey, writing as Philalethes, reveals as much:

> *And truly it is not our intent to make the Art common to all kind of men, we write to the deserving only; intending our Books to be but as Way-marks to such as shall travel in these paths of Nature, and we do what we may to shut out the unworthy: Yet so plainly we write, that as many as God hath appointed to this Mastery shall certainly understand us, and have cause to be thankful unto us for our faithfulness herein. This we shall receive from the Sons of this Science, whatever we have from*

others: therefore our Books are intended for the former, we do not write a word to the latter.

By encrypting the details of their craft in this manner, alchemists protected medical and industrial chemical trade secrets and ensured that only the worthy would ever gain access to the Philosophers' Stone. For the uninitiated, alchemical research often proved to be a frustratingly labyrinthine exercise. Modern private and professional research aimed at decoding alchemy's entire corpus of materials and methods remains ongoing after centuries of effort.

Properties and Applications
The Philosophers' Stone was not naturally occurring, but rather a compound substance resulting from skilled artisanship. The alchemist's role in the Stone's confection was to prepare the materials properly and combine them in the correct proportions. Once prepared, sealed in a vessel and fired at exactly the correct temperature regimen, the Philosophers' Stone was said to create itself by means of a self-synthesising process; attention need only be paid to the degrees of fire. The deceptive simplicity of the process was enticing; yet discovering the exact ingredients presented a seemingly insurmountable obstacle to the uninitiated. Some alchemists became obsessed with rediscovering these secrets, to the point of insanity.

Some tantalising success stories encouraged determined souls to continue by presenting descriptions of the finished product. An alchemist who achieved success at confecting the Stone was called an adept. One of the earliest European adepts to provide a description of the finished Philosophers' Stone was the 14th century French alchemist, Nicholas Flamel, who wrote:

When the noble metal [gold] was perfectly prepared, it made a fine powder of gold, which is the Philosophers' Stone.

The primary ingredient in the archetypal Philosophers' Stone recipe is always alchemically prepared gold. George Starkey would echo Flamel three centuries later, writing that:

> Some Alchemists who are in search of our Arcanum seek to prepare something of a solid nature, because they have heard the object of their search described as a Stone.
>
> Know, then, that it is called a stone, not because it is like a stone, but only because, by virtue of its fixed nature, it resists the action of fire as successfully as any stone. In species it is gold, more pure than the purest; it is fixed and incombustible like a stone, but its appearance is that of very fine powder ...
>
> It does not exist in Nature, but has to be prepared by Art, in obedience to Nature's law. Its substance is in metals; but in form it differs widely from them, and in this sense the metals are not our Stone.
>
> Thus, you see that though our Stone is made of gold alone, yet it is not common gold.

The above quotations clearly identify the primary ingredient as gold, yet traditional descriptions of the Stone's color, density, structure, and even scent are at great variance with those of gold. The Philosophers' Stone has been described as a heavy vermilion powder; a ruby colored waxy substance; a deep red translucent crystal; a saffron or yellow powder, and even a golden colored liquid. It has been described as possessing an odour of sea salt, or having no scent at all. Although descriptive accounts of these indicators may at first seem contradictory, they are each quite accurate in that what these depictions portray – *a product at various stages of refinement.*

Depending upon the intended therapeutic or industrial application of the product, the Philosophers' Stone needed to undergo further refinement. Alchemists used the basic Philosophers' Stone product to create either a tonifying / rejuvenating *universal medicine* in liquid form – known as *aurum potabile* – or a *powder of projection* that facilitated the manufacture of alchemical gold from copper, lead, mercury, or a combination thereof. In some cases, the Stone was used for wet-gilding silver with a layer of metallic gold. It is also the most likely candidate for the secret ingredient required in order to create the rose-colored stained glass, which adorned the great Gothic cathedrals.

Inherent Dangers

Confecting the Philosophers' Stone was fraught with dangers, both inside and outside the laboratory. The Stone required a laboratory, but due to widespread prohibitions on alchemy, if one was not a court-sponsored alchemist it meant working in an illegal laboratory. Clandestine alchemical labs, as well as many operating under court patronage, featured built-in charcoal furnaces posing the first obvious severe health threat. The toxic nature of various chemicals and gases associated with alchemy had not been fully understood, and alchemists handled these substances with little or no protective protocols. Alchemists routinely inhaled noxious fumes, such as mercury, and tested their products by tasting or ingesting them; in effect acting as human guinea pigs.

In addition to severe health risks, alchemists also faced legal dangers. For clandestine laboratories, the danger of being discovered was an ever-present source of concern. Hauling large daily loads of charcoal into small private quarters – not to mention the noxious odours emanating from such labs – would have attracted unwanted public attention. Desperate characters suffering from disease also targeted alchemists, believing that they possessed secret cures. The mere hint that an alchemist may be in possession of the Philosophers' Stone placed him in grave danger as Starkey, writing as Philalethes, reveals:

> *It was only a short time ago that, after visiting the plague-stricken haunts of a certain city, and restoring the sick to perfect health by means*

> *of my miraculous medicine, I found myself surrounded by a yelling mob, who demanded that I should give to them my Elixir of the Sages; and it was only by changing my dress and my name, by shaving off my beard and putting on a wig, that I was enabled to save my life, and escape from the hands of those wicked men. And even when our lives are not threatened, it is not pleasant to find ourselves, wherever we go, the central objects of human greed ...*
>
> *I know of several persons who were found strangled in their beds, simply because they were suspected of possessing this secret, though, in reality, they knew no more about it than their murderers; it was enough for some desperate ruffians, that a mere whisper of suspicion had been breathed against their victims. Men are so eager to have this Medicine that your very caution will arouse their suspicions, and endanger your safety.*

As will be detailed in subsequent chapters, alchemical gold differs from genuine elemental gold. Nevertheless, the ruling elite would not necessarily have fully understood the difference. A fundamental reason for the prohibition of alchemy was the state's belief that if the Philosophers' Stone was successfully achieved, alchemists could potentially tap into an unlimited source of wealth and thus destabilize currency markets and existing power structures. This attitude was first recorded during the late 3rd century CE, in an edict by the Roman Emperor Diocletian who feared that Alexandrian alchemical gold could pass for genuine currency and thus empower those who opposed him. He ordered all materials in the Empire "which treated of the admirable art of making gold and silver" destroyed. Diocletian's edict was the first, but not final, criminalization of alchemy. During the 14th century, the Dominican *Directorium inquisitorum*, the inquisitors' directory, lists alchemists among magicians and wizards.

The perception that the Philosophers' Stone could generate unlimited wealth and immortality was all too enticing for those in power, and is one

INTRODUCING THE STONE

of the reasons for alchemy being patronized and pursued by Europe's ruling elite and religious figureheads. For this reason, court sponsored alchemists achieved somewhat of a celebrity status. If unsuccessful, alchemists risked punishment for either being a fraud or withholding secrets. Far from guaranteeing protection, successfully achieving the Philosophers' Stone presented its own unique set of perils. If success was even suspected, the alchemist in question faced interrogation, torture, imprisonment, or even execution; those in power would stop at nothing to acquire and harness the power of the Philosophers' Stone. Greed enticed con artists, charlatans and impostors to seek out the Philosophers' Stone and, upon inevitably failing, to use its reputation to defraud enthusiastic yet unwary patrons. This resulted in a very precarious and unstable environment for genuine alchemists operating with good intentions and endowed with a refined skill set.

Despite these dangerous pitfalls, the Philosophers' Stone continued to captivate some of the greatest intellectual minds of Europe. Alchemy survived public scandal, legal prohibition, political intrigue, and religious persecution. It evolved into the early modern *chymistry*, which played a foundational role in the development of pharmacy and modern chemistry. By the 19th century, alchemists were responsible for the discovery of a great many substances. Alchemy had played the role of proto-science for chemistry, iatrochemistry, pharmacology, and physics, among others, and had contributed to the formulation of the scientific method. Alchemy also profoundly influenced the development of psychology, historiology, astronomy, astrology, and the occult sciences. As the various sciences became more compartmentalized alchemy seemed to fade into the background, with the art surviving among secret societies and fraternities or via individual transmission. Thousands of alchemical texts survive, many detailing in tantalisingly veiled language the secrets of confecting the Philosophers' Stone. The search continues even today.

The Rediscovery

The Philosopher's Stone was an actual material substance; a tangible product synthesized by an impressive number of well-respected historical personalities – alchemists, philosophers, physicians, and the like. The question has never been whether the Philosophers' Stone was real, but rather what exactly is the molecular structure of the Stone (considered in the light of modern chemistry, pharmacology and industry) and how can it be reproduced?

This is not a treatise regarding alchemy *per se*. Rather, this study focuses upon the reproducibility of a genuine Philosophers' Stone substance according to traditional processes. The primary goal was to synthesize the exact substance traditionally understood to be the Philosophers' Stone, using materials and methods detailed in alchemical texts, and to analyze the history and evolution of the chemistry involved.

The study was limited to alchemical authors that were generally considered authentic – that is, the great adepts of alchemy in each tradition. The methodology was to find a common denominator inherent in the published syntheses of these alchemists and begin research from there. To do so, required an alchemical approach – *solve et coagula*. Hundreds of alchemical texts were examined for commonalities linking Maria Hebrea's descriptions circa 200 CE with those of European alchemists writing between the 12th to 20th centuries. It was a quest for an archetypal materials list and methodology profile that stood the test of time. The search resulted in the identification of the standard ingredients and chemical reactions common to most recipes:

1. **Gold** – *featured as the primary and most obvious of the three basic ingredients; the gold described is always refined to a high purity, then reduced to the finest particle size possible before being used;*

2. **Antimony** – *occurs naturally as a sulphide ore today called stibnite; in former times stibnite was called antimony, whereas purified metallic antimony was*

called regulus and, earlier in history, flowers of antimony; both stibnite and purified antimony are used at specific stages in the archetypal recipe;

3. *Flux / Menstruum – the final constituent was far more difficult to distinguish; it is a substance with the capability of dissolving gold without violence or corrosion, yet has the additional capacity to crystallize, congeal or coagulate under the right conditions; described as being metallic in nature, yet translucent and viscous.*

The primary challenge was to identify the flux / menstruum. Gold and antimony are considered by alchemists to be related. The flux / menstruum is described as being intimately related to, and having the same nature as, both gold and antimony. The greatest and most guarded secret to confecting the Philosophers' Stone is the identity of the flux, more commonly called the *universal solvent*. Complication arises because different substances were used at different periods throughout history. These substances ranged from a unique salt during the Alexandrian period, to the salt-saturated urine of Islamic alchemy, to the unique chemical compounds developed and utilized throughout European alchemy. Put simply, the flux / menstruum is the *Key* to the whole art, and knowledge of it unites brothers in an invisible fraternity of adepts spanning centuries.

Cross-referencing and carefully analyzing original sources resulted in reproducibility of the various flux / menstruum chemical reactions according to the materials and methods presented in alchemical texts. Various indicators were scrutinized, such as appearance, vapour temperature, viscosity, ability to dissolve gold without violence, etc. Once working samples of each were reproduced, the chemistry was scrutinized and compared to the reactions described in alchemical texts in order to discern historical variations. It became apparent that the flux / menstruum was achieved by a variety of techniques, which offered alchemists a margin for divergence in materials and methods. So ended the *solve* aspect of the research. What was next required was the *coagulation* of a vast body of arcane data, sourced from centuries of alchemical literature, into a coherent narrative of the evolution of the chemistry associated with the creation of the Philosophers' Stone and its valuations and applications.

Recipes – written by alchemists who shared a common approach, in accordance with an archetypal technique of uniting gold and antimony via a flux / menstruum – were gathered. Comparisons between various recipes provided further insight into the core process that ultimately enabled reproducibility. By the end of the 17th century, alchemists had developed comprehensively diverse techniques, yet the primary ingredients and finished products remained essentially equivalent to those of the Alexandrian adepts' so many centuries earlier. Achieving reproducibility resulted in the writing of chemical equations for the Philosophers' Stone, and to fully elucidating the subject of alchemical gold and transmutation from a contemporary scientific perspective. A comprehensive understanding of the Stone's molecular structure was fundamental towards creating a hypothesis for a therapeutic mechanism of action *in vivo*. The following treatment reveals the origins and evolution of Alexandrian, Islamic and European methods to achieve the same single chemical compound – the Philosophers' Stone.

BOOK 1

Origins & Evolution

From this, we need to consider the beginnings of things:

how they grow in one another and are as a single thing;

and how each has its function, in order to make the corpus complete.

Paracelsus

Chapter 1 – Sophic Substances

Listen, then, while I make known to you
the Grand Arcanum of this wonder-working Stone,
which at the same time is not a stone, which exists in every man,
and may be found in its own place at all times.
The knowledge which I declare is not intended for the unworthy,
and will not be understood by them.
But to you who are earnest students of Nature God will,
at His own time, reveal this glorious secret.

Eirenæus Philalethes

TRIA PRIMA – SULFUR, MERCURY AND SALT

His [the Adept's] knowledge cannot be acquired in a few months, no, nor even in a few years – and yet when once it is acquired it may be communicated to another, who has made himself a little acquainted with the old philosophers, in a few hours.
    – Sigismund Bacstrom, 18th century

The archetypal Philosophers' Stone synthesis is deceptively easy in theory. In an attempt to conceal this simplicity and other trade secrets, alchemists became masters of encryption. They accomplished this primarily by creating obstacles using wordplay. Once the preparatory work has been finished, the remaining work of confecting the Philosophers' Stone was so simple that it has been likened to women's work and child's play. Alchemists routinely comment upon the simplicity of confecting the Stone – one of the Art's most guarded secrets – once the experimenter is fully familiar with the synthesis:

BOOK 1 – ORIGINS & EVOLUTION

> *... the worke is no hard labour to him that knows it.*
> *– Ibn ar-Tafiz (Artephius), 12th century*

> *... sayth of the Art that it is so easy that it may be learned in the space of 12 houres & be brought to action in eight days.*
> *– Anonymous, 17th century*

One of the methods employed to protect the secret of the Stone's simplicity was sophisticated discourse. Bernard Trevisan openly admits to the use of wordplay in this capacity:

> *The thing [confecting the Stone] is very easy, yea soe easy that if I should tell it in plaine words you would scarce believe it. Al the difficulty consists in our words & our meaning in them.*
> *– Bernard Trevisan, 15th century*

Alchemist and physician George Starkey echoes Bernard in his laboratory notes for the *Diana Denudata*, an unfinished work that he intended to publish as part of the Philalethes writings:

> *For it is only this one thing which is secret, & hidden, which if it should be told plainly, with the pondus and regimen, even fooles would deride the Art, it is soe easy, being indeed but the worke of women, & the play of Children, obscured and vayled by the Antient Sages by al meanes possible, being indeed an Arcanum not fit to be divulged.*
> *– George Starkey, 17th century*

He did publish the same sentiment under his pseudonym in other texts:

> *Also another Philosopher sayth that if the Art were told in plaine words, even very fooles would laugh at it. Hence it is called the worke of women & the play of Children. – Eirenæus Philalethes, 17th century*

More than mere analogy, the recipe for the Philosophers' Stone paralleled Egyptian template recipes for bread and beer. European alchemists apparently considered the duties of baking and brewing

CHAPTER 1 - SOPHIC SUBSTANCES

as work for women and children, but they are aslo works of chemistry. The archetypal bread recipe consists of three basic ingredients – flour (body), yeast (soul) and a medium in the form of water (spirit). Similarly, the archetypal beer recipe consists of three ingredients – grain (body), ferment (soul) and water as a medium (spirit). Bread, once placed in the oven, was self-creating, requiring no further work other than observation and tending to the temperature. In a similar fashion, ancient beer brewing was more a work of observation than labor. This understanding fed builders of the Egyptian empire and provided food and beverage to the gods in the form of bread and beer offerings. When alchemists commented upon the parallels between baking, brewing and creating the Philosophers' Stone, they did so in recognition of a uniting principle; an operative template that was both practical and efficient.

The bronze-craft and glass making technologies were mother and father of the Philosophers' Stone recipe. Two of the earliest recorded recipes for the Philosophers' Stone are preserved in the teachings of a respected Jewish craftswoman named Maria Hebrea living and teaching in Alexandria sometime around 200 CE. Writings attributed to her reveal Maria to have been an expert bronzesmith who used the Philosophers' Stone as the key ingredient in *Corinthian bronze*-craft. Her recipes feature only three ingredients all derived from the mineral-metal kingdom and each ingredient related to the other.

The recipe for the Philosophers' Stone follows the three-ingredient template recipe for bread baking and beer brewing. Like bakers and brewers – whose work resulted in a tremendous diversity of legitimate expressions of bread and beer in varying forms – alchemists would eventually develop a diversity of different techniques and modifications stemming from the archetypal Philosophers' Stone recipe. The three ingredients common to every Philosophers' Stone synthesis are gold, antimony and a fluxing agent, yet they are encrypted using a staggering number of different cover-names and descriptives, a small sampling of which are presented below.

Gold	Antimony	Salt / Menstruum
Au – metallic gold calx	Sb_2O_3 – flowers of antimony	NH_4Cl – sal ammoniac
Au – metallic gold calx	$Sb_2O_2SO_4$ – glass of antimony	$SbCl_3$ – butter of antimony
Fixed and non-flammable	Volatile and unchanged	Volatile and changeable
First Principle	Second Principle	Third Principle
Red Man	White Woman	Spirit of Life
Red Gumm	White Gumm	Lunaria
Sulfur	Mercury	Salt
Salt	Mercury	Sulfur
Body	Spirit or Soul	Soul or Mediator
Sol	Luna	Mercury
Sun	Moon	Venus or Sirius
Father	Mother	Wind
Old Man	Steel, Our Lead, magnet	Sword (Chalybs)
Perfect Body	First Water	Second Mercurial Water
Father of Life	Fountain of Life	Center of Life
Latten, Rebis, Lion's Blood, Hermaphrodite, Androgyne, Sophic Sulfur (gold-antimony glass or alloy)		Our Vinegar Mercury Sublimate

Doubled Sophic Mercury, Mercurius Vitæ,
the White Stone, Hermes
(flowers of antimony + sal ammoniac),
Eudica, Eagle's Gluten, Azoth, Ignis Aqua,
Fire-Water, Sophic Mercury
(butter of antimony)

The number of cover-names can be disconcerting to the uninitiated, but they actually refer to specific chemical substances or compounds that

CHAPTER 1 – SOPHIC SUBSTANCES

were perfectly comprehensible to other adepts. An alchemist familiar with confecting the Stone is able to identify cover-names in context as Bacstrom reveals:

> *The most ancient first principles were* Sulphur *and* Mercury, *and to these in process of time was added* Salt ... *[These three cover-names] ... were mere* terms of art, *and very often had no relation whatever to the substances now known by these names.*
> – Sigismund Bacstrom, 18th century

Robert Tauladanus, an alchemist writing during the 16th century, provides a neat overview of the process:

> *According to the testimony of all philosophers there are three parts belonging to the Elixir, viz., soul, body, and spirit.*
>
> 1. *The* soul *is nothing else but the ferment or the form of the Elixir [antimony].*
> 2. *The* body *is the paste or matter [gold].*
> 3. *The third part of the Stone is the* Spirit *[salt / menstruum].*
>
> *If this mediator [flux] were taken away the* soul *[antimony] could never be centrally and permanently united with the* body *[gold] ... This* spirit *is called by the Philosophers: Heaven, Dissolving Mercury, Menstruum, Azoth, Quintessence and a hundred other names.*
> – Robert Tauladanus, 16th century

Tauladanus is revealing that the flux not only aids dissolution, it also permanently binds gold and antimony into a single compound substance. A fluxing agent is usually a salt product of some form. In Egypt, two naturally occurring fluxing salts, natron and sal ammoniac, were commonly used in both metallurgy and glass making. A flux such as natron allows for dissolution at lower temperatures and can aid solidification. The fluxing agent, sometimes referred to as the mediator, was the *Key* to the entire art and its greatest and most guarded secret. It is also an ingredient that could vary somewhat as we will explore later.

Aside from bread and beer analogies, alchemists employed other imagery involving trinities such as the Father, Son and Holy Spirit. Often the egg, with its sulfurous yolk, was also used as an analogy:

> *As an egg is composed of 3 things, the shell, the white, and the yolk, so our Philosophical Egg [Philosophers' Stone] composed of a body, soul, and spirit. Yet in truth it is but one thing, a trinity in unity and unity in trinity –* Sulphur, Mercury, and Arsenic [sophic salt].
> – *Wolfgang Dienheim, late 16th-17th century*

The first commonality encountered when researching the Philosophers' Stone is the pervasive notion that there are only three ingredients required to confect the Stone. In order to reproduce the Philosophers' Stone authentically, it is very important to identify the materials involved accurately. One of the great alchemist-physicians of the Islamic Empire, Ibn Sīnā (Avicenna), identifies the three components but in a veiled manner in keeping with tradition:

> *I advise you to* work only in Sol, and Luna [gold and antimony] and [sophic] Mercury, *because the whole benefit of the Art consists in them.*
> – *Ibn Sīnā (Avicenna), 10th century*

The following excerpts have been selected from alchemists largely regarded to be adepts, meaning they have succeeded in confecting the Philosophers' Stone or they were in direct communication with someone who has. The quotations that follow serve to reinforce the fact that the Philosophers' Stone is synthesized with only three carefully prepared ingredients:

> *Compose therefore our most secret stone from these three things and nothing else, for in no other things are contained that which so many seek after. This amalgama or natural composition, when managed in*

CHAPTER 1 – SOPHIC SUBSTANCES

the right manner, you may say in truth is but one thing – our stone. This whole composition is a mixture whose price and value is inestimable. This is our Brass mentioned in the Turba. – Anonymous

Know ye that no true tincture can be made but from our Brass, that is from our confection which is made of three things. Employ these and you must get the mineral stone.
– The Turba Philosophorum, 12th century

They say that the three things are of one nature, of one matter and essence, one water and one root – and they verily tell the truth.
– Nicolas Flamel, 14th century

When three are made one, in the form of a congealed substance, then it hath in it a true tincture ... – Bernard Trevisan, 15th century

Three species and no more than three enter our work ... Three species only are necessary in this work which Count Bernard has plainly enough indicated where he says: "Our work is made of one root [gold] and two mercurial substances [antimony and salt / menstruum], crude but pure, extracted out of their mines." By the Root the Count means ... the mature sulphur which is in gold perfectly digested ... This one principle determines and glorifies the other two, which are therefore called superficial principles. – Georgius Agricola, 15th-16th century

The three principles of the Universal are but one thing – the true spirit of mercury and Anima sulphuris, with the Spiritual salt, united under one heaven and dwelling in one body.
– Basil Valentine, 15th-16th century

One thing [flux] containeth and conjoins the medicine, two compose it [gold + antimony] therefore three are joined in one body [Philosophers' Stone]. – Lorenzo Ventura, 16th century

The father of the Stone is Sol [gold], the mother is Luna [antimony], and the wind [flux] carries it in its belly – that is the Sal Alcali called by the

> *Philosophers Sal acumu nivenum, the vegetating salt hidden in the body of magnesia [antimony salt].*
> — *Michał Sędziwój (Sendivogius), 16th-17th century*

> Seek three in one things, and one in three. *Open these and shut them up again and you have the whole art. Solve et coagula.* The Spirit [flux] will give the soul [antimony] to the body [gold]. *The Spirit attracts the soul and returns it to the dead body [reduced gold], and* at length the three remain united. — *Johann Siebmacher, 16th-17th century*

SOPHIC SULFUR
Gold Calx (Au)

The first principle, the ground work and foundation of the whole art is Gold – common pure gold, without any ambiguity or double meaning.
This is 'Our Sulphur'.

— *Sigismund Bacstrom*

To alchemists, gold was the king of all metals due to its indestructible, incorruptible and immortal nature. The only problem with gold, according to the alchemical theory of organic geology, was that gold had ceased growing due to its having already reached perfection – it had no growth or regenerative principles left to impart. The alchemical theory of organic geology viewed sulphide ores (such as cinnabar, stibnite, vitriol, galena, orpiment, and realgar) as growing or ripening into metals. Base metals would evolve towards the final perfection of gold according to their inherent balance of sulfur and mercury principles.

> ... *the father of the Stone is Sol [gold].*
> — *Michał Sędziwój (Sendivogius), 16th-17th century*

The alchemical work required the infusion of a new growth / reproductive principle into the body of gold. In order to enable such infusion, gold's very weak remaining soul and spirit needed to be killed or destroyed – only then would the gold be fit to receive the transplant of spirit-soul

CHAPTER 1 – SOPHIC SUBSTANCES

from its donor, antimony. However, before this could happen, the gold had to be purified and reduced to a powder that was then referred to by a number of cover-names, one of the most popular being *Sophic Sulfur*.

> *Since the alchemists can obtain* this sulphur *they can rejoice. This* is the foundation of their Universal Medicine ...
> — Johann Joachim Becher, 17th century

> *Some of the philosophers say* they used gold as the basis for their work, but they first reduced it to a sulphur by dissolving it ...
> — J.W. Hamilton Jones, 20th century

Alchemists believed that metals could be reduced to their first nature by the process of dissolution, followed by the creation of a powder or calx. We now know that what alchemists called calx were elemental metals at a very fine state of division or a metal salt or ash. In any case, once they had achieved pure calx, the gold was considered *philosophical*, meaning reduced to its original nature and fit for alchemical use.

> *We truly do dissolve Gold that it may be reduced into its first nature,* which is to say [a type of] Mercury.
> — Muhammad ibn Zakariyā al-Rāzī (Rhazes), 9th-10th century

Gold calx is technically an elixir. In Greek, the language of Alexandrian alchemy, the word used to denote *elixir* was xēre / xêrion meaning *medicinal powder*. By the 6th century, *elixir* was also referred to as *pharmakon*, implying therapeutic value. In Arabic, the language of Islamic alchemy, *iksir* is derived from either Greek or from Fujian-Yangzhou dialects of Chinese *ek-chi / yik-chi* meaning *primordial essence*. The prefix *al-* (meaning *the-*) was added arriving at *al-iksir*, or elixir, in English. Gold needed to be reduced to *xēre / iksir* first in order to achieve therapeutic or transformative potential.

> *Gold being the most noble among metals,* the most compact, perfect, and fixed, if it be dissolved and separated in most little parts it becomes

> *spiritual and volatile* like the Mercury and that by reason of its heat; *and then it hath a Tincture without end,* and that tincture is called the hot masculine sperm. – Ibn Sīnā (Avicenna), 10th century

An early technique to achieve gold calx was to dissolve gold by combining saltpetre (potassium nitrate; KNO_3) with sal ammoniac (ammonium chloride; NH_4Cl) to create an acid capable of reducing gold to a fine powder. Often, due to an excess of ammonia, the gold would take on an olive green tint. Gold in this form was known by the cover-name *Green Lion*. The cover-name Green Lion also applied to antimony in various forms or in fusion with gold, indicating that antimony was a Green Lion in the form of immature or unripe gold.

Once gold has been transformed into Sophic Sulfur, a.k.a. Sophic Gold Calx, it must next be married to *Sophic Mercury*. This process is known as the First Marriage or creating the *Rebis, Hermaphrodite* or *Androgyne*. The *Rebis* is prepared by casting the gold elixir into molten antimony, sometimes referred to as Sophic Mercury, in order to achieve a *gold-antimony glass* or alloy. Where Sophic Sulfur is male, Sophic Mercury is female, hence the imagery of marriage or dual sexuality.

> *Sol, which is our sulphur,* is reduced into [a type of] Mercury by [sophic] Mercury. – Roger Bacon, 13th century

> [Sophic] *Sulphur doth, in his work, supply the place of the male,* and whosoever undertakes the transmutation art *without it, all his attempts will be in vain ...* – George Starkey (Philalethes), 17th century

The Second Marriage is the process whereby the *Rebis* is sealed in a vessel and concocted over low heat. The product undergoes a slow equilibrium chemical reaction over time to arrive at the final product known to alchemists as the Philosophers' Stone.

> *You must make a living and incombustible water, and then* congeal or coagulate it with the perfect body of Sol. – Ibn ar-Tafiz, 12th century

CHAPTER 1 - SOPHIC SUBSTANCES

For an in-depth study of alchemical gold, see *Gold Elixirs: A Cross-Cultural History of Therapeutic Gold* by the author.

SOPHIC MERCURY
Prepared Antimony (Sb; Sb_2O_3; $Sb_2O_2SO_4$)

The second [principle] is Mercury, not common quicksilver, however, but that substance to which the Philosophers have given the name of 'Our Mercury', 'Our Diana', 'Our Moon', 'Our Luna', 'Unripe Gold', and many other names.

— Sigismund Bacstrom

The identity of the *prime material* was one of the great mysteries of alchemy. It has been described as a common item found underfoot or in the marketplaces (in ancient desert cultures) and of having little or no value. European alchemists described the prime material as a product of underground caverns or mines. Maria's reference to it as *Our Lead* is one of the earliest uses of a cover-name in alchemy; referring to a naturally occurring sulphide ore called stibnite (antimony trisulphide; Sb_2S_3). When Maria described stibnite as *Our Lead,* she was comparing it to the other lead, another naturally occurring sulphide ore called galena (lead sulphide; PbS). Pioneering scientist and alchemical adept, Roger Bacon, clearly interprets alchemical lead to mean stibnite:

BOOK 1 – ORIGINS & EVOLUTION

> *Stibium or Antimony,* **as the Philosophers say, is composed of a Noble Mineral Sulphur, which** *they accounted to be the black secret Lead of the Wise.* **– Roger Bacon, 13th century**

> *... Antimony, which is better than common lead,* is called the PHILOSOPHERS LEAD, *or their* SECRET LEAD; *of so many named, but known of few ...* – Rudolf Glauber, 17th century

Both stibnite and galena were household items in antiquity, ground fine and used as antiseptic powder or as an eye cosmetic known as *kahâl*. Stibnite has been used by desert communities across the Sahara, Mesopotamia and Persia for around 5,000 years. When stibnite ore is reduced to remove its sulfur, the result is purified antimony. Alexandrian technique involved roasting the stibnite ore until a fine greyish to brilliant white powder called *flowers of antimony* is achieved. European alchemical technique favoured metallurgical reduction in order to create pure metallic antimony called *star regulus*.

The star regulus of antimony refers to pure metallic antimony after it has been isolated from its sulfur matrix. It is shiny and metallic like silver, yet very brittle and displays a crystalline pattern likened to a star. The star regulus – metallic antimony – appears like fine steel, and due to this association, one of the cover-names for pure metallic antimony is *chalybs*. The etymology for *chalybs* may originate from the Greek *kalyptō* meaning *to cover, conceal* or *hide*. An alternative etymology is that it originates from an Indo-Iranian tribe famed for their metallurgy, known as the Iron or Kalybs tribe. *Chalybs* eventually was adopted by Latin to mean *steel*. King Arthur's sword, *Caliburn*, is named for this association.

> *There is granted unto us* one metallic substance which hath a power to consume the rest, for it may be considered their water and mother. *Yet there is one thing, namely the radical moisture of the sun and moon [gold and silver in this instance] that withstands it, and is bettered by it. That I may discover it to you,* it is called chalibs [sword], or Steel ...
> – *Michał Sędziwój (Sendivogius), 16th-17th century*

CHAPTER 1 - SOPHIC SUBSTANCES

It is also likened to quicksilver (*argent vive*) because, like quicksilver, antimony dissolves gold easily and at a much lower temperature than the standard melting temperature of gold.

> *Antimony is a mineral participating of saturnine parts and has in all respects the nature thereof.* This saturnine antimony agrees with gold and contains in itself argentum vivum ... and *gold is truly swallowed up by this antimonial argent vive* ... – Ibn ar-Tafiz (Artephius), 12th century

> *Here, my son, thou must* understand Luna metaphorically, *and not according to the letter ... by* Luna is understood mercury or the prime matter ... *and not mercury vive [quicksilver], as the sophisters suppose. For the first matter of metals is not mercury vive.*
> – Incertus Macrocosmus, 17th century

Once antimony and gold are alloyed, the antimony can then be removed by vapourization. This process leaves behind purified gold. In alchemical terms, this process 'kills' the dual soul of gold and leaves the body behind, soulless. For this reason, antimony, as gold's assassin, is sometimes referred to by the cover-name *chalybs* meaning *sword*. An additional cover-name for antimony is *magnesia* or *magnet*, due to its uncanny capacity to combine with other metals, especially gold.

> ... in Antimony also there is a Spirit, **which effects whatsoever in it, or can proceed from it, in an invisible way and manner, no otherwise, than** as in the Magnet **is absconded a certain invisible power, as we shall more largely treat in its own place,** where we speak of the Magnet.
> – Basil Valentine, 15th-16th century

The following passage from *The Turba Philosophorum* addresses the reduction of stibnite to metallic antimony, the text encrypted by the use of four cover-names:

> *Leave, therefore, manifold and superfluous things, and take* quicksilver *[antimony], coagulate in the* body of magnesia, *in* kuhul *or in the* sulphur

> *which does not burn [all referring to stibnite] ...*
>
> *– The Turba Philosophorum, 12th century*

Because of antimony's affinity to gold, its ability to join so easily with gold and to purify it in a manner similar to that in which quicksilver does, alchemists believed these substances to be intimately related:

> *The whole secret of our preparation is, that you take that mineral which is next of kin [related] to gold and to mercury ...*
>
> *– George Starkey (Philalethes), 17th century*

Antimony however, differed from mercury in one significant aspect, which allowed it to win out over quicksilver as a philosophical substance. Quicksilver was considered completely devoid of the sulfur principle. Antimony was viewed as inherently containing both mercury and sulfur principles. The mercury principle was cold and humid and the sulfur principle hot and dry, and both were believed to be embodied in antimony. From an alchemical perspective, this meant that antimony was much closer to ripening or maturing towards becoming gold than was quicksilver. In addition, like gold, metallic antimony resists corrosion by acids. The spirit and soul – or the growth and regenerative principles – inherent in antimony served as a perfect dual soul transplant donor into soulless gold calx.

> *Antimony is endued with all the four first qualities; it is cold and humid, and against it is hot and dry, and accommodates it self to the four Seasons of the year, also it is volatile and fixed.*
>
> *– Basil Valentine, 15th-16th century*

Early European descriptions of metallic antimony appear in writings attributed Vannoccio Biringuccio during the 16th century. Nearly two centuries later, Dutch physician and founder of clinical teaching, Dr. Herman Boerhaave sums up the properties of antimony referring to it as a *menstruum* and reflects upon its use in alchemy:

CHAPTER 1 - SOPHIC SUBSTANCES

This term [menstruum], indeed, was at first appropriated to the solvent for the Philosophers' stone, but afterwards came to be applied generally to all solvents ... All minerals of a metalline nature are solid menstrua, and especially Antimony which dissolves metals with as much ease as fire thaws ice ... This is certain, that nothing is better suited to alter the nature of metals than Antimony. Whence I cannot but suspect the Adepts made use of antimony as a menstruum in the preparation of their stone ... – Herman Boerhaave, 17th-18th century

In the Middle East, antimony has a far more ancient past reaching back to the dawn of recorded history and civilization. The Sumerians called copper *ur-udu*, stibnite *liš-a-bár*, and bronze *zabar*. The earliest examples of tin-bronze originate from Ur in Mesopotamia circa 3500-3200 BCE.

Substance	Earliest Known Use	Region
Tin	3500 BCE	Unknown
Antimony	3750 BCE	Caucasus Mountains
Carbon	3750 BCE	Mesopotamia / Egypt
Iron	5000 BCE or earlier	Egypt
Silver	5000 BCE or earlier	Middle East
Gold	6000 BCE or earlier	Middle East
Lead	7000 BCE	Near East
Copper & Stibnite	9000 BCE	Anatolia / Mesopotamia

Arsenical bronze, also called antimonial bronze, was created by reduction of antimony-bearing copper arsenates, the earliest artifacts of which were discovered on the Iranian plateau and date from sometime between 4000 and 5000 BCE. Ceremonial antimonial bronze was created at Tulaylat al-Ghassul near the Dead Sea with antimony imported from the Caucasus Mountains as early as 3750 BCE. This type of ancient bronze was known in Greek as *aes-stimmi* or by the cover-name *molybdochalkon*. Early alchemists must have understood the relationship between copper and antimony because the alchemical symbol for antimony, when inverted, is the symbol for copper. Combined, the symbols for antimony and copper form one of the earliest symbols for brass / bronze.

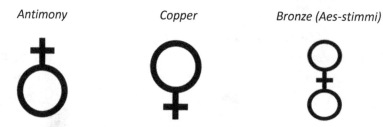

By the time Maria Hebrea began working with stibnite and antimony in Alexandria, antimony's practical use had already been long established as part of civilization's health, beauty and trade-crafts for millennia.

Ancient Health, Beauty and Tradecrafts
Desert cultures from Morocco in the west to Afghanistan in the east, and all cultures in-between, used stibnite and galena as eye cosmetics to reduce harsh desert glare and indicate social status, beauty and wealth. Bacteria-carrying desert flies and contaminated water were causes of eye infections for ancient desert dwellers. Modern research reveals that these ancient eye cosmetics acted as an antiseptic for the prevention and treatment of eye diseases such as blepharitis, trachoma, chalazion, pterygium, cataract, conjunctivitis, ectropion, and trichiasis. As such, these eye cosmetics served as an important form of disease prevention.

Apart from the varied number of alchemical cover-names used to encrypt stibnite and antimony in alchemical texts, these substances also occur in native languages across the globe. The most ancient is Sumerian *šembi* meaning *stibnite makeup*, whereas *gùnu* or *gùn* is the verb form meaning *to apply it*. In Alexandria, stibnite was called *stimmi*, *stibi* or *antimony* in Greek, and later *stibium* in Latin. The Egyptian word for stibnite or galena as a cosmetic is *mśdmt*, in Hebrew it is *kahâl*, a loan word from Assyrian *guḥlu* meaning eye-paint, and on the Indian subcontinent, it is called *kājal*. Indian alchemists named their synthetic black mercuric

CHAPTER 1 - SOPHIC SUBSTANCES

sulphide (HgS), a substance crucial to many alchemical processes, after *kājal* calling it *kajjali*. The Arabic word for stibnite- or galena-based eye cosmetics today is *koḥl*. *Koḥl* was used in the town of *Mascara*, Algeria, as a cosmetic to lengthen eyelashes. *Mascara* is just one modern term originally associated with stibnite that is still in use today. Stibnite or galena served as the *prime material* or base substance of these cosmetics to which an admixture of organics such as medicinal herbs, flowers or essential oils would be added. Recipes were inheritances passed on from one generation to the next.

Koḥl appears in the Bible as *kahâl* in Ezekiel, *pûkh* in II Kings and *stibium* in Jeremiah; each referring to an eye cosmetic. Among Muslims of both sexes, *koḥl* is an important eye protector and is known alternately as *surma* and *isthmid* depending upon the region. Its use in Islam is described as *Sunnah* (*Holy Act*) in Abu Dawud Tib 14, and Tirmidhi Tib 9. Stibnite-based *koḥl* is used in Morocco where it symbolizes the *Kaaba*, Islam's holy black stone at Mecca. The Egyptians knew it as *mśdmt* and incorporated it into funerary rites where it was applied during the mummification process, with beautiful vials of *mśdmt* accompanying the dead in the tomb for use in the afterlife. Egyptian priests used *mśdmt* in religious rites in their temples, applying it to the eyes of statues of their gods. Archeological evidence reveals that everyone – male and female in every strata of Egyptian society – wore a thick layer of black or green *mśdmt* around the eyes daily. Different qualities of *mśdmt* worn by Egyptians indicated social status.

Egyptian *mśdmt* was available in black and green colors. Common black *mśdmt* was traditionally made from galena, but also from rare imported stibnite, which remained popular in Jewish and Islamic cultures. *Mśdmt* was also worn in ancient Canaan, Armenia, Persia, and Afghanistan where large natural stibnite deposits occurred. Stibnite was a trade good imported to Alexandria via overland and sea trading routes, a record of which occurs in a trading journal known as the *Periplus of the Erythraean*

BOOK 1 – ORIGINS & EVOLUTION

Sea. *Periplus* in Greek literally means 'a sailing around' and is the rough equivalent of the captain's log for a merchant vessel. Alexandria was the major trading hub or clearing house connecting Roman trade routes in the west to India at the far eastern end of the route during the 1st century CE. The *Periplus* describes the many trade goods exchanged between the Roman Empire and India, one of which was antimony:

> There are imported into this market-town [Indian harbor of Barigaza] wine, Italian preferred, also Laodicean and Arabian; copper, tin, and lead; coral and topaz; thin clothing and inferior sorts of all kinds; bright-coloured girdles a cubit wide; storax, sweet clover, flint glass, realgar, **antimony, gold and silver coin,** on which there is a profit when exchanged for the money of the country; **and ointment, but not very costly and not much …**
>
> — Periplus of the Erythraean Sea, Chapter 49, 1st century

The antimony-copper combination so important to early Bronze Age metallurgy finds an odd parallel in Egyptian cosmetics. Both stibnite and galena served as the basis for black *mśdmt*, red *mśdmt* was rare naturally occurring antimony oxysulfide (Sb_2S_2O) whereas green *mśdmt* was produced from a prized copper ore called malachite ($Cu_2CO_3(OH)_2$). Malachite was mined in the Sinai at Timna, a site once believed to be the location of King Solomon's mines. Aside from metallurgy and cosmetics, these minerals were also widely used in ancient ceramics and glass industries as colors, dyes and pigments; chemical technologies already highly advanced by Maria Hebrea's time.

Stibnite was a household substance vital to the health and hygiene of the earliest civilizations. Stibnite and antimony are also mentioned in classic medical writings such as *The Ebers Papyrus* and the works of Celsus, Pliny, Dioscorides, Hippocrates, and others. Modern research suggests that *koḥl* boosts the body's production of nitric oxide – a chemical known to stimulate the immune system and fight bacteria. European alchemist-physicians were latecomers to the scene, with Roger Bacon's antimonial alchemy during the 13th century. By the time Basil Valentine began popularizing antimony a few centuries following Bacon, it had already been in use in Egypt, Mesopotamia, Persia and the Indian sub-continent in some form or another for more than 4,500 years.

Stibnite was a *universal substance*, important to metallurgy, cosmetics, health and hygiene, bronze and glass making, dye and pigment manufacture, priestly rites, and the Philosophers' Stone in Egypt. When viewed from its historical and cultural frame of reference, it seems natural that work to confect the Philosophers' Stone begins with this important and useful *prime material*.

The First Perfection

> And *by the Star to him they are brought near ...*
> This substance is *stellate ... This is our steel, our true hermaphrodite:*
> This is *our moon ... our unripe gold ... Old Saturn's son ...*
>
> – Sir George Ripley, 15th century

The *First Perfection* refers to the initial process of antimony purification. In the Alexandrian tradition, this meant creating *flowers of antimony* [ἄνθο μονιον; ánthos̱ monion], literally meaning *fixed flowers*, whereas in some forms of European alchemy the first step was to isolate the star regulus. As *flowers of antimony*, the powder's transition from black to grey or white indicated purity. Antimony in its pure metallic form displays characteristic crystalline patterns likened to the shape of a star. To alchemists, this was proof that this amazing substance was of celestial origin. They named it *regulus*, meaning *little king*, after the brightest star in the constellation Leo and one of the brightest in the night sky – Alpha Leonis. In Arabic, it is known as *Qalb[u] Al-'asad*, meaning *heart of the lion*. Alchemically, *antimony is the heart of gold*. Writes Zosimus:

> *Go to the waters of the Nile; there you will find a stone which has*

> *a spirit; take it, cut it in two; put your hand in its interior and draw out its heart: because its soul is in its heart.*

In alchemy, lions and kings symbolize gold and sometimes antimony. Antimony was believed to contain the *seed* or *sperm of gold* and, because of this, was considered intimately related to gold and to the Sun; it was the crown prince, unripe or immature gold – a king in the making. As an immature lion, the star regulus is depicted as a *Green Lion* to indicate its unripe state, which differentiates it from gold as a *Golden Lion*, and highly purified and reduced gold powder as the *Red Lion*. Antimony is also sometimes referred to as *Son of Saturn*, lead being the metal typically associated with Saturn. This is the European equivalent of Maria's use of the cover-name, *Our Lead*, to indicate stibnite.

> ... *the* Stone called the Philosopher's Stone, comes out of Saturn.
> – Johann Isaac Hollandus, 16th century

> *Saturn is our Philosopher's Stone, and our Latten [alloy], out of which our Mercury and our Stone is extracted* with small Labour, little Art and Expense, and in a short time. *– Johann Isaac Hollandus, 16th century*

> *The substance which we first take in hand, is a mineral similar to Mercury,* which a crude sulphur does bake in the Earth; *and is called Saturn's Child,* which indeed appears vile to sight, but is glorious within; it is sable coloured, with argent veins appearing intermixed in the body, whose sparkling line stains the connate sulphur; it is wholly volatile and unfixed, yet taken in this native crudity, it purgeth all the superfluity of Sol ... *– Eirenæus Philalethes, 17th century*

Researchers generally credit Italian metallurgist Vannoccio Biringuccio as the first person to describe a procedure for isolating antimony during the 16th century, yet Biringuccio himself credits alchemists as originators of the technique. The ancient Greeks described an antimony-copper alloy called *aes-stimmi*, which indicates an early Greek use of antimony in metallurgy and specifically bronze-craft. This is reinforced by the fact that Pliny the Elder distinguished between male and female forms of antimony; stibnite being male and antimony being female.

Philosophers' Mercury, unlike elemental mercury (Hg), does not occur naturally. Purified antimony must be separated from its sulfur matrix before it is suitable for use. An alchemist begins the preparatory processes to confect the Stone only after achieving purified antimony, a stage known as *passing the first gate* or *first degree*.

> *The mercury of the Philosophers is not found in the earth, but must be prepared by art ... and this Mercury has a power to unlock, kill and revive metals ... – Michał Sędziwój, 16th century*

> *... this [antimony] is the first Key, this is the principle part of the whole Art, this opens to you the first Gate, this will also unlock the last, which leads to the Palace of the King. – Theodore Kirckring, 17th century*

The Stone of the First Degree

> *In the first degree the Stone is called Adrop, Philosophical lead, Antimony. ... The Philosophers' Lead is not Lead Ore but the stellated regulus of Antimony.*
>
> *– Incertus Macrocosmus, 17th century*

Antimony isolated from its sulfur and in a purified form was known as the *Stone of the First Degree* or *Stone of the First Order*. Depending upon the chemical technology of the period, stibnite reduction to antimony was performed in various ways at different times throughout Western alchemy's history. Varied approaches to reduce stibnite to pure metallic antimony do not necessarily imply an altered recipe, but rather a different operational variation of the archetypal recipe using the same basic materials to achieve an identical product. We will explore these processes in detail in subsequent chapters.

1. **Alexandrian and Islamic alchemy** – Maria and her contemporaries calcined mśdmt / kaḥâl / ḳoḥl, a household item easily acquired in the markets and

bazaars at very little cost. It was available in powdered form, yet even in mineral form, it could be easily ground. Mineral sulfur (S) boils at 444.6 °C and sublimes easily, whereas metallic antimony (Sb) boils at 1,587 °C, thus allowing for separation. By calcining (pan-frying) mśdmt / kahâl / ḳoḥl at the correct temperature for the proper duration, volatile sulfur is removed and thus stibnite is efficiently reduced to *flowers of antimony* ($Sb_2O_3 \cdot Sb_2S_3$) via a simple low-tech procedure. This method is called *calcination*. Islamic alchemists also adopted this method.

2. **European alchemy** – Europeans eventually favoured the metallurgical technique of reducing stibnite with iron. Iron bonds with the sulfur present in stibnite resulting in what alchemist called *green vitriol* (iron sulphide; FeS), thus freeing metallic antimony from its sulfur bonds. The antimony would then be repeatedly purified using a fluxing agent to achieve a very pure star regulus of antimony and Mars; iron being the metal that alchemists associated with Mars. Alchemists mistakenly believed that the use of iron potentiated antimony. This method is known as *separation*.

The Chalibs [sword] of Sendivogius is the Regulus of Antimony and Mars, which is the first and the coagulated Mercury of the Philosophers; but it must be highly pure ... The philosophers' Mercury, which dissolves Gold and Silver, is a dry mercury otherwise it could not be coagulated with the perfect metals. – Multum in Parvo, 17th century

It is possible to identify the origins or stylistic influences of a Philosophers' Stone recipe by carefully analyzing the stibnite reduction method employed. Regardless of the technique used, the desired result is always a purified form of antimony.

Purified antimony was regarded as the *essence* extracted from stibnite or ḳoḥl. Islamic alchemists expressed this as *al-ḳoḥl*, which was transliterated from Arabic to Latin and eventually into English as *alcohol*. Originally, it meant *extracted essence*, with the extraction of antimony from stibnite serving as the archetypal model. Later, when distilling spirits of wine became popular among alchemists, the distillate was referred

CHAPTER 1 – SOPHIC SUBSTANCES

to as *alcohol of wine*, or *spirit of wine* (*spiritus vini*) where *alcohol* eventually came to indicate a *distilled spirit*. According to this etymology, *al-koḥl* is the *First Perfection* towards confecting the Philosophers' Stone.

SOPHIC SALT, EUDICA, SPIRIT, WATER
Antimony Trichloride (SbCl$_3$)

The third [principle] is what they call their 'Secret Fire', 'Our Mercurial Water', 'Dissolving Water', 'Fire against Nature', 'Spirit of Life', 'The Moon', 'The Priest', etc.

– Sigismund Bacstrom

Secret Fire – Ignis Aqua

The identity of the universal solvent, fluxing salt or menstruum that could both dissolve and coagulate was one of the most carefully guarded secrets of alchemy. Rediscovering this universal solvent presented an obstacle of great magnitude that must be overcome by every alchemist hoping to achieve the Philosophers' Stone. It was typically described as a *mercurial water*, convincing some alchemists that quicksilver was the actual substance being described. However, if we study carefully, adepts always cautioned against this interpretation.

> *It is a mistake to suppose that you can work miracles with a clear limpid water extracted from mercury.* **Even if we could get such a water, it would be of no use ...** *– Bernard Trevisan, 15th century*

Misinterpretation of *Mercury of the Philosophers* to mean quicksilver (or mercury of the vulgar) dates much further back than 15th century mediæval Europe. An early indication of this confusion appears in the 4th century writings of Synesius where he clarifies that *Sophic Mercury* is neither elemental mercury, nor a product derived from it. Toussaint, in a commentary on a Synesius fragment, identifies *Sophic Mercury* as a type of water, referring to it by six different cover-names:

> *I will nevertheless endeavour to make myself be better understood by these essential words of the Abbot Sinesius, who says that the Mercury of the Philosophers is not the Mercury of the vulgar, nor of the Mercury of the vulgar wholly; and I to speak much more clearly than he, I tell you, that it is no more the Mercury of any metal, [that is to say common quicksilver] but the Mercury of philosophers, the Mercury of Metals; the pontic water, the most sharp vinegar, the fire, and the viscous humidity of the philosophers. – Alexandre Toussaint, 17th century*

Purified antimony was the *Mercury of the Philosophers*. It may seem contradictory that *Sophic Salt* is also sometimes referred to as *Mercury of the Philosophers* or *Sophic Mercury* as in the above excerpt. To alchemists, however, this was not a contradiction because they viewed their solvent as antimony in its highest state of perfection. In other words, the liquid solvent was the seed, sperm or the very essence of antimony. This is an advanced form of *al-koḥl*; the essence of stibnite in its most refined or exalted state. Quicksilver was indeed used since ancient times to purify and reduce gold. Later, it would be experimented with and used successfully to create a version of the Philosophers' Stone, but this must be regarded as an offshoot or side-branch of alchemy not in accord with the archetypal recipes originating in Alexandria.

CHAPTER 1 - SOPHIC SUBSTANCES

By manipulating antimony in just the right way, alchemists arrived at a group of substances that ended with a fiery liquid solvent that could dissolve all the metals known to alchemy, including gold and silver at a fine particle size. This substance was known in Latin as *Ignis Aqua* (*Fire Water*) or by cover-names such as *Secret Fire* and *Universal Solvent*.

> *The error in this work is chiefly to be attributed to ignorance of the true fire, which is one of the moving principles that transmutes the whole matter into the true philosophers' stone ... In a short time that fire, without any laying on of hands will complete the whole work ... Therefore seek out this fire with all thy industry for having once found it thou shalt accomplish thy desire, because it performs the whole work, and is the true Key of all the Philosophers, which they have never yet revealed. – Ibn ar-Tafiz (Artephius), 12th century*

The mystery surrounding the solvent was extremely complex for many reasons. Initially, one must discover the method to distil antimony to its pure white salt. Once this had been achieved, alchemists often further refined the product via a complex series of redistillations in order to arrive at a stabilized, clear liquid solvent. The solvent must then be capable of dissolving gold; yet in order to do so the gold needed to have been perfectly prepared. Any errors in this multi-step process and all ended in failure. According to many recipes, successful creation of the solvent was the foundation upon which the entire confection of the Philosophers' Stone rested.

> *Our living water [flux] therefore is a fire which burns, breaks and mortifies the gold more than elementary fire; and the more the gold is mixed with our living water and scoured therewith, in a gentle heat, the more is it torn asunder, centrally opened and attenuated by our fiery living water. – Arnaldus de Villa Nova, 13th century*

As Arnaldus indicates in the passage above, the liquid must be capable of bringing gold to an even finer state of division. Both antimony and this solvent, being derived from the same source, are capable of doing so. Modern proponent of alchemy, R.W. Councell, commenting upon Dr. Sigismund Bacstrom's work, provides an insightful clue that the solvent is a product of distillation; an 800 year old echo of Artephius. Pure and white describes antimony salt, whereas clear fluid refers to the salt's molten state or its re-distillate. In addition, Councell provides a few more of the numerous cover-names by which it may be identified:

> The whole of this secret is ... Antimony, and a mercurial [i.e. antimonial] sublimate ... Our moist fire, by dissolving and subliming that which is pure and white [butter of antimony], casts forth or rejects, its faeces or filth, like a voluntary vomit ... The pure and white substance ascends upwards, and the impure and earthly remains fixed in the bottom ...
> – Ibn ar-Tafiz (Artephius), 12th century

> ... the philosophers' clear fluid mercury is a distilled liquid. It is, therefore, a separation from something ... Their [alchemists'] virgin and blessed water is also named Bird of Hermes, Vessel and Seal of Hermes ...
> – R.W. Councell, 20th century

Sophic Mercury is always understood as being able to dissolve gold gently, like melting ice or butter, without corrosion or violence. When alchemical texts describe *water extracted from mercury*, or *mercurial sublimate* as Artephius does, their cover-names are referring to a unique substance known as *butter of antimony*. It is *mercurial* because antimony is the identity of *Sophic Mercury*. It is a *sublimate* because it is a product of *sublimation*, a term synonymous with *distillation* in the past. It is described as a *mercurial sublimate*, indicating its properties, as opposed to the chemical ($HgCl_2$) which was generally referred to as *corrosive sublimate* by alchemists.

Butter of Antimony – Salt

One of the greatest challenges to reproducibility is decoding the language and imagery in which alchemical substances and processes are

CHAPTER 1 – SOPHIC SUBSTANCES

encrypted. To complicate matters further, some alchemical authorities appear to contradict each other openly. In the two passages below, the *Key* to the whole work is first identified as a *salt*, but then as a *water*:

> *... though they describe only sulphur and mercury [gold + antimony] yet without salt, they could never have attained to this work, since salt is the Key and beginning of this sacred science ...*
> – Michał Sędziwój (Sendivogius), 16th century

> *Our Philosophical mercurial water is the Key whereby all coagulated, fixt and unfixed metallic and mineral bodies are radically and physically dissolved and reduced into their first principle. This mercurial water has been kept very secret by all the philosophers, as the secret of the whole art ... How this philosophical water or fire, water of mercury, is to be prepared, the philosophers have carefully hidden.*
> – Nodus Sophicus Enodatus, 17th century

What classical alchemical texts reveal is a process rather than a paradox. By what process is it possible to transform metallic antimony to an antimony salt and from there to a clear, fiery thick liquid? This mystery kept European alchemists' fires burning for centuries in pursuit of a solution. A process which could do so would demonstrate that the salt and the water (cover-names: salt and dew), both paradoxically described as the *Key* in many texts, each derive from and are related to antimony. In effect, they were searching for a *mercurial water*, derived from a *mercurial salt*, originating from antimony and imbued with the power to both dissolve and coagulate.

> *When the Philosophers speak of mercury, understand our [sophic] mercury; by the water understand Mercury sublimated from its proper salt and coagulated into a salt.* – Benedictus Figulus, 17th century

Sophic Salt is the mediator between *Sophic Sulfur* (gold) and *Sophic Mercury* (antimony). By distilling this special salt, a solvent is created that dissolves fine gold-antimony glass calx first into a liquid, after which it then solidifies the composition into the Philosophers' Stone.

> *Sulphur is the father of life, Mercury is the fountain of life,* the salt is the center of life ... *The constant companion of [sophic] Sulphur is [sophic] Mercury: they never quit each other, for the one needs the other. But the* Salt preserves what Sulphur and Mercury produce. Thus is Salt the true copulator of Sulphur and Mercury.
> – *Johannes de Monte Raphaim, 18th century*

The following excerpt from Edward Kelly provides a clue that helps to identify *Sophic Salt*. The salt alchemists referred to as *butter of antimony* matches this description perfectly. *Butter of antimony* is today known as antimony trichloride ($SbCl_3$), a soft solid at room temperature that becomes molten at a gentle heat, crystallizes when it cools and will burn the skin. Antimony trichloride could then be optionally redistilled into a more refined menstruum.

> *The water of the sages adheres to nothing except homogeneous substances.* It does not wet your hands if you touch it, but scorches your skin, and frets and corrodes every substance with which it comes in contact, *except gold and silver – it would not affect these until they have been dissipated and dissolved by spirits in strong waters – and with these it combines most intimately.* – *Edward Kelly, 16th century*

Just like ordinary household butter melted in a saucepan, when *butter of antimony* is gently heated to 73.4 °C or higher, it melts into a clear thick translucent liquid.

> *The Philosophers have written much of* a vaporous water, which they have called the fire of Wisdom, *and they have said that this is material or elemental, but [yet] an essential or supernatural fire,* sometimes called a Divine fire – this is our aqua Mercurii, *which is excited by the help of common external heat, administered by Art.*
> – *Johann Siebmacher, 16th century*

CHAPTER 1 – SOPHIC SUBSTANCES

Watery Azoth – Dew

These are not fables. You will touch with your hands, you will see with your own eyes, the Azoth, the Mercury of Philosophers, which alone will suffice to obtain for you our Stone.

— *Heinrich Khunrath*, Amphitheatrum

The term adopted for such a universal menstruum throughout Western alchemical history was *Azoth*. *Azoth* was the agent of transformation by which the Philosophers' Stone would self-synthesize. The word *Azoth* employed by Maria and Zosimus in early Alexandria, Jābir and other Islamic alchemists, remained an important alchemical term right up to the 17th century. The word itself is an encryption of the properties of molten *butter of antimony*. The encryption lies in the meanings traditionally associated with letters of the ancient Phoenician and Hebrew alphabets.

Azoth is an acronym composed of the first letter of Greek and Hebrew alphabets, with the letter A symbolising oneness, omnipotence and transformation from the crude to the spiritual. The letter Z, originally meaning *spear*, was closely linked to the number 7 and signified spiritual struggle. The next letter is Omega from Greek meaning *the end*, followed by the letter Tau, the final letter of the ancient Phoenician and Hebrew alphabet, symbolising life and rebirth. In the case of Azoth, the spelling suggests:

A + Z + Ω + T/h

All in one → Struggle to a spiritual state (distillation) → Ending → Rebirth

The word is also related to the *Ain Soph* (ultimate substance) of kabbalah. *Azoth* was Arabicized to *al-zā'ūq* and later to Andalusian Arabic *az-zāwūq* meaning *the mercury*, or more accurately *Philosophers' Mercury*. The goal of European alchemists was to decode this occult formula encrypted within a cover-name. Paracelsus provides individual cover-names for the substance in its fiery, airy, watery, and earthly stage of formulation; the liquid stage being *Azoth*.

> This spirit in its *fiery* form is called a *Sandarac [stibnite]*, *in the aerial a kybrick [flowers of antimony]*, *in the* watery *Azoth [butter of antimony]*, *in the* earthly *Alcohoph and Aliocosoph [regulus and algarot].*
> – *Paracelsus, 15th-16th century*

Incertus reconfirms that *Azoth* is extracted from antimony and he identifies it by the same cover-name used by Artephius during the 12th century ... *Vinegar of the Philosophers*:

> Distilled Vinegar is not the *Vinegar of the Philosophers. Their most sharp vinegar is the secret Fire, which extracts the essence from antimony, that is from the Regulus of Antimony and Mars [metallic antimony] and forms Azoth.* – *Incertus Macrocosmus, 17th century*

Azoth was considered the *All*-in-one, due to the belief that it was a perfect combination of Aristotle's four elements – earth, air, water, and fire. *Azoth* began as earth in the form of purified antimony. The salt was then distilled, thus creating a vapour or air that condensed into *Ignis Aqua* – fiery water. This process was symbolized by the following formula:

Earth + Air + Water + Fire = Azoth

Earth	=	Adam, origins and the letter	A
Air	=	Spirit, struggle and the letter	Z
Water	=	End or final form and the letter	Ω
Fire	=	Rebirth, renewal and the letter	T/h

CHAPTER 1 - SOPHIC SUBSTANCES

According to the encryption, *Azoth* is an earth distilled into a fiery corrosive liquid; is derived from a single substance and possesses inherent regenerative properties. The word *Azoth* may be a derivative or variation of Maria Hebrea's *Ade-Zethet*. The symbol for *Azoth* came to represent concepts of a *principle body*, along with concepts of *absorption, assimilation, permeation* or *saturation*, and *reverberation*. *Azoth* is a composition of Aristotle's four elements or essences, known in Greek as the *Tetrasomia*, to which a fifth is later added. The fifth element represents the *æther* and is known to alchemists as the *Rebis*, and will be detailed in the next section. When the *Rebis* is joined with *Azoth*, the fifth element facilitates the creation of a new composite known to alchemists as the *Quintessence* – better known as the Philosophers' Stone.

Even after having arrived at this understanding, the process of creating the solvent still presented numerous additional challenges.

> Antimony is a Mineral made of the Vapour of the Earth changed into Water, which Spiritual Sidereal Transmutation is the true Astrum of Antimony; which Water, by the Stars first, afterward by the Element of Fire, which resides in the Element of Air, is extracted from the Elementary Earth, and by Coagulation formally changed into a tangible Essence ... – Basil Valentine, 15th-16th century

> Look upon that despicable thing [stibnite], whereby our Secret is opened. For it is a thing which all know well, and he that knoweth it not, will hardly or never find it: the wise man keepeth it, and the fool throws it away, and the reduction is easy to him that knows it. But, my Son, it is the greatest secret to free this Stone, or Mercury vive, from its natural bonds, wherewith he is tied by Nature, that is, to dissolve and reduce it into its primogenial Water; for without this be done, all will prove but labour lost: for else we should not be able to sever and extract the true Spirit or Watery substance, which dissolveth the Bodies. – Anonymous

BOOK 1 – ORIGINS & EVOLUTION

Mastering the process of creating *butter of antimony* and redistilling it became a labyrinthine trap that ensnared many would-be adepts. Alchemical texts often completely omit any operative details for creating or redistilling *butter of antimony*. Texts that address *butter of antimony* generally do so only with regard to descriptions. Even the foremost authority on working with antimony, Basil Valentine, was very cryptic in his writings on the subject. What was important was to achieve *butter of antimony* and, optionally, a redistilled menstruum able to dissolve gold calx and later crystallize or coagulate the concoction into the final product. Alchemical authors protected the secret by employing a mind-numbing variety of cover-names:

> *The philosophers have called this maid and blessed water by many thousands of different names in their books. They call it heaven, a heavenly water, a heavenly rain, a heavenly thaw, a May thaw, water of Paradise, an aqua fortis and an aquam Regis, a corrosive aquafort, a sharp vinegar and liquor, also Quintam essentiam vini, a waxy green juice, waxy mercurium, green water and Leonem viridis, quicksilver, menstruum or blood. They also call it urine and horse piss, milk and virgin's milk, water of arsenic, silver, Luna or Lunae water, woman, a female seed, a sulphuric steam and smoke, a fiery, burning spirit, a deathly all-penetrating poison, a dragon, a scorpion which eats its young, a hellish fire of horse dung, a sharp salt, sal armoniacum, a common salt, a lye, a viscous oil, the stomach of an ostrich which eats and digests all things, an eagle, a vulture and hermetic bird, a vessel and Sigillum Hermetis, a melting and calcinating oven, and innumerable other names of animals, birds, plants, waters, juices, milks and blood, etc. They have used all these names and written of it figuratively in their books. They have suggested that such a water is made of these things, with the result that all ignorant people who have searched for it in these things, have not found the desired water. – Anonymous*

Alchemists found this amazing substance to be a key that seemingly enabled the Philosophers' Stone to self-synthesize –the master *Key* that

CHAPTER 1 - SOPHIC SUBSTANCES

unlocked the mysteries of alchemy's most sought-after prize. Rediscovering the solvent was extremely difficult by design. Alchemists typically describe alchemical dissolution of gold via a transparent fluid that resembles dew, interpreted here as butter of antimony:

> ... we state that our fire is [a chemical] very similar to the artificial fire of the chemists, and seeks nothing more in nature than that same fire, into which it also transfers and converts itself with the greatest readiness and without resistance. And indeed gold is wholly fire, and, just like fire, is dissolved by fire not into a mercurial liquor (as some think), nor by aqua fortis (as our enemies incorrectly insist) but into and through that sappy, transparent and fluid substance that most closely resembles the celestial dew. – Gerhard Dorn

Alchemical authors used omission, vagueness, contradictions, and innumerable cover-names in order to keep the secret of keys such as this always just out of reach, while at the same time communicating success to other adepts. For those uninitiated, only the very persistent or extremely insightful experimenters ever rediscovered this fundamental key. Adepts created more cover-names for their salt / menstruum than any other single substance; a practice first introduced in Alexandria or possibly much earlier.

> Behold, now I have doubled mercury in my possession: Now I own it – white lily, powder of adamantine, chief central poison of the dragon, spirit of arsenic, green lion, incombustible spirit of the moon, life and death of all metals, moist radical, universal dissolving nutriment, true menstruum of the philosophers, which without doing any harm reduces metal to first matter. This is the true water for sprinkling, in which the living seeds of metal inhere, and from which other metals can be produced ... – Solinus Saltzal

Paracelsus recognized the value of *butter of antimony* to alchemy and developed a legend around this salt in the form of *Elias Artista*. He concealed in his legend that alchemy is nothing more than the *Artist* and his mastery of *salts*. This is actually rather accurate. According to Alexandrian methodology only *sal ammoniac*, by which all of their work was accomplished, is required to confect the Stone. While Islamic and European traditions employed an assortment of salts in the preparation of their reagents, generally they used only two salts in the actual confection of the Stone – *sal ammoniac* and *butter of antimony*. The secret to confecting the Philosophers' Stone has always relied upon the identification and the manipulation of these salts.

Paracelsus encrypted this hidden truth in the name *Elias Artista*. *Elias Artista* was said to be an ancient wandering alchemist who would appear in the future to reveal the secrets of alchemy to the world. The name is an anagram that, when rearranged, clearly makes his point:

Elias Artista = Artista i Sale (Artist of the Salt)

Liquid Antimony Menstruum
Throughout the tradition of European alchemy, *butter of antimony* remained the crowning glory of menstruums used to confect the Stone. European alchemists during the early modern period further experimented with redistillations of *butter of antimony*. The various historical techniques for reducing stibnite yielded either *flowers*

CHAPTER 1 – SOPHIC SUBSTANCES

of antimony (antimony ash; $Sb_2O_3 \cdot Sb_2S_3$), *glass of antimony* (antimony oxysulfate; $Sb_2O_2SO_4$) or the star regulus (metallic antimony; Sb) as precursors. In the European tradition, any one of these was then reacted to yield *butter of antimony* (antimony trichloride; $SbCl_3$) by which various methods are available. Once *butter of antimony* had been achieved, it was optionally distilled into *liquid antimony menstruum* by several possible procedures. These manipulations resulted in a number of new substances of interest to chemistry, first produced by European alchemists. European alchemists may have created complex oxychlorides, antimonic acid or stibine as byproducts. If so, these remained either secret or unidentified substances until being characterized by modern chemists.

- *Flowers of antimony*   antimony trioxide ($Sb_2O_3 \cdot Sb_2S_3$)
- *Glass of antimony*   antimony oxysulfate ($Sb_2O_2SO_4$)
- *Butter of antimony*   antimony trichloride ($SbCl_3$)
- *Powder of Algarotti*   antimony oxychloride ($SbOCl$; $Sb_4O_5Cl_2$)

Alexandrian and Islamic alchemists reduced readily available stibnite to *flowers of antimony* by calcination, whereas later Europeans preferred to begin with the star regulus of antimony. *Flowers of antimony* and *star regulus of antimony* are very different starting materials, each requiring specific methods by which to achieve *butter of antimony*. *Powder of Algarotti*, a byproduct of working with *butter of antimony*, played an increasingly important role, going on to become one of the first alchemical substances distributed widely throughout Europe on a commercial basis. *Powder of Algarotti* was the identity of alchemical Hermes, Cleopatra's Serpent, Zosimus' *divine water* fixed to Mercury and Synesius' White Stone of Hermes. *Liquid antimony menstruum* required skill and understanding in order to achieve it in its pure form.

Millennia of metallurgy and glass making informed the expertise of Maria Hebrea and other Alexandrian alchemists with the use of various menstruums and fluxing agents. Alexandrian alchemists used *sal ammoniac* to distil *butter of antimony* as their *Sophic Salt*. Maria's methodology, as presented to King Aros, clearly indicates a 1st to 3rd century technology for creating *flowers of antimony* by calcination using the very same technique detailed by Basil Valentine in his landmark work on the subject, *Triumphal Chariot of Antimony*, published during the 16th century. *Sal ammoniac* decomposes to create vapourous and liquid hydrogen chloride responsible for the chemical reaction that ultimately results in the Philosophers' Stone. It was a salt that turned into a fiery vapour; its action responsible for the chemical reaction between gold, antimony and chlorides.

One of the innovations in Islamic alchemy is the use of distilled or salt-saturated *urine menstruum*, a widespread ancient industrial technology used in fabric dyeing and leather tanning that dates back to prehistory. Al-Razi made great use of *urine menstruum* in his brand of alchemy, but its use is first mentioned by Pseudo-Democritus during the late 2nd century in his *Physika kai Mystika*.

Many of the techniques employed by European alchemists evolved from the period between the 12th and 17th centuries and cannot claim ancient origins. Operative details confirm European technological advances, yet the desired result was always the same according to a European archetypal process – the reduction of stibnite, creation of *butter of antimony* and its transformation to a liquid solvent by applied heat or redistillation. Although European technical innovations advanced control and efficiency, it is also possible that the Alexandrian archetypal recipe suffered unnecessary over-complications. By carefully studying a broad spectrum of alchemical writings with an eye towards the salt / menstruum and the various processes employed to create it, we can gain greater insight into the technological and cultural heritage of each particular Philosophers' Stone synthesis preserved in alchemical texts.

CHAPTER 1 - SOPHIC SUBSTANCES

Not Alone – and Best Supporting Actor Goes to ...

European antimony was sourced from mines in France, Spain, Hungary, and Italy. Hungary is mentioned as a source of quality antimony in numerous alchemical texts. Antimony has unique properties that make it useful as an alchemical substance. Just like water freezing to ice, antimony expands when it cools and solidifies. Its crystals contain an unusual amount of empty space similar to water, causing it to resist contraction when it cools. With antimony in a molten state, gold alloys with it easily and at a much lower temperature than is normally required to melt gold. As the alloy cools, it serves to break gold down to a fine state of division. A *gold-antimony alloy* is also very brittle upon cooling, which allows for easy grinding to powdered form. These properties have fascinated early alchemists and modern scientists alike:

> *... it [antimony] may be reputed one of the seven Wonders of the World*
> *– Basil Valentine, 15th-16th century*

> *If chemical elements could win Oscars, antimony would get my vote for Best Supporting Actor. – Richard I. Gibson, 20th century*

This is high praise for a metal that is not a precious metal or even a true metal. Not only is antimony gold's lover, wife and assassin, it is also sometimes referred to by cover-names such as *prostitute, whore, harlot* or *friend of metals* due to its usefulness in metallurgical processes apart from alchemy. It serves as a hardener when alloyed with other metals, making them stronger and corrosion resistant. There is a good chance the lead posts in your car's battery are actually a lead-antimony alloy. Antimony was employed in the manufacture Britannia metal, pewter and printing type metal to increase firmness and smooth edges. Since the time of the Gutenberg printing press in the 15th century until movable type printing was recently replaced by modern technologies, a special alloy created by Johannes Gutenberg himself relied upon antimony as a key metal of choice along with lead, tin and a bit of copper to manufacture his moveable type metal. Another unique property of antimony is that, just like gold, it does not tarnish or corrode. When antimony is used in the manufacture of bell metal and cymbals, it enhances resonance and prevents corrosion.

Alchemists' fascination with antimony and their belief that it was celestial in origin is certainly understandable. It is not actually a metal, nor is it a non-metal. It is not crystal in the common sense yet displays a crystalline, star shaped pattern and behaves in a similar manner to water crystals by expanding upon cooling. Antimony is about as rare as silver and sits between metals and halogens on the periodic table of elements, which classifies it as a *metalloid* and, as such, a semiconductor. The name antimony is aptly chosen from Greek where *anti + monos = not alone*. Throughout metallurgical history, antimony has only rarely ever been used alone.

Basil Valentine, alchemist and expert on the subject of antimony, must have intuited its identity as a metalloid rather than a metal when he wrote:

> *Antimony is no other than a Fume, or (as I may otherwise call it) a Mineral Vapour, which is genited from above by the Stars.*
>
> – Basil Valentine, 15th-16th century

The importance of stibnite, antimony and its role in confecting the Philosophers' Stone is encapsulated in the following passage. Interpretations [in square brackets] are those of the author.

> *There is a stone and yet it is no stone [metalloid] wherein the whole art lieth concealed [in stibnite].* **Nature has formed it but has not brought it to perfection [of flowers, regulus or glass].** *You will not find it above ground: it groweth only under the foundation of the mountains [where it is mined]. In this subject lies the whole art.* **Whosoever hath the fumes or vapours [distillate; butter of antimony] of this thing and the golden splendour of the Red Lion [gold-antimony glass], with a highly pure**

CHAPTER 1 - SOPHIC SUBSTANCES

[sophic] mercury [butter of antimony], and knoweth the red sulphur [gold-antimony glass] in this composition, he has the foundation of the whole art. – Michał Sędziwój (Sendivogius), 16th century

LATTEN, BODY, EARTH, THE REBIS
Gold-Antimony Glass (Au·Sb$_2$O$_2$SO$_4$)

*The first [principle] being well purified,
and the second [principle] properly prepared,
they are then joined together, and the compound which is called Rebis
is then reduced to powder and mixed with the third.
Thus are all the three principles united in proper proportion.*

– Sigismund Bacstrom

The Hermaphrodite or Androgyne
The *Rebis* is the *Stone of the Second Degree* or *Stone of the Second Order*. Operatively speaking, the secret to the *Rebis* is one of proportion. The ratios of gold to antimony vary depending upon which recipe the experimenter is following, or from which point in history the recipe originates.

Creating the *Rebis* is a two-step process, according the above passage from Bacstrom:

1. First, a union is created by fusing refined and reduced gold calx with purified antimony in just the right proportion to make the gold brittle. This process is sometimes referred to as vitrification – literally to make a glass.
2. Once cooled, the gold-antimony alloy is then ground to a very fine powder and combined with the third principle, butter of antimony or liquid antimony menstruum, in the proper proportions.

The *Rebis* was a term used by alchemists meaning *dual thing* and denotes a union of opposites into a single substance believed to be greater than the sum of its two parts. It is likened to a hermaphrodite in the sense of the conjoining of male Hermes and female Aphrodite into an androgynous divine being beyond duality. It is also known as the sacred chemical marriage between the Sun King (Sol; gold) and the Moon Queen (Luna; antimony); their intercourse resulting in a divine androgynous offspring. A number of emblems were adopted to represent the *Rebis*, the most notable being the two-headed eagle with outstretched wings.

> *These two are Sol [gold] and Luna [antimony]. You will never obtain perfection unless Sol [gold] and Luna [antimony] be united into one body.* – Lorenzo Ventura, 16th century

> *The true matter has been named by various appellations by the philosophers, though in truth it is one thing, Rebis ... the hermaphrodite.*
> – Incertus Macrocosmus, 17th century

The two-headed eagle as a symbolic representation is found in ancient civilizations that stretch back to the earliest human settlements ranging from India to Persia, Mesopotamia, Egypt, Turkey, and Armenia. The earliest image of the *Rebis* can be traced back in history to Sumeria around the beginning of the Bronze Age, circa 3800 BCE. In India, the *Rebis* appears as the mythological double-headed Eagle known as *Gandaberunda*.

In pre-dynastic Egypt, long before any pyramids had been built, populations living along the Nile River Valley worshiped a Mother Earth goddess represented in stone carvings as a two-headed bird. Very little is known about this civilization, but what is clear is that these representations were common. Archeological research has unearthed many tombs containing relics in the form of a double-headed bird deity.

The *Rebis* was worshiped by the ancient Kenites as a Sun god at Mount Sinai, then known as the Wilderness of Kadesh or sometimes simply as the Wilderness. This god was an extension of an earlier Hittite storm god also appearing in the form of a double-headed eagle. This is likely

CHAPTER 1 – SOPHIC SUBSTANCES

the majestic and powerful Lord of thunder and lightning alluded to in the twenty-ninth psalm in the *Bible*, a psalm of David:

ז קוֹל־יְהוָה חֹצֵב לַהֲבוֹת אֵשׁ *7 The voice of the Lord flashes forth flames of fire.*

ח קוֹל יְהוָה יָחִיל מִדְבָּר יָחִיל יְהוָה מִדְבַּר קָדֵשׁ *8 The voice of the Lord shakes the wilderness; the Lord shakes the wilderness of Kadesh.*

ט קוֹל יְהוָה יְחוֹלֵל אַיָּלוֹת--וַיֶּחֱשֹׂף יְעָרוֹת וּבְהֵיכָלוֹ-- כֻּלּוֹ אֹמֵר כָּבוֹד *9 The voice of the Lord makes the deer give birth and strips the forests bare, and in his temple all cry, "Glory!"*

The double-headed eagle image later became a very important symbol to the Roman Empire and Christianity, and served as a powerful emblem in heraldry and Scottish Rite Freemasonry.

> *Pure gold is of most difficult and hard fusion, but with Antimony it melts in a moderate fire ... If you mix Gold with Regulus of Antimony the Gold forgets its pristine stubbornness in the fire, and now melts, like lead or tin, in a small heat. – Georgius Agricola, 15th century*

The passage from Agricola above draws attention as to why alchemists and metallurgists of all ages were fascinated by the reaction between gold and antimony. Ordinarily, gold melts at 1,064.18 °C. Antimony, however, melts at 630.63 °C. Metallurgists observed the phenomenon that when gold is added to molten antimony it melts into a fusion with antimony at a much lower melt temperature than required for gold alone. This provided an important operative advantage in that it reduced the amount of heat and fuel normally required to smelt gold.

BOOK 1 – ORIGINS & EVOLUTION

Lion's Blood and Eagle's Gluten

> It is neither mineral nor metal,
> but the original Mother and First Matter of all minerals and metals.
> It is nothing else but the Lion, with his coagulated blood,
> and the gluten of the White Eagle.
>
> – Paracelsus

Paracelsus suggested that the Philosophers' Stone was nothing more than a combination of *Lion's Blood* and *Eagle's Gluten*. *Lion's Blood* can refer to a number of alchemical substances, all of which serve the exact same purpose of acting upon gold in such a way that it is more easily reduced and coagulated. Although Paracelsus was referring only to the first example below, the term *Lion's Blood* has been applied to the following substances:

1. **Latten / Rebis** – *gold-antimony glass or alloy ground to the smallest particle size possible;*
2. **Gold calx** (Au) – *purified and reduced to the smallest particle size possible;*
3. **Anhydrous gold trichloride** ($AuCl_3$) – *a red crystalline gold salt achieved by drying gold trichloride monohydrate to the anhydrous form.*

The discovery of aqua regia by Jābir ibn Hayyān in the 8th century CE allowed alchemists to achieve the gold salts listed above with little effort. Prior to the 8th century, Philosophers' Stone recipes required gold and antimony to undergo a much longer reaction time using various fluxing salts, such as the Alum of Spain mentioned by Maria in the passage below.

> *Sol [gold] or Luna [silver in this example] must be calcined philosophically with the first water [molten glass of antimony] that the perfect body [of the precious metal] may be opened and become porous to enable the second mercurial water [butter of antimony] to have the readier ingress.* – Khālid ibn Yazīd (Calid), 7th-8th century

CHAPTER 1 - SOPHIC SUBSTANCES

... you shall call the Stone our Latten, and call the Vinegar Water, wherein our Stone is to be washed; this is the Stone and the Water whereof the Philosophers have wrote so many great Volumes.
— *Johann Isaac Hollandus, 16th century*

Paracelsus used the cover-name *Eagle's Gluten* to refer to *butter of antimony*. The eagle is also a reference to *sal ammoniac*, derived from a reading of Herodotus' *Histories*, Book IV, where he describes *ammonium chloride* salts found at Augila (*Eagle*), the second stop on the trans-Saharan caravan route. Paraceslus' cover-name is a strong indicator that he knew *sal ammoniac* was originally used to create *butter of antimony*. The reaction to create *butter of antimony* first entered Europe through translations of Morienus, Artephius popularized it and Basil Valentine immortalized it in his Second Key. Alchemically speaking, the term *eagle* came to imply volatization.

The eagle is a predator able to devour many kinds of prey. *Sal ammoniac* was widely used as early as the 1st century CE as a flux, solvent and volatization agent. *Glūten* is Latin for *glue* or *binding substance*. Today we use the term *gluten* in reference to the mixture of proteins found in foods processed from grains that give dough its elastic texture. Again, we come across a parallel with bread baking, in that the fluxing agent has traditionally been equated to the leavening agent in bread. Although Paracelsus was originally referring to *butter of antimony*, the term *Eagle's Gluten* can refer to the following substances:

1. **Aqueous Sal ammoniac** *(NH_4Cl) – the fluxing salt used in early Alexandrian archetypal recipes;*
2. **Butter of antimony** *($SbCl_3$) – an antimony salt, antimony trichloride, achieved by a number of different methods;*

The lion is the king of earthly predators and the eagle the king of aerial predators. By merging these two kings into a type of hybrid, it was as if the heavens and the Earth had been joined into a single sacred being.

BOOK 1 – ORIGINS & EVOLUTION

Alchemical imagery often depicts a lion with eagle's wings known to the Greeks as a *Griffin*. *Griffin* imagery suggests vitrified gold, as the *Rebis* or *Lion's Blood* (fixed earth principle) undergoing union with *Eagle's Gluten* as *butter of antimony* (volatile water principle). The dual nature of the *Griffin's* composition means that the *Griffin* technically qualifies as a *Rebis*, but usually it refers to the matrix that will undergo reaction resulting in the Philosophers' Stone.

In the writings of Flavius Philostratus and Pliny the Elder, *Griffins* are portrayed as the guardians of gold and able to quarry and unearth gold from hidden terrestrial depths; a legend likely arising due to gold deposits found close to fossil beds along Issedonian trade routes. Greeks considered the *Griffin* to be the king of all creatures earthly or aerial, and as such, they were responsible for guarding the world's treasures.

Once again, this very ancient image originates in Sumeria. The Mesopotamian *Rebis-Griffin* deity was named Zu, Anzu, Igig, or Imdugud and along with being depicted as a double-headed eagle, was depicted as a lion-eagle hybrid or *Griffin*. Travelling Greek scholars studying in Mesopotamia likely adopted the legend and image, incorporating it into their own mythology and teaching it upon their return. It appears that alchemical imagery of the *Rebis*, *Lions* and *Eagles* can all be accounted for as stemming from the Sumerian Zu mentioned above. Ancient Sumerian and Akkadian mythology depicts Zu as the divine storm-bird that stole the Tablets of Destiny – clay tablets bearing decrees of the divine – and secreted them away to an unknown mountaintop. This theme is revisited millennia later in the story of Moses retrieving the tablets from Mount Sinai in the presence of an ancient metallurgical Kenite god who, in earlier traditions, took the form of a double-headed eagle.

The symbol of the double-headed eagle or the lion-eagle hybrid is deeply embedded in the collective consciousness of the world's great cultures.

CHAPTER 1 - SOPHIC SUBSTANCES

That these symbols date back to the dawn of Bronze Age civilizations and correspond so well with metallurgical and alchemical processes is compelling evidence – although speculative and circumstantial – that the Philosophers' Stone may claim far more ancient origins than Græco-Roman Alexandria, as generally postulated by historians of metallurgy, alchemy and chemistry. An Indo-Iranian Chalcolithic or early Bronze Age metallurgical culture is a fair candidate for the Stones' prehistoric origins.

Metallurgy and bronze technology played a pivotal role in society at a time in history when the wealth and strength of a city-state was dependent upon artisans producing valuable trade goods, providing armour and weapons to a standing army, and creating artistic furnishings to adorn palaces and temples. Trade secrets were guarded and complete ideologies in line with religious belief systems were infused by artisans into their craft traditions.

Cultural intercourse between Sumerians, Akkadians, Babylonians, Assyrians, Hittites and others engendered a sharing of related cultures, pantheon of gods, belief systems, and technologies. One of the technologies shared by Mesopotamian cultures was metallurgy as it relates to antimonial bronzes and gold technology. If we understand the imagery of the Zu as pertaining to metallurgy, the history of the Philosophers' Stone is quite ancient and can be viewed as preserving Sumerian metallurgical imagery related to the early Bronze Age. It remains to be proven whether *Rebis-Griffin* imagery represents Sumerian or earlier Indo-Iranian bronze-craft or not, but what is very clear is that alchemists understood these images to be directly representative of confecting the Philosophers' Stone.

The Chemical Wedding

> *Make the marriage between the Red husband and the White wife and thou shalt have the mastery.*
>
> – *Muhammad ibn Zakariyā al-Rāzī (Rhazes)*

The *Chemical Wedding* is the process of finishing the Philosophers' Stone, which upon completion is known as the *Stone of the Third Degree* or *Stone of the Third Order*. The three orders refer to antimony passing through three stages: 1) purification; 2) union, and 3) fixation. *Lion's Blood* and *Eagle's Gluten* are synonymous with the Red Man and the White Woman. The process of creating *Lion's Blood* or *Rebis* (*gold-antimony glass* or alloy) alloy and combining it with the *Eagle's Glue* (*butter of antimony*), – a mixture symbolized by the *Griffin* – was also known as the *Chemical Wedding* alluded to by al-Rāzī in the passage above.

This allegory would remain popular with European alchemists in general, and the Rosicrucian order in particular. The *Chemical Wedding* is immortalized in the allegorical tale known as *The Chymical Wedding of Christian Rosenkreutz* written during the 15th century CE. The story

CHAPTER 1 – SOPHIC SUBSTANCES

is reputed to be the inspiration for the development of the various Rosicrucian societies that followed. It fuses alchemical processes with the rite of Passover, resulting in a curious union of alchemy and religion. The wedding is indeed a marriage of two chemical substances.

The white woman, if she be married to the red man, presently they embrace, and embracing are coupled. **By themselves they are** *dissolved* **and by themselves they are** *brought together, that* **they** *which were two, may be made as it were one body.* **–** *Masculinus, in Arnaldus de Villa Nova's Rosarium Philosophorum, 13th-14th century*

... the Red Man *[gold] and the* White Woman *[glass of antimony] be made one,* **Spoused with the Spirit of Life** *[butter of antimony] to live in love and rest. – Sir George Ripley, 15th century*

Our art is nothing else but an equal mixture of the powers of the elements, of heat and dryness – cold and humidity; a natural equality; a union of the man [rebis] and his wife [butter of antimony] ...
– Michał Sędziwój (Sendivogius), 16th century

The beginning of this Art is One only thing [The Stone] composed of two substances – a fixt [Rebis] and an unfixed [Butter of Antimony]. **One is the seed, the other is and remains the Mother. The one is the** *Red fixed servant [gold; Sun], the other is the* White Wife *[antimony; Moon]. One is the [Sophic]* Mercury, *the other is the [Sophic] sulphur.*
– Incertus Macrocosmus, 17th century

The entire mass comprised of a matrix of gold, antimony and *butter of antimony* should have the consistency of a dough or soft clay with no excess moisture. The mass is then placed into a digestion vessel and sealed airtight. The digestion vessel is then subjected to low heat and the mass need only be monitored for color indicators that signal the alchemist to adjust the temperature accordingly. Final union is achieved when the chemical reaction results in the Philosophers' Stone. Today it is possible to reproduce this reaction extremely efficiently using ratios based upon the molecular weight of each ingredient; a luxury unavailable to alchemists of old.

The color indicators are a result of a slow equilibrium chemical reaction, chloride oxide ion exchange, which takes place due to a gradual increase in heat. This is called a *regimen* and proceeds according to a traditional seven stages characteristic of the *Ars Magna* (Great Art). The *Ars Brevis* (Brief Art) to confect the Stone is by far the most ancient of Philosophers' Stone recipes. It likely originated as shared Bronze Age metallurgical technology uniting Tartessos, Mesopotamia, Persia, Egypt, and Phoenicia, and possibly Bactria.

ALL PATHS LEAD TO THE STONE

Via Sicca – Dry Path

The Dry Path (*Via Sicca*) is also known as the Brief Art (*Ars Brevis*). It is a metallurgical process where the entire work is performed in a crucible heated by a furnace or over a strong fire. This path is simple, elegant and can be completed in very short time. The fascinating history behind this path is extremely ancient and, based upon available data, can be attributed to Semitic Akkadians from Mesopotamia or Semitic populations that migrated to Egypt from the Mediterranean. As suggested earlier, the *Via Sicca* is far more ancient and widespread than current research indicates. A key to this path was the fluxing salt. The Dry Path, which was used by Moses, as I show in the next chapter, was known to Maria Hebrea.

Via Humida – Wet Path

When we think of the Philosophers' Stone and the archetypal synthesis to confect it, the *Ars Magna* (Great Art) is what is generally being referred to with all of its imagery, colors, planetary, and zodiacal correlations. This is also referred to as the Wet Path (*Via Humida*) and cannot be traced any further back than Maria Hebrea's practice in Alexandria circa 200 CE.

CHAPTER 1 - SOPHIC SUBSTANCES

It appears to be an adaptation of the Dry Path where metallurgical furnace and crucible work was replaced by slow heating and digestion techniques possibly originating from medicine, perfume and fragrance craft technologies. Interestingly, the earliest renowned alchemists in recorded history were a group of Alexandrian craftswomen active between the 1st to 3rd centuries CE. These founding mothers of alchemy assumed legendary Jewish, Egyptian and Ptolemaic pseudonyms. Aside from Maria Hebrea, the writings of Isis the Prophetess and Cleopatra the *chrysopœian* survive to the present as fragments in various alchemical compilations. The works of Alexandrian alchemy's grand matriarchs reveal an accurate and complete template or archetypal theory and operative methodology to confect the Stone.

During the 1st to 3rd centuries CE in Alexandria, mineral acids had not yet been characterized. Fluxing salts held the key to early Alexandrian Philosophers' Stone Wet Path syntheses. Maria graciously reveals that her secret salt was *sal ammoniac*. In order to create a reaction between gold and antimony oxysulfates using low heat, a volatile chloride salt is required to serve as a catalyst. *Sal ammoniac* was a key component to chemical technologies in early Alexandria, and employed to create *divine water* and the Stone via the Wet Path. In Maria's time, *sal ammoniac* was sourced from the Temple of Jupiter in Ammonium located at the Siwa Oasis in Libya, or from the Augila Oasis further along the trans-Saharan caravan route.

By the 7th century CE, Alexandrian alchemical texts authored by Stephanos of Alexandria reveal a shift in emphasis away from the industrial application of the Philosophers' Stone as a metallurgical and glass coloring agent towards deeper discussions of astronomy, medicine and esoteric gold making as a unified philosophy and path to personal redemption. Stephanos identified a number of plant substances and their ashes in relation to alchemy. He also employed the word *pharmakon* –

which he termed *ash* – as synonymous with *elixir*. Stephanos listed two applications of his ash:

1. *Ash which is used by a physician in healing;*
2. *Ash which is used in the art of making gold.*

His method of *chrysopœia* was consistent with earlier Alexandrian alchemists. Egyptians practiced gold making by projecting powders or ashes (metal-oxides) into molten copper. It was around the time of Stephanos that trade in Chinese, Indian alchemical imports reached its peak, and Indian medicinal metallic ashes, called *bhasma*, were available in Alexandrian markets as tonifying or rejuvenative elixirs.

Stephanos' disciple was a monk named Morienus. Around the end of the 7th century CE, alchemy was transmitted to an Islamic prince living In Damascus named Khālid ibn Yazīd who sent for the best teacher of alchemy to serve as his private tutor; Morienus being considered the best available teacher at the time. Although a Christian, Morienus' spiritual and alchemical beliefs were in harmony with the beliefs of followed by Prince Khālid. The Prince's alchemical teachings were eventually transmitted to a Sufi named Jābir ibn Hayyān, Islam's extremely influential alchemical pioneer. Jābir's incredibly important contribution to alchemy was the discovery of *aqua regia*, a highly corrosive double acid able to dissolve gold. *Aqua regia* opened the door to creating the various gold salts used in alchemy. From this point onward, various salts and their mineral acid distillates would continue to play a vital role in the development of new variations of the archetypal Philosophers' Stone recipe. When we see a synthesis for confecting the Stone that requires *aqua regia* or the distillation of mineral acids, we can generally date the recipe to the 8th century CE or later. This chemical technology is the heritage of the Islamic alchemical tradition, rather than the Alexandrian.

CHAPTER 1 - SOPHIC SUBSTANCES

Once the Philosophers' Stone reached Europe, recipes to confect it evolved into a great variety of approaches. Newer and stronger acids, the ongoing identification of a wide variety of chemical substances and intense experimentation all contributed to an outpouring of alchemical texts. Operative technique to create the Philosophers' Stone in Islamic Andalusia derived from Morienus and recorded by 12th century Islamic alchemist, Artephius, in his *Secret Book*, revealed the *Secret Fire* and *Vinegar of the Philosophers* as *butter of antimony* created via antimony and sal ammoniac. Much of Artephius' book consists of descriptions and deserved praise of this amazing solvent that would become the mainstay of European alchemy for nearly 600 years afterwards.

Artephius' solvent was understood and appreciated by Paracelsus who discovered a unique method of reducing stibnite directly to *butter of antimony*, resulting in a byproduct that he called *cinnabar of antimony*. Paracelsus confected the Philosophers' Stone using his *Eagle's Gluten*, a technology that originated with Morienus. Vittorio Algarotti, having been influenced by Paracelsus, manipulated *butter of antimony* to achieve *Algarot powder*. In a recipe attributed to Synesius in the 4th century, in addition to those of Cleopatra and Zosimus, the product of their recipes is *Algarot powder*, suggesting a separate Alexandrian point of origin prior to its rediscovery in Europe. The fascinating aspect of this powder is that it offered the possibility of an 'instant' Philosophers' Stone. When *Algarot powder* is 'fermented' with the proper measure of gold calx, gold salts or other metal via fusion, the result is the Philosophers' Stone.

One of the most sophisticated, yet direct approaches to confecting the Stone comes from writings attributed to the European alchemist Basil Valentine, who presented the entire synthesis in detail. He included the use of *aqua regia*, ether and careful distillation to achieve gold salts. He also employed *butter of antimony* using Artephius' recipe and combined it with the *Rebis*, which he called *vitrified gold*, to achieve the Philosophers' Stone. Some see Basil Valentine's work as the crowning of European alchemical formulation. His gold purification and reduction technique deserves further research in order to determine his candidacy as the European discoverer of a method to achieve dry gold nanoparticles.

Regardless of which approach one takes, the foundational principle of joining gold to antimony via the medium of a salt / menstruum is present in all genuine Philosophers' Stone recipes. The original Alexandrian template or blueprint to confect the Philosophers' Stone unites all forms of Western alchemy, whether discussing the earliest Alexandrian methods; Islamic alchemy's use of *urine menstruums*, mineral acids and distillation; European advances that employ a variety of chemical reagents and methodology that is more sophisticated; Dry Path or Wet.

The Red Thread

The identity of the basic Philosophers' Stone from a chemistry perspective is a species of gold-antimony salt. Gold-antimony salt is achieved by a process of ion exchange. For the process to work, ion donors are required to donate chlorides, oxychlorides and oxysulfides to a gold-antimony recipient. A number of substances may be used effectively to create the reaction between gold, antimony and chloride ions. The earliest was Alexandrian alchemy's sal ammoniac (ammonium chloride; NH_4Cl). Salt-saturated or distilled *urine menstruums*, rich in chlorides and ammonia, came into vogue in the Islamic tradition, followed by *butter of antimony* (antimony trichloride; $SbCl_3$) which donated both antimony and chloride ions. Europeans experimented with *redistilled butter of antimony. Liquid antimony menstruum* serves as a donor of antimony and chloride ions. Alchemists correctly attributed vitality and regenerative principles to their salts and menstruums. In retrospect, the ion exchange potential qualifies these substances as philosophic.

CHAPTER 1 – SOPHIC SUBSTANCES

Upon analyzing the various recipes or paths, it soon becomes clear that what is considered a path may be understood as an alchemist's personal preference or methodology by which he prepared his reagents. The preparation of all reagents artisanally from scratch is an aspect of alchemy that largely differentiates alchemists from chemists. Reagent preparation is also an area where wide latitude of options, variations and personal choice come to play a crucial role. Gold must be purified and reduced, antimony separated from its sulfur, salts crystallized and distilled and in some cases redistilled. Each of these processes may be performed via a wide selection of methods involving metallurgy and/or chemistry. An alchemist viewed his options and decided upon which techniques to employ. He then combined them and thus formed a unique designer formula; a personalized recipe or path. The various paths are each a collection of preparation processes that reflect the preferences of the individual alchemist. This has given rise to the notion that there are many different paths to create the Philosophers' Stone.

Most alchemy researchers have concentrated upon understanding and individually reproducing each path. This approach may be understood as a *process-based approach*. To make an analogy, each of these paths may be likened to an individual and unique gemstone or bead of varying colors, value and lustre strung together on the necklace of alchemy via a unifying, hidden red thread. This treatment is an attempt to characterize the red thread and reveal it as an underlying principle that unites all the various paths to the Philosophers' Stone. The aim is to reveal an archetypal blueprint or template that, when overlaid with any valid Philosophers' Stone recipe, demonstrates how success is achieved by adhering to a set of inviolable fundamental principles of confection. From this perspective, there is only one path to confect the Philosophers' Stone, yet achieved in a variety of ways.

Another interesting revelation was that each canonical recipe that adhered to the basic principles theoretically results in the exact same species of product. One of the details rarely mentioned in any canonical recipe is the exact ratio measure of reagents to confect the Stone. In recipes that do provide measures or ratios, it can be difficult to discern the measurement as indicating mass or volume, or the historical system of weights and measures. Some recipes for both reagent preparation and Stone confection were inefficient, creating waste or byproducts that needed to be separated by decanting, distillation or evaporation, whereas the more efficient recipes result in a quality end product free of waste or byproducts. The chemical reaction between gold, antimony and chloride ions is at the core of the Philosophers' Stone confection. Sigismund Bacstrom, likely due to having been directly initiated into Rosicrucian technique, understood these fundamentals. This allowed him to analyze and provide insightful commentary upon a vast collection of recipes. Approaching the Stone's confection from an in-depth understanding of the red thread of underlying principles necessary for a successful end product may be understood as a *product-based approach*.

By beginning with a principle or product-based approach, the alchemist becomes freed from the dogma of canonical recipes developed by other alchemists and can begin to develop a unique and personal path to the Philosophers' Stone. This is precisely how canonical recipes were originally developed, by the process of innovation with an aim towards efficiency born of a deep understanding of the fundamental principles, the archetypal template, to confect the Stone. Maria Hebrea's recipe may be regarded as the prototype for the metallurgical *Ars Brevis* and her personal innovation, the original chemistry-based *Ars Magna*. From the perspective of modern chemistry, Maria's *Ars Magna* is a magnificent example of efficiency and elegance. European innovations resulted in a new twist on the *Ars Magna* when Morienus and later Artephius introduced *butter of antimony* developed from Maria's innovative use of sal ammoniac to distil *divine water*. Morienus' influence resulted in a new European Philosophers' Stone template that was followed by great alchemists such as Paracelsus, Heinrich Khunrath, Basil Valentine, Sigismund Bacstrom and many others. Yet the archetypal recipe – a chemical reaction between gold, antimony and chloride ions – remained intact. Regardless of the individualized methods employed to prepare the reagents, the final product of Alexandrian, Islamic and European foundational recipes is the archetypal Philosophers' Stone as a species of gold-antimony oxysulfide.

A completely separate methodology prevalent in Islamic and European alchemy is the use of mercury in elemental and chloride forms. Alchemists adopting the technique of employing mercury to confect the Stone were called *mercurialists*. This practice was never part of Judeo-Egyptian Alexandrian technique, although mercury was widely used in ancient Egypt for gold reduction and as a Babylonian-style fire gilding agent. Islamic alchemy made use of mercury in alchemy and medicine, but they imported much of the theory and application from India – a culture with a far more ancient and widespread use of mercury, and one of the most comprehensive branches of calcined mineral and metal rejuvenative medicines ever known. The innovation of mercurial alchemy can be traced to China. Indian alchemical tradition credits a Chinese alchemist with exporting the technical use of mercury from China to India. The archetypal recipe of *mercurialists* centers on mercuric oxide by first alloying prepared gold and antimony followed by amalgamating the alloy with mercury. This mixture undergoes a process of oxidization in a sealed vessel, catalysed by heat. If we consider the archetypal Philosophers' Stone to be a combination of gold and antimony united by a general menstruum, then mercurial alchemy must be considered valid. However, if we base this interpretation upon wet chemistry ion exchange as being the archetype derived from an Alexandrian template, then we must think of mercurial alchemy as a valid yet separate and subsequent descendant on the alchemical family tree. This branch may be viewed as an artifact of the union – instituted by Islamic alchemists – between

Eastern elixir alchemy and the Western industrial and esoteric gold making tradition. The history and evolution of mercurial alchemy will be presented in a separate book by the author.

As alchemy evolved and transitioned into its various expressions, the notion of a Philosopher's "Stone" evolved as well. Islamic alchemist, al-Razi employed a basic template approach to create several "Stones" along with a general test to gauge success. Paracelsus developed a number of alchemical "Stones" in his brand of European alchemy known as Spagyrics, which may have been at least partially influenced by al-Razi. Renowned modern alchemist Dennis William Hauk describes a sampling of different alchemical "Stones" (alluded to in the Foreword) from personal experience based on a lifetime devoted to the subject:

> For an adept-alchemist making stones is no big deal. Most adepts produce five or more during their active years. One can get a good idea of what it takes to create a stone by reviewing the process of confecting a vegetable stone.
>
> Most alchemists would follow the spagyric method set down by Paracelsus in the sixteenth century. It involves isolating and recombining the Three Essentials of Sulfur, Mercury, and Salt at the astrologically appropriate times.
>
> The plant's soul (or Sulfur) is obtained by steam-distilling the oil from the crushed plant. The remaining parts of the plant are digested and fermented, and the resulting solution is rectified to get the alcohol or spirit of Mercury, which is its life force. The residue of stems and leaves left behind is the plant's dead body. This material is dried and incinerated to ash. The ash is ground fine, dissolved in water, and filtered. The liquid is then evaporated, which produces a fine, white crystalline Salt. This is the plant's true body.
>
> The separated Three Essentials are now recombined. First, the crystallized Salt is roasted for a week. Then the Salt is ground again and spread in a warm dish outdoors to absorb atmospheric moisture.

Alternatively, a few drops of fresh dew can be added. This whole process is repeated two more times.

Next, the Salt is saturated with its oily Sulfur and placed in an incubator. After a week, the Salt is reanimated by saturation with the alcohol spirit of its Mercury. After a period of digestion, the solution is distilled by cohobation, in which the condensate is poured again and again upon the matter left at the bottom of the vessel. Finally, the original plant is resurrected in an exalted form known as the elixir.

To create the stone, the elixir material is evaporated, ground, gently calcined, reanimated, and distilled again by cohobation. To cure the stone, it is digested in an incubator for another 6-12 months. The cured stone is said to have fundamental healing power at the level of soul that affects the whole body. It will be able to draw out and restore the soul (sulfuric oil) from any other macerated plant by soaking them together in alcohol.

Because of astrological considerations, creating a vegetable stone can take 2-6 years or more and usually requires the preparation of more plant material, oil, and alcohol to feed the process. It is an ongoing relationship that requires a lot of dedication on the part of the alchemist. It should be noted that the alchemist's intuition, based on his or her relationship with the substance to be transformed, is what guides the work. Quantities, timing, and the sequence of operations are all subject to change based on this deeper connection. It ain't chemistry.

Another way to create a vegetable stone is known as the alchemical method, which involves taking the material through the stages of transformation set down by the ancient masters. This method is also how the mineral stone is created.

In mineral alchemy, there are three traditional ways to create the Philosopher's Stone. The first is the Wet Way using lead acetate. The second is the Dry Way using antimony and the Star Regulus. And the third way is known as the path of Mercury.

Since the authors have already covered the Wet and Dry paths, I will briefly summarize the path of Mercury. Many do not take this path because of the dangers of working with mercury. It emits odorless fumes that attack the brain and nervous system.

Many alchemists believed mercury was the First Matter, and like the First Matter, they felt it was influenced by human consciousness. Mercury seems to flow and vibrate with a mind of its own, which is why alchemists called it quicksilver and viewed it as the living metal Argent Vive.

The work begins with the purification of common mercury. Today, it is distilled three times to purify it, but in the past, it was usually washed in rainwater and sea salt. When the salt blackened, it was rinsed out, and the process repeated until the salt darkened no more. Next, an equal amount of sea salt was added to the mercury, and the mixture was saturated with vinegar. After mixing thoroughly, the salt and vinegar were rinsed out for the final time.

The purified mercury has to be reanimated with Celestial Fire to awaken its generative power. There are several ways to do this, but the simplest is using the Seed of Gold. Powdered gold is added to the purified mercury and digested for six months. The mercury is animated by absorbing the seed of gold.

Another way is to use an alloy of antimony, silver, and copper known as the Lunar Venusian Regulus. It is amalgamated to the purified mercury and distilled. This process is repeated seven times, which produces an animated mercury.

Still another way to animate mercury is by Cinnabar (mercury sulfide). Equal amounts of sulfur and purified mercury are blended until the mixture turns black. Then an equal weight of iron filings is mixed in. The mixture is transferred into a retort and distilled, and the vapor condenses into mercury. The process is repeated seven times to obtain the animated mercury.

CHAPTER 1 – SOPHIC SUBSTANCES

Once we have the animated mercury, we can proceed to that grand moment in alchemy – the Sacred Marriage. Mercury is the Queen in this royal marriage, and gold is the King. Their offspring is the Philosopher's Child, which will mature (or cure) into the Philosopher's Stone.

The animated mercury and pure powdered gold are mixed together to form an amalgam. This is sealed hermetically in a glass vessel and digested until it passes through the three stages of Black, White, and Red. The temperature at the start is 50°C. When the mass turns black, it is raised to 65°.

In about three months, an iridescent sheen will appear on the surface, which marks the Peacock's Tail. In another nine months or so, the matter will begin to turn white. Increase the heat to 130°C for a few months, until the matter displays a yellow color that soon turns reddish. Increase the heat to 200°C for another two months until a deep red color shows. Then break open the glass to retrieve the Red Stone.

To use the Red Stone to transmute metals, it must undergo a further process known as ceration to give it a waxy consistency. The amalgam is squeezed through a cloth and the softer material that comes through is poured into a glass vessel. Digest the material for three months at each of the heats in the previous step. Then, break open the vessel and remove the waxy material, which can now be projected onto metals.

The projection power of the Red Stone can be increased significantly in a process referred to as multiplication. Use part of the Red Stone prepared prior to ceration and add ten parts animated mercury. Once again, seal it in a glass vessel and digest it according to the color-based procedure outlined above. This time, the colors will progress much faster through the cycle.

This process can be repeated seven times. It is said that if this is done more than seven times, the Stone will begin to emit light and become unstable. If all works out, the Stone can be ground into a powder that is cast into molten metals to transmute them to a nobler state.

I should mention a fourth tradition known as the path of the First Matter. It begins with the seemingly impossible task of capturing the First Matter, which is such a mystery even alchemists have a hard time talking about it. That is probably why the main text for this work is a series of wordless drawings known as the Mutus Liber or "Silent Book."

Towards the end of my apprenticeship, I assisted extracting First Matter from black dirt, although I am still not clear about what we accomplished. The dirt around Vienna was apparently not good enough, so we drove all the way to the Black Forest in southwestern Germany to fetch the blackest dirt we could find. Then began an eight-year process of filtering, digesting, fermenting, and distilling the products over and over again. The materials added over the years included many liters of dew, various kinds of young sprouts, thatch, but no metals.

I was only present a few months before I had to return to America. I can only describe it as psychophysical work in which one's attitude can affect the outcome. Once the First Matter is achieved, it is treated like a kind of animated mercury, and the work continues with the Four Elements and Three Essentials. I returned to observe the progress of the work a couple times, but the final products were never shared with me.

Whether or not the First Matter can be captured in the lab, I cannot say. I do believe it is an important ingredient in the Philosopher's Stone that is represented by the universal mercury or Azoth. I was taught First Matter was easiest to retrieve from black dirt, while some traditions say urine or other rejected substances should be used. But no matter where it is obtained, the First Matter is the same for everything.

TOOLS OF TÉCHNI – GLASSWARE AND FURNACE

The Griffin's Egg

The *Philosophers' Egg,* sometimes referred to as the *Griffin's Egg,* is conceptually a triple-entendre in that it refers to a substance, a vessel and the philosophical macrocosmic counterpart to the vessel. From the passage below, we can see that the egg may serve as an analogy for

CHAPTER 1 - SOPHIC SUBSTANCES

the Philosophers' Stone where masculine fixed gold assumes the role of the egg yolk; the egg white is feminine and volatile antimony, with the eggshell representing final conjunction of the two as a finished product.

> *From this Egg, three things must be considered; namely, the Yolk, the White, and the Shells; and this last is only and altogether necessary for Philosophers, which is called the end of the Egg; that is, the last part, rejoicing in perfection, having the likeness of a little Mountain, and also generated between Male and Female; which when it is perfectly calcined it exceeds all Earths whatsoever in whiteness and subtility, enduring the greatest fire embracing Tincture, and desiring a metal in Nature, which is hardly believed of Workers in the Art, unless they being overcome by experience the Mistress of things, that they be compelled to confess and admire it. – St. Dunstan, 17th century*

While this is a valid explanation for the *Philosophers' Egg*, it is not what is typically referred to in alchemical texts when the *Philosophers' Egg* is discussed. The egg is symbolic of the womb of creation. To alchemists, the Philosophers' Stone incubates and develops upon incubation of the "heat of a hen sitting on her eggs" in the vessel, meaning gentle heat. It is alternately referred to as the *ostrich's stomach* or represented in alchemical iconography as coffins or tombs.

As mentioned previously, the *Griffin* is the lion-eagle hybrid symbolising the union of fixed gold (lion) and volatile antimony (eagle). The *Griffin's Egg* is the womb or digestion vessel in which this embryonic mixture matures into the Philosophers' Stone. Just like any egg, the digestion vessel must be sealed airtight and remain so throughout the entire process. This seal is known as the *Seal of Hermes*. We retain this imagery in modern parlance when we use the term *hermetically sealed* to describe airtight packaging.

Two vessels are used to confect the Philosophers' Stone. The first is the *digestion vessel* in which the amalgam of gold, antimony and flux is reacted. Bacstrom provides a detailed description of a digestion vessel in the passage below:

> *The glasses are globular digesting glasses, of five, or at most six, inches in diameter, with a neck five or six inches in length and wide enough to admit your thumb or at least your middle finger.* You must make stoppers of fine grained, very dry Oak, to fit nicely into the necks which should be ground in the inside a little way so as to form a nice round hole for the stopper: the stopper should go into the neck one inch deep, and should leave one inch above the neck.
>
> – Sigismund Bacstrom, 18th century

Bacstrom's *digesting glass* is equivalent to a modern 500 ml long neck, round bottomed or Florence boiling flask. Bacstrom's writings suggest performing as many as five or six reactions simultaneously. The amount of amalgam traditionally occupies one-third of the volume of the flask. From this, we can infer that Bacstrom was a relatively wealthy man. The amount of gold in combination with antimony and flux required to meet this volumetric criterion is tremendous. The alternative is that Bacstrom worked according to Cleopatra, Zosimus and Synesius' method of first creating butter of antimony via digestion or the White Stone without gold, and adding the gold later.

CHAPTER 1 - SOPHIC SUBSTANCES

Once the Philosophers' Stone reaction is complete, it is ground to a powder, joined with additional menstruum and reacted a second time. This process, known as either *fermentation* or *multiplication*, takes place in a much smaller vessel.

> *The globes for multiplication are about two inches in diameter, with necks three or four inches long and wide enough to admit a finger.*
> *– Sigismund Bacstrom, 18th century*

Bacstrom's *multiplication globe* is equivalent to a 100 ml long neck, round bottomed or Florence boiling flask. Laboratory glassware in former times was not as strong as modern standard borosilicate products. The threat of exploding distillation and digestion vessels was always an issue, and the desire for stronger glassware is apparent in Bacstrom's writings.

> *All the glasses should be made of considerable strength – at least one eighth of an inch in substance. – Sigismund Bacstrom, 18th century*

Athanor – Mountains and Ovens

Alchemy, metallurgy and mining were interrelated. Alchemists adhered to a theory of organic geology where cosmic influences descended into the Earth and formed fluids believed to act as naturally occurring solvents, which then reacted with sulfur and mercury in varying proportions, generating metals that ultimately matured into gold over time. Alchemists attempted to reproduce this process in their ovens.

> *It is not only by the common operations of mining and digging in the profundities of the earth that it is possible to obtain Gold. It is quite within the powers of Art to imitate Nature in this matter, for Art perfects Nature in this as in many other things. – Martinus Rulandus*

Mining seemed to vindicate the alchemists' organic geology theory in a number of different ways. Alchemists gleaned insight into the nature of metals and their apparent formation within the Earth directly from miners. Observation revealed that water and a viscous mineral fluid or sludge were ever-present in the depths of mines. It was assumed that the source of this fluid was aerial vapours and precipitation that percolated down into the Earth.

> ... *In the caves of the mountains where the workers labour and dig our gold or silver,* a white oil drips out and when it has disappeared in the ground in which there is this Cohyle or the seed and the beginning of the gold, *there will be something glowing from the earth like a tear or like a white blood, and like a tear of a plant or a grapevine when they are cut, and it is similar to drops of light water in its seeping out, and after a day or night it will coagulate and be similar to the saliva of the mouth or the milk or water foam. And after a certain time when you see it, you will find it slightly reddish and this redness will increase every day, and when it is redder than coagulated blood, but not yet hard as stone, but soft and like a salve and cream, then* the gold in it is completed, but not yet stable *in the heat of the fire, and it will not be stable until it coagulates and becomes similar to a hard rock ...* – Jacob Juran

The further one penetrates into a mine, the hotter and often wetter the mine becomes. The mine was the womb of Mother Earth and it was here that these celestial fluids mixed and mingled with sulfurous fumes and mercurial vapours to create a thick vitriolic liquid known to alchemists and miners as *gur*. This *gur* was believed by early modern alchemists and mining authors to indicate the presence of a mineral ore that would eventually coagulate into a metal. The fact that soft sulphide ores are present at mines in larger quantities than precious metals was taken as evidence that these sulphide ores were in the process of maturing into gold.

Alchemists believed that it was possible to artificially reproduce *gur* and mimic the processes taking place inside Nature's furnace deep within mountains and mines. The process of confecting the Philosophers' Stone in the laboratory served as a model to both reproduce and observe this natural Earth process in miniature and in a significantly accelerated

CHAPTER 1 - SOPHIC SUBSTANCES

timeframe. The natural process of metal formation and maturation to gold could be reproduced by an alchemist in a short time in a laboratory, rather than over millennia in Nature. An alchemist who could duplicate Nature in this manner was seen as having gained mastery over the mineral-metal kingdom and the Earth's inner workings. This was nothing less than magic and was considered a great secret revealed only to initiates of the tradition.

> *Magister Degenhardus, Lullius and Matthesius, in his Serpa Concione 3, write that* the material of the metals should be like buttermilk before it hardens into a metallic form, and that it can be spread like butter. They call it GUR, and I have found it myself in mines where Nature has made lead. And *if one is also able to make such a material here above the earth,* then that should be a sure sign not only that one has the right Materia, but also that one is undoubtedly on the right path. *– J. Grashof*

Miners ultimately shared their magic with metallurgists and smiths. In the following passage, Mircea Eliade encapsulates the relationship between mining, metallurgy and sacred sciences:

> *There would appear to have existed therefore, at several different cultural levels (which is the mark of great antiquity), a close connection between the art of the smith, the occult sciences (shamanism, magic, healing, etc.) and the art of song, dance and poetry ... One element nevertheless is constant – that is* the sacredness of metal and ... all mining and metallurgical operations ... *It is in any case significant that in contrast to pre-agricultural and pre-metallurgical mythologies, where, as a natural prerogative, God is the possessor of the thunderbolt ... in the myths of historic peoples, on the other hand (Egypt, the Near East and the Indo-Europeans), the God of the hurricane receives these weapons – lightning and thunder – from a divine smith.*

During the beginnings of civilization, metal-crafts were as sacred as the mysterious processes taking place within the bowels of the Earth. Their furnace was likened to a mountain in which metals were transformed.

The smith's craft was not viewed as merely symbolic, but rather metallurgy was a secret and mysterious technology forbidden to outsiders. Metallurgy provided trade goods and weapons extremely important to the development, strength and stability of Bronze Age city-state kingdoms.

Alchemy was eventually viewed as the culmination of all mineral and metal technologies applied either therapeutically or industrially. In order for an alchemist to claim mastery, he first needed to have mastered the ability to reproduce the most difficult natural process; the organic growth of minerals into metals and their final maturation into gold.

> *Therefore it is to be known, that Nature hath her passages and veins in the Earth, which doth distill Waters, salt, clear and turbulent. For it always observed by sight, that in the Pits, or Groves of Metals, sharp and salt Waters do distil down. While therefore those water do fall downwards, (for all heavy things are carried downwards) there are sulphureous vapours ascending from the center of the Earth, that do meet them. Therefore if the waters be saltish, pure and clear, and the sulphureous vapours pure also; and that they embrace one another in their meeting, then a pure Metal is generated ...*
>
> *– Joan Baptista van Helmont*

The effort to imitate Nature as described above, the alchemist attempted to replicate natural conditions, as he understood them. He began by creating an analog for the Earth and its subterranean processes in the form of a furnace called an *athanor*.

The oldest form of the word *athanor* is the Greek *athanatos* meaning *non-dying* or *eternal*. The ongoing geological processes deep within mountains occurred over æons and, as such, were considered eternal. Islamic alchemists arabicized the word to *al-tannūr* or *at-tannūr*, meaning *baker's oven*. It was shaped like a tower or mountain with a domed roof

CHAPTER 1 – SOPHIC SUBSTANCES

and was able to maintain a steady non-dying or eternal heat analogous to the depths of a mountain.

Philalethes (George Starkey) gives a description of his favourite type of stove and ends by advising that a good stove should match the following criteria:

1. *... firstly, it must be free from draughts;*
2. *... secondly, it must enable you to vary the temperature, without removing your vessel;*
3. *... thirdly, you must be able to keep up in it a fire for ten or twelve hours, without looking to it [in order to sleep]. Then the door of our Art will be opened to you; and when you have prepared the Stone, you may procure a small portable stove, for the purpose of multiplying it.*

The athanor was primarily used for metallurgical processes such as purifying and reducing gold, calcining stibnite to create *flowers of antimony*, creating the star regulus, and to perform Maria Hebrea's *Ars Brevis* (Short Dry Path).

Professor William R. Newman, in *Gehennical Fire*, provides unique insight into the difficulties of acquiring an athanor as late as the 17th century. In the passage below, Professor Newman is describing Robert Boyle's challenges in this regard:

> *Obtaining a reliable furnace in the mid-seventeenth century could take on the appearance of a quest for the Holy Grail. Robert Boyle, in a much-quoted letter to his sister Katherine, bemoans the difficulties that he encountered in conveying a "great earthen furnace" to his laboratory. After having it "transported a thousand miles by land," Boyle received "that grand implement of Vulcan ... crumbled into as many pieces, as we into sects." If Boyle, one of the best-connected men in England, could have such problems obtaining a proper furnace, one can imagine the difficulties that beset Starkey.*

In *The Work with the Butter of Antimony* by Mr. Hand:

> ... if the first distillation required the heat of three wicks in the lamp, the second will not require more than two.

Between the 17th and 19th centuries, alchemical furnace descriptions and diagrams appeared in the following publications:

- **1650** – John French's Art of Distillation: round furnace, lamp furnaces and makeshift types with pictures;
- **1670** – LeFebre's Compleat Body of Chemistry: philosophical lamp furnace with diagram (pictured);
- **1689** – Johann Joachim Becher's Tripus Hermeticus Fatidicus, seu I Laboratorium Portable;
- **1731** – Peter Shaw: pirated Becher's ideas, published An Essay for Introducing a Portable Laboratory;
- **1803** – Joseph Black's Lectures on the Elements of Chemistry: improved portable furnace sold until 1912.

The Three Baths – Water, Ash and Sand

Once all the preparatory work had been completed and the *Griffin* embryo had been sealed in its egg, a different heating method was required. Water, ash, dung and sand baths offered reliable control over temperature and were used in all Western alchemical traditions. European alchemists employed a number of techniques, the most interesting of which was horse dung. As late as the 18th century, horse dung was used to indicate temperature control. Sigismund Bacstrom provides very clear instructions for using horse dung to indicate reaction temperatures in degrees Fahrenheit:

CHAPTER 1 - SOPHIC SUBSTANCES

The greatest heat is in the middle, where it is generally from 140 to 150 degrees: less heat is round the staves where it varies from 90 to 100 and from that to 120 degrees.

Your horse dung must be procured before, as it takes sometimes five, six or more days before it ferments and gives the necessary heat. This is soon discovered by the steam arising from it, and by the thermometer buried in it nine or ten inches deep.

You must have two hogsheads or casks, in order to prepare a second before the fermenting heat has entirely left the first; which heat seldom lasts longer than three weeks; as your work must never become cold one single moment. You must cover the top with clean straw pretty thick, and also all round the casks, especially in winter, or the work will be too cold and your operation will be very much retarded, if not fail.

When reading ancient alchemical works, especially when alchemists address stoves, fires, temperature management, etc. their frustration becomes apparent in the writing. It is difficult for those of us who have never lived without electricity to imagine the struggle to maintain a constant fire with an exact temperature day in and day out for extended periods; not to mention working with horse dung, charcoals or wood fuels with uneven burning temperatures.

Fortunately, all modern scientists need to do is pre-adjust the thermocouple's digital setting for the heating mantle to the exact temperature required and monitor the reaction. Sigismund Bacstrom provides a clue towards the proper digestion flask heating technique when he instructs experimenters to:

... bury them [digestion flasks] in the bath all round the cask [filled with horse dung], where the gentlest heat prevails, deep enough that only the upper part of the neck and stopper, that which is luted, may be in sight or level with the surface of the bath.

He is describing what today might be called a full coverage zippered heating mantle or full submersion in an oil or sand bath, leaving just the neck of the digestion flask exposed. The reason for doing so is to keep the body of the flask warm and the neck of the flask cooler towards the upper part creating a current of convection. The vapours ascend the exposed cooler neck, condense and descend the wall of the glass. By connecting the heating mantle to an uninterruptible power supply, an analog for an *athanor* or the baths is created.

The Seal of Hermes

During the Bronze Age, proto-alchemical procedures were mainly metallurgical. Originally, the alchemical seal so often mentioned in alchemical texts referred to a layer of fluxing salt that sealed the metals in the crucible from the ambient atmosphere. This process is associated with the *Ars Brevis* used by early metallurgists and alchemists.

In Alexandria, Maria's innovative *Ars Magna* made use of lower temperatures over a longer period. A feature of this new synthesis was that the digestion vessel must be sealed airtight and remain so until the process was complete. Alexandrian use of sal ammoniac resulted in volatile hydrogen chloride gas that could easily escape an improperly sealed vessel and ruin the reaction. This airtight seal was known as the *Seal of Hermes* and its use in the *Ars Magna* paralleled the seal created by fluxing salts in the metallurgical Dry Path.

Bacstrom describes the procedure he employed in the 18th century to hermetically-seal his digestion flask:

> *The globe glass is immediately to be shut* with a stopper made of oak, which fits nicely, in order that the superfluous remaining humidity (the phlegm) may, during the putrefaction, penetrate and evaporate through the pores of the oak. *The joining of the neck and the stopper must be luted* with something that is able to resist outward warmth and moisture.

CHAPTER 1 - SOPHIC SUBSTANCES

He then describes in detail his technique for ensuring an airtight seal by securing the neck of the digesting vessel:

> ... with strips of linen pasted round them, and harpsichord wire wound over the linen, with a varnish over the whole, and with lute where the stopper joins the upper brim of the neck. This lute may be sealing wax dropped on (all round) by a burning candle, or rosin, or quick drying varnish, thickened with filings of iron, or any other good luting that can withstand warmth and moisture; but take care to keep the bottom of the stopper in sight, by not covering the neck of the glass quite so low as to hide it.

The reader should now have an introductory understanding of the Philosophers' Stone in a general historical context, and be familiar with the fundamental ingredients and equipment required to confect the Philosophers' Stone. The following section will delve in-depth into the Stone's origins and evolution from classical antiquity up until the early modern period. For readers interested in getting directly to work, skip ahead to Section 3 – Materials and Methods.

Pray, Read, Read, Read, Re-Read, Work, and Discover – Heinrich Khunrath

Chapter 2 – Moses and the Golden Calf

כ וַיִּקַּח אֶת-הָעֵגֶל אֲשֶׁר עָשׂוּ, וַיִּשְׂרֹף בָּאֵשׁ, וַיִּטְחַן, עַד אֲשֶׁר-דָּק; וַיִּזֶר עַל-פְּנֵי הַמַּיִם, וַיַּשְׁקְ אֶת-בְּנֵי יִשְׂרָאֵל.

Exodus 32:20 He took the [golden] calf that they had made and burned it with fire and ground it to powder and scattered it on the water and made the people of Israel drink it.

כא וְאֶת-חַטַּאתְכֶם אֲשֶׁר-עֲשִׂיתֶם אֶת-הָעֵגֶל, לָקַחְתִּי וָאֶשְׂרֹף אֹתוֹ בָּאֵשׁ, וָאֶכֹּת אֹתוֹ טָחוֹן הֵיטֵב, עַד אֲשֶׁר-דַּק לְעָפָר; וָאַשְׁלִךְ, אֶת-עֲפָרוֹ, אֶל-הַנַּחַל, הַיֹּרֵד מִן-הָהָר.

Deuteronomy 9:21 Then I took the sinful thing, the [golden] calf that you had made, and burned it with fire and crushed it, grinding it very small, until it was as fine as dust. And I threw the dust of it into the brook that ran down from the mountain.

Jasher 82:19 And Moses came to the camp and he took the [golden] calf and burned it with fire, and ground it till it became fine dust, and strewed it upon the water and gave it to the Israelites to drink.

MOSES – THE FIRST ALCHEMIST

In the mountainous wilderness of the ancient Sinai desert, a legendary event occurred. This event, which has captivated man's imagination for centuries, offers fascinating implications for the study of both religion and alchemy. This event described in the narrative of *Moses and the golden calf*, constitutes the earliest written record of the manipulation of gold to be ingested by humans. In addition to the accounts of the three major religions of the Holy Land – Judaism, Christianity and Islam – the narrative is also detailed in a work of Jewish literature, the Book of Jasher (or, traditionally, the Book of the Just Man). The *Moses and the golden calf* narrative succinctly encapsulates the alchemical process of confecting ingestible gold. The ability to break gold's metallic bonds, reduce it to a powdered form and ingest it as a form of spiritual sacrament or psychospiritual medicine are touchstone topics in the tradition of alchemy. With reference to the alchemical *solve et coagula* (dissolution and coagulation), burning gold in a fire would render it molten, thus signifying the *solve* stage. Congealing the molten gold to a solid form, brittle enough to grind to a powder, falls within the *coagula* stage. Yet gold does not burn: it melts and is malleable rather than brittle.

Jewish sage-mystics maintain that the story of Exodus, and indeed any Biblical narrative, is but a fractal slice of a multi-dimensional whole and must be interpreted at a number of levels in order to reveal the greater truth hidden within. However, for the purpose of alchemical discussion this treatment must remain bound to a literal interpretation, two-dimensional as it may be, as a valid approach to understanding the metallurgical elements presented in the *Moses and the golden calf* narrative. The crux of the issue revolves around the chemical technology required to reduce gold to a powder a millennia-and-a-half before the Common Era (CE); the therapeutic value of such a powdered gold product, and how the secret of its confection may have served to strengthen the material and spiritual wealth of Israel. Alchemists have never questioned *whether* Moses actually confected powdered gold, but rather *how* he did so. Alchemists not only realized the plausibility of the technology implied in the narrative, but also the therapeutic potential, whether such is physical, psychological or spiritual. Moses' technique has been a driving force of alchemy from the time of the early Alexandrian alchemists of the 2nd century CE until the 18th century European alchemists and the chemists of the 19th century. As the first man in any written historical account to create a powder from gold, and to preside over its ritual ingestion, Moses is naturally accorded the honor of being the first alchemist. Moses, in company with Hermes / Thoth, represents the Jewish, Greek and Egyptian origins of alchemy in the Western tradition sometimes described as the Mosaico-Hermetic Art.

One of the earliest treatments of Moses can be found in *The Turba Philosophorum, or, Assembly of the Sages*, an alchemical classic that entered Europe with Islam during the 12th century CE. Europeans were

CHAPTER 2 - MOSES AND THE GOLDEN-CALF

first exposed to *The Turba Philosophorum* as an Arabic or Hebrew treatise. Raphael Patai, in his comprehensive work, *The Jewish Alchemists*, describes the scene:

> When Moses' turn comes to speak, he explains the various names of quicksilver, and quotes "the Philosopher" as having said that the quicksilver which tinges gold is the quicksilver out of cinnabar (argentums vivum cambar), and that this is the "magnesia," while the quicksilver of the auripigmentum or orpiment is the sulphur which ascends from this mixed compound material. "You must, therefore," he says, "mix that thick thing with the 'fiery venom,' and let it putrefy, and diligently pound it until a spirit is produced which is hidden in that other spirit; then it will become a tincture for everything that you wish."

The Turba Philosophorum provides a simple and elegant hypothesis, which postulates the use of cinnabar ore. The Egyptians sourced cinnabar from the Giza Plateau. Mercury derived from cinnabar was used in extractive metallurgy during the time of Moses. Based on the exquisite purity of the gold and the quality of artisanship found in Tutankhamen's tomb, it is evident that gold technology was very advanced at this time. That Moses would have stocked up on cinnabar or mercury would have demonstrated remarkable foresight and planning. For this reason, and others, it is unlikely that Moses' technique involved cinnabar or mercury. Delving deeper, Occam's razor carves a much more likely scenario.

Swarna bhasma, the hallmark gold elixir of the Indian alchemical tradition, is created by mechanical grinding or sublimation, thus verifying that it is possible to either hand-grind elemental gold to achieve a dark red powder with particles in the nanometer size range or, like another Indian gold elixir – makaradhwaja – achieve the same via sublimation with purified cinnabar or mercury.

The product that results from mercury gold processing is synonymously known as gold lime, gold sponge or gold calx. It varies in appearance from fluffy beige to a fine black powder. If processed even further it achieves a deep reddish-brown nanoparticulate form. This answers the question as to whether Moses' feat was possible or not, but how he did it remains the question. Although Patai presents a summary of the arguments on the subject ranging throughout the 17th to 19th centuries CE, what becomes clear from the debate is that a consensus has yet to be reached.

Dr. Ken Gilleo published the article 'The Alchemy of Nanotechnology' in 2006. Dr. Gilleo's brief treatment of the subject of *Moses and the golden calf* reminds us that the topic is still being considered during the 21st century by modern academics. He approaches the subject by questioning:

> ... is there a chance that chemists and their predecessors, the Alchemists, were making "Nano-stuff", long before the big Nano-boom? Let's step back to seek out the Nanotech pioneers.
>
> Just how old is Nano-gold? The early metalworkers in Egypt and Mesopotamia, who invented the first lead-free solders probably made nano-gold. The Christian Bible provides convincing evidence and suggests that Moses may have been the father of Nanotech. Moses destroyed the infamous Golden Calf with an intense fire, created powder, and made up a potable mixture of the gold in water. The Israelites were made to drink the liquid that was probably a colloidal suspension of nano-gold.

Elsewhere the article states that ...

> ... Nano-metallurgy played a role in medicine and religion. Both gold and silver nano-particles were used in Biblical times and probably much earlier.

In order to understand how Moses achieved medical, and potentially industrial, gold nanotech approximately 3,500 years before Science had identified it or had even developed a language for communicating the

CHAPTER 2 - MOSES AND THE GOLDEN-CALF

concept, it is important to examine the advanced metallurgy being practiced in Egypt and the Sinai by the Kenite tribes in the land of Midian during the time of Moses. To understand the complex dynamic of the region and its inhabitants during Biblical times more clearly, it serves to gain a greater perspective of the land of Midian and the Kenite metallurgist-priesthood.

The Land of Midian

The land of Midian takes its name from Midian, a son of Abraham from his second wife, Keturah:

א ויסף אברהם ויקח אשה ושמה קטורה   *Genesis 25:1* Now Abraham took another wife, whose name was Keturah.

ב ותלד לו את זמרן ואת יקשן ואת מדן ואת מדין--ואת ישבק ואת שוח   *2* She bore to him Zimran and Jokshan and Medan and Midian and Ishbak and Shuah.

This marriage likely led to a league of interrelated tribes occupying the region known as Midian. Both the place and its people are mentioned in Jewish, Christian and Islamic traditions. Based on archeology and religious texts, it is generally understood to have a geographical horizon stretching from northwestern Arabia on the east shore of the Gulf of Aqaba, extending into the Sinai Peninsula at least as far as Serabít el-Khâdim.

The land of Midian is significant to the *golden calf* narrative due to it being where Moses spent 40 years in voluntary exile after murdering the Egyptian.

טו וַיִּשְׁמַע פַּרְעֹה אֶת-הַדָּבָר הַזֶּה, וַיְבַקֵּשׁ לַהֲרֹג אֶת-מֹשֶׁה; וַיִּבְרַח מֹשֶׁה מִפְּנֵי פַרְעֹה, וַיֵּשֶׁב בְּאֶרֶץ-מִדְיָן וַיֵּשֶׁב עַל-הַבְּאֵר.   *Exodus 2:15* When Pharaoh heard of it, he sought to kill Moses. But Moses fled from Pharaoh and stayed in the land of Midian. And he sat down by a well.

It was in Midian that Moses met Jethro, the high priest of the land and later his father-in-law (once he had married Jethro's daughter, Zipporah).

כא וַיּוֹאֶל מֹשֶׁה, לָשֶׁבֶת אֶת-הָאִישׁ; וַיִּתֵּן אֶת-צִפֹּרָה בִתּוֹ, לְמֹשֶׁה. *Exodus 2:21* And Moses was content to dwell with the man, and *he gave Moses his daughter Zipporah.*

Moses would have become completely familiar with the region's geography, industry, trade, and intertribal politics. Being adopted into Egyptian royalty and having proven himself as a skilled military commander, Moses was ultimately accepted by Jethro and gained access into the metallurgical and other mysteries of Jethro's priesthood. Jethro held one of the most powerful and important positions within Midianite territory, yet he was Kenite rather than Midianite.

Of Indo-Iranian ethnicity, the Kenites had migrated into the region from South Asia in great antiquity. As a people, they play an important role in the history of ancient Israel. Like many other ancient Indo-Iranian tribes, the Kenites were coppersmiths and metallurgists. The etymology of Kenite, from a Biblical perspective, is derived from Tubal-Cain, founder of metallurgy. Religious texts mention the Kenites living in or around Canaan as early as the time of Abraham. The term *caina* in Chaldean signifies a goldsmith, whereas *cainiah* or *caina* in Syriac signifies a brass / bronze smith. Kenites in the region were associated with copper.

כב וְצִלָּה גַם-הִוא, יָלְדָה אֶת-תּוּבַל קַיִן--לֹטֵשׁ, כָּל-חֹרֵשׁ נְחֹשֶׁת וּבַרְזֶל; וַאֲחוֹת תּוּבַל-קַיִן, נַעֲמָה. *Genesis 4:22* Zillah also bore *Tubal-cain; he was the forger of all instruments of bronze and iron. The sister of Tubal-cain was Naamah.*

יח בַּיּוֹם הַהוּא, כָּרַת יְהוָה אֶת-אַבְרָם--בְּרִית לֵאמֹר: לְזַרְעֲךָ, נָתַתִּי אֶת-הָאָרֶץ הַזֹּאת, מִנְּהַר מִצְרַיִם, עַד-הַנָּהָר הַגָּדֹל נְהַר-פְּרָת. יט אֶת-הַקֵּינִי, וְאֶת-הַקְּנִזִּי, וְאֵת, הַקַּדְמֹנִי. כ וְאֶת-הַחִתִּי וְאֶת-הַפְּרִזִּי, וְאֶת-הָרְפָאִים. *Genesis 15:18-21* On that day the Lord made *a covenant with Abram, saying, "To your offspring I give this land, from the river of Egypt to the great river, the river Euphrates, the land of the Kenites, the Kenizzites, the Kadmonites, the Hittites,*

CHAPTER 2 – MOSES AND THE GOLDEN-CALF

כא וְאֶת־הָאֱמֹרִי, וְאֶת־הַכְּנַעֲנִי, וְאֶת־הַגִּרְגָּשִׁי, וְאֶת־הַיְבוּסִי. {ס} the Perizzites, the Rephaim, the Amorites, the Canaanites, the Girgashites and the Jebusites."

During the time of the Exodus, Jethro and his clan lived in close proximity to Mount Horeb and Serabít el-Khâdim.

א וּמֹשֶׁה, הָיָה רֹעֶה אֶת־צֹאן יִתְרוֹ חֹתְנוֹ--כֹּהֵן מִדְיָן; וַיִּנְהַג אֶת־הַצֹּאן אַחַר הַמִּדְבָּר, וַיָּבֹא אֶל־הַר הָאֱלֹהִים חֹרֵבָה. *Exodus 3:1* Now Moses was keeping the flock of his father-in-law, Jethro, the priest of Midian, *and he led his flock to* the west side of the wilderness and came *to Horeb, the mountain of God.*

Although Jethro is clearly described in religious texts as a Midianite priest occupying a high position among the Midianites, and as living within Midianite territory, the same texts also make it clear that he was a Kenite.

טז וּבְנֵי קֵינִי חֹתֵן מֹשֶׁה עָלוּ מֵעִיר הַתְּמָרִים, אֶת־בְּנֵי יְהוּדָה, מִדְבַּר יְהוּדָה, אֲשֶׁר בְּנֶגֶב עֲרָד; וַיֵּלֶךְ, וַיֵּשֶׁב אֶת־הָעָם. *Judges 1:16* And *the descendants of the Kenite, Moses' father-in-law, went up with the people of Judah from the city of palms into the wilderness of Judah, which lies in the Negeb near Arad, and* they went and settled with the people.

The Kenite region abutted the southern border of Judah and considerable cultural intercourse took place between the two cultures. By the time of King David's reign, the Kenites were settled among the tribe of Judah. David shared the spoils of war with the Kenites and other allies.

כו וַיָּבֹא דָוִד אֶל־צִקְלַג, וַיְשַׁלַּח מֵהַשָּׁלָל לְזִקְנֵי יְהוּדָה לְרֵעֵהוּ לֵאמֹר: הִנֵּה לָכֶם {ר} בְּרָכָה, מִשְּׁלַל אֹיְבֵי יְהוָה. {ס} **כז** לַאֲשֶׁר בְּבֵית־אֵל {ס} וְלַאֲשֶׁר {ר} בְּרָמוֹת־נֶגֶב, {ס} וְלַאֲשֶׁר בְּיַתִּר. {ס} **כח** וְלַאֲשֶׁר {ר} בַּעֲרֹעֵר {ס} וְלַאֲשֶׁר בְּשִׂפְמוֹת, {ס} וְלַאֲשֶׁר {ר} בְּאֶשְׁתְּמֹעַ. {ס} *1 Samuel 30:26-31* When *David came to Ziklag, he* sent part of the spoil to his friends, *the elders of Judah, saying, "Here is a present for you from the spoil of the enemies of the Lord." It was* for those in Bethel, in Ramoth of the Negeb, in Jattir, in Aroer,

כט וְלַאֲשֶׁר בְּרָכָל, {ס} וְלַאֲשֶׁר {ר} בְּעָרֵי הַיְּרַחְמְאֵלִי, {ס} וְלַאֲשֶׁר, בְּעָרֵי הַקֵּינִי. {ס} ל וְלַאֲשֶׁר {ר} בְּחָרְמָה {ס} וְלַאֲשֶׁר בְּבוֹר-עָשָׁן, {ס} וְלַאֲשֶׁר {ר} בַּעֲתָךְ. {ס} לא וְלַאֲשֶׁר, בְּחֶבְרוֹן; וּלְכָל-הַמְּקֹמוֹת אֲשֶׁר- {ר} הִתְהַלֶּךְ-שָׁם דָּוִד, הוּא וַאֲנָשָׁיו. {פ}

in Siphmoth, in Eshtemoa, in Racal, in the cities of the Jerahmeelites, in the cities of the Kenites, in Hormah, in Bor-ashan, in Athach, in Hebron, for all the places where David and his men had roamed.

The role that Jethro and the Kenites played during the Exodus was later remembered by Saul:

ו וַיֹּאמֶר שָׁאוּל אֶל-הַקֵּינִי לְכוּ סֻּרוּ רְדוּ מִתּוֹךְ עֲמָלֵקִי, פֶּן-אֹסִפְךָ עִמּוֹ, וְאַתָּה עָשִׂיתָה חֶסֶד עִם-כָּל-בְּנֵי יִשְׂרָאֵל, בַּעֲלוֹתָם מִמִּצְרָיִם; וַיָּסַר קֵינִי, מִתּוֹךְ עֲמָלֵק.

1 Samuel 15:6 Then Saul said to the Kenites, "Go, depart; go down from among the Amalekites, lest I destroy you with them. For you showed kindness to all the people of Israel when they came up out of Egypt." So the Kenites departed from among the Amalekites.

Midianite Gold

Whether the narrative is an accurate portrayal of an historical act or not is not as important as the fact that whoever wrote the narrative understood gold technology. To have enough gold to create a statue implies that the Israelites were in possession of gold.

לה וּבְנֵי-יִשְׂרָאֵל עָשׂוּ, כִּדְבַר מֹשֶׁה; וַיִּשְׁאֲלוּ, מִמִּצְרַיִם, כְּלֵי-כֶסֶף וּכְלֵי זָהָב, וּשְׂמָלֹת. לו וַיהוָה נָתַן אֶת-חֵן הָעָם, בְּעֵינֵי מִצְרַיִם--וַיַּשְׁאִלוּם; וַיְנַצְּלוּ, אֶת-מִצְרָיִם. {פ}

Exodus 12:35-6 The people of Israel had also done as Moses told them, for they had asked the Egyptians for silver and gold jewelry and for clothing. And the Lord had given the people favor in the sight of the Egyptians, so that they let them have what they asked. Thus they plundered the Egyptians.

Along with the gold plundered from the Egyptians, the possibility remains that Moses knew of another source of gold in the Sinai. To explore this possibility, it is necessary to investigate the lands that Moses had

CHAPTER 2 - MOSES AND THE GOLDEN-CALF

traversed. His first and most important destination was the Wilderness of Shur.

כב וַיַּסַּע מֹשֶׁה אֶת-יִשְׂרָאֵל מִיַּם-סוּף, וַיֵּצְאוּ אֶל-מִדְבַּר-שׁוּר; וַיֵּלְכוּ שְׁלֹשֶׁת-יָמִים בַּמִּדְבָּר, וְלֹא-מָצְאוּ מָיִם. *Exodus 15:22 Then Moses made Israel set out from the Red Sea, and they went into the wilderness of Shur. They went three days in the wilderness and found no water.*

Interestingly, the earliest Biblical mention of any metal was gold from Havilah, as recounted in Genesis. This passage refers to the gold of Amalekites in the Wilderness of Shur.

יא שֵׁם הָאֶחָד, פִּישׁוֹן--הוּא הַסֹּבֵב, אֵת כָּל-אֶרֶץ הַחֲוִילָה, אֲשֶׁר-שָׁם, הַזָּהָב. *Genesis 2:11 The name of the first [river] is the Pishon. It is the one that flowed around the whole land of Havilah, where there is gold.*

The Wilderness of Shur was located in the Sinai and inhabited by Midianite, Kenite, Amalekite, and Ishmaelite tribes. Their territories often bordered or overlapped each other. The majority of the Sinai Peninsula was inhabited by Amalekites in the north and Midianite tribes to the south, extending eastward into Arabia.

The Sinai is not traditionally associated with gold, but Deuteronomy of the Greek Septuagint seems to indicate otherwise:

οὗτοι οἱ λόγοι οὓς ἐλάλησεν μωυσῆς παντὶ ισραηλ πέραν τοῦ ιορδάνου ἐν τῇ ἐρήμῳ πρὸς δυσμαῖς πλησίον τῆς ἐρυθρᾶς ἀνὰ μέσον φαραν τοφολ καὶ λοβον καὶ αυλων καὶ καταχρύσεα *Deuteronomy 1:1 These [are] the words which Moses spoke to all Israel on this side Jordan in the desert towards the west near the Red Sea, between Pharan Tophol, and Lobon, and Aulon, and the gold works.*

The Midianite territory extending eastward into northwestern Saudi Arabia was rich in gold and mineral deposits. This area was known as Ophir, and gold from Ophir was well known and commented on by Jewish scholars. However, by the time of Roman conquest, the area had been depleted. Renowned British explorer Sir Richard Burton printed the following quote on the inner leaf of an adventure journal:

> "We have the authority of Niebuhr, that the precious metals are not found or known to exist in Arabia, which has no mines either of gold or silver." – Crichton's History of Arabia, ii. 403

Burton, in true character, set out to prove otherwise and eloquently recounted his explorations in *The Gold Mines of Midian and the Ruined Midianite Cities*, published in 1878. This was followed in 1879 by a two-volume set, *The Land of Midian*, in which he sets the record straight by comprehensively detailing the lost Midian cities and gold mines of Ophir. The valleys cutting into the high plain of Nedched contained the remains of silver, copper and gold mines. Commenting on the region's depletion:

> The Land of Midian is at present, in fact, much like California, when the pick and fan men had done their work: she is still wealthy, but her stage is that when machinery must take the place of the human arm.

Burton does concede however to the bounty of the region in ancient times, explaining that:

> To the [Jewish] explorers of the Mining Cities of Midian, the most interesting part of the story is the quantity and variety of metals produced by the land – gold, silver, copper, tin, and lead!

ח וְאֶת-מַלְכֵי מִדְיָן הָרְגוּ עַל-חַלְלֵיהֶם, אֶת-אֱוִי וְאֶת-רֶקֶם וְאֶת-צוּר וְאֶת-חוּר וְאֶת-רֶבַע-- חֲמֵשֶׁת, מַלְכֵי מִדְיָן; וְאֵת בִּלְעָם בֶּן-בְּעוֹר, הָרְגוּ בֶּחָרֶב.

Numbers 31:8 They killed the kings of Midian with the rest of their slain, Evi, Rekem, Zur, Hur, and Reba, the five kings of Midian. And they also killed Balaam the son of Beor with the sword.

כא וַיֹּאמֶר אֶלְעָזָר הַכֹּהֵן אֶל-אַנְשֵׁי הַצָּבָא, הַבָּאִים לַמִּלְחָמָה: זֹאת חֻקַּת הַתּוֹרָה, אֲשֶׁר-צִוָּה יְהוָה אֶת-מֹשֶׁה.

31:21-2 Then Eleazar the priest said to the men in the army who had gone to battle: "This is the statute of the law that the Lord has commanded Moses: only the

CHAPTER 2 - MOSES AND THE GOLDEN-CALF

כב אַךְ אֶת-הַזָּהָב, וְאֶת-הַכֶּסֶף; אֶת-הַנְּחֹשֶׁת, אֶת-הַבַּרְזֶל, אֶת-הַבְּדִיל, וְאֶת-הָעֹפָרֶת. *gold, the silver, the bronze, the iron, the tin, and the lead...*

In his fascinating treatment of the gold and metal mining industry of the Midianites, Burton reveals a booming trade existing in antiquity, which sets the precedent for a gold trade route through the Sinai. In the ancient world, Egypt produced the majority of the Mediterranean region's gold output. Egypt colonized Nubia for its mines, some reaching to more than 100 meters in depth and with a combined operational output of more than four million pounds of gold. Egypt's lust for gold was unquenchable and the Egyptians imported it from wherever possible, including the Midian mines of Ophir.

The *Jewish Encyclopedia* acknowledges the possibility of an Arabian location for Ophir, but places it a bit further south along the Peninsula. Its compilers were perhaps unaware of Burton's research findings at the time of writing.

> ... *there is ample justification for the view which prefers the western coast of Arabia,* **especially as there are a number of references in the ancient authors to the rich gold of the southwestern coast of Arabia. According to Agatharchides,** *these mines contained pieces of gold as large as walnuts:* **but this metal was of little value to the inhabitants, and iron and copper were worth two and three times as much.** *It is hardly probable that Solomon and Hiram would have sent their ships past Yemen to fetch gold from the Gulf of Persia, which was much farther away.* – 1906 Jewish Encyclopedia

Egypt certainly would have exploited this source of Midianite gold. Ophir gold likely entered Egypt via existing Sinai trade routes dating back to 3000 BCE. One of the major routes, well known in Biblical times, began in Egypt at Heliopolis or Memphis and traversed the Sinai to the port city of Aqaba. From here, it continued north connecting to Petra, Babylon and eastward to Bactria and the Indian sub-continent.

> יז נַעְבְּרָה-נָּא בְאַרְצֶךָ, לֹא נַעֲבֹר בְּשָׂדֶה וּבְכֶרֶם, וְלֹא נִשְׁתֶּה מֵי בְאֵר: דֶּרֶךְ הַמֶּלֶךְ נֵלֵךְ, לֹא נִטֶּה יָמִין וּשְׂמֹאול, עַד אֲשֶׁר-נַעֲבֹר, גְּבֻלֶךָ.
>
> *Numbers 20:17* Please let us pass through your land. We will not pass through field or vineyard, or drink water from a well. *We will go along the King's Highway. We will not turn aside to the right hand or to the left until we have passed through your territory."*

As a veritable pipeline of commerce, it was known as the King's Highway, as mentioned in the passage above. The King's Highway developed primarily because of Egypt's love of Lebanon cedar from the north, and lapis lazuli and other exotic goods originating in Asia to the Far East.

The problem that arises is that there must have been some form of trading station, treasury or mined ore refinery within the Sinai to accommodate Midian gold and other mined goods. Indeed, there exists just such a place in the form of a highly protected and unusual mining temple / workshop, which will be discussed in detail. This temple / workshop existed within Midianite territory overseen by a Kenite priest and was likely staffed with Midianite or Kenite laborers and smiths. It remained important to many dynastic successors during Egypt's long history. Jethro's position as high priest would have placed him in control over the major temple in the region, the Hathor Temple at Serabít el-Khâdim. A better understanding of this temple will help clarify the chemical and metallurgical technology in use at the Hathor Temple by Jethro and his miners and smiths, and the temple's importance to Egypt.

THE REFINERY OF HIDDEN GOLD

The Biblical narrative of *Moses and the golden calf* explains that the gold utilized for the *golden calf* statue was jewelry that had been melted. Whether the statue was solid gold or gold laminated over a wooden frame is irrelevant. In either case, the gold first needed to be melted into a usable form. This highlights two very important considerations:

CHAPTER 2 - MOSES AND THE GOLDEN-CALF

1. *Goldsmith technology implied that the goldsmith in question either worked in the temple-craft tradition for the Pharaoh, meaning that his craft was overseen by an Egyptian priesthood within a temple complex, or the priest and smiths were actually of another tradition with full working knowledge of the process;*
2. *A facility would have been required with a furnace that could reach the 1,064 degrees Celsius / 1,948 degrees Fahrenheit minimum temperature required to melt gold.*

An existing facility figures quite nicely into the story at this point. The *golden calf* episode takes place as Moses returns from the mountain. Biblical accounts clearly describe two mountains by name, which are often undifferentiated or explained as a single mountain with dual names by some Biblical researchers. These are named Mount Sinai and Mount Horeb / Choreb. If a facility near a few prominent mountains in the Sinai could be found, this would support the argument that the Israelites were capable of the difficult task of smelting gold by taking advantage of such a facility.

Professor Sir W. M. Flinders Petrie was an English Egyptologist who pioneered the use of systematic methodology in archaeology. He has been called 'the Father of Egyptian Archaeology' and held the first chair of Egyptology in the United Kingdom. On 03 December 1905, Petrie mounted a research expedition to the Sinai with backing from the Egypt Exploration Fund. He was charged with the task of "promotion of surveys and excavations for the purpose of elucidating or illustrating the Old Testament narrative." On 11 January 1906, he laid eyes on a fascinating temple that would allow him to do as he had been charged. The intriguing temple complex before Petrie was the Temple of Hathor at Serabít el-Khâdim, and it had been there since before the Great Pyramids of Giza had been built.

The temple – a revealing addition to Egyptian archaeology – not only fulfills the geographical criteria of Exodus; it also added to the archaeological record:

- the first recorded Semitic inscriptions;
- a pre-existing Semitic temple complex;
- an industrious mining and manufacturing facility complete with workers' barracks and dwellings;
- a metallurgical workshop;
- and evidence of the beginnings of the first phonetic alphabet based on Semitic language – Proto-Sinaitic.

The temple lies south of the King's Highway and was in full operation during the time of Moses. It was an incredible archaeological find that remains largely uncelebrated to this day. Petrie and his research assistant, Lina Eckenstein, were able to account for every stage of the Exodus in detail, thus transforming the story from Biblical narrative to history backed by archaeology and a profound understanding of Sinai's natural terrain. The findings of Petrie and Eckenstein are so relevant to our topic that they are worth examining in some detail.

That the temple contained a foundry and metallurgical facilities is typical given that other prominent and more celebrated temples, such as the Edfu Temple, maintained smelting and metallurgical facilities. Commenting on Hatshepsut's shrine to Sneferu, Petrie laments the poor condition of the Hathor Temple remains:

> There is no other such monument known which makes us regret the more that it is not in better preservation.

As far as the geographical criteria from Biblical accounts, the match is staggering in its accuracy. The temple complex is just a few miles from three major peaks in the region: Jebel Saniya (possible Mount Sinai), Jebel Ghorabi (possible Mount Horeb / Choreb), and the highest and most

CHAPTER 2 – MOSES AND THE GOLDEN-CALF

picturesque in the region, Serabít el-Khâdim. Mount Sinai and Mount Horeb are so closely associated in the *Bible* that they appear to be interchangeable. The region was not only considered holy or sacred – as a temple, mining operation and gold processing workshop, it would have been off-limits and highly protected by the Pharaoh and resident high priest, with severe penalty for even approaching.

In addition, the name Serabít el-Khâdim itself provides a tantalising clue to the question regarding gold processing in the region. The etymology of the name Serabít el-Khâdim is still uncertain due to the many valid possibilities. Aside from Arabic, the etymology could be either ancient Egyptian or Semitic. In ancient Egyptian, *Sar = exalted, bit = mines* and *Khetem* signifies *fortress*, giving us the title, *Fortress of the Exalted Mines*. The Hebrew transcription of *Serabít* is *Tsarephat,* which carries the quite literal meaning of *workshop for refining gold*. Hebrew for *Khâdim* is *Katem* meaning *hidden gold*. Combined, the fascinating linguistic possibility is that Serabít el-Khâdim is an Arabic transliteration of Hebrew *Tsarephat Katem* with the suggestive meaning, *The Refinery of Hidden Gold*.

Oddly, it was not Petrie but Eckenstein who, many years after the expedition, presented the radical proposition that the area matched Biblical accounts in her book *A History of Sinai*, published in 1921. Part of her argument was based on archaeological evidence of the presence of a pre-existing Semitic inhabitation at the time of the Exodus. She also drew attention to the two deities worshipped at the Hathor Temple, the first being Hathor who was worshipped by Egyptians as a goddess of fertility and giver of life, and the second being Sopdu who was viewed as a protector, war god of the east and son of Sopdet (regarded by some as the alter-ego of Isis).

The oldest trace of occupation at the Hathor Temple includes a statue dating back to the reign of Sneferu circa 2500 BCE. Petrie concludes:

> Thus, from the history of the place known to later ages, and from the hawk found here, we must credit Sneferu with having opened these mines.

The Hathor Temple was perceived by Egyptian royalty as extremely important. This is evidenced by records of patronage by every Pharaoh from the Twelfth to Twenty-sixth dynasties. Each Pharaoh enlarged the complex, beginning with Amenemhat III in 1833 BCE and concluding with inscriptions from the Ramses VI reign, which ended in 1156 BCE. Although not continuous occupation, this temple boasts an incredible active history of more than 1300 years.

Archaeological Irregularities

Petrie was struck by a number of archaeological anomalies, the first of which was the temple itself, as explained by Eckenstein:

> These buildings were all worked in the red sandstone of the place, and were decorated with figures and hieroglyphs in the formal style of Egypt. The arrangement and the disposition of the buildings have nothing in common, however, with the temples of Egypt.

The bizarre layout of the temple was hiding another enigma in the form of 50 tons of white powder that Petrie presumed to be potash (potassium carbonate or K_2CO_3), remaining from what he deduced to have originally amounted to several hundred tons. He estimated the powder's origin to predate the Eighteenth dynasty – that is, before the Exodus. This incredible discovery presented quite an enigma.

Some temples are built upon a foundation bed of fine white sand, but Petrie would have been aware of this. Petrie's writings reveal a very astute eye for mineral substances and he would have easily been able to discern the difference between sand and ash. Smelting produces a dense black slag of which there was none in this part of the mining operation but plenty elsewhere near the mines; a remarkable discovery

CHAPTER 2 - MOSES AND THE GOLDEN-CALF

in its own right, indicating that resident metallurgists could achieve temperatures high enough to smelt copper offsite.

Petrie considered that the ashes were perhaps a result of extracting alkali ash; this based on a precedent found at a site in Jerusalem. He thoroughly examined the ashes in an attempt to explain the anomaly. He recounts his efforts twice:

1. Though I carefully searched the ashes in various parts, and though my men would have preserved anything noticeable, we did not find aught but pottery.
2. Though I carefully searched these ashes in dozens of instances, winnowing them in a breeze, I never found a fragment of bone, or anything beyond clean ash.

Whatever Petrie's ash was, it was obviously held in high esteem by either the Egyptians or local priests and artisans. He discounts it being commercial alkali ash due to a lack of vegetable sources to burn. However, the Hathor Temple was the Pharaoh's project linked to Egypt by boat and donkey caravans. Temple inscriptions indicate that in some cases five tons of food was delivered overland each day to the Hathor Temple by donkey-trains of up to 500 animals strong. All necessities needed to continue production at a temple as important as the Hathor Temple were made available. With this understanding, the logistics of supplying fuel or vegetable sources for potash production would have presented only a minor obstacle.

Potash has been used from the dawn of history for bleaching textiles and, later, in glass production. Extremely abundant in the Earth's crust, potassium is a major plant and crop nutrient, and an ingredient in the manufacture of soaps, glass, ceramics, chemical dyes, and medicines. An ancient method of making potassium carbonate was to leach wood ashes with water and then evaporate the solution, in the result being a white residue. Potash combined with silica from sand constitutes the primary ingredients for ancient glass production. In India, alkali ash known as *kshara* is a fundamental ingredient in the process of creating Ayurvedic organo-metallic medicines, and is especially important to the Siddha branch of medicine. Siddha tradition claims that the practice originated from beyond the seas in a time of great antiquity. The technical use of *kshara* to produce organo-metallic and metallo-therapeutic medicine is still being practiced today by modern Indians in Tamil Nadu. Likewise, the ash could have been powdered natron, a naturally occurring sodium carbonate, bicarbonate and chloride combination salt used for mummification, as a fluxing agent in metallurgy and to make Egyptian and, later, Roman soda-lime glassware. In all later alchemical traditions, the identity of various fluxing salts, and liquid menstruums created from such salts, was regarded as the mysterious *Key* and great secret of the Art.

Another irregularity identified by Petrie was a series of water cisterns located in the section he called the *Hathor Hanafieh*, described as the most imposing part of the temple. The *hanafieh* is the central water basin where the faithful wash to purify themselves before worship in Middle- and Near-Eastern religions. This universal practice is found in Egyptian temple rites, among Muslims prior to prayer and Jewish priests before the door of the tabernacle, and in the ritual immersion of the Essenes and the Christian counterpart, baptism. The anomaly arises upon considering that, by tradition, there is only a single central *hanafieh* – if, in fact, that was what the cisterns of the Hathor Temple actually were. Aside from this main basin, Petrie found three other large cisterns, which he surmised to have served the purposes of a unique and unusually complex ritual:

> Such a series of ablutions [ritual cleansing] must have belonged to a complex ritual; each applying to a different part of the body.

CHAPTER 2 - MOSES AND THE GOLDEN-CALF

The cisterns only become an anomaly if one rigidly holds to the idea that the additional cisterns – or any of the cisterns for that matter – were used for ritual bathing. Upon acknowledging the possibility that the cisterns might also have served as large leaching tanks for the creation of alkali salts, another avenue of possibility regarding the utility of these unusual basins reveals itself. Petrie comments on the unusual placement of the cisterns:

> *It is noticeable that only one of these tanks is before the outer door of the temple, while three of them are in the finest buildings of the temple ...*

He explains that a centralized *hanafieh* or ritual cleansing area is the established custom, and then downplays the observance in the Egyptian tradition, explaining that cleansing was simply a matter of routine:

> *We do not find this multiplication of washings defined in the early Jewish ritual, where it is only said that the priests were washed before the door of the tabernacle, at the laver [first detailed in Exodus 30:18] ...*

> *The principal court of a mosque is that of the hanafiyeh, which is an octagonal basin in the midst of the court, usually surrounded with pillars.*

> *The ablutions [ritual cleansings] were not a preliminary cleansing before any religious ceremony could be worthily performed; but they were part and parcel of the acts of religion in the temple itself.*

If the cisterns were indeed meant to serve as *hanafieh*, this implies an unusually complex cleansing ritual unique to the Hathor Temple. However, one must also consider Petrie's position as one possible interpretation alongside others. By Petrie's own account, the temple contained several hundred tons of ash at one point in time. Since the Hathor Temple has been well established by Petrie himself as a bustling foundry, mining and manufacturing facility, and metallurgical workshop in an area named *the Place for Refining Hidden Gold*, the idea that large cisterns may have served a more industrial purpose is at least plausible. This could mean that the *hanafieh* was situated just outside the main workshop, with the reaction vessels inside as Petrie describes. Upon reflection, one finds contemporary examples of just such a setup in modern chemistry labs and hospital operating rooms in the form of a clean room at the entrance of the workspace.

Another anomaly is that of the single baker mentioned. The population of the Hathor Temple during operation was large, meaning a considerably large number of mouths to feed. Although at times up to five tons of food was delivered daily, kitchen staff remain unmentioned except for a single baker. From this, it is possible to surmise that this baker was either extremely busy, or he served some other more important role that deserved special mention and could be performed or supervized by a single artisan. Petrie comments on this curious fact:

> The cook or baker, pesy, is named in one part. It is curious that there were not more men employed in the preparation of food, as it takes one man's time to bake and cook for twenty or thirty men, so that a dozen or twenty cooks would have been required for an ancient expedition.

The possible role of a single high ranking baker as relates to the Hathor Temple and the subject of ingestible gold will be explored later in this chapter, as the subject deserves special consideration and a section of its own.

A further anomaly is that of the men who *give birth to minerals*. The highest ranking of the technical artisans at the Hathor Temple were titled *mes en āati*. Petrie understood the etymology of the title, yet needed to venture a guess at what this particular title actually signified:

> The root of the title is, of course, mes, birth; and the derived meanings of meses, to draw a figure, and mesmes, planning or invention ...

CHAPTER 2 – MOSES AND THE GOLDEN-CALF

Petrie clarifies further:

> *It has been thought that this was the title of a mason or stone-cutter, but it is far more important than that ...*

Without a suitable answer, Petrie then reasoned that the men who *give birth to minerals* might have been prospectors. I might add that if artisans were manipulating copper ores and gold into a new gold-bronze alloy, this title and the high rank accompanying it would aptly fit the position.

Scribes and monument masons are detailed on stelae personnel accounts, yet a few stand out as unique and deserve mention. Petrie makes note:

> *... there were 50 "people of the temple of Amen" ... There are also two obscure titles which can only be discussed at greater length than we can do here.*

In *Researches in Sinai*, Petrie discusses minute details of his observations at great length in his own unique way, including things that have nothing to do with archaeology such as nomadic desert lifestyle, daily camp life, flora and fauna, etc. It is rather odd that he noticed something important in the archaeological record that he was unwilling to discuss in any detail or share in a public forum. A mention of "people of the temple of Amen" would seem to indicate either a group of priests or artisans, but the inscription does not specify which.

A later tablet depicts the figure of Ramses I, founder of the Nineteenth dynasty, described as *Prince of Every Circuit of the Aten,* which struck Eckenstein as notable for the reason that Atenism was believed to have died out after the death of Akhenaten and Tutankhamen. Eckenstein commented on this, explaining that the inscription:

> ... suggests that the adherents of the religious reformer [Akhenaten] after his downfall sought and found a refuge at the relatively remote centre of Serabit.

Ramses I also mentions making a delivery of goods to the priests of his new refuge. A likely assumption can be made that these donations fell directly under the control of the head priest of the temple. Although it is an inscription of the Nineteenth dynasty, it provides our first archaeological proof that gold was actually delivered to the Hathor Temple from the Pharaoh whose royal inscription reports sending a shipment to:

> "... my Mother Hathor, mistress of [mafkat] turquoise," silver, gold, royal linen, and things numerous as the sand. And they brought back to the king wonders of real [mafkat] turquoise in numerous sacks, such as had not been heard of before.

Eckenstein interpreted the inscription *mfkA.t* to mean turquoise. This provides us with a tantalising new development in our story. It appears that the Pharaoh would deliver a veritable treasury of precious metals, fine cloth (certainly meant for residing priests rather than laborers) and other valuables with the expectation of receiving valuable *mafkat* in return. This implies that resident priests of the Hathor Temple maintained control over an incredibly valuable commodity to be traded, in a manner of speaking, for silver and more importantly gold. The question whether or not gold was delivered and at least stored at the Hathor Temple seems to have been answered conclusively in the affirmative by Pharaoh Ramses I himself who carved it in stone. Moses would have had access to a tremendous amount of gold during the Exodus.

The Hathor Temple's Role and Scope of Operation

Is it possible that Moses – member of the royal household, Egyptian priest, family member of a Kenite high priest and Midianite ruling class,

CHAPTER 2 - MOSES AND THE GOLDEN-CALF

having himself lived in the Sinai for forty years – would have been unaware of the existence of such a facility? Moses must have known of the Hathor Temple. This answer can be ascertained by looking at the scope of the operation according to the archaeological record. Petrie provides insight:

> The season for carving records was, then, from the middle of January to the middle of May, but generally not later than March. This just accords with our experience of the climate. In January the weather is cold, but good for active work; by April the heat begins; in May it is almost unbearable in the valleys.

Records reveal that Pharaoh mounted fully outfitted royal expeditions to work and worship at the Hathor Temple, requiring a level of organization verging on military. Such expeditions would normally arrive in the spring after the rains and conclude as the weather became too hot a few months later. Petrie estimated that expeditions such as these were undertaken every other year. Names of the important members of the expedition were engraved on stelae with crudely scratched graffiti-style inscriptions of laborer names at the bottom, possibly indicating that the local labor force remained behind after the main party returned home. An example of these detailed records provides some insight into the monumental scale of the operation that is indicative of Egyptian organizational ingenuity:

- *Expedition leader – 'Seal Bearer of the God[-King]' or general;*
- *Twenty-five grades of general officials;*
- *Eleven various titles for local mining management officials;*
- *Eight classes of technical artisans;*
- *Nine varieties of laborers making up most of the population;*
- *Guarded storehouse with treasurer, guards, storekeeper-banker;*
- *Interpreters to aid communication between the Egyptians and laborers speaking Retennu, Aamu (Syrian) and Semitic (Hebrew, Arabic, Aramaic, etc.);*

- *Inspectors (no description of what they were responsible for inspecting);*
- *Up to six boats for sea transport, delivering cargo every second day at the height of operation;*
- *Great donkey trains of up to 500 animals, transporting up to five tons per day at the height of operation.*

This much activity in the heart of the Sinai would have drawn considerable attention and been known to the local tribes who likely provided the labor force, as indicated by graffiti and the need for interpreters. What is not recorded is actually more important. Curiously, no records exist for the priesthood at the Hathor Temple, nor any mention of what their duties may have been, a possible indication that resident priests served as full-time caretakers during the interval between mining expeditions, and were not considered actual expedition members.

The omission of priesthood initially seems to contradict the archaeological evidence that the Hathor Temple was an operational place of devotional worship. However, it is very unlikely that such an important temple was without priesthood or sat vacant for any length of time. It has already been firmly established that the resident priesthood was provided with cloth and precious metals by order of the Pharaoh, indicating a resident population. Artifacts such as carved stelae records of expedition personnel strongly suggest that Moses, as a member of the royal household, would have been fully aware of such an undertaking. As a royal, he likely would have participated at some point. This is evidenced by records of various members of the Eighteenth dynasty royal household worshipping at the Hathor Temple.

The inscriptions at the Hathor Temple add tremendously to our understanding of the Sinai during the time of Moses. An incredible volume of the stelae and monuments, and the temple design itself, are

CHAPTER 2 - MOSES AND THE GOLDEN-CALF

in a style completely alien to that of the Egyptians, yet featuring elements common to Semitic places of worship. The nature of the temple is one of devotion to light, life and bounty rather than the sepulchral. Carved inscriptions often depict a devotional scenario, sometimes accompanied by a detailed description of the setting engraved by the scribe at the bottom.

Two inscription motifs stand out at the Hathor Temple – the word *mafkat* and conical *white-bread* offerings. The first mention of *mafkat* in the Egyptian language is associated with the Hathor Temple and is found on the tablet of Dadkara describing the *Fkat* country (*fkat* is short for *mfkA.t*). Until Petrie's expedition, the term *mafkat* primarily indicated a location. At the time of Petrie's expedition, it presented an enigma and remains so today. It has been mistakenly written as *mfkzt* that indicates an error reading the fourth character, which looks very similar to the number 3 or cursive Z.

A more accurate spelling is *mfkA.t,* which communicates a variety of meanings. Gerald Massey, poet and self-styled Egyptologist of Petrie's era, provides insight into Hathor, Horus, hawks, and *mafkat*:

> *Horus on his pedestal or papyrus is a figure not to be mistaken ... The hawk or vulture on the pedestal or papyrus (uat) was indefinitely older than the human type of Horus the child in Egypt. Horus as the hawk or vulture ... was the child of Hathor; and these two, Hathor and Horus, were the divine mother and child. The gold hawk of Horus is connected with the Egyptian mines, whilst precious metals and stones, especially the turquoise, were expressly sacred to the goddess Hathor.*

Petrie noted that hawk figurines were some of the oldest artifacts found at the Hathor Temple. The golden sparrowhawk (*Accipiter nisus*) was symbolic of Hathor's golden child, Horus. Massey further explains that *mafkat* was a non-specific description of precious stones and minerals roughly equivalent to "bounty of the earth":

> The Egyptian goddess Hathor, as a form of the Earth-mother, was the mistress of the mines, and of precious stones and metals, called mafkat. It was here she gave birth to the blue-eyed golden Horus as her child, her golden calf or hawk of gold. The Egyptian labourers who worked the mines of the turquoise country in the Sinaitic peninsula were worshippers of this golden Hathor and the golden Horus. These two are the divinities most frequently invoked in the religious worship of the Egyptian officers and miners residing in the neighbourhood of the mafkat mines. ... Lastly, as Mafekh or Mafkhet is a title of Hathor, as mafekh is an Egyptian name for the turquoise, for copper and other treasures of the mines, as well as of Hathor ...

He continues by explaining that the east – the location of the Hathor Temple in relation to Egypt – was important to solar mythology as the location of the rising Sun and the star Sirius. This accorded with Akhenaten's new religion, as the icon that represented Atenism was the solar disk casting its rays downward. *Pyramid Texts* refer to a "field of *mafkat*" which makes sense when understood in the context of *mafkat* as depicting a holy realm to the east being related to the Sun god, birth and eternal life as a place of *coming forth*. Petrie explains:

> But in the mythos the place of coming forth had been given to the sun god in the east, and this became the holy land in the solar mythology which has been too hastily identified by certain Egyptologists with Arabia as the eastern land.

Hathor of Mafkat might just as easily be translated as *Hathor of the Place of Coming Forth*, and in a mineralogical context, *mafkat* may be understood as *Mineral Which Has Come Forth*. This might explain *mafkat* used synonymously to mean *precious stone* in general and *turquoise, malachite* and *other copper ores* specifically. It has also been used to describe green plants, papyrus, and an unknown mineral, the color of sunrise, Wadi Maghara, and even Sinai. *Mafkat* is also sometimes synonymous with the word *joy*, but could indicate joy associated with

CHAPTER 2 - MOSES AND THE GOLDEN-CALF

receiving bounty; the joy of the Sun's rays at dawn or that of bringing forth new life, such as Hathor giving birth to Horus. The word *mafkat* and the idea of *coming or bringing forth* appear to be closely related mining and industry in the Sinai.

Since the focus of mining in the region was turquoise, Petrie and Eckenstein settled on turquoise as their interpretation of the word and it remains the mainstream definition used by Egyptologists to date:

There is always a difficulty in reading aright the names of precious stones. Mafkat was often rendered as malachite, and it needed the turning over of the rubbish heaps at Serabit and the discovery of turquoise, in order fully to establish the nature of the stone that was the product of the area ...

The inscriptions of *mfkA.t* on stelae and walls at the Hathor Temple were directly related to the industry of the temple. Another form of representative inscriptions and carvings also relates to a product of the temple, which is introduced briefly here. In Egyptian worship, a very special food offering is presented, often by the Pharaoh himself, to a deity or God. It is a conical shaped offering much like an upside-down ice cream cone fitting in the palm of one's hand. Images of these conical offerings depict a type of sacramental bread called *conical white-bread*, and can be found on temple walls at a number of sites throughout Egypt. They are associated with the act of offering, represent eternal life and relate to Sirius, Sopdu and the first day of the year as Sirius rises with the Sun.

Conical white-bread offerings at the Hathor Temple are to be expected. What is very unusual is the high concentration of depictions in the region. Not only were these *conical white-bread* offerings being baked at the Hathor Temple, they were exported to Egypt for use in ceremonies and rituals. Archaeological evidence shows that the Hathor Temple was producing bread offerings for very important occasions as late as 654 BCE, as Eckenstein explains:

> *... in the list of temples which made gifts to Nitocris on the occasion of her adoption by the Pharaoh Psamtek I, about the year b.c. 654, a gift was made of a hundred deben of bread by the temples of Sais, of Buto, and by "the house of Hathor of Mafkat" i.e. Serabít ...*

The interesting part of the above quote is that the bread was measured by weight / value rather than number of loaves. This weight / value indicator, *deben*, is normally used for items with a trade value such as gold, silver, copper, grain, etc. The word was also used to represent both value and weight in a similar manner to Great Britain's pound sterling.

The weight of *deben* during the time of Psamtek I was roughly 91 grams, for an overall weight of temple-crafted bread amounting to approximately 91 kilograms / 200 pounds. Either that is a lot of bread or the number represents value. The amount of gold by weight in a gold *deben* of the time was 12 grams. As a measurement of value, 100 gold *deben* at 12 grams of gold each equals a total value of 1.2 kilograms of gold. One may interpret this passage to indicate the amount of bread offered or the value of the offering; either way this was no ordinary bread.

Conical white-bread offerings, common to the Hathor Temple, appear repeatedly in engravings at many locations at Serabít el-Khâdim. Scribal notes describe the conical objects being offered in these scenes clearly as *white-bread*. Upon recalling the anomaly of the single baker mentioned earlier as an artisan on expedition personnel stelae, the role of a very noteworthy baker-artisan appears now to have a plausible explanation. Used for offerings, temple-crafted bread was a highly valued and very expensive commodity. If the production of this unique bread was under the supervision of high-ranking baker, it implies that there was something very special about every aspect of the bread – even more so when we take into consideration than *conical white-bread* offerings were gifts from the Pharaoh and associated with eternal life. *White-bread* was

CHAPTER 2 - MOSES AND THE GOLDEN-CALF

also related to Hathor, Horus, Sopdu, the east as a holy realm, the star Sirius, the Sun, gold, birth and a coming forth: essentially everything that the Hathor Temple stood for could be symbolized and embodied in *conical white-bread*.

Conical white-bread depicted in carvings appears to be unleavened bread baked on a conical stone form. They are depicted as slightly irregular, as opposed to the perfect triangle, in the hieroglyph that represents them. The logical explanation for this irregularity is that the carvings genuinely depict the irregular shape of an actual baked good. To further the connection with the Hathor Temple, two conical stone forms were found in the ruins as Petrie describes:

> ... in the portico were two conical stones of about 6 inches (15 cm) and 9 inches (22.5 cm) in height respectively.

The hieroglyph for Sirius is a triangle drawn in roughly the same proportions as *conical white-bread* offerings: a triangle measuring approximately nine units high and four units at the base. The *Sothic* (Greek: Sirius) *Triangle* containing a smaller inner triangle of the same proportion, is the hieroglyph to mean *give, offer, appoint, cause, permit, grant,* etc. and implies *to bring forth*. The *Sothic Triangle* combined with the *Ankh* forms the hieroglyph for *to give, offer or bring forth everlasting life*, commonly found in conjunction with *conical white-bread* offerings. The hieroglyph for *to be satisfied or content* is depicted as a *conical white-bread* offering in a shallow tray placed squarely on an offering mat.

The two unusual motifs – *mafkat* as a *Substance or Compound Which Has Come Forth*, and *conical white-bread* inscriptions and their physical counterpart as temple-crafted *white-bread* representing the *Bringing Forth of Eternal Life* – are aspects unique to the Hathor Temple insofar as they appear at this temple in a significantly higher concentration than elsewhere in Egypt.

The Importance of Petrie and Eckenstein's Work

Clearly, the Hathor Temple was profoundly important to Egypt and the land of Midian. Petrie and Eckenstein's research is crucial to establishing the industry and significance of the region during the time of the Exodus, and has provided important contributions to other areas of study. Petrie's discovery of a Proto-Sinaitic script marked the transition from a written Egyptian pictographic language to the crude beginnings of our modern phonetic alphabet.

> I am disposed to see in this one of the many alphabets which were in use in the Mediterranean lands long before the fixed alphabet selected by the Phoenicians.

Professor Orly Goldwasser, head Egyptologist specialising in languages at the Institute of Archaeology at the Hebrew University in Jerusalem, confirms the importance of Serabít el-Khâdim as the birthplace of the Western alphabet. In her article *Canaanites Reading Hieroglyphs*, she explains:

> Serabit el-Khâdim may have been a natural cultural site for the creation of the new script. A melting-pot, a place of a "plurality of cultural contexts," ...

She refers to Proto-Sinaitic as the "script of the poor" and explains that it would be quite some time before it was fully adopted by the Greeks:

> Regarding the Protosinaitic script and its Protocanaanite successors as the "script of the caravans," or, better, the "script of the poor," ...

> All this would have to wait until the official establishments of the 9th century [BCE] states adopted the "script of the poor" and made it the new official script of the Near East. With its adoption by the Greeks, it became the script of most Western civilizations.

CHAPTER 2 - MOSES AND THE GOLDEN-CALF

That the origins of our modern alphabet can be traced back to the Hathor Temple is fascinating in itself and a substantial contribution by Petrie. However, his main purpose was to elucidate the story of Exodus. In this regard, Petrie commented on the question as to whether the Israelites would have encountered the Hathor Temple:

> *The argument that the Israelites would not have travelled down to the region of the Egyptian mines has no force whatever. The Egyptians never occupied that mining district with a garrison, but only sent expeditions; at the most these were in alternate years, and in the times of Merenptah only once in many years. Hence, unless an expedition were actually there in that year, no reason existed for [the Israelites] avoiding the Sinai district.*

Petrie then did something rather unexpected: he attempted to correct Biblical interpretation, which may have been the reason why his *Researches in Sinai* has never received the same accolades as his other revered publications, and to date remains his most uncelebrated work. He argues the impossibility of Sinai accommodating the number of Israelites listed in the Bible, and explains this as a misinterpretation of a single word – *alaf* – as likely resulting from poor editing during the 9th century BCE or sometime during the post-exilic age.

> *Is there any emendation of a likely error which could yield a probable form for the original record? The total number stated in Exodus is the result of the census of separate tribes stated in Numbers. ...The word alaf has two meanings, "a thousand" and "a group" or "family." Hence the statement in words of 32 alaf, 200 people, might be read as 32,200, or as 32 families, 200 people. The statement is ambiguous, and is one which is, therefore, peculiarly liable to corruption of the original meaning in editing an earlier document. We have at least a working hypothesis that the "thousands" are "families" or tents, and the "hundreds" are the total inhabitants of those tents.*

BOOK 1 – ORIGINS & EVOLUTION

This was a bold statement at the time and not well received. He devoted a whole chapter to the subject, and then described Moses from the hypothetical first person perspective of an Egyptian. This could have been regarded as a sacrilegious experiment by both Jews and fundamental Christians of his day, and very risky.

> Finally, let us sketch out what a cultivated Egyptian would have said of the Israelites' history, so far as we can at present understand it. One of them who had been well educated by us had run away into the desert, and settled in Sinai. Seeing that the land there was sufficient to support his kindred, he came back and tried to get permission for them to go on a pilgrimage to a sacred mountain. This was refused; but many troubles of bad seasons, and a plague at last, so disheartened us that, in the confusion, some thousands of these tribes escaped into the wilderness. They safely crossed the shallows of the gulf, but a detachment of troops following them was caught and swept away.

Petrie's characterization of Moses and the Israelites revealed a very human figure embroiled in the social drama of an oppressed group of approximately 5,000 to 6,000 people fleeing under the leadership of a high-ranking, Egyptian educated rebel. He summarizes his work:

> We have now seen how far some fresh information from Sinai, and some new examination of our documents, may lead us in understanding the history of Israel. Probably no party will be content with such results; but such seem to be the data for future dealing with this question. Only a few more points have been cleared of the many which are yet in darkness. But if we can at least gain some clearer view of the limits of uncertainty, of the problems yet to be solved, and of the checks from other sources which limit the bounds of speculation, we shall not have made this review of the subject in vain.

Eckenstein's *A History of Sinai* is a masterpiece of Biblical archaeology in which she accounts for the minutest details found in the Exodus narrative with the trained eye of a professional observer possessing

CHAPTER 2 - MOSES AND THE GOLDEN-CALF

firsthand experience of living in the region for extended periods. Her research is so compelling that it must stand as the prevailing explanation until the subject is furthered by a more accurate depiction backed by archaeological evidence. Her landmark elucidation of the events of Exodus read like a play-by-play account distinguished by great accuracy concerning distances, terrain, sources of water, etc. She summarizes her work succinctly:

> ... the study of the episode [Exodus] reviewed in the light of modern research, reveals an unexpected accuracy [referring to logistics], and once more shows that tradition is of value in proportion to our power of reading it aright.

From the perspective of archaeology presented by Petrie and Eckenstein, it is plausible that the Exodus occurred sometime during the Eighteenth dynasty. Moses was raised as an Egyptian member of the royal household, likely holding title as a priest. Moses was fully familiar with the Hathor Temple, having lived as a family member of the Kenite priest and tribal leader in Midianite territory. In addition, the Hathor Temple was likely Moses' and the Israelites' primary destination in Sinai.

As to whether Moses would have had access to a workshop in Sinai *en route* that would facilitate the manipulation of gold; overwhelming archeological evidence affirms this. In addition, Petrie also proved that copper smelting was being performed offsite at temporary furnaces based on copper slag evidence. These furnaces would have needed to reach a minimum temperature of 1,984 degrees Fahrenheit / 1,085 degrees Celsius to melt copper; higher than what is required to melt gold. Petrie further revealed that gold was physically delivered to the Hathor Temple by a Pharaoh. This argues very strongly in favour of the possibility that Moses had access to the materials and technological means to accomplish the task depicted in the *golden calf* narrative. Although not specifically mentioned in religious texts, the chance that either Moses or Jethro were unaware of such monumental industry in the region is highly unlikely. With their leadership qualities, the alternative scenario is that they were probably in charge of the operation.

HOW DID MOSES DO IT?

That secret gold and copper metallurgy was taking place in an Egyptian Kenite temple in Midian territory of the Sinai at the time of the Exodus is profoundly significant but not without precedent – antimonial bronze technology had existed in the southern Levant for over two millennia by the time of Moses. The main question remains: how did Moses dissolve gold only to coagulate it to a brittle mass before grinding it to a fine powder using metallurgical, as opposed to chemical, techniques? The answer becomes clear upon recalling the importance of *mśdmt / kahâl / ḳoḥl* to the Egyptians, Hebrews and other desert dwellers, as addressed in an earlier chapter.

Paracelsus, a European alchemist working during the 16th century, succinctly explained the entire operation. The process he describes is deceptively simple, effective and employs only metallurgical, as opposed to chemical, techniques. Although he makes use of purified antimony, this process will work with stibnite ore just as well. All Moses would have needed is a metallurgical facility, a *golden calf* and a supply of local *mśdmt / kahâl / ḳoḥl*; all of which were abundantly available to him in Sinai.

Paracelsus' Metallurgical Technique – 16th Century
Where Paracelsus learned this technique is a mystery, as he remained silent on the subject. It is possible that he was inspired by the *golden calf* narrative or that he arrived at it experimentally. The technique involves straightforward metallurgy and presents the most reasonable hypothesis for Moses' operative technique. This process came to be known as the *Ars Brevis* (Quick Art) or, alternately, *Via Sicca* (Dry Path). Paracelsus

CHAPTER 2 – MOSES AND THE GOLDEN-CALF

seems to have been the first European alchemist to write so clearly about it. The great mysterious early 20th century French alchemist, Fulcanelli, was the last proponent of any renown to champion the *Via Sicca* to create the Philosophers' Stone. The recipe below was translated from Paracelsus' *Sphaera Saturni* and preserved by Dr. Sigismund Bacstrom:

> *Take the* kerib ☉ *[gold] & throw him before the* wolf ♃ *[stibnite] that it may devour him, which, however, causes a long struggle. When the King* ☉ *is apparently devoured,* make a large roaring △ *[fire], & the* wolf will be devoured *also.*

> *When the* Lion ☉*, the* Red Lion*, or the king* ☉ *is or* remains conqueror*, at last his internal spirit by this battle fortified, & his eyes are* Luminous like the ☉ *[sun]; but do not let the matter rest here, says Paracelsus.*

Paracelsus is describing the process of bringing gold into fusion with antimony, followed by an increase in temperature to the point that the antimony vapourizes, thus leaving highly purified gold in the crucible.

> *Many do think this means the purgation of* ☉ *by* ♃ *but there's a hidden sense in it.*

> *Take* fine ☉ & melt it *in a strong new* ☾ *[crucible], by the blast with 3 parts of pure* ☍ ♃ ♂ *[marital regulus of antimony], &* keep blowing the △ *[fire], until the* ☍ *[antimony] is vanished.*

The recipe calls for an additional three parts of pure metallic antimony to molten gold, and allowing the antimony to vaporize. Paracelsus adds some operative details, explaining that it does not matter whether one uses stibnite or purified antimony, but to continue to add three parts each time the previous addition vaporizes, while keeping the gold in a molten state the entire time.

> *Continue adding gradually some fresh* ☍ ♃ ♂ *[antimony] or even crude* ♃ *[stibnite] & continue blowing & melting, so that* the ☉ remains in

> *constant fusion, until the* ♂ *[stibnite], or its* ☿ *[regulus; antimony] has carried away the whole body of the* ☉*, to all appearance, which is effected in 10, 11, or 12 times, when there remains our* ☉*, our* ℞ *[tincture; Philosophers' Stone], transparent Red like a ruby.*

The effect of repeated vaporization of stibnite or antimony over molten gold is the suspension of gold in what is known as *glass of antimony*. The process is known to alchemists as *vitrification* or glass making. Because the gold has achieved a nanoparticle size, this contributes to the deep red color of the glass. This glass will resemble a crystal or translucent stone and remains our best candidate for the Philosophers' Stone of Moses.

Sigismund Bacstrom's Metallurgical Technique – 18th Century
Two centuries after Paracelsus, Rosicrucian adept Sigismund Bacstrom reopened the experiment successfully as recorded in *A Compendium of Alchemical Tracts II* and included a few more additional details:

> *Take three parts or scruples of pure* ☿ *[antimony regulus], & one part or scruple of fine* ☉ *[gold], without alloy. Melt this in a covered* ♆ *[crucible], & let stand in the* △ *[fire], covered, until the* ☿ *is evaporated & the* ☉ *remaining alone. When cold, you will find the gold look paler than before.*

> *Melt this gold again with three parts* ☿ ♂ ♂ *[antimony] & let it stand in the* △*, in a covered* ♆*, in constant fusion, or you do nothing, until the* ☿ *is again evaporated.*

Bacstrom explains that each time this operation is performed gold reaches a finer state of division with a very miniscule amount of antimony remaining behind fused with the gold. He then proceeds to list the weight specifications and number of fusions for the entire synthesis matching those of Paracelsus in the synthesis above:

> *Repeat this operation (each time melting three parts* ☿ *with your one part of* ☉ *by weight) eleven times, or even twelve times.*

CHAPTER 2 - MOSES AND THE GOLDEN-CALF

Bacstrom then provides a few operative details to ensure success and to protect the health of the operator:

Each fusion must last in a strong heat, until the Chalybs [antimony] is fairly evaporated, but you must NOT BREATHE THESE ANTIMONIAL FUMES. *The little furnace should stand under a chimney.*

What follows is an explanation as to how the gold has become vitrified, along with color indicators in order to gauge success:

... 3 scruples will remain fixed with the one part or 1 scruple of spermatic or regenerated ☉, & will, & must, I think, appear RED, *or at least, of a deep orange colour, although the deeper the red, the more rich in △, or tincture.*

The materials and methods presented above involve metallurgical skills well within the capabilities of metallurgists in the Bronze Age and, specifically, copper smelters and smiths at Serabít el-Khâdim. What is not addressed above is a use for an alkali ash or fluxing salt. If the product above is placed into a crucible and fused with the correct amount of alkali ash, natron or even Dead Sea Salt, the gold-antimony alloy will fuse with the salt in a process known as vitrification to create a crystalline structure containing nanoparticles of gold and antimony within a crystalline salt structure. This ruby colored gold-antimony salt crystal can be ground easily and will dissolve in liquid to achieve a colloidal suspension of gold-antimony nanoparticles.

The process also works with silver. Gold or silver reduced and crystallized via repeated vapourization of local *mśdmt / kahâl / ḳoḥl* results in two products that would each later be classified by alchemists as a Philosophers' Stone. Alchemists from later traditions would associate these two incredible stones with the two stones, the *Urim* and *Thummim*, on the breastpiece of judgment crafted for Aaron in the wilderness. If sand or quartz (silica dioxide or SiO_2) is used as the fluxing agent in place of salt and fused at around 1,725 degrees Celsius, a unique type of glass is achieved: one clear translucent and the other ruby-red. In antiquity, these would have been considered sacred gemstones produced by a magico-religious process. If a high enough temperature could have been achieved in order to create the glass products mentioned above, a plausible hypothesis for the identity of the *Urim* and *Thummim* reveals itself. These two stones are mentioned eight times in the *Bible*, generally signifying spiritual authority. The two stones are employed to arrive at decisions.

ל וְנָתַתָּ אֶל-חֹשֶׁן הַמִּשְׁפָּט, אֶת-הָאוּרִים וְאֶת-הַתֻּמִּים, וְהָיוּ עַל-לֵב אַהֲרֹן, בְּבֹאוֹ לִפְנֵי יְהוָה; וְנָשָׂא אַהֲרֹן אֶת-מִשְׁפַּט בְּנֵי-יִשְׂרָאֵל עַל-לִבּוֹ, לִפְנֵי יְהוָה--תָּמִיד. {ס}

Exodus 28:30 And *in the breastpiece of judgment you shall put the Urim and the Thummim,* and they shall *be on Aaron's heart, when he goes in before the Lord. Thus Aaron shall bear the judgment of the people of Israel on his heart before the Lord regularly.*

It remains unclear whether or not Moses fused his product with salt. The product in either case will be a brittle product capable of being easily ground. This unique type of red gold product matches the descriptions of Parvaim gold mentioned in 2 Chronicles:

ו וַיְצַף אֶת-הַבַּיִת, וַיְחַר וֶאֱלִישָׁמָע, וֶאֱלִיפָלֶט.

2 Chronicles 3:6 He *adorned the house with settings of precious* stones. *The gold was gold of Parvaim.*

Rafael Patai lists two types of gold mentioned in Jewish sources as possessing a unique red coloration:

Muphaz gold – *Rabbi Patroqi, the brother of R. Drosa, said in the name of R. Abba bar R. Buna: It looked like sulphur set on fire. ...*

Parvayim gold – *R. Shim'on ben Lazish said: It is red like the blood of the bull [par]. And some say that it brings forth fruit ...*

CHAPTER 2 - MOSES AND THE GOLDEN-CALF

(in Mishna Yoma 4:4), "Every other day its gold was yellow, on this day [on Yom Kippur] *it was red"* **– this is the** Parvayim gold which is like the blood of bulls.

The descriptions of Parvaim gold tantalisingly beg the question as to whether this type of red gold, described as "red like the blood of the bull", might be better described as being the red blood of the *golden calf*.

Regardless of whether Moses vitrified his product with a fluxing agent or not, he created a ruby red crystallized gold product that was ground to a fine powder. Greeks and Alexandrians would later term this powder *xēre / xêrion*, simply meaning powder in Greek. Greek *xēre / xêrion* was transliterated into Arabic *iksir* and finally *al-iksir*. Moses created the world's first powdered gold *elixir* by the process above. Later, in our exploration of Maria Hebrea's alchemy in early Alexandria, we will cover additional methods of working with this technique and further explain its tremendous industrial value. The work of Paracelsus and Bacstrom establish plausibility of the biblical account of Moses' metallurgical feat with the implications that Moses may have been the first recorded alchemist in the western tradition. If Paracelsus and Bacstrom's process accurately reflects Moses' methodology, it may have served as the proto-archetypal recipe foundational to Maria Hebrea's Judeo-Egyptian alchemy that developed centuries later in Alexandria.

Who Taught Moses His Alchemical Technique?
Where did Moses learn the skills to perform such an alchemical feat approximately 3,500 years ago? There are three possible sources for technology of this sort:

1. *He learned it from Joseph in Egypt;*
2. *He learned it from Egyptian temple craftsmen, as was his royal right;*
3. *He learned it from Jethro at the Hathor Temple at Serabít el-Khâdim in the Sinai.*

BOOK 1 – ORIGINS & EVOLUTION

Biblical Joseph (Yusuf) was sold into slavery, only later to become the most powerful man in Egypt after Pharaoh. Aside from confirming that Midianites traded with Egypt, possessed precious metals and traded in slaves, the passage below describes poor conditions by which Joseph arrived in Egypt:

> כח וַיַּעַבְרוּ אֲנָשִׁים מִדְיָנִים סֹחֲרִים, וַיִּמְשְׁכוּ וַיַּעֲלוּ אֶת-יוֹסֵף מִן-הַבּוֹר, וַיִּמְכְּרוּ אֶת-יוֹסֵף לַיִּשְׁמְעֵאלִים, בְּעֶשְׂרִים כָּסֶף; וַיָּבִיאוּ אֶת-יוֹסֵף, מִצְרָיְמָה.
>
> *Genesis 37:28* Then Midianite traders passed by. And they drew Joseph up and lifted him out of the pit, and sold him to the Ishmaelites for twenty shekels of silver. They took Joseph to Egypt.

Yuya has been identified by scholars as a possible candidate for Biblical *Yusuf*. Eckenstein uses the term *Syrian* to describe Yuya's daughter Thyi's Semitic ethnicity:

> The reader will recall that Queen Thyi also was of Syrian origin, and that Amen-hotep III and Thyi were the parents of Amen-hotep IV, better known as Akhen-aten, the great religious reformer of Egypt.

Egyptologists have concluded that Yuya's true origins remain unclear and that his mummy analysis reveals that he was taller than average with unusual features and appearance not typically Egyptian, thus indicating a possible foreign origin. Even his name was uncharacteristic for an Egyptian. Yuya's mummy was the only Egyptian mummy to have his hands in a prayer position under his chin, Semitic features and a beard in the style of ancient Israelites.

> כד וַיֹּאמֶר יוֹסֵף אֶל-אֶחָיו, אָנֹכִי מֵת; וֵאלֹהִים פָּקֹד יִפְקֹד אֶתְכֶם, וְהֶעֱלָה אֶתְכֶם מִן-הָאָרֶץ הַזֹּאת, אֶל-הָאָרֶץ, אֲשֶׁר נִשְׁבַּע לְאַבְרָהָם לְיִצְחָק וּלְיַעֲקֹב.
>
> כה וַיַּשְׁבַּע יוֹסֵף, אֶת-בְּנֵי יִשְׂרָאֵל לֵאמֹר: פָּקֹד יִפְקֹד אֱלֹהִים אֶתְכֶם, וְהַעֲלִתֶם אֶת-עַצְמֹתַי מִזֶּה.
>
> *Genesis 50:24-26* And Joseph said to his brothers, "I am about to die, but God will visit you and bring you up out of this land to the land that he swore to Abraham, to Isaac, and to Jacob." Then Joseph made the sons of Israel swear, saying, "God will surely visit you, and you shall carry

CHAPTER 2 - MOSES AND THE GOLDEN-CALF

כו וַיָּמָת יוֹסֵף, בֶּן-מֵאָה וָעֶשֶׂר שָׁנִים; וַיַּחַנְטוּ אֹתוֹ, וַיִּישֶׂם בָּאָרוֹן בְּמִצְרָיִם. {ש} *up my bones from here." So Joseph died, being 110 years old. They embalmed him, and he was put in a coffin in Egypt.*

Yuya was Amenhotep III's chief advisor, holding the posts of King's Lieutenant and Master of the Horse. His official title was *Father of the God*, possibly in reference to his being Amenhotep III's father-in-law. Yuya was also Akhenaten's grandfather, his daughter Tiye being Akhenaten's mother. The history of Yuya correlates quite readily with the description of Biblical Yusuf, but falls short in one regard: Yuya's mummy was found in Egypt, whereas Yusuf's remains were in Moses' possession as he left Egypt.

יט וַיִּקַּח מֹשֶׁה אֶת-עַצְמוֹת יוֹסֵף, עִמּוֹ: כִּי הַשְׁבֵּעַ הִשְׁבִּיעַ אֶת-בְּנֵי יִשְׂרָאֵל, לֵאמֹר, פָּקֹד יִפְקֹד אֱלֹהִים אֶתְכֶם, וְהַעֲלִיתֶם אֶת-עַצְמֹתַי מִזֶּה אִתְּכֶם. כ וַיִּסְעוּ, מִסֻּכֹּת; וַיַּחֲנוּ בְאֵתָם, בִּקְצֵה הַמִּדְבָּר. *Exodus 13:19-20 Moses took the bones of Joseph with him, for Joseph had made the sons of Israel solemnly swear, saying, "God will surely visit you, and you shall carry up my bones with you from here." And they moved on from Succoth and encamped at Etham, on the edge of the wilderness.*

This discrepancy between historical Yuya and Biblical Yusuf can be reconciled upon allowing for the possibility that the authors of Exodus employed the coded term *bones of Joseph* to mean that Moses was carrying with him a hidden knowledge or tradition that originated with Yusuf / Yuya in Egypt. Thus, an Egyptian-Hebrew source of Moses' alchemical knowledge can be postulated.

Being raised as Egyptian royalty would have provided Moses with many opportunities to learn metallurgical technologies or even to be in charge of such industry. Jethro's position as a Kenite high priest in Midianite territory, abounding in copper mines and a temple associated with metallurgy, provides an ample 40-year window of additional opportunity for Moses to have learned from Jethro.

BOOK 1 – ORIGINS & EVOLUTION

Religious and Cultural Implications

A question that must be addressed is this: by making the children of Israel drink a solution containing gold nanoparticles, was Moses punishing them, or providing a unique psychospiritual medicine or sacrament now lost to history. Modern research on the therapeutic value of gold nanoparticles may help us to answer the question. In 1998, an article appeared that addressed gold's potential as a nootropic (i.e. smart drug) and nervine (i.e. therapeutic treatment for nervous system disorders). A few years later in 2000, researchers demonstrated that gold nanoparticles exhibited anti-anxiolytic (i.e. stress reducing), anti-depressant and anti-cataleptic (i.e. versus physical or psychological disorders such as epilepsy, schizophrenia, etc.) actions in mice with a wide margin of safety. The study also revealed immunostimulant actions without any discernible negative side effects. In 2006, Harvard Medical School published an article on gold's biological mechanism of action in treating autoimmune diseases such as juvenile diabetes, lupus and rheumatoid arthritis. Finally, in 2010, researchers demonstrated that gold nanoparticles end up nesting in the mid-brain; specifically the thalamus and hypothalamus, hippocampus and, to a lesser degree, the cerebral cortex.

The sub-trace amounts of antimony are diaphoretic, resulting in heavy sweating – this alone is therapeutic in harsh desert conditions. If Moses had indeed brought his product into salt fusion, the salt content simply adds much needed electrolytes and minerals to the therapeutic profile of Moses' elixir. Metallotherapeutic and organometallic research is ongoing in modern laboratories and remains a fascinating subject of investigation.

Patai in *The Jewish Alchemists* explains that many centuries after Moses, the concept of just such an elixir was well known to rabbis of the Talmud:

> ... *there is a kind of gold which is so soft that it can be spun like a thread and stretched like wax, and another which is like burning sulfur in*

CHAPTER 2 - MOSES AND THE GOLDEN-CALF

appearance. ... The other great quest of the alchemists – to be able to produce a miraculous substance or essence that could give health and youth, and banish death – was also known to the rabbis of the Talmud. They called it in Hebrew sam hayyim, and in Aramaic samma dihayya, that is, "elixir of life," and believed in its actual or possible existence. The references to it in rabbinic literature are few and indirect, but clear enough to render the presence of such a belief indubitable.

Is it possible that Moses knew of gold's therapeutic value? Was the ingestion of gold an ancient Egyptian, Kenite or Semitic sacramental act? Was Moses' gold product baked into ancient Egyptian and Jewish sacred breads? The answers to these questions may never be resolved. What remains clear is that the story of Moses and the *golden calf* is so plausible that it must be considered an historical probability when viewed from an informed perspective.

There is every indication that the powder created by Moses is what the Egyptians industrially called a *powder of projection*. Ancient Egyptians used powders that were *projected*, meaning placed in a crucible with copper oxides or molten metal as a technique for alloying and coloring in contrast to gilding technologies of the Babylonians – a technical distinction that distinguished Judeo-Egyptian alchemy. These metallic powders and salts when applied to bronze craft enable a level of reproducibility, with resultant high quality alloys.

These powders could be dissolved or suspended in water and ingested as Moses had the children of Israel do, baked into valuable sacramental breads or combined with fine copper to create a *gold-antimonial bronze* that would later come to be known as *Corinthian bronze*. The *powder of projection* must have been the identity of the mysterious *Mafkat* immortalized and carved in stone on sacred stelae; the enigmatic *Mineral of Coming Forth* created by the men who *give birth the minerals*.

The copper mines provide a hint as to the ultimate purpose of the Hathor Temple. The artisans serving under the high priest, most likely Jethro, were transforming gold into *powder of projection*. The resulting *gold-antimonial bronze* created via the *powder of projection* was the original *alchemical gold*. We find mention of this unique bright golden bronze described as being as precious as gold in the Book of Ezra:

וּכְפֹרֵי זָהָב עֶשְׂרִים, לַאֲדַרְכֹנִים **כז** *Ezra 8:27* ... *twenty bowls of gold worth a thousand darics, and two vessels of fine bright bronze as precious as gold.* אָלֶף; וּכְלֵי נְחֹשֶׁת מֻצְהָב טוֹבָה, שְׁנַיִם-- חֲמוּדֹת, כַּזָּהָב.

By the 1st century CE, this unique bronze sold for a higher price than its equivalent weight of pure gold. Moses and the early Jewish alchemists of Alexandria were not creating gold, but rather a unique bronze trade good even more valuable than gold.

Corinthian bronze would later become a major trade good produced and exported from Alexandria until at least the late 3rd century CE. Approximately 1,500 years after the Exodus, *Corinthian bronze* crafted by Jewish artisans in Alexandria was used to adorn the Second Temple of Jerusalem. If Moses were indeed in possession of the technology to create *Corinthian bronze* so early in history, it would have provided a source of immense wealth for the struggling new nation. The operative simplicity of its creation would have demanded extreme secrecy. *Corinthian bronze* would have provided an incredibly powerful legacy to be taken to the Holy Land and stewarded by Kings David and Solomon.

Chapter 3 – Maria Hebrea's Tincture

*One becomes two, two becomes three,
and by means of a fourth, achieves unity.
Thus two are but one.*

Maria Hebrea

THE MOTHER OF ALCHEMY

Of the many alchemists known to the history of Western alchemy, Maria Hebrea may be suggested as the greatest of them all. Maria is believed to have lived in Alexandria sometime around 200 CE. Alchemists were introduced to one of the most ancient texts regarding the Philosophers' Stone and its confection through writings attributed to Maria, recorded by the Alexandrian alchemical anthologist Zosimos of Panopolis who preserved her brand of alchemy around 300 CE. Maria is credited as the founder of Western alchemy as Raphael Patai explains in *The Jewish Alchemists*:

> *There is historical evidence, on the other hand, that among the Hellenistic alchemists there were several Jewish adepts, one of whom, Maria the Jewess, was considered by them the founder of their art. She is constantly referred to by Zosimos and others as the ultimate authority in both the theory and the practice of alchemy,* **and her teachings, in many cases in the form of pithy aphorisms, are quoted as if they had been prophetic announcements.**

Zosimos, and the great Islamic alchemist Jābir ibn Hayyān, plus countless other alchemists respected Maria most of all, and with good reason. A strong distinction must be made between the myriad chemical technologies being developed in Alexandria as compared to alchemy. An array of chemical technologies existed in Maria's Alexandria that would later become incorporated into alchemy as it transitioned into *chymistry*. These technologies include perfume manufacture, glass production, pigment and dye manufacture, and herbal medicines, among others. What makes Maria so fascinating is that she specialized in a unique art known as *chrysopœia*, or alternately *aurifaction* – this as opposed to the *aurifiction* of the Babylonian gilding tradition of Pseudo-Democritus. In this treatment, *chrysopœia* and *aurifaction* are used synonymously to indicate Maria's brand of alchemy.

Aurifiction was an attempt to falsify gold via processes described in the Leyden and Stockholm Papyri. In contrast, *aurifaction* was the art of creating high quality bronze, indistinguishable from genuine gold, via Maria's unique processes. Maria came from a centuries old tradition that worked copper into the most exquisite bronzes in history, created in Alexandria and exported as valuable trade goods around the classical world. Maria was the head of an artisan guild in an art that was highly revered and honored; a tradition she referred to simply as *the Art* (or *téchnī*; τέχνη). Women seem to have enjoyed tremendous respect and egalitarian social status in general and in the Art of *chrysopœia* in particular. These grand matriarchs of Alexandrian *chrysopœia* held professional trade names such as Isis, Cleopatra, Madera, and others whose writings reveal a deep understanding of the Art.

Maria was Jewish; a topic well researched by Patai who provides a number of details about Maria the person, including the lineage of her art. Arabic sources imply that Maria was trained at Memphis in Egypt in the tradition of the great Persian magus and brother of Xerxes, Ostanes, who taught there centuries earlier. Hermodorus mentions Ostanes as one of the names common in a supposed line of magi that ran from Zoroaster down to Alexander's conquest. Roman historian Pliny the Elder, writing during the 1st century CE, was skeptical of such an ancient succession, but held the view that the Ostanes who had accompanied

Xerxes on his expedition to Greece was indeed the first man to write the classics on magic studied by the Greeks in Alexandria.

Ostanes was considered one of the fathers of alchemy, medicine, mineralogy, and botany. Arab alchemical author al-Habīb presents Maria as a contemporary of Pseudo-Democritus, and the teacher of King Aros, although she may have lived a century after Pseudo-Democritus. Pseudo-Democritus lived in 1st to 2nd century Alexandria and authored an alchemical classic entitled *Physical and Mystical Matters*. In it, he describes his particular tradition as "An art, purporting to relate to the transmutation of metals, and described in a terminology at once Physical and Mystical" – a fitting summation of Alexandrian *chrysopœia* in general and Maria's approach in particular.

When Alexander founded Alexandria in 332 BCE, he invited Jews to reside there and assigned them the Jewish quarter, a walled city within a city as large as the Greek quarter and home to the largest Jewish community in the world at the time. Alexandrian Jews formed an integral part of the Jewish community in the Near East, recognized the authority of the High Priests and paid tribute to the Temple at Jerusalem. They were governed by a Jewish council and laws, ruled by 71 elders and an *Ethnarch* (Greek: leader of the people). Although Alexandrian Jews retained their culture and religion, they incorporated Greek as a second language. Many of Alexandria's artisans were Egyptian Jews. The language of the earliest Alexandrian alchemical works may have been Greek, but the most respected artisans were Jewish, as Pseudo-Democritus reveals:

> *It was the law of the Egyptians that nobody must divulge these things in writing ... The Jews alone have attained a knowledge of its practice, and also have described and exposed these things in a secret language.*

What is never divulged is whether these Jewish artisans were Gnostic Christians, Therapeutae or members of the Jewish diaspora – all considered Jewish at the time. Compare Pseudo-Democritus' passage above with the 1st century Jewish Philosopher Philo of Alexandria's description of the Therapeutae below:

> *For they read the holy scriptures and* draw out in thought and allegory their ancestral philosophy, since they regard the literal meanings as symbols of an inner and hidden nature revealing itself in covert ideas.
>
> — Philo, paragraph 28

Maria felt that her Tincture contained principles of incorruptibility, growth and regeneration that perfected copper alloys, and that this process paralleled similar processes at the human level. She believed that creating the Tincture of the Philosophers, or Ios in Greek was sacred to Jews and she intended to keep it that way, as Patai explains:

> Maria Hebrea intended to limit the knowledge of the Philosophers' Stone to "the race of Abraham" that is, to Jews.

Patai bases this upon two statements attributed to Maria, one of which appears in a number of Greek manuscripts:

> Do not touch the Philosophers' Stone with your hands; you are not of our race, you are not of the race of Abraham.

The second statement consists of Olympiodorus' commentary upon a similar statement made by Maria, which appears in Zosimos' writings:

> If you are not of our race, you cannot touch it, for the Art is special, not common.

Patai reveals that Zosimos viewed the Jews and Maria particularly, as authoritative sources for ancient alchemical theory and practice, in contrast to the Egyptians and Greeks who were not as forthcoming with their secrets. He seems to have regarded sacred knowledge in each tradition as equally valid, however.

CHAPTER 3 – MARIA HEBREA'S TINCTURE

There are two sciences and two wisdoms: that of the Egyptians and that of the Hebrews, which latter is rendered more sound by divine justice. The science and wisdom of the best dominate both: they come from ancient centuries. – Zosimos

The Therapeutae were a contemplative and chaste group of Jewish mystics and healers living along the borders of Lake Mariout just south of Alexandria. They followed the teachings of the five books of Moses, the prophets and psalms in addition to mystical teachings regarding numerological formulae and allegorical religious interpretation, and claimed a healing tradition far superior to what was available in the cities. They were a disciplined group who, according to Philo, were:

> ... the best of a kind given to 'perfect goodness' that 'exists in many places in the inhabited world'

On the Sabbath, a day very important to the Therapeutae, they gathered in a meetinghouse – men and women separated by an open partition – and listened to religious discourses. Every seventh week they enjoyed a banquet by serving one another, after which they spent the night singing hymns together. No one is certain whether Maria was a member of this group, but what is certain is that she was very well respected, skilled at her art, a leader in her community, and brought a very tangible sense of depth and meaning to her practice. If Maria were a member of the Therapeutae community, the overwhelming respect for her talent and character would appear to have been even more justified. Although it is impossible to be certain, there is a strong likelihood that Maria was at least familiar with the Therapeutae, if not actually a member.

Alternately, historical Maria Hebrea may have been Miriam II, daughter a Jewish high priest residing in Alexandria who later became third wife to King Herod (Aros) the Great, or possibly pseudo-named after her. Josephus recounts the story of their wedding:

BOOK 1 – ORIGINS & EVOLUTION

> *There was one Simon, a citizen of Jerusalem, the son of one Boethus, a citizen of Alexandria, and a priest of great note there; this man had a daughter, who was esteemed the most beautiful woman of that time; and when the people of Jerusalem began to speak much in her commendation, it happened that Herod was much affected with what was said of her; and when he saw the damsel, he was smitten with her beauty, yet did he entirely reject the thoughts of using his authority to abuse her, as believing, what was the truth, that by so doing he should be stigmatized for violence and tyranny; so he thought it best to take the damsel to wife. And while Simon was of a dignity too inferior to be allied to him, but still too considerable to be despised, he governed his inclinations after the most prudent manner, by augmenting the dignity of the family, and making them more honorable; so he immediately deprived Jesus, the son of Phabet, of the high priesthood, and conferred that dignity on Simon, and so joined in affinity with him [by marrying his daughter].* – Josephus, Antiquities of the Jews, Book XV, Chapter 6

This scenario suggests that the earliest and most reknown and regarded alchemist may have been Miriam II, daughter of Simon Boethus, Jewish High Priest of Alexandria and later Jerusalem, inheritor of Judeo-Egyptian occult secrets of the highest order including gold-making technology. Maria became a Jewish Queen by marriage to Herod the Great from 23 BCE to 5 CE. A gold-maker of such high socio-political status would under-standibly view the Art of Chrysopœia as a Jewish legacy. This explains the gold-making work of another matriarch of Judeo-Egyptian alchemy in Maria's tradition, named after Herod's fourth wife Cleopatra of Jerusalem. He and Cleopatra controlled the lucrative trade in asphalt sourced from the Dead Sea, and copper mines on Cyprus leased from Rome. The enigmatic alchemist named Archelaus may have been named after Herod's son, Herod Archelaus.

One of the most admirable qualities of Maria Hebrea is her ability to transmit information using few words. She was not only a skilled artisan, but also a talented designer of practical laboratory equipment;

some of which can still be found in modern laboratories today. Writes Patai:

> *Maria constructed and described various ovens and apparatuses for cooking and distilling made of metal, clay and glass ... She considered glass vessels especially useful, because "they see without touching."*

It must be understood that Maria was expert at manufacturing her Tincture and applying it to create alchemical gold. As was her tradition, she veiled her materials and methods in a layer of cover-names, but always in such a way that an astute and persistent mind was afforded the opportunity to understand her. This exploration of Maria's processes focuses specifically upon the archetypal Brief and Great recipes to achieve the Philosophers' Stone, which stood as the foundation for later Alexandrian, Islamic and European traditions.

Three of Maria's core teachings offer direct insight into the origins of applied alchemy:

1. **Maria's maxim** – *a complete outline of the confection of the Philosophers' Stone by any method in just 16 words;*
2. **Dialogue of Maria and Aros** – *a comprehensive theoretical outline of Brief and Great Philosophers' Stone recipes;*
3. **Gold making from 1st to 3rd century Alexandria** – *transmuting copper alloys into alchemical gold, i.e. Corinthian bronze.*

The following sections will address the major cover-names used by Maria as an aid to understanding her work. This chapter attempts to reveal the origins of the Philosophers' Stone in the most simple and direct terms, complete with modern balanced chemical equations to aid reproducibility of Maria's *Ars Brevis* and *Ars Magna* – the two archetypal Philosophers' Stone recipes upon which most other variations are based. By understanding the basic theory of confecting the Philosophers' Stone Maria's way – magnificently presented in the three works addressed in this treatment – the doors to understanding most other paths or techniques for confecting the Stone will be flung wide open. Maria's work provides a point of origin from which to trace the evolution of methods for achieving the Tincture of the Philosophers, the archetypal Philosopher' Stone, during the first centuries CE and continuing up until the more recent alchemical greats, Sigismund Bacstrom and Fulcanelli, in the 18th and 20th centuries respectively.

In order to understand Maria's art, and the origins of Western alchemy in general, it is necessary to become familiar with the various cover-names Maria employed to veil her secrets from the uninitiated. Maria explains clearly, consistently and without waver that her art consists of four elements, which she describes as 'Four Stones':

1. *The Black Stone (stibnite; Sb_2S_3) – this mineral was sold as powdered eye cosmetics, as detailed in a previous chapter. It was inexpensive and sold in the markets of Alexandria. One could end up stepping on msdmt / kahâl / kohl that had dropped upon the ground. It was available everywhere, found in every household and used by both males and females of all ages. This is Maria's First Stone.*

2. *The White Stone (flowers of antimony; antimony ash; $Sb_2O_3 \cdot Sb_2S_3$) – Maria worked with powders. To achieve flowers of antimony, Maria simply pan roasted ready powdered msdmt / kahâl / kohl in open air or with quicksilver. The sulfur melts and then vaporizes at a relatively low temperature. As sulfur melts or reacts with quicksilver, the mass first turns reddish before becoming yellow, then white. This color regimen has been a cause of confusion for many alchemy researchers struggling to understand which comes first: red or white or vice versa. By continued stirring at increased temperature, the remaining metallic antimony oxidizes into a brilliant white fluffy powder called flowers of antimony. Maria describes the process as "Whitening the [first or black] Stone." This is Maria's Second Stone.*

3. *The Red Stone (Latten; $Au \cdot Sb_2O_2SO_4$) – this is the product similar to the Moses recipe presented in an earlier chapter. The Moses method involves continually feeding stibnite to molten gold until a fusion occurs, after as many as 12 such feedings. Maria created her version of the Latten as a single*

CHAPTER 3 – MARIA HEBREA'S TINCTURE

fusion of gold and flowers of antimony. It is an alloy created between gold and flowers of antimony to make the gold a brittle calx. Maria describes flowers of antimony *as feminine or white, and gold as masculine or red; the combination being likened to a red and white opium poppy flower.* When gold is purified and particle-size reduced, it achieves a reddish-brown color and is a soft heavy powder. Maria uses the cover-name gum to describe a soft powder. This is Maria's Third Stone.

4. *The Violet Stone (the Tincture; $Au \cdot Sb_2S_2O$) – the Tincture is the Philosophers' Stone, originally called* vafaí, Ios or Ios of Gold. *In Greek,* Ios *means* venom, medicine, arrow, purple, tingeing agent *among others. Its verb form,* iosis, *meant* to tinge. *Purple was the color of Phoenician royalty, rulers during antiquity. It was Saturn and Lilith's sacred color, this being the origin of its use by emperors and kings to denote descent from Lilith and the Titans. Tyrian purple fabric dye was produced by the Phoenicians at Tyre from the fresh mucus secretion of the spiny-dye murex snail (Bolinus brandaris), which was worth its weight in silver.*

Iosis *or purpling in the alchemical sense, referred to gold receiving its royal purple robe as a sign of perfection.* Ios *eventually was superseded by the term* Elixir *in Islamic alchemy. Maria completes the Stone by fixing* Latten *with a finely powdered fluxing salt. In the* Ars Brevis, *the fusion is created in a crucible at a temperature just high enough to allow powdered* Latten *and the powdered salt to achieve fusion. This removes any water or oxygen trapped in the salt matrix, causing it to appear as a red gold-antimony salt crystal. When this is heated, it will melt like wax yet produce no fumes. This is Maria's Fourth Stone.*

Maria pioneered a second technique known as the Ars Magna. *In order to achieve fusion of the three principles, Maria united the fluxing salt with flowers of antimony to achieve divine water, which comprises the female dual soul principle. She then unites the female spirit-soul with the male body, which is prepared gold-antimony glass calx. In the* Ars Magna, *the fusion takes place in a glass vessel over low heat, allowing the three ingredients to liquefy and achieve salt fusion. Over time, the heat is raised in order to remove any water or oxygen trapped in the salt matrix, causing it to appear* as a red gold-antimony salt crystal. The result is the same product as above and, when heated, will melt like wax yet produce no fumes.

Powder of projection is the same Tincture, Maria's Fourth Stone, fused with antimonial bronze called *molybdochalkon* to create a unique *gold-antimonial bronze* known as *Corinthian bronze*, and as an additive to glass or rock-crystal to create artificial rubies and diamonds – these being the primary industrial trade uses for the Philosophers' Stone in Maria's Alexandria. Variations of her unique bronzes additionally were surface treated with acids to dissolve outer layers of base metals to create unique liver-purple patinas highly valued throughout the Roman Empire. Alexandrian *chrysopœians* viewed this fascinating change from copper to gold, or creating purple patinas as transmutation.

Maria also explains that the entire process can be completed with silver in place of gold, in order to create a white Stone. The white Philosophers' Stone is a silver-antimonial salt crystal. She suggests that this salt, when combined with lead and/or mercury, creates artificial silver by coagulating the soft metals, understood today as operating in the same way a mercury dental amalgam hardens; antimony acting as a metal hardening agent. There were many other ways of manipulating these products and applying them industrially, but the following analysis will concentrate only upon the fundamentals of Maria's art.

Maria's maxim

In her maxim, Maria accurately describes – in just 16 words – how the Art unfolds. Her maxim is a general outline or theoretical overview of the basic steps involved in confecting the Philosophers' Stone.

> *One becomes two* – Stibnite ore is the one *First Stone* or prime material, also called the One by Cleopatra, Zosimus, Synesius and others. It reduces to two elements, antimony and sulfur – but this is not what Maria is referring to. When stibnite is reduced to ash by pan roasting, the sulfur vaporizes and the remaining metallic antimony oxidizes. The

CHAPTER 3 - MARIA HEBREA'S TINCTURE

ash retains sulfur dioxide, silica and in some cases, iron and free sulfur in trace amounts. These grey to brilliant white powders are collectively known to alchemists as flowers of antimony. *Maria refers to this process as "whitening the stone" and correctly asserts that the process can be performed in a very short time. The First Stone is stibnite, which is converted to the Second Stone, flowers of antimony; thus one becomes two.*

Two become three – *The* flowers of antimony *are then mixed with purified gold calx, the result is known by the cover-name* Latten. *Latten is the chemical marriage that results in the first union of opposites; thus, these two-things become the Third Stone, and two becomes three.*

And by means of the third and fourth, achieves unity – *the fourth is a medium related to both gold and antimony that dissolves them both, keeps them in flux and also serves to coagulate or crystallize the end product. Maria describes this medium as "the white, clear and much honored Herb, which is found on low hills"* – *a description she derived from reading Herodotus. The white fluxing salt for her* Ars Brevis Via Sicca *(Brief Dry Path) was vapour or dew from during the night, crystallized under the Egyptian Sun; thus making it related to both the Sun and the Moon and viewed by Maria as being naturally fixed, although it is also of a very volatile nature. The fluxing salt for her* Ars Magna Via Humida *(Great Wet Path) was sal ammoniac. Thus, the Third Stone, when fused with this fourth substance* – *the fluxing salt* – *achieves unity as the Fourth Stone.*

Thus two are but one – *In Maria's fast and slow methods, the salt medium brought both gold and antimony (the two) into a homogenous solution that later crystallized or solidified into a fused compound metal salt (the* One*). The entire art is based upon dissolving gold and antimony, homogenising them via liquification followed by coagulation. As Maria writes, "Join the male and the female and you will find what is sought." European alchemists summarized this process with a simple maxim,* Solve et Coagula.

DIALOGUE OF MARIA AND AROS

Sometime between the 1st and 3rd centuries CE, a renaissance of philosophy, science and technology occurred and spread from Alexandria throughout the Roman Empire. The most comprehensive single rendering of the Art of *chrysopœia* and the Philosophers' Stone in any tradition is preserved in the form of a dialogue between Maria and King Aros (Herod the Great or Iro̱di̱s), traditionally assigned to this period. The name Aros likely arose from a series of mis-transliterations from Greek *Iro̱di̱s o Mégas* (Ηρώδης ο Μέγας) to Arabic *Arodos al-Kabir* (هيرودس الكبير) and from there ultimately to Latin and other European languages.

The Dialogue of Maria and Aros was first published in Basel in 1593, and again during 1610 in a two volume Latin work entitled *Artis auriferae quam chemiam vocant* – a Latin translation, likely from al-Habib's Arabic version derived from an unidentified Greek source. Michael Maier believed that the original was in Hebrew, which would explain some misspellings and poor transliteration; unfortunately, the original remains to be found. The adaptation that follows is the author's interpretation based upon various sources. The foundational principles and operative details remain intact along with the conventional name 'Aros'. A number of translations exist for those interested in researching alchemical literature (see Appendix B for Latin excerpt from *Artis auriferae ...*).

This dialogue is a complete and thorough course on creating the *Tincture of the Philosophers* – or Ios, the original Philosophers' Stone. A student wishing to master confecting the Stone could study only this dialogue, forsaking all other alchemical writings, and upon interpreting it correctly gain mastery of the Stone's confection. Maria first details the process for confecting the Philosophers' Stone via the *Ars Brevis* – the original metallurgical method – in a pedagogical, precise and generous manner. The order suggests that it is the older of the two methods.

CHAPTER 3 - MARIA HEBREA'S TINCTURE

She follows this by revealing her secret technique, the *Ars Magna*, which she appears to have pioneered as her own innovation based on a chemical reaction requiring distillation. She describes her materials and methods using gently veiled yet appropriate cover-names, providing valid clues in order to enable astute students of the Art to solve her puzzles. She is generously clear with her presentation and even re-emphasizes the *prime material* (or *prima materia*) in the second lecture of the dialogue. Throughout, she delivers subtle hints and indicators to aid the operator and finishes with clues for turning base metals into precious products. In short, *The Dialogue of Maria and Aros* is a masterpiece.

Maria's Ars Brevis

Aros the Philosopher had a meeting with Maria the Prophetess the Sister of Moses, and approaching to her, he paid her respect and said to her: Oh Prophetess, I have truly heard many say of you that you whiten the Stone in one day.

And Maria said: *Yes, Aros, even in a part of one day.*

The beginning of the dialogue is a trap. It would appear that Aros is referring to the Philosophers' Stone, but this is not the case. One of Maria's technical advances in *chrysopœia* was the use of *flowers of antimony*. Aros is asking about her technique. The cover-term for her technique is *whiten[ing] the [First] Stone*. The technique would later be called *calcination* and is based on ancient technological precedent.

Aros said: *Oh Lady Maria, when will the york be which you affirm? How shall we whiten and afterwards add blackness?*

Whitening the Stone is the initial step. Adding blackness is a reference to the final process and can signify two specific operations: either the creation of *Corinthian bronze* by combining the Philosophers' Stone with antimonial bronze via fusion, or the pickling process that creates a unique patina. What Aros is really asking of Maria is to explain the process in its entirety.

> *[Maria quoting Hermes] "**Likewise Hermes** [said]": Have you heard of the Stone mixed of saffron and white, and [that its] pulverization evinces separation, which whitens the gem's core [κέντρο γεμ; kéntro gem]?*

Here Maria is attempting to ascertain whether Aros is already familiar with the process of working with stibnite, which she grinds and calcines to a fine white-ish powder. It is a qualifying test to see if Aros knows the *prime material*, stibnite (*mśdmt / kahâl / koḥl*; Sb_2S_3). Stibnite is Maria's *prime material*, in other words her First Stone – the Black.

> ***Then answered Aros***: *It is as you say, oh Lady but in a long time.*

Aros knows of the process but believes it takes a long time to accomplish.

> ***Maria answered***: *Hermes mentioned in all the books of his that Philosophers whiten the Stone in one hour of the day.*

Maria, in her own unique way, is either explaining to Aros that she is not the inventor of the technique – by crediting Hermes with the discovery – or implying the use of philosophers' mercury in the process. The process she is alluding to is omitted from the dialogue. Here is an example of Maria providing an opportunity for discovery. Powdered *flowers of antimony* is Maria's Second Stone – the White. The simplified chemistry for *whitening the stone* is as follows:

> *whitening the stone* via calcination = $2\ Sb_2S_3 + 9\ O_2 = 6\ SO_2 + 2\ Sb_2O_3$

> ***Then answered Aros***: *Oh how excellent that is!*

> ***Said Maria***: *It is most excellent to the person unaware of it.*

CHAPTER 3 - MARIA HEBREA'S TINCTURE

Aros said: Oh Prophetess, if in the presence of all people [humanity] there are 4 Elements [Tetrasomia; earth, air, water, fire], he [Hermes] said they could be completed, and complexed, and their fumes coagulated, and retained in one day, so long as they follow [...preparation; are fully prepared].

Maria cleverly introduces her materials list under the cover-name *four elements* – stibnite (black), antimony (white), gold (yellow / red), and fluxing salt (creates a violet end product); not to be confused with the Four Stones. After a basic ingredient has been purified or prepared by fire, it is referred to by the cover-name *fume*. Maria is explaining to Aros that once each of the basic ingredients has been prepared, they can be combined and the process of completing the Philosophers' Stone via the *Ars Brevis* can be completed in a single day. The combination of the four elements – the *Tetrasomia* – results in a fifth essence. The fifth essence was later termed the *Quintessence*, and is synonymous with the *Elixir* and the *Philosophers' Stone* or *Stone of the Wise*.

Maria said: Oh Aros, by God, if your understanding was not firm, you would not hear these words from me, so long as the Lord filled my heart, by grace of divine will.

Nevertheless: Here is the Recipe – alum of Spain – [i.e.] white gum and red gum – which is [red] Philosophers' sulfur, their Sol [☉; gold] and greater Tincture – and marry gum with gum by a true union. Do you understand, asked Maria? Certainly Lady [replied Aros].

Alum of Spain is a cover-term for calcined prime material, stibnite, identified several centuries later by Synesius as *Roman antimony*. Egypt was a primary exporter of *alum* since around 500 BCE, and therefore it is doubtful that an imported type of alum was the pre-Arabic term used. The original cover-term was likely *chalcanthum*, or more specifically *misy* or *sory* to suggest a sulfide or sulfate ore to encrypt the prime material.

> **Maria said:** *Make it [the compound] flow like water; vitrify this water [convert it to glass] for a day, a fixed body of two components.* **Liquefy by the secret of nature in a philosophic vessel.** *I don't understand, oh Lady [replied Aros].*

Maria sometimes uses the cover-name *gum* synonymously with *fume* to indicate a prepared powder, but also in the context of a liquid or solid compound reagent such as a glass or salt. Having provided a synopsis of the entire process for confecting the Stone, Maria then provides instructions for creating gold-antimony glass – *Latten* (Au·Sb$_2$O$_2$SO$_4$). *Latten*, a glass or alloy understood in alchemy to be a hermaphroditic or androgynous body or marriage, is Maria's Third Stone. The two technological advances presented by Maria include Hermes' *whitening the Stone* process and her own *calcination-vitrification* process for *Latten*, explained to Aros below in further detail:

> **At any rate, said Maria:** *Watch the fume carefully that none of it escape,* **with a gentle fire** *such as is the measure of the heat of the sun in June or July. Stay near the vessel and carefully observe, how black to red begins in less than three hours of the day, and* the fume penetrates the body, the spirit binds [it] and will become like milky wax, melting and penetrating. *And that is the secret.*

Maria begins with a description of *whitening the Stone* – calcination at relatively low temperature. She accurately details the color indicators to look for as the black stibnite is calcined to *flowers of antimony*, followed by fusion with fine gold powder to create red *Latten*. According to her description, *whitening the Stone* followed by creating *Latten* is performed sequentially as a combined operation.

In the following passage, Maria hints at the flux process using a fluxing salt. Although not specifically identified in the Moses technique, the presence of 50 tons of alkali ash or fluxing salt at Serabít el-Khâdim is a strong indicator that the artisans at the Hathor Temple finished their

CHAPTER 3 – MARIA HEBREA'S TINCTURE

Stone in a similar manner to Maria's technique described in the following passages. Maria is the first to include this technological innovation as a standard part of her method of operation – the third so far.

Aros said: Do not say [I can't believe] that it will always be thus.

Maria replied: Aros, there is a more marvelous thing, of which did not exist among the ancients, nor available to them as anything medicinal. And that is: Take the white herb, bright and much honored, occurring on small mounds [hills; hillocks], grind it fresh in the hour [of its collection], and this is the true body that does not flee from the fire.

In the passage above, Maria introduces her fluxing salt by the cover-term *white herb, bright and much honored*. Her odd statement of secrecy that it was not anything "medicinal" is a hint that the *white, bright and much honored herb* is actually a mineral and not a medicinal plant material; a clue often misunderstood or misinterpreted. She seems to suggest that the use of sal ammoniac as an operative agent is her own innovation. Maria is revealing to Aros that the identity of her *much honored herb* was sal ammoniac; honored for reasons detailed later in this chapter. She generously provides operative clues explaining that the sal ammoniac should be powdered, is stable to 338 °C ("does not flee from the fire") and is a flux that instantly reacts with metal oxides present in *Latten*.

And Aros said: Is not this the true Stone?

And Maria said: Yes truly. But others do not know of this regimen with its quickness.

Aros said: And then what?

In the passage below, Maria presents the finishing process of joining the salt to *Latten* via projection over low heat, a technique known as *vitrification* where oxide-chloride exchange results in a metal chloride.

European alchemy would later refer to this technique as *detonation*. Maria's fluxing salt is her fourth principle and is presented as an operative trade secret. The heat must be enough to catalyse the reaction, but not so hot that it vapourizes the salt or antimony. She indicates that the ability to fuse *Latten* and fluxing salt into a homogenous body is the secret of the Art. Once the product cools, it takes the form of a crystallized product – the Philosophers' Stone.

> ***Maria said****: Make a glass of this sulfur and mercury,* ***and*** *they are the two fumes comprised of two Lights,* ***and*** *project therein the completion [fulfilling ferment] of the Tincture* ***and*** *spirits, according to the true weight, …*

Flowers of antimony and purified gold calx are the *two fumes* mentioned in the passage above. Maria's use of the cover-term *two Lights* is in reference to Genesis 1:16, indicating the Sun and the Moon.

טז וַיַּעַשׂ אֱלֹהִים, אֶת-שְׁנֵי הַמְּאֹרֹת הַגְּדֹלִים: אֶת-הַמָּאוֹר הַגָּדֹל, לְמֶמְשֶׁלֶת הַיּוֹם, וְאֶת-הַמָּאוֹר הַקָּטֹן לְמֶמְשֶׁלֶת הַלַּיְלָה, וְאֵת הַכּוֹכָבִים.

Genesis 1:16 *And God made the* two great lights *– the greater light to rule the day and the lesser light to rule the night – and the stars.*

This is the earliest textual alchemical equation where the Sun = gold and the Moon = antimony. Maria refers to these by additional cover-names: *Kibrich* and *Zubech*. *Kibrich* is a mistransliteration of Arabic *kibrīt* (كبريت) meaning *sulfur* derived from Greek *theío* (θεῖο) meaning *sulfur* or *sulphide ore*, but to alchemists is used as a pun and interchangeably with *theía* (θεία) meaning *divine* or *God*. The words *kibrich* and *zubech* in the Latin text are a mistransliteration from Arabic *kibrīt* (كبريت) and *zībaq* (زئبق), meaning simply *sulfur* and *mercury* respectively, into poor Latin. In this scenario, *gold = sulfur* and *flowers of antimony = mercury*. The first union is a union of gold (sulfur) + antimony (mercury), the product of which is then considered *sophic sulfur*. Gold-antimony glass is then joined to the fluxing salt, often referred to by the cover-term *sophic mercury*.

CHAPTER 3 - MARIA HEBREA'S TINCTURE

The fluxing salt is described via the cover-terms *fulfilling or ferment of the Tincture and of the Spirits*. Maria describes the reaction using the cover-name, *vitrify*, roughly meaning *to fuse*. She omits the process of finely powdering *Latten* beforehand, yet divulges a very subtle operative clue regarding *Latten* with the words "and when it [*Latten*] is perfect..." She is hinting that *Latten* must first be at the right temperature before the powdered salt is *thrown or projected* onto *Latten*. Once the reaction is complete, the product is then removed, allowed to cool and is then pulverized:

> ... *then pulverize it, and put it into the fire*, *and you will see marvels brought about by them [the compound].* ...

The final change is performed by grinding the product to a powder and liquefying it a final time. This process helps remove any residual moisture, free oxygen, ammonia, or chlorides. It is an ancient metallurgical form of purification.

> ... *The entire regimen depends on the temperature of fire.* Oh how wonderful, how it [the Stone] is moved from colour to color in less than an hour of the day, until it reaches the goal of red and white [white and red according to Bacstrom's interpretation from German]. Extinguish the fire, and let it cool, and open it [the vessel], ...

When Maria says, "before it becomes red and white," she is not saying "red *and then* white," but is instead referring to an equal fusion between the red and white principles as indicated by the next passage. Bacstrom interpreted the passage according to Zosimos' traditional regimen of black, white, yellow, and red color stages in sequential order; as opposed to Maria's fusing of red and white gums.

In other writings attributed to Maria, the final product is called *Ios* meaning *violet* or Tincture, but this jargon evolved early in alchemical history to the term *Elixir*. When the final product is ground to a fine

135

powder, it is called Ios – Tincture, the Elixir or the Philosophers' Stone – Maria's Fourth Stone. She presents specific color indicators for her product. When the stone is created in a crucible via *projection-detonation* the product may not necessarily be uniform in color.

> ... *and you will find a bright pearly body in the color of wild poppy mixed with white [red and white mentioned previously], and it is waxy, melting and penetrating, and* its golden color falls upon one thousand two hundred *[can be projected upon or tinge 1,200 parts]. That is the hidden secret. Then Aros kneeled and prostrated.*

When Aros prostrates, this can be understood as a subtle indicator or hint given by Maria that the exposition is finished; this concludes Maria's instruction on the *Ars Brevis* method of confecting the Tincture. She adds a final clue, without expressly saying it, that this finished Stone is the *powder of projection*.

Any Philosophers' Stone recipe that calls for confecting the Stone via metallurgical technique, employs the use of *flowers of antimony*, or employs the technique of combining *Latten* with a fluxing salt in a crucible is most likely a descendant of the archetypal recipe preserved in Maria's *Ars Brevis*.

Template for **Maria's *Ars Brevis***:

1. **Black** – begin with the *prime material; mśdmt / kahâl / ḳoḥl* (stibnite; Sb_2S_3) – the First Stone;
2. **White** – calcine (i.e. pan roast) stibnite to *flowers of antimony* (antimony ash; $Sb_2O_3 \cdot Sb_2S_3$) – the Second Stone;
3. **Red** – fuse purified gold calx with *flowers of antimony* to create *gold-antimony glass* ($Au \cdot Sb_2O_2SO_4$) – the Third Stone;
4. **Violet** – detonate sal ammoniac with powdered *Latten*, grind to create the *Tincture / Elixir* – the Fourth Stone.

CHAPTER 3 – MARIA HEBREA'S TINCTURE

Maria's Ars Magna

In the passage below, Maria indicates that she will begin a new lecture on another process. This lecture comprises one of the most fascinating parts of *The Dialogue of Maria and Aros*. In the following passage, she introduces her fourth innovation. This process is none other than the archetypal Philosophers' Stone recipe known as the *Ars Magna*, the template upon which nearly all subsequent Philosophers' Stone recipes in alchemical history were based.

> **Maria said to him**: Raise your head Aros, because *I will abbreviate it for you, take that bright body strewn on mounds [small hills], which is not subject to putrefaction or movement. Here is the Recipe – and grind it with gum Elsaron, and unite the two powders, for gum Elsaron is a seizing body, and grind it all.*

In the *Ars Brevis*, Maria began her lecture with a discussion of calcining the prime material to *flowers of antimony*; here she begins with sal ammoniac. Sal ammoniac cannot be conquered by putrefaction in its natural state. It crystallizes or vaporizes. The most obvious cover-name, *gum Elsaron*, is derived from a passage in *The Song of Solomon* that mentions the *rose of Sharon*, or *Chavatzelet HaSharon* (חבצלת השרון), used here to indicate *flowers of antimony*. Whether Maria was actually referring to the flower *lily of the valley* (*Convallaria majalis*) remains uncertain, yet it was routinely depicted in European alchemical emblematic art. It was named *majalis* after Maia, mother of Mercury-Hermes. Maria is generously allowing for Jewish identification of *flowers of antimony*.

> א אֲנִי חֲבַצֶּלֶת הַשָּׁרוֹן, שׁוֹשַׁנַּת הָעֲמָקִים. *Song of Songs 2:1* I *am a rose of Sharon, a lily of the valleys.*

This lecture presents an innovative method in which she explains the use of sal ammoniac and her *Ars Magna* technique. Her abbreviation or synopsis is in reference to the method of confection post-preparation, and to the length of time required to complete the entire synthesis, which she addresses at length towards the end of the dialogue. The technical details by which she *"unite[s] the two powders"* remain unclear – Maria could have united her powders via metallurgical technique or via chemical reaction. Each of these options merits further consideration.

Metallurgical process: Maria could have created a reaction between the two powders in a crucible. Sal ammoniac works by reacting with the metal oxides present in *flowers* or *glass of antimony* to form a fluid metal chloride salt. The chlorides of sal ammoniac are exchanged with the oxides in *flowers of antimony*, resulting in crude antimony trichloride – a very tricky type of reaction later termed *detonation*.

Chemical reaction: Maria most likely created a chemical reaction between the powders via distillation. Sal ammoniac decomposes to form ammonia and highly corrosive hydrogen chloride gas, which reacts with *flowers of antimony*, resulting in crude antimony trichloride.

F. Sherwood Taylor pointed out in 1930 in an article entitled 'The Origins of Greek Alchemy' that Zosimos credited Maria with the first distillation apparatus. He wrote: *"... it appears likely that she perfected the apparatus for distillation of liquids in a form so efficient as to have suffered little alteration in two millennia."* In 1949 Stapleton, Lewis and Taylor identified a description of *Divine Water* distillation recorded in a Maria fragment preserved by Islamic alchemist ibn Umail:

> **Maria also said**: *The 'Water' which I have mentioned is an Angel and descends from the sky, and the earth accepts it on account of its [the earth's] moistness ...*
>
> **Commentary by ibn Umail**: *She meant by this the 'Divine Water',* **which is the 'Soul'. She named it 'Angel'** *because it is spiritual, and because that 'Water' has risen from the earth to the sky of the Birba [i.e., from the*

CHAPTER 3 – MARIA HEBREA'S TINCTURE

bottom to the top of the Alembic]. And as for her statement '[The Water] descends from the sky', she meant by this its return to their Earth.

Another Maria fragment appearing in the *Book of Morienus* addresses her *whitening the Stone* process, yet the passage lends itself to an alternate interpretation as the creation of an antimony distillate. The cover-term *body well prepared* suggests *flowers of antimony* via calcination. This is followed by the phrase *extraction of the spirit from the body*; a reasonable decoding of which is that *spirit = distillate*, and *body = flowers of antimony*.

> ... *when* the white spirit or milk has been made and raised, *and the residual* body well prepared, so that its darkness or blackness is removed and that foul odour withdrawn from it, *the* extraction of the spirit from the body *has been accomplished. And at the moment of conjunction of that spirit with that body, great wonders will be seen. On this point, the philosophers who gathered before Maria said to her: "fortunate are you, Maria, for the splendid mystery hidden from many has been revealed to you." – Morienus, 7th century*

The mention of *Soul* and its description as spiritual is consistent with an antimony compound as a spirit or distillate. The above quotes and commentaries support the possibility that Maria had discovered how to create *butter of antimony* via distillation or redistillation. Zosimos relates a process from a Jewish source in *On the Evaporation of Divine Water that Fixes Mercury*, for distilling *divine water* using a Tribikos distillation unit in less than a single day. This is likely a reference to Maria.

Regardless of which technique she may have used, the archetypal template of uniting gold, antimony and chlorides remains intact. If Maria united her powders via the metallurgical or chemistry-based processes above, the staggering implications are that she discovered *butter of antimony* approximately 500 years before Morienus introduced it to Khālid and 1,000 years before Artephius wrote about it. She omits any details regarding her ingredient ratios. Assuming that Maria was indeed creating antimony trichloride, the chemistry might appear as follows:

Maria's Divine Water $\quad Sb_2O_3 + 6\ NH_4Cl + 3\ H_2O = 2\ SbCl_3 + 6\ NH_4OH$

Maria uses the cover-names *finely powdered fixed body* (to indicate sal ammoniac), *gum* (to indicate a fine powder) of *Elsaron* (the *rose of Sharon* being a flower) in order to describe *flowers of antimony*. Combined, these represent a unified *dual-soul* (soul + spirit), ethereal lunar female principle.

> ... *Approach therefore [draw near; pay attention] – it all melts. If you project [the husband] upon this wife, it [the liquid] will be as distilling water [water dripping], and when it cools, it coagulates [freezes] and becomes a unified body,* **and make projection of it [into molten copper], and you will behold wonders.**

She is describing butter of antimony, the product of uniting her two powders, and the phenomenon in which it melts at 73.4 °C yet becomes eutectic upon uniting it with Latten and remaining liquid at a relatively low temperature. She refers to Latten by the cover-name *her spouse*. Maria is very generous with regard to well-considered cover-names. Gold is generally considered the corporeal solar male principle in alchemy; it is the king of metals. In this case, the cover-term *spouse* identifies a chemical marriage between the *gold-antimony glass* (red male) and the antimony salt matrix (white female), which unites at elevated heat and solidifies as it cools. This is understood as infusing female dual-soul into male body resulting in a perfected hermaphroditic or androgynous product – a body + soul-spirit union.

> ... *Oh Aros that is the hidden secret of the Art [schools; guilds]. And know that the* **two aforementioned fumes [prepared alchemical reagents] are the roots of this Art, they are the white Sulfur and fused powder, but the**

CHAPTER 3 - MARIA HEBREA'S TINCTURE

fixed body is from the Heart of Saturn which preserves the tincture, and the campus of wisdom or of the Art.

White Sulfur is the cover-name for *flowers of antimony* or *ground gold-antimony glass* depending on the context. The fixed body is sal ammoniac (NH$_4$Cl); to Jews in antiquity Saturn held sacred associations paralleling Zeus to Greeks or Amun to the Egyptians. In the passages that follow, Maria is clear that the f*ixed body from the Heart of Saturn* is her sacred salt. The Latin *humid calx* refers to gold that has undergone fusion with molten glass of antimony, in otherwords according to Maria, *"... it [the compound is] made to flow like water, then vitrified ..."* and *"... melted by the secret of nature ..."*

Maria uses very specific verbiage to describe sal ammoniac and rightly so, as it is the *Key* to success in the *Ars Magna*. The beauty of Maria's linguistic technique is its accuracy. Indeed, any number of salts may be used in the *Ars Brevis* to create an archaic gold-antimony salt fusion. Sigismund Bacstrom successfully achieved reproducibility of Maria's *Ars Brevis* using saltpetre as a substitute for Maria's local natron or sal ammoniac. However, in her *Ars Magna* recipe specifically, sal ammoniac is crucial to creating the fluxing salt employed to produce her Tincture.

> *... And the Philosophers called it by many and all names [this fixed body], taken from small mounds [hills; hillocks] this bright white body.* These are the reagents of this Art, *some [of which are] prepared and some found on small hills [hillocks].* And know Aros, although sages have not mentioned it by name in any thesis-dissertation, except that the Art cannot be completed unless by it. And in this Art there are nothing but wonders.

The *parts that are prepared* are the cosmetic *mśdmt / kahâl / ḳoḥl* and gold. The part *found on small hills* is sal ammoniac, the key to Maria's Judeo-Egyptian *chrysopœia*. Hers is the original archetypal Philosophers' Stone recipe that evolved from metallurgy to chemistry, known later in Europe as the *Ars Magna*. It is a chemical reaction, in a sealed digestion vessel, between sal ammoniac, *flowers of antimony* and gold calx, resulting in the archetypal Philosophers' Stone according to the following equation:

> Sophic Sulfur (gold) + Sophic Mercury (flowers of antimony)
> + Salt Mediator (sal ammoniac / divine water)
> = Ios, Tincture, Philosophers' Stone

> **Initiation**: *and in this [work] are 4 stones, and their regimen is true as I have said. And this is primary, the compulsory study [of] Ade and Zethet [Azoth], by allegory [as found in] Hermes' books and Teachings.*

> *Philosophers have always indicated a long regimen, and have pretended that the work consists of various things that [in fact] were unnecessary to perform the work. And they [pretend to] take a year to perform the magistery; but only for hiding it from the ignorant, until they can comprehend it, because the Art is not completed except with gold; which is a great and divine secret. And those who hear our secret cannot verify it because of their ignorance. Do you understand Aros?*

The four stones are Black (stibnite; Sb_2S_3), White (antimony; $Sb_2O_3 \cdot Sb_2S_3$), Red / Yellow (*Latten* / gold-calx; $Au \cdot Sb_2O_2SO_4$), and Violet / Tincture. In veiled language, Maria confirms to Aros that she has fulfilled his request to explain her method from beginning to end. Here she addresses the lengthy regimen typically associated with the *Ars Magna* as being a ploy to hide the technique from the unworthy. The great and divine secret here, echoed by Isis in another alchemical dialogue, is that the gold content of the Stone is the active ingredient to create alchemical gold from copper. Maria's *Ars Magna* is the *Great Art*, but it is not necessarily a lengthy one.

> **Aros said**: *Certainly. But tell me about that vessel, without which the work will not be completed.*

CHAPTER 3 – MARIA HEBREA'S TINCTURE

Mary said: It is the vessel of Hermes, although Stoics [Philosophers] have concealed it, and it is not a vessel used for magic or divination [νεκρομαντία], but is the measure of your fire.

What Maria does not address anywhere in the dialogue is that the *Seal of Hermes* is a metallurgical cover-name for fluxing salt, because it forms a seal over molten metal in a crucible. Here she is addressing the *Ars Magna* only, which takes place in a sealed glass digestion flask instead of a crucible. This sealed flask is generally understood to be the secret to success. Maria addresses this error in understanding and offers an alternate perspective.

Her cover-term *vessel of Hermes* is associated with the *measure of fire*. The *fire* Maria is referring to is the *secret fire* inherent in her fluxing salt – namely the salt's potential for chemical reaction upon heating to create butter of antimony – *divine water*, philosophical Mercury (Hermes). Decomposition of sal ammoniac generates ammonia + hydrochloric acid vapour, according to the following equation:

$$NH_4Cl = NH_3 + HCl$$

As metal chlorides form, the entire mass liquefies over heat. It is for this reason that Maria's *Ars Magna* was termed *Via Humida* (or Wet Path) by later European alchemists. The great *Key* to the chemical reaction is the amount of salt used – in other words, the *measure of fire*. The efficiency of the chemical reaction depends heavily upon accurate ingredient ratios. The result is a body or *vessel* in the form of a compound substance composed of three principles. Hermes was named *Thrice-Great* because he was seen as the embodiment of three great bodies of wisdom – alchemy, astrology and magic. Maria likens the Philosophers' Stone to Hermes Trismegistus because, like him, the Stone embodies a trinity – *body-soul-spirit* in the form of *gold-antimony-salt*. She is suggesting that the salt is the master *Key* and secret of the Philosophers' Stone.

> **To which Aros said:** *Oh Lady you have obeyed the Society of Scholars: Oh Prophetess, have you found the secret of the Philosophers, who maintain in their books that one can make the Art from one body [i.e. a single substance]?*
>
> **And Maria said:** *Certainly. Hermes did not teach that because the root of our Art [school; guild] is the unbreakable body, venom and a toxin that kills all bodies and pulverizes [powders] them, and its odor solidifies quicksilver.*

Just in case Aros has not clearly understood the *prime material*, Maria graciously provides a refresher. She employs the cover-name *root of our art* in the passage above to indicate common commercial *mśdmt / kahâl / ḳoḥl* (stibnite; Sb_2S_3). When the finished silver-based Stone is used to create alchemical silver from quicksilver, the antimony content – meaning the *fume* of the *prime material* – serves to solidify or coagulate mercury in a manner similar to the way a dental amalgam hardens. She also clarifies that stibnite is the source of *divine water*:

> **And she said:** *I swear to you by eternal God, that when that venom is dissolved until it becomes a subtle [Divine] water, no matter how the solution is done, it solidifies Mercury [quicksilver] into Luna [silver] with the power of truth, and falls into the Throne of Jupiter [tin], and forms [silvers] itself into Luna [silver].*

The fixed metals responsible for providing the silver or golden colors of *Corinthian bronze* are elemental silver and gold. In the passage above, Maria provides a clue towards the creation of alchemical silver by combining the Philosophers' Stone with elemental mercury (Hg), tin (Sn) or a combination of both.

> *... There is a science of all bodies [substances], but the Stoics because of life's shortness and the work's duration have concealed this, and they found tingeing [coloring; potent] elements, and they increased them.*

CHAPTER 3 – MARIA HEBREA'S TINCTURE

And all Philosophers teach those things rather than the vessel of Hermes, because it [the vessel of Hermes] is divine, and, by the Lord's wisdom, hidden from the Gentiles. And those who do not know it are ignorant of the true regimen because of their ignorance of the vessel of Hermes.

Maria criticizes Philosophers who focused on different methods of coloration as opposed to Jewish chemical technology. She explains that the *vessel of Hermes*, *secret fire* and *divine water* in the form of butter of antimony was the treasure and great mystery hidden from the gentiles. *Divine Water* during the early Alexandrian period was synonymous with *Hermes*, called *Mercury* (not to be confused with quicksilver) in Latin and *Philosophers' Mercury* in traditions that followed. Maria's implication was that the *vessel of Hermes* was a secret known only to Judeo-Egyptians who, during her time, were the stewards of the Art of *chrysopœia*. Herein rests the primary difference between the chemical technologies of the Pseudo-Democritan Babylonian style of *aurifiction*, preserved in Alexandrian alchemical texts such as the Leyden and Stockholm Papyri, and Maria's unique style of *aurifaction*. Her operative technique of working with sal ammoniac may be the deciding factor for the differentiation between Alexandrian chemical technologies in a general sense and authentic alchemy.

Template for **Maria's *Ars Magna*:**

1. **Prepare**: the two *fumes / gums* – *flowers of antimony* and purified gold calx;
2. **Unite**: the *flowers of antimony* (soul; $Sb_2O_3 \cdot Sb_2S_3$) with sal ammoniac (spirit; NH_4Cl) to create the dual-soul;
3. **Join**: the dual-soul with the dual-body (*gold-antimony glass* calx; $Au \cdot Sb_2O_2SO_4$) and seal in a glass digestion vessel;
4. **Digest**: at ascending temperature, until the powders liquefy and then crystallize into a reddish-purple solid.

Maria's *Ars Magna* technique for creating her Tincture is a magnificent feat of efficient chemistry. Any recipe that calls for confecting the Stone by creating a chemical reaction between antimony and gold via ion exchange as a single reaction in a sealed glass digestion vessel, or employs the use of sal ammoniac as the primary fluxing-salt, is likely a descendant of the archetypal recipe preserved in Maria's *Ars Magna*.

In both of Maria's recipes, the composition of her Philosophers' Stone is a combination of gold, antimony and salt, which would later come to be known collectively by the Latin, term *Tria Prima*. Paracelsus, a European alchemist working during the 15th century, encrypted the identity of the *Tria Prima* with the cover-names sulfur, mercury and salt. Maria also referred to compounds by pairs of cover-terms *white gum* ↔ *red gum*, *kibrich* ↔ *zubech* or *two fumes that contain two lights* (sun and moon) in general and additionally by specific cover-names for each of the two compounds such as *Gum Elsaron*. These two compounds found fullest expression in the 7th century writings of Morienus and commentary by Sir Isaac Newton where they are known by the cover-terms *latten* ↔ *eudica*, *body* ↔ *blood* or *earth* ↔ *water*. When alchemy was transmitted to the Islamic Empire, these two compounds became known as *sophic sulfur* ↔ *sophic mercury*. Centuries later in the European tradition, Paracelsus referred to this pair of compounds by the cover-terms *lion's blood* ↔ *eagle's gluten*.

THE HERB, OASIS AND ALCHEMICAL GOLD

The Much-Honored Herb

Herodotus, writing during the 5th century BCE, was known to educated Alexandrians as the *father of history*; his works widely studied and revered for centuries. Maria was certainly educated in the *Histories* written by Herodotus when she encrypted the identity of her fluxing agents under a number of various cover-names, two of which appear to be sourced directly from *Histories*, Book IV Chapters 181-185. Her

CHAPTER 3 - MARIA HEBREA'S TINCTURE

generosity lies with the fact that anyone educated at the time would be familiar with her verbiage as sourced directly from *Histories*. Upon reviewing the cover-names employed in *The Dialogue of Maria and Aros*, a direct connection becomes clear.

> ... *because* the Art will not be completed but by it – *Maria Hebrea*
>
> 1. The spirit that binds;
> 2. A fixed body that preserves the tincture;
> 3. The white herb, bright and much honored, occuring on small mounds (hills; hillocks), ground fresh;
> 4. The true fixed body that does not flee from the fire;
> 5. That bright body, strewn on mounds (small hills; hillocks); not subject to putrefaction;
> 6. Fixed body from the Heart of Saturn, which preserves the tincture;
> 7. Bright white body.

Later in alchemical history, this enigmatic substance would come to be known by the cover-name *Eagle*, supposedly because it is volatile. A better reading of Maria's source reveals an alternate possible etymology. The following excerpt from Herodotus' *Histories*, 4:181-5 is an account of the established trans-Saharan caravan route connected by a series of oases; a close reading of which clears up three long-standing mysteries of alchemy.

> *Farther inland than these is that Libyan country which is haunted by wild beasts, and beyond this wild beasts' land there runs a ridge of sand that stretches from Thebes of Egypt to the Pillars of Herakles. At intervals of about ten days' journey along this ridge* there are masses of great lumps of salt in hillocks: *on the top of every hillock a fountain of cold sweet water shoots up from the midst of the salt: men dwell round it who are farthest away toward the desert and inland from the wild beasts' country. The first on the journey from Thebes, ten days distant from that place, are* the Ammonians, who follow the worship of the Zeus of Thebes; *for, as I have before said, the image of Zeus at Thebes has the*

head of a ram. They have another spring of water besides, which is warm at dawn, and colder at market-time, and very cold at noon: and it is then that they water their gardens: as the day declines and the coldness abates, till at sunset the water grows warm. It becomes ever hotter till midnight, and then it boils and bubbles; after midnight it becomes ever cooler till dawn. This spring is called the spring of the sun.

At a distance of ten days' journey again from the Ammonians along the sand ridge, *there is a hillock of salt like that of the Ammonians, and springs of water, where men dwell; this place is called Augila [αετός; Eagle]; it is to this that the Nasamonians are wont to come to gather palm-fruit.* **After ten days' journey again from Augila** *there is yet another hillock of salt and springs of water and many fruit-bearing palms, as at the other places;* **men dwell there called Garamantes, an exceeding great nation,** *who sow in earth which they have laid on the salt.*

After another ten days' journey from the Garamantes *there is again a salt hillock and water; men dwell there called Atarantes ... After another ten days' journey* *there is again a hillock of salt, and water, and men dwelling there. Near to this salt is a mountain called Atlas, the shape whereof is slender and a complete circle; and it is said to be so high that its summit cannot be seen, for cloud is ever upon them winter and summer. The people of the country call it the pillar of heaven. These men have got their name, which is Atlantes, from this mountain. It is said that they eat no living creature, and see no dreams in their sleep.*

I know and can tell the names of all the peoples that dwell on the ridge as far as the Atlantes, but no farther than that. **But this I know, that the ridge reaches as far as the Pillars of Herakles and beyond them.** *There is a mine of salt on it every ten days' journey,* **and men dwell there.** *Their houses are all built of the blocks of salt;* **for even these are parts of Libya where no rain falls; for the walls, being of salt, could not stand firm if there were rain.** *The salt there is both white and purple.*

CHAPTER 3 - MARIA HEBREA'S TINCTURE

Nearly two-and-a-half millennia after Herodotus, T.E. Lawrence (Lawrence of Arabia) in his *Seven Pillars of Wisdom* describes a similar timeless scene when he wrote about a stopover in a desert salt hamlet:

> *Nebk, to be our next halt, had plentiful water, with some grazing. Auda had appointed it our rallying place, because of the convenient nearness of the Blaidat, or 'salt hamlets'. In it he and Sherif Nasir sat down for days, to consider enrolling the men, and to prepare the road along which we would march, by approaching the tribes and the sheikhs who lived near.*

These fascinating excerpts reveal each desert spring or well – called a *salt hamlet* by Lawrence –an area exists where a concentration of salt occurs. Herodotus describes the second hill of salt at Augila (αετός; Eagle) as being *a hillock of salt like that of the Ammonians*. The famed salt of the Temple of Amun is sal ammoniac (NH_4Cl). This similar salt at Augila (and elsewhere) is easily explained upon considering that camels are watered and rested in a central area at such salt hamlets. Salt deposits were identified by Herodotus at each salt hamlet on the trade route, approximately ten days' journey from each other. The first two oases encountered upon leaving Alexandria were the Siwa Oasis and Augila; both sources of sal ammoniac according to Herodotus.

The primary source of sal ammoniac nearest to Alexandria was the Temple of Amun / Zeus / Jupiter at Siwa Oasis, with the second found at a location that translates to *Eagle*. Furthermore, Herodotus is explicit that these salts can be sourced from small hills or hillocks. This identification of salt deposits was available to the educated class familiar with Herodotus' *Histories*, taught for approximately 700 years prior to the time Maria drew upon it. The problem arises when we consider that camel urine contains no ammonia, but is extremely rich in hippuric acid. The question regarding ammonium salts from camel urine was addressed by Bernard Read of Beijing Union Medical College in his 1925 article, 'The Chemical Constituents of Camel Urine'. Camels drawn to watering holes are induced by the scent of urine to urinate at the same location. Enzymes acting upon fresh urine initiate ammoniacal fermentation of the hippuric acid present, resulting in ammonia-rich salt deposits, thus explaining unrefined sal ammoniac sourced from desert salt hamlets.

The three mysteries of Alexandrian, Islamic and European alchemy answered by the above include:

1. The identity of Maria's fluxing salt central to Alexandrian alchemy;
2. The purpose of urine and urine distillate common to Islamic alchemy;
3. The etymology and chemical identity of Eagle salt in the Islamic and European alchemical traditions.

It is possible, therefore, to identify with certainty Maria's *much-honored herb*, Islamic alchemy's *urine menstruums* and European alchemy's *Eagle salt* directly from Maria's original source. Maria knew sal ammoniac as *halas ammoniakôn* (ἅλας ἀμμονιακων) and derived at least two of her cover-names directly from Herodotus' discussion of ammonium salt deposits at Siwa and Augila. Pliny the Elder, a Roman philosopher, army and navy commander and author writing during the 1st century CE, corroborates accounts of sal ammoniac being derived from northeastern Sinai and the Libyan Desert during the Roman period in the following passage:

> King Ptole-mæs discovered salt also in the vicinity of Pelusium [Πηλούσιον], *when he encamped there; a circumstance which induced other persons to seek and discover it* in the scorched tracts that lie between Egypt and Arabia, *beneath the sand. In the same manner, too, it has been found in the thirsting deserts of Africa, as far as the oracle of Hammon, a locality in which the salt increases at night with the increase of the moon.* The districts of Cyrenaica [Κυρηναϊκή] are ennobled, too, by the production of hammoniacum, a salt so called from the fact of its being found beneath the sands [ἄμμος] there.
>
> – Pliny the Elder, The Natural History, 31:39

CHAPTER 3 - MARIA HEBREA'S TINCTURE

Arrian of Nicomedia was a Greek historian, official, military commander, and philosopher writing during the 2nd century; primarily remembered for his history of Alexander's conquest. In his section detailing Alexander's trip to the Siwa Oasis, Arrian recounted sal ammoniac collection, packaging in woven palm leaf baskets, and its perceived value as a luxury good or religious offering.

> *In this place also natural salt is procured by digging, and certain of the priests of Ammon convey quantities of it into Egypt.* **For whenever they set out for Egypt** *they put it into little boxes plaited out of palm, and carry it as a present to the king, or some other great man.* **The grains of this salt are large, some of them being even longer than three fingers' breadth; and** *it is clear like crystal.* **The Egyptians and others who are respectful to the deity, use this salt in their sacrifices, as it is clearer than that which is procured from the sea. Alexander then was struck with wonder at the place, and consulted the oracle of the god.**
> – *Arrian*, Journey of Alexander, Chapter IV, 3-5, 'The Oasis of Ammon'

γίγνονται δὲ καὶ ἅλες αὐτόματοι ἐν τῷ χωρίῳ τούτῳ ὀρυκτοί· καὶ τούτων ἔστιν οὓς ἐς Αἴγυπτον φέρουσι τῶν ἱερέων τινὲς τοῦ Ἄμμωνος. ἐπειδὰν γὰρ ἐπ' Αἰγύπτου στέλλωνται, ἐς κοιτίδας πλεκτὰς ἐκ φοίνικος ἐσβαλόντες δῶρον τῷ βασιλεῖ ἀποφέρουσιν ἠεῖ τῳ ἄλλῳ. ἔστι δὲ μακρός τε ὁ χόνδρος 'ἤδη δέ τινες αὐτῶν καὶ ὑπὲρ τρεῖς δακτύλουσ' καὶ καθαρὸς ὥσπερ κρύσταλλος: καὶ τούτῳ ἐπὶ ταῖς θυσίαις χρῶνται, ὡς καθαρωτέρῳ τῶν ἀπὸ θαλάσσης ἁλῶν, Αἰγύπτιοί τε καὶ ὅσοι ἄλλοι τοῦ θείου οὐκ ἀμελῶς ἔχουσιν. ἐνταῦθα Ἀλέξανδρος τόν τε χῶρον ἐθαύμασε καὶ τῷ θεῷ ἐχρήσατο.

Synesius of Cyrene was the Bishop of Ptolomais during the 4th century CE. He pursued higher education at Alexandria, where he became a devoted disciple of the famous Hypatia. Synesius is credited for preserving a recipe for creating the White Philosophers' Stone that relied upon sal ammoniac.

BOOK 1 – ORIGINS & EVOLUTION

> *There is, I swear by Holy Hestia, with us a native salt of the earth, which is less far south, than the distance to the North Sea, the salt of Ammon, we call it. The mineral is crumbly developing under a crust, which, the crust which conceals it when you remove it, the underlying earth may be easily furrowed deeply with hands as with a hoe. Furthermore now that it is dug up, that is to say the salt, the very sight of which, as well as the taste is a most enjoyable delight.*
>
> – Synesius of Cyrene, Letter 148, to Olympius (The Good Life)

> *Est, per saeram vestem juro, est apud nos terrestre sal, quod minore intervallo ab Austro distat, quam ab Aquilone mare; hoc ammonis sal a nobis appellatur. Lapide hoc a friabili alitur atque tegitur, quem crustae in modum insidentem cum detraxeris, facili negotio manibus ae sarculus altius proscindi subjecta tellus potest. Id porro quod effoditur, sal est, cum visu ipso, tum alia gustandi voluptate jucundissimum:*

In 2008, an article was published, 'Ammonia in the environment: From ancient times to the present', in which two considerations regarding the Siwa Oasis were presented. The first was the consideration that the temple would have had considerably large dung fires that would have included offerings of local salt, resulting in ammonium chloride rich soot that could have been identified by the priesthood. The second finding was that in a detailed study on rain chemistry, the Nile delta showed very high levels of precipitation ammonia with Siwa having the second highest concentration in rain. Moreover, the ammonia values at Siwa coincided with equally high values of chloride. The Siwa Oasis turned out to be quite rich in environmental ammonium chloride.

Siwa Oasis and the Temple of Amun

The ancient name for the Siwa Oasis was Ammonium and it was one of Egypt's most isolated settlements. According to the High Priest residing there during Alexander's conquest, it was a place where Egyptians were banished by oppressive Persian rulers. Centuries later, the Romans would do the same. Greek historian Diodorus Siculus, writing during the

CHAPTER 3 – MARIA HEBREA'S TINCTURE

1st century CE, describes the Oasis in *Bibliotheca Historica* (or *Historical Library*) Book 17: 50, as Maria would have experienced it:

> The land where this temple lies is surrounded by a sandy desert and waterless waste, destitute of anything good for man. The oasis is fifty furlongs in length and breadth and is watered by many fine springs, so that it is covered with all sorts of trees, especially those valued for their fruit. It has a moderate climate like our spring and, surrounded as it is by very hot regions, alone furnishes to its people a contrasting mildness of temperature.

Siculus goes on to detail the temple and the oracle's method of divination, yet makes no mention of sal ammoniac. It was mentioned by Herodotus during the 5th century BCE and described as a thriving industry as late as the 4th century CE by Synesius, indicating that salt production remained uninterrupted for nearly a millennium and probably much longer by the time of Synesius' writing.

As explained by author Ahmed Fakri in *Siwa Oasis*, hieroglyphs on the inner temple wall provide the name of the original builder as *Wenamun – the Great Chief of the Deserts*. Wenamun is depicted wearing an ostrich feather, symbolising that he was of Libyan ancestry. Fakri suggests that Wenamun succeeded his father as Great Chief of the Desert and built the temple during his rule.

Wenamun appears in an ancient papyrus held at the Moscow Museum known as the *Travels of Wenamun: The Journey of Wenamun to Phoenicia*. It was once believed that Wenamun embarked upon his journey around 1100 BCE, but it was revealed in 1924 – through cross-referencing Hebrew sources – that the actual date was around 400 BCE. This explains why Herodotus makes no mention of a temple in his accounts of the Oasis. The papyrus offers a fascinating glimpse into the life of a powerful sea captain and merchant trading in the Mediterranean.

Wenamun is also clearly depicted as a priest and an official. The tale is told in the first person and the account reads like a captain's log:

> ... the day on which Wenamon, the Senior of the Forecourt of the house of Amon, [lord of thrones] of the Two Lands, set out to fetch the woodwork for the great and august barque of Amon-Re, king of the gods, which is on [the river and which is named:] `User-het-Amon'.

A negotiation with a trader of Lebanese cedar provides insight into the stern character of Wenamun and his views regarding his god Amun-Ra:

> You are stationed [here] to carry on the commerce of the Lebanon with Amon, its lord. As for you saying that the former kings sent silver and gold – suppose they had life and health! Now as for Amon-Re, king of the gods – he is the lord of this life and health, and he was the lord of your fathers. They spent their lifetime making offering to Amon. And you also – you are the servant of Amon!

Wenamun's chiefdom at the Siwa Oasis, and the temple he founded, would go on to play a key role in a dramatic turning point in history.

During Alexander's time, three great oracles existed in the region of the eastern Mediterranean; the Oracles of Delphi and Dodona in Greece, and the Oracle of Amun at the Siwa Oasis. Oracles in classical antiquity were believed to be divine portals through which one could directly communicate with the particular god of the temple. Such oracles were consulted by royalty for prophecies, predictions and wise council. Under oppressive Persian rule, all three oracles were known to each other and likely in communication. The Egyptian god Amun-Ra was viewed by the Greeks as the Egyptian representation of their god Zeus-Pitar. The Greeks changed the name to Zeus-Amun, later called Jupiter-Hammon by the Romans.

Amun was chief of the Egyptian gods associated with the Egyptian overthrow of the oppressive 15th Dynasty rulers, the Hyksos. Henceforth

Amun was viewed as a protector deity, and eventually associated with the chief solar deity Ra; he thus became Amun-Ra. The tribes of the Libyan Desert worshipped a guardian deity who appeared as a ram that watched over their flocks. The Egyptians identified this ram deity as Amun-Ra. The Egyptian word *amun* originally meant *to shepherd* or *to feed*, but also means *the hidden*. This ram-headed protector God was viewed as symbolic of Alexander's role as liberator and protector by the Egyptians and Alexander himself. In order to legitimize his reign, Alexander had to first visit the great oracles in order to gain confirmation that he was a demi-god as Heracles and Perseus were, and thus of their lineage.

Alexander had already been declared *Son of Ra* during a sojourn to Heliopolis and Memphis, but this would be reconfirmed at Siwa where he was declared *Son of Zeus-Amun*; ram-headed protector deity or a *serpent* according to Alexander's biographer, Arrian of Nicomedia. Accounts of this arduous journey 400 kilometres into the Libyan Desert are recorded by historians Diodorus, Rufus, Arrianus, and Plutarch. The story is also immortalized on an inscribed column known as the *Stela of the Banishment of Menkheperre*, now located in the Louvre Museum in France.

Alexander posed questions to the deity and received favourable answers and a special gift from Menkheperre, the High Priest of Amun-Ra. Respected alchemy researcher and author, Dennis W. Hauck, presents the story as preserved in the tradition of European alchemy:

> *Convinced it was his destiny to reveal the ancient secrets, Alexander immediately headed across the Libyan desert to an ancient temple at Siwa near where the tomb was located. According to Albertus Magnus and others, that is where Alexander found the Emerald Tablet.*
>
> *Alexander took the tablet and scrolls he found in the tomb to Heliopolis, where he placed the scrolls in the sacred archives and put the Emerald*

> Tablet on public display. Construction of the city of Alexandria to house and study the Hermetic texts was begun immediately, and he assembled a panel of priests and scholars to prepare Greek translations. According to esoteric historian Manly P. Hall, the mysterious Emerald Tablet caused quite a stir. One traveller, who had seen it on display at Heliopolis, wrote: "It is a precious stone, like an emerald, whereon these characters are represented in bas-relief, not engraved. It is esteemed above 2,000 years old. The matter of this emerald had once been in a fluid state like melted glass, and had been cast in a mold, and to this flux the artist had given the hardness of the natural and genuine emerald, by his art."

Herodotus' *Histories* was written more than a century prior to Alexander's liberation of Egypt from the Persians. Alexander must have been introduced to sal ammoniac during his trip to Siwa Oasis because, according to Herodotus, the Oasis was known for this type of salt long before Alexander's arrival. Alexander is said to have identified a sal ammoniac cave deposit in a region that was plagued by underground burning coal seams in what is now Tajikistan, the implication being that he was familiar enough with this unique salt to identify it during his campaign in Central Asia. According to noted present-day alchemist and author, Rubellus Petrinus, these salts from Tajikistan were exported to the West under the name *sal tartari* (*subterranean salt* or *salt of the Tartars*, i.e. Mongolians). The term sal tartari would later come to denote potassium carbonate, also known as pearl ash, first characterized in early modern Europe.

The secret method of the Temple of Amun's refined sal ammoniac manufacture was first revealed to the Académie Royale des Sciences in 1716 by Siccard, and later to the Royal Society in detail by Linneus in 1760. During Siwa's lush spring each year, animals were set to graze upon unique legume pastures that were presumably rich in nitrogen and salt. Throughout March and April, peasants burnt the dried dung from the livestock as fuel, and then collected the soot from their chimneys.

CHAPTER 3 - MARIA HEBREA'S TINCTURE

Narrow-mouthed glass vessels coated in clay were filled with the soot and heated above a furnace. Purified sal ammoniac sublimate collected at the vessel's opening, which was then removed.

If the above legend of Alexander and the *Emerald Tablet* is in any way describing factual events, then it is possible that the story is an allegory concealing that Alexander was introduced to a chemical technology or industry revolving around sal ammoniac, which he then introduced to Alexandria. At this stage in history, it was customary to transmit an historical event clothed in the form of legendary allegory. This being the case or not, the Siwa Oasis and its temple inspired a certain mystique and awe in the ancient world. That Maria made use of sal ammoniac is not surprising. This amazing salt was applied to a number of chemical technologies in Egypt including fabric dyeing, leather tanning, glass manufacture, metallurgy, food preparation, and medicine, among others. It was a powerfully magic substance sourced from an honored and important temple and associated with the solar deity Amun-Ra.

Upon reading the various histories that mention Siwa Oasis, Maria's use of the cover-terminology – *the white, clear and much-honored Herb, which is found on low hills* – seems appropriate on a number of levels. It was *much-honored* due to its source at the Temple of Amun and the Augila Oasis, and a salt *found on low hills* matching the descriptions of ammonia-rich urine salts present at caravan salt hamlets. It is therefore possible to ascertain with a degree of certainty that Maria's secret salt was, indeed, sal ammoniac. Experimentation with sal ammoniac provided Maria with a substance that could be innovatively applied to her already ancient Art. It provided the potential for chloride-oxide ion exchange, which would eventually characterize production of *Divine Water*. Maria stood at the pivotal point in alchemical history where operative metallurgical techniques of *aurifaction / chrysopœia* transitioned to chemistry-based methods: the smith had become the chemist.

BOOK 1 – ORIGINS & EVOLUTION

Alexandrian Alchemical Gold

Maria's *chrysopœia* was a two-step process. Creating the Tincture was only half the job; the other half being the industrial application of the Tincture to create alchemical gold. Maria created her Tincture with either gold or silver; the resulting elixirs known as the Red Stone and the White Stone respectively by later alchemists. With these two Stones, she produced alchemical gold and silver products. According to Zosimos, she also made a unique four metal mixture that she referred to as *Our Lead*, and ingeniously applied a liver-purple patina to her silver product by a process known as *iosis*. Throughout her writings, as recorded by Zosimos, Maria primarily worked with copper, yet she was equally familiar with lead, mercury and tin for the manufacture of alchemical silver.

Pliny the Elder provides an interesting clue regarding the products that Maria was creating in his *Naturalis Historia*, Book 34: 3. Pliny describes copper-alloys produced in Alexandria and elsewhere that correspond with Maria's work almost perfectly.

> **Naturalis Historia *34: 3 – The Corinthian Brass [Bronze]***
> *Formerly a mixture was made of copper fused with gold [oreikhalkos] and silver, and the workmanship in this metal was considered even more valuable than the material itself; but, at the present day, it is difficult to say whether the workmanship in it, or the material, is the worst. Indeed, it is wonderful, that while the value of these works has so infinitely increased, the reputation of the art itself is nearly extinct. But it would appear, that in this, as in everything else, what was formerly done for the sake of reputation, is now undertaken for the mere purpose of gain.*
>
> *For whereas this art was ascribed to the gods themselves, and men of rank in all countries endeavoured to acquire fame by the practice of it ...*

Pliny introduces a copper product alloyed with precious metals (Greek, ορείχαλκος *oreikhalkos*; Latin; *aurichalcum*), known as *mountain copper* or *gold-copper* to indicate a gold-based bronze. Pliny is correct to assert

CHAPTER 3 – MARIA HEBREA'S TINCTURE

great antiquity to *aurichalcum* due to its earliest mention in a Homeric hymn by Hesiod in the 7th century BCE. *Aurichalcum* is also mentioned in the writings of Cicero, Plato and Virgil prior to Pliny's description. He appears to lament how a once respected artisanal tradecraft had degenerated to crass commercial industry. Since Pliny wrote of *Corinthian bronze* nearly two centuries prior to Maria's time, this was likely the case during the period in which she operated. There is a strong possibility that Maria and her contemporaries were in control of a financially successful enterprise. Scholars have always puzzled over the fact that Pliny had never mentioned alchemy even once in all of his writings, but perhaps he was actually describing the Art in its original form, as it existed in Pliny's day: the *Corinthian bronze* industry.

The only genuine Corinthian vessels, then, are those which these men of taste metamorphose, sometimes into dishes, sometimes into lamps, or even into washing basins, without any regard to decency.

... the white variety, approaching very nearly to the splendour of silver, and in which that metal forms a large proportion of the compound; a second kind, in which the yellow colour of gold predominates; and a third, in which all the metals are mixed in equal proportions.

This last is highly prized for its colour, which approaches to that of liver, and it is on this account that it is called 'hepatizon:' it is far inferior to the Corinthian metal, but much superior to the Æginetan and Delian, which long held the first rank.

Maria created her alloys by combining *molybdochalkos* (antimonial bronze) with either the White or Red Elixir in exact proportions, in order to create a very fine silver or gold antimonial alloy described by Pliny as silver or gold-alloyed *Corinthian bronze*. She created an additional product matching Pliny's description of a third form of *Corinthian bronze* "in which all the metals are mixed in equal proportions." The unique liver-colored bronze, *hepatizon*, was crafted by treating *Corinthian bronzes* with sulfur in order to affect a purple-black patina. (See Appendix C for detailed process)

Maria's products match the three *Corinthian bronze* products described by Pliny almost verbatim. Numerous sources mention *Corinthian bronze* from great antiquity up until the 10th century CE. It should not surprise us that the early Alexandrian artisans – generally accepted as the first alchemists by mainstream researchers – were actually a guild of bronze artisans who had evolved the Art from bronze metallurgy to chemistry, based upon a new technical application of sal ammoniac. Their unique bronze work, *Corinthian bronze*, was the original alchemical gold.

Corinthian bronze was likely the fine bright bronze described in the Book of Ezra, thus indicating that the bronze as valuable as gold known to Ezra in the 5th century BCE was still available during the 1st century CE.

כז וּכְפֹרֵי זָהָב עֶשְׂרִים, לַאֲדַרְכֹנִים אָלֶף; וּכְלֵי נְחֹשֶׁת מֻצְהָב טוֹבָה, שְׁנַיִם--חֲמוּדֹת, כַּזָּהָב.

Ezra 8:27 ... *20 bowls of gold worth 1,000 darics, and two vessels of* fine bright bronze as precious as gold.

This "fine bright bronze as precious as gold" may be the "gleaming metal" mentioned in the Book of Ezekiel.

ד וָאֵרֶא וְהִנֵּה רוּחַ סְעָרָה בָּאָה מִן-הַצָּפוֹן, עָנָן גָּדוֹל וְאֵשׁ מִתְלַקַּחַת, וְנֹגַהּ לוֹ, סָבִיב; וּמִתּוֹכָהּ--כְּעֵין הַחַשְׁמַל, מִתּוֹךְ הָאֵשׁ.

Ezekiel 1:4 As I looked, behold, a stormy wind came out of the north, and a great cloud, with brightness around it, and fire flashing forth continually, and in the midst of the fire, *as it were gleaming metal.*

Jewish historian Josephus, writing in the 1st century CE in his *Antiquities of the Jews*, explains that the craftsman of the Temple of Solomon was an expert in precious metals and brass.

> Now Solomon sent for an artificer out of Tyre, whose name was Hiram; he was by birth of the tribe of Naphtali, on the mother's side, (for she was of that tribe,) but his father was Ur, of the stock of the Israelites.

CHAPTER 3 - MARIA HEBREA'S TINCTURE

This man was skillful in all sorts of work; but his chief skill lay in working in gold, and silver, and brass; by whom were made all the mechanical works about the temple, **according to the will of Solomon.**

Book 8, Chapter 3: 7 describes vessels in the Temple of Solomon made of a bronze that was like gold in beauty:

He also made a brazen altar, **whose length was twenty cubits, and its breadth the same, and its height ten, for the burnt-offerings.** *He also made all its vessels of brass, the pots, and the shovels, and the basons; and besides these, the snuffers and the tongs, and all its other vessels, he made of brass, and such brass as was in splendour and beauty like gold.*

In Book 15, Chapter 11: 5, Josephus makes further reference to 162 massive pillar caps, called *chapiters,* **which he described as follows:**

Their chapiters [pillar-caps] were made with sculptures after the Corinthian order, *and caused an amazement [to the spectators], by reason of the grandeur of the whole.*

Although Josephus does not explicitly state that the pillar caps were *Corinthian bronze,* the mention of a sculpture described as Corinthian is intriguing. It is reasonable to connect Maria's Art, a *Corinthian bronze* industry and the Alexandrian Jewish community to Josephus' accounts of brass or bronze that exhibited the "splendour and beauty like gold" and the story of the *Beautiful Gate* or *Nicanor Gate* of Herod's Temple in Jerusalem. In Josephus' *War of the Jews*, Book 5, Chapter 5: 3, he describes the *Beautiful Gate* as being more beautiful than gold and more costly. He also connects *Corinthian brass* to two distinct colors – silver and gold – and comments upon their thickness. *Corinthian bronze* would naturally be much thicker than pure silver or gold gilding found elsewhere.

> *Now nine of* these gates were on every side covered over with gold and silver, *as were the jambs of their doors and their lintels; but* there was one gate *that was without the [inward court of the] holy house,* which was of Corinthian brass, and greatly excelled those that were only covered over with silver and gold ... *Now the magnitudes of the other gates were equal one to another;* but that over the Corinthian gate, which opened on the east over against the gate of the holy house itself, was much larger; for its height was fifty cubits; and its doors were forty cubits; and it was adorned after a most costly manner, as having much richer and thicker plates of silver and gold upon them *than the other.*

The *Beautiful Gate* mentioned in the Old and New Testament, separated the inner court from the outer court of the Gentiles on the east. It was estimated to have been approximately 75 feet high and extremely heavy.

> καί τις ἀνὴρ χωλὸς ἐκ κοιλίας μητρὸς αὐτοῦ ὑπάρχων ἐβαστάζετο, ὃν ἐτίθουν καθ' ἡμέραν πρὸς τὴν θύραν τοῦ ἱεροῦ τὴν λεγομένην Ὡραίαν τοῦ αἰτεῖν ἐλεημοσύνην παρὰ τῶν εἰσπορευομένων εἰς τὸ ἱερόν·
>
> *Acts 3:2* And a man lame from birth was being carried, whom they laid daily at *the gate of the temple that is called the Beautiful Gate [Nicanor Gate] to ask alms of those entering the temple.*

> καὶ εἶδεν πᾶς ὁ λαὸς αὐτὸν περιπατοῦντα καὶ αἰνοῦντα τὸν θεόν· ἐπεγίνωσκον δὲ αὐτὸν ὅτι οὗτος ἦν ὁ πρὸς τὴν ἐλεημοσύνην καθήμενος ἐπὶ τῇ ὡραίᾳ πύλῃ τοῦ ἱεροῦ καὶ ἐπλήσθησαν θάμβους καὶ ἐκστάσεως ἐπὶ τῷ συμβεβηκότι αὐτῷ.
>
> *Acts 3:9-10* And all the people saw him walking and praising God, and recognised him as the one who sat at *the Beautiful Gate [Nicanor Gate] of the temple,* asking for alms. And they were filled with wonder and amazement at what had happened to him.

The Jewish Virtual Library preserves a fascinating legend associated with Nicanor's gates:

> *This gate was one of the best known of the gifts made to the Temple and "miracles were performed in connection with the gate of Nicanor and his memory was praised" (Yoma 3:10). Of these miracles the Talmud states: "What miracles were performed by his doors? When* Nicanor went to Alexandria in Egypt to bring them, *on his return a huge wave*

threatened to engulf him. Thereupon they took one of the doors and cast it into the sea but still the sea continued to rage. When they prepared to cast the other one into the sea, Nicanor rose and clung to it, saying 'cast me in with it.'" The sea immediately became calm. He was, however, deeply grieved about the other door. As they reached the harbour of Acre it broke the surface and appeared from under the sides of the boat. Others say a sea monster swallowed it and ejected it out onto dry land. Subsequently all the gates of the Sanctuary were changed for golden ones, but the Nicanor gates, which were said to be of bronze, were left because of the miracles wrought with them. But some say that they were retained because the bronze of which they were made had a special golden hue. R. Eliezer b. Jacob said, "It was Corinthian copper which shone like gold" *(Yoma 38a).*

Jewish sage-mystics would likely advise us to decipher the hidden meaning encrypted in the passage above, yet it is possible to glean a few literal facts from the passage relevant to the history of *Corinthian bronze* in Alexandria. The first is that the gate was a gift by Nicanor who personally sourced them in Alexandria and was accompanying them in transit to Jerusalem. They were bronze, specifically described as Corinthian copper, and "shone like gold." They obviously displayed great mystique and filled onlookers with admiration.

The reconstruction of Moses' technique and purpose in antiquity finds artisanal life in Herod's renovation of Solomon's Temple and industrial purpose in classical Alexandrian *chrysopœian* bronze-craft. Maria's techniques and innovations were based upon ancient Bronze Age precedents appearing in Jewish culture and religion. Maria had transformed ancient Bronze Age metallurgical techniques into chemistry made possible by an honored secret ingredient, sal ammoniac. The Bronze Age had evolved into an embryonic period of chemical technologies with Maria occupying center stage as the drama unfolded.

Alexandria during Maria's time had developed into an incredibly cosmopolitan center of trade, scientific learning and artistic expression that surpassed Athens or Rome. Talented men and women of all fields flocked to its great markets and libraries in order to make their fame and fortune. Since Ptolemy I founded the great edifices of Alexandria, magnificent libraries and universities were erected to house one of the largest collections of writings the world had ever seen. Learning had become synonymous with the city where mathematics, cosmology, astronomy and astrology, magic, religion, and philosophy all coexisted and often intermingled harmoniously until Roman domination, followed shortly thereafter by strict Christian oppression.

Fall of the Gold Makers

It was approximately 267 CE, when the Romans assassinated Odenathus, the king who succeeded in recovering the Roman East from the Persians and restored it to the Empire. The king, however, had bequeathed his rule to his charismatic wife, Zenobia of Palmyra. Zenobia raised an army loyal to her and took on the Roman Empire, determined to bring the eastern lands of Syria, Palestine, Egypt, and large parts of Asia Minor under her rule. Historian Edward Gibbon, in his 18th century *History of the Decline and Fall of the Roman Empire*, describes Zenobia as a beautiful yet powerful and charismatic warrior-queen:

> *Zenobia is perhaps the only female whose superior genius broke through the servile indolence imposed on her sex by the climate and manners of Asia.* She claimed her descent from the Macedonian kings of Egypt, equalled in beauty her ancestor Cleopatra, and far surpassed that princess in chastity and valour. Zenobia was esteemed the most lovely as well as the most heroic of her sex. She was of a dark complexion, *(for in speaking of a lady these trifles become important.)* Her teeth were of a pearly whiteness, and her large black eyes sparkled with uncommon fire, tempered by the most attractive sweetness. Her voice was strong and harmonious. Her manly understanding was strengthened and adorned by study. *She was not ignorant of the Latin tongue, but*

CHAPTER 3 – MARIA HEBREA'S TINCTURE

possessed in equal perfection the Greek, the Syriac, and the Egyptian languages.

She had inured her constitution to fatigue, disdained the use of a covered carriage, generally appeared on horseback in a military habit, and sometimes marched several miles on foot at the head of the troops. The success of Odenathus was in a great measure ascribed to her incomparable prudence and fortitude.

Based upon Gibbon's description, Zenobia was a fearless woman with the endurance and stamina of a soldier and the allure of a goddess. Zenobia achieved her goals and assumed the title Queen of the East. Although she claimed Macedonian ancestry, Henry Milman, writing in the late 18th century, explains that Zenobia was Jewish. Alexandria seemed to flourish under her rule and the climate would have favored *Corinthian bronze* production and trade throughout Zenobia's queendom.

These circumstances would prove to be short-lived with the arrival of Roman Emperor Aurelian who was determined to regain control of Zenobia's lands and reunify his empire. Battles raged throughout areas of Alexandria where the Royal Library was a casualty of fire; the Serapæum escaping destruction. By default, the Serapæum Library and the Cæsareum Library allowed for the continuation of Alexandrian literary and scientific learning. Aurelian eventually captured Zenobia and triumphantly paraded her in golden shackles through the streets of Rome. Gibbons describes the scene:

The beauteous figure of Zenobia was confined by fetters of gold; a slave supported the gold chain which encircled her neck, and she almost fainted under the intolerable weight of jewels. **She preceded on foot the magnificent chariot, in which she once hoped to enter the gates of Rome. It was followed by two other chariots, still more sumptuous, of Odenathus and of the Persian monarch. The triumphal car of Aurelian (it had formerly been used by a Gothic king) was drawn, on this memorable occasion, either by four stags or by four elephants.**

Aurelian was eventually kind to Zenobia, providing her with an elegant villa not far from the capital. She became a respected member of Roman society with her daughters marrying into Roman nobility.

Few surviving records have emerged regarding the *Corinthian bronze* trade after Aurelian regained control of Alexandria, but it likely continued unaffected for another few decades after he retook the city. The most severe blow to the *chrysopœia / aurifaction* tradition of the period came with Diocletian's securing control of Alexandria in 298 CE.

Diocletian attempted to bring the Egyptian tax system in line with Roman standards, which led to a revolt in the region. These tax reforms certainly would have affected the wealthy Jewish guild controlling the lucrative *Corinthian bronze* industry. It is likely that Diocletian was concerned that a revolt would be financed by *Corinthian bronze* makers, based upon an unusual edict issued a few years before entering Alexandria, as the Byzantine encyclopedia the Suda reveals under the section titled *Chimeía*, meaning alchemy:

> *[Chimeía] The preparation of silver and gold. Diocletian sought out and burned books about this. [It is said] that due to the Egyptians' revolting behavior Diocletian treated them harshly and murderously. After seeking out the books written by the ancient [Egyptians] concerning the alchemy of gold and silver, he burned them so that the Egyptians would no longer have wealth from such a technique,* **nor would their surfeit of money in the future embolden them against the Romans.**

> Χημεία: ἡ τοῦ ἀργύρου καὶ χρυσοῦ κατασκευή, ἧς τὰ βιβλία διερευνησάμενος ὁ Διοκλητιανὸς ἔκαυσεν. ὅτι διὰ τὰ νεωτερισθέντα Αἰγυπτίοις Διοκλητιανῷ τούτοις ἀνημέρως καὶ φονικῶς ἐχρήσατο. ὅτε δὴ καὶ τὰ περὶ χημείας χρυσοῦ καὶ ἀργύρου τοῖς παλαιοῖς αὐτῶν γεγραμμένα βιβλία διερευνησάμενος ἔκαυσε πρὸς τὸ μηκέτι πλοῦτον Αἰγυπτίοις ἐκ τῆς τοιαύτης προσγίνεσθαι τέχνης μηδὲ χρημάτων αὐτοὺς θαρροῦντας περιουσίᾳ τοῦ λοιποῦ Ῥωμαίοις ἀνταίρειν. ζήτει ἐν τῷ δέρας.

CHAPTER 3 - MARIA HEBREA'S TINCTURE

It is a safe assumption that Diocletian did not have a vendetta against alchemists, nor was he attempting to save a naïve population from silly notions of gold making. His edict appears to have been clearly directed towards the Art of Maria's tradition – the *Corinthian bronze* industry and those involved in it. Diocletian's edict to destroy all works and books addressing the *Chimeía* of gold and silver *Corinthian bronze* making resulted in a tremendous loss of literature on the subject. Those texts that survived appear to have been collected and preserved in the writings of Zosimos. Criminalization of the Art was a deathblow for *chrysopœian* guilds and artisans working in the tradition of Maria, Cleopatra, Isis, and others. If the Philosophers' Stone was to survive as a tradition, it needed to undergo a transmutation that would both disguise and evolve the Art to a new form.

This new system of alchemy would need to protect the Art from association with illegal gold making by removing the main ingredient – gold. It was a transmutation from Maria's industrial chemical gold making technology into an abstract cosmological model representative of philosophical concepts at the heart of Pagan, Christian, Jewish, Islamic, and Gnostic ideologies. It was the birth of a new purpose for the Philosophers' Stone as a visual model to elucidate sacred philosophy; the Stone's new primary function within Hermetic, Gnostic and Neoplatonic philosophical traditions.

This necessary shift from a chemical technology to abstract philosophical model was a direct result of Diocletian's new tax increases and criminalization of *chrysopœia*. Evidence of this transition is found in alchemical writings reflecting the next stage of Alexandria's turbulent history. An example of the change in technique and purpose is preserved in a Philosophers' Stone recipe attributed to Synesius, Bishop of Ptolemais and disciple of Hypatia, presented in the following chapter. These changes marked the end of original industrial alchemy and the beginnings of alchemy's second form as an esoteric or spiritual model by which to approach philosophical truths.

Chapter 4 – White Stone of Hermes

*Look at the divine water which gives them drink and the air
that governs them after they have been given a body in a single being.
... One is the serpent, which has its poison according to two compositions.
... One is All and through it is All and by it is All
and if you have not All, All is nothing.*

Chrysopœia of Cleopatra

Throughout classical antiquity, innumerable figures stand out as having played an important role in laying the scientific, cultural and political foundations upon which the edifices of modernity have been built. Great astronomers, mathematicians, philosophers, poets, artists, conquerors, and kings have influenced society long before Christianity gained a foothold as the Western world's universal authority on all matters pertaining to God and creation. At virtually the precise moment that the transition towards enforced Christianization took place, Hypatia struggled to keep Greek philosophy alive. She was known to all in Alexandria as simply *the Philosopher*. As her most devoted student, Synesius is intimately bound to the legacy of Hypatia. Hypatia's tragedy places the next evolution of the Philosophers' Stone in historical context.

Hypatia was an accomplished and free-spirited universal genius who influenced the greatest scholars and most prominent religious and political personalities of her time. A visit to Hypatia – if not to study, then only to meet her – was a rite of passage for dignitaries arriving in Alexandria. Great men sought her affections, leaders her council and nearly every intellectual Alexandrian her teachings. She taught anyone who wanted to learn – rich or poor, Christian or Pagan – without discrimination. Because of this, her students remained fiercely devoted to her. One such devoted student was Synesius, bishop of Ptolemais who described her in a letter as:

> A genuine guide in the mysteries of philosophy – Synesius, Letter 137

Under her father's influence Hypatia developed a brilliant intellect counter-balanced with physical athleticism in which she excelled at sailing, horsemanship and hiking. She was primarily a mathematician, but in addition, she taught an eclectic mix of cosmology, astronomy and astrology and, according to some, alchemy and other mysteries. Even when other Pagan schools underwent persecution, Hypatia's school was allowed to continue. At one point Hypatia's works could be found in most great libraries of the ancient world. As a teacher at the forefront of ancient science and mystery, she would certainly have been aware of Alexandria's *chrysopœia* history at the very least, and likely a proponent of the quasi-religious or spiritual brand of alchemy preserved by Synesius.

Hypatia was born into philosophy. Her father, Theon Alexandricus, was a great mathematician and last keeper of the Royal Library at the Musæum in Alexandria. She received formal education in Athens and Italy before establishing herself in Alexandria as teacher of philosophy *par excellence*. Theon would live to see his daughter take over his position as Alexandria's leading mathematician, but she would never teach in the Serapæum as her father did. In 391 CE, a conflict broke out between Pope Theophilus and Alexandria's non-Christian community. The Christian faction forced a standoff with the non-Christians besieged in the Serapæum. The emperor offered a pardon and safe passage for the non-Christians but ordered the Serapæum, its contents and library destroyed, as Byzantine Church historian Socrates Scholasticus, writing during the early 5th century, recounts in *The Ecclesiastical History*:

> Seizing this opportunity, Theophilus exerted himself to the utmost ... he caused the Mithraeum to be cleaned out ... Then he destroyed the Serapeum ... and he had the phalli of Priapus carried through the midst

CHAPTER 4 - WHITE STONE OF HERMES

of the forum ... the heathen temples ... were therefore razed to the ground, and the images of their gods molten into pots and other convenient utensils for the use of the Alexandrian church ...
— *Socrates Scholasticus, 5th century*

Hypatia's school was a political assembly hall in a marketplace called the Agora where she taught mathematics, astronomy and philosophy as was allowed until the new Christian faith, in edicts read aloud to the populace, proclaimed one subject after another forbidden. As a result, these subjects would come to be known as the mysteries or the occult sciences and, by necessity, were taught clandestinely in order to avoid persecution. The crux of Hypatia's philosophy – personal and direct knowledge of the divine principle and its manifestations, i.e. gnosis – was at odds with dogmatic Christian beliefs.

For political more than religious reasons, Hypatia found herself embroiled in a power struggle between two strongly opposed schools of Christianity vying for influence and control. As a non-Christian wielding tremendous influence over influential Christian leaders, she fell victim to circumstances and an impassioned Christian mob. Hypatia was ambushed in her chariot on the way home, shamefully stripped of all her clothing as a form of humiliation, and brutally dragged through the streets of Alexandria to the Caesaræum where she was murdered. The mob flayed her skin from its skeleton with potshards and bits of broken abalone shell and burnt her remains outside the city walls. Her works would later be gathered from the library collections around the Byzantine Empire and incinerated. What little is known of Hypatia's life and work comes mainly from letters and correspondence preserved by her most devoted student, Synesius. The details of her death are well known:

... they completely stripped her, and then murdered her by scraping her skin off with tiles and bits of shell. After tearing her body in pieces, they took her mangled limbs to a place called Cinaron, and there burnt them.
— *Socrates Scholasticus, 5th century*

BOOK 1 – ORIGINS & EVOLUTION

> *And they tore off her clothing and dragged her ... through the streets of the city till she died. And they carried her to a place named Cinaron, and they burned her body with fire.* – John of Nikiû, 7th century

The earlier period of classical antiquity was marked by a cultural and intellectual climate that fostered free and open scientific inquiry and philosophical expression that resulted in the great Babylonian, Egyptian and Greek traditions. There are those who view Hypatia's tragic murder as symbolic of the end of this Golden Age. Although the Alexandrian philosophical tradition would continue for some time after Hypatia's death, it limped forward under heavy Christian oppression in the form of anti-Pagan laws, proclamations and social pressure. Science and philosophy were non-Christian traditions and eventually criminalized, punishable by property confiscation, exile or death. Many of the great discoveries of classical antiquity were lost, only to be recovered centuries later when the intellectual climate became more favourable.

Like Hypatia, the Philosophers' Stone was a product of Alexandria, its legacy also subject to tumultuous political, religious and cultural changes. These changes influenced and transformed how the Stone was valued and what it meant to those who preserved the Art. The Philosophers' Stone began as a revered chemical technology responsible for a thriving and lucrative bronze industry. In the early 4th century, *chrysopœia* historian, encyclopedist and alchemist Zosimos began transforming the Stone into an icon or study model in order to elucidate Neoplatonic, Gnostic and especially Hermetic philosophies that collectively came to characterize the Alexandrian philosophical tradition. A greater understanding of the social, political and religious climate throughout the 4th and 5th centuries enables the suggestion of an hypothesis regarding how and why recipes for, and literary expressions of, the Stone underwent modifications during this period.

CHAPTER 4 – WHITE STONE OF HERMES

Astonishment and Wonder

From the 1st century CE, through Maria's period and right up to the time of Hypatia's tragedy, Alexandrians had been conditioned to associate Pagan temples with astonishment and wonder. Upon arriving at a temple, great doors would ceremoniously open without anyone touching them. Once inside, the gods spoke, thunder roared, statues cried, danced and even levitated. The wealth of a temple rested upon the number of worshippers it could attract. If a temple could demonstrate that the god/s were present as an invisible force behind visible phenomena, its wealth and influence grew. Alexandrians had come to expect a visible demonstration that validated their belief.

This approach to temple magic, while reaching its highest degree of sophistication in Alexandria between the 1st and 4th centuries, actually dates back to ancient Egypt where speaking tubes secreted inside statues brought Egyptian gods to life. Competition between Pagan temple worship and Christianity had steadily increased with the temples becoming increasingly reliant upon secret machine made 'miracles' in order to retain their followers in the face of emergent Christianity and Gnostic philosophies. The driving force behind temple technology was Heron of Alexandria, a mathematician, physicist and engineer who lived and taught during the 1st century CE.

In his landmark work, *Pneumatics*, Heron details several devices designed specifically for use in various temples throughout the city. These amazing machines operated on hot air, steam, vacuum, water, wind, and even magnetism to create fantastic illusions for expectant temple crowds. Heron's devices were not just gimmicks, but rather fantastic examples of early propulsion technology and mechanical engineering originating from the mind of a highly educated technical genius.

Heron is believed to have served as the director of the school known as the Musæum – Greek for *Temple of the Muses* – adjacent to the Royal

Library where he was the most well-known and last recorded member. Under Heron, the Musæum became one of the world's first polytechnic institutes. He is credited for inventing the first steam engine, surveying instruments, weapons and military machines, as well as the mechanical magic behind temple wonders. He was called *Michanikos*, Greek for *engineer* and origin of the modern word *mechanic*. In total, Heron invented more than 100 machines and devices that were largely responsible for Alexandrians associating religious experience with genuine observable phenomena. These fascinating marvels left the worshipper in a state of astonishment and wonder.

It should therefore come as no surprise to learn that even rational philosophers such as Hypatia and her students used models as tools to better elucidate philosophical truths. The process of confecting the Philosophers' Stone in a glass vessel – dry powders liquefying, changing color and magically transforming into a new stable substance before one's eyes – was a powerful example of just such a model. Today chemical reactions such as these still hold fascination. The Art that began as an advanced industrial technology, having become temporarily criminalized, preserved and reinvented itself as a viable cosmological / philosophical model.

After Diocletian criminalized the Art, emphasis shifted from its industrial value to importance as a cosmological / philosophical model disassociated from illegal *aurifaction*. This transformation required some specific changes in the recipe to confect the Stone and its presentation. Any detail that could tie the recipe or product to the Art of *aurifaction / chrysopœia* would place the artist in great jeopardy during the period of Diocletian's ban. The Tincture's connection to gold needed to be downplayed, resulting in the removal of elemental gold from the recipe – at least in appearance. The recipe itself became cloaked in garments of Neoplatonic, Gnostic and Hermetic verbiage and symbolism.

CHAPTER 4 - WHITE STONE OF HERMES

Diocletian did not last very long but Zosimos, writing around 300 CE, protected the tradition of Alexandrian *chrysopœia* from any further such oppression by collecting, codifying and preserving alchemical literature that survived Diocletian's bonfires. He distinguished the soul and body of a metal, and recognized the parallel between the behaviour of matter and events in the human psyche. In the writings of Zosimos, the Art began its transition from chemical technology to a cosmological model of philosophical ideas and spiritual development.

What is within is also what is without – Zosimos of Panopolis, c. 300 CE

CHRYSOPŒIA OF CLEOPATRA

Alchemical-Hermetic writings and fragments attributed to an early-Alexandrian alchemist named Cleopatra, preserved by Zosimos, reflect the period of transition from industrial technology to cosmological / philosophical model. Entire volumes could be written detailing parallels between confecting the Philosophers' Stone and the Hermetic, Gnostic and Neoplatonic philosophies prevailing in Alexandria at the time. In detailing the history and evolution of technique, this chapter will focus upon three extant recipes in which the chemistry and esoteric symbolism appear to be closely related if not identical.

One of the more interesting Philosophers' Stone recipes is a collection of fragments known as the *Chrysopœia of Cleopatra*. The title of the work refers to gold making, yet the language clearly indicates that the process was understood to represent the *One* of Hermetic and Gnostic philosophy symbolically. The word *chrysopœia* suggests that the recipe and product were linked to *Corinthian bronze* technology prior to Diocletian's edict. However, the language clearly presents the process and product as a

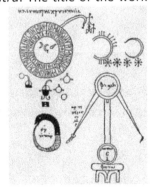

175

philosophical model or allegory for spiritual development. Since most of Cleopatra's work is in the form of diagrammatic sketches, it lends itself to a number of interpretations. An interesting possibility is that Cleopatra's recipe existed prior to the Diocletian edict in practice, yet survived as a practical model after the ban on the Noble Art.

Cleopatra's diagram contains five groupings of sketches accompanied by a few phrases. Each sketch denotes a chemical process and a correlative philosophical concept. Two glass vessels are pictured, one is a distillation vessel with a furnace underneath and the other is an evaporation vessel.

Ouroboros: A very recognisable alchemical image is the serpent devouring its tail, known in alchemy as the *Ouroboros*; the beginning and end, or cyclic transformation of the *One*. The words in the center of the *Ouroboros* read "One is All." Philosophically speaking the *One* – also called the *Father* or *Goodness* – was considered the supreme divine  principle, ultimate truth, source, or cause from which all things originate according to Alexandrian Gnostic philosophies. Gnostics called it the *Light* and considered it pure energy, completely devoid of matter, and the unmanifest source of all manifest creation. It is an assertion of a cosmic uniting principle and cyclic nature of duality, i.e. growth and regeneration. Alchemically, the source from which the Philosophers' Stone originates – through which it achieves perfection and matures into the Stone – is centered on stibnite / antimony represented by the serpent with its venom.

Stibnite is likened to a black winged dragon or serpent. When *flowers of antimony* are achieved via roasting, the volatile sulfur is removed as sublimated sulfur dioxide fumes. The result is white *flowers of antimony*. The entire process is symbolized in alchemical art as a black winged dragon (stibnite) losing its wings (volatile sulfur) and becoming a white

CHAPTER 4 – WHITE STONE OF HERMES

serpent (*flowers of antimony*). The *Ouroboros* is a symbolic representation of the chemical process of manipulating crude antimony ore – mśdmt / kahâl / ḳoḥl – to more refined forms.

Chemically speaking, the Philosophers' Stone is the *One* because it incorporated various ingredients into a single unified substance. Maria created a reaction between three primary ingredients – *flowers of antimony* + gold calx, via the medium of sal ammoniac – to create the *One*. If the gold was omitted in order to comply with Diocletian's edict, it is still possible to create a reaction between *flowers of antimony* + sal ammoniac to form a new compound. Since Cleopatra equates the serpent with the *One*, it indicates that her Stone is a manipulation of antimony.

Sophic Mercury / Lead: The large circular emblem in the upper left corner has two concentric circles with three symbols in the center, typically interpreted as gold, silver and mercury according to the Hermetic maxim, "Its father is the Sun and its mother is the Moon. Thus the wind bore it within it and the Earth nourished it." *It* takes center 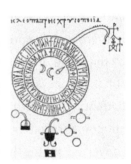 position in the diagram as Sophic Mercury, meaning an antimony product. Liquid Sophic Mercury must then be converted to a dry pristine white powder, which can be then be fused in a crucible with either gold or silver to create the Red or White Philosophers' Stone respectively. According to this interpretation, the diagram is symbolic of the Philosophers' Stone in the tradition of Maria's *Ars Magna* insofar as it includes precious metals.

A second interpretation suggests *Sophic Mercury / Lead* as the primary product. The central larger waning crescent with a handle at the bottom is generally interpreted as Mercury / Hermes, yet can also be interpreted as symbolising Saturn, the child of the sky god Uranus and the Earth goddess Gaia. The larger Saturn crescent-sickle occupies the central position as lead, not the ordinary but rather the *Philosophers' Lead* or *Secret Lead of the Wise*; cover-names identifying an antimony product as *mercury of Saturn*. In European alchemy, antimony is known by the cover-name *Son of Saturn*. In this context, the *Philosophers' Lead* can be fused with either gold calx or silver calx to finish the Stone, thus explaining the lunar and solar symbolism. This is further evidenced by the inscription that reads:

> *One is the serpent* **which has its tingeing venom according to** *two compositions.*

The end-product, the *One*, is clearly identified by Cleopatra as synonymous with the serpent. It is also composed of two things, interpreted here as sal ammoniac and *flowers of antimony*. The *One* is, therefore, the product of the reaction between these two, symbolized by the sickle of Saturn. The process of working with the *One* begins with stibnite (antimony trisulfide) which is calcined to *flowers of antimony* (antimony trioxide). The *One* is achieved by a chemical reaction with sal ammoniac resulting in *divine water* (antimony trichloride), which is finally converted to the Hermetic White Stone (antimony oxychloride), also called Sophic Mercury or Sophic Lead. It is the manipulation of a single prime material through its different phases.

One, the serpent = Antimony, artisanally manipulated

The *One* that remains constant throughout these reactions is the essence of antimony. The next inscription reads as follows:

> *One is All* **and** *through it is All* **and** *by it is All* **and** *if you have not All, All is nothing.* – Cleopatra the Alchemist

CHAPTER 4 – WHITE STONE OF HERMES

Compare the above to the passage below, in which Zosimos' perspective on *divine water* accords with Cleopatra's nearly verbatim, suggesting that they were both discussing the same chemical compound:

> This [divine water] is the great and divine mystery, the object of the search. **For** this is the all, **and** the all is from it **and** the all is through it. Two natures, one being; the one draws the one, the one rules the one.
> – Zosimus

All refers to the philosophical notion of totality or unicity underlying phenomena. This can be taken quite literally by decoding the cover-word *One* as correlating to the dissolving and coagulating properties of antimony; the sickle symbolizes death and regeneration as it cuts the harvest making way for new growth. The fragments suggest a modified Philosophers' Stone recipe with little evidence for the initial use of gold. The result is a white composite identified by Cleopatra as the serpent or the *One*. The central symbol takes pride of place in the diagram. Whether it represents Sophic Mercury (via a modified symbol for Mercury) or Sophic Lead (as the sickle of Saturn) is irrelevant, as these can be considered synonymous. The *One* is composed of two ingredients – resulting in thick translucent *divine water* known as *Sophic* or *Philosophers' Mercury / Hermes*, which is then converted by a second manipulation to a pristine white powder called *Sophic* or *Philosophers' Living Doubled Mercury / Hermes*.

Spirit and Soul: Three images appear in the upper right corner – two crescents and four asterisks. Throughout alchemical history, the asterisk symbolizes sal ammoniac. The two crescents – one radiant and the other plain – represent the female lunar antimony element before and after the reaction with sal ammoniac. *Flowers of antimony* + sal ammoniac represent combined female elements of soul + spirit. The soul and spirit become unified into the feminine lunar dual-soul, which then only needs to be joined to the solar male element, gold calx, in order to create the finished Philosophers' Stone. The diagram suggests ingredient proportions. Three lines above also appear below and the five radiating lines represent the five metals that can be later fused with the Hermetic White Stone – gold, silver, copper, iron and tin.

Dialogue of Cleopatra: In an accompanying text known as the *Dialogue of Cleopatra and the Philosophers*, Cleopatra draws a parallel between plant growth and confecting the Philosophers' Stone.

> *Look at* the divine water which gives them drink *and the* air that governs them after they have been given *a body in a single being.*

The use of the term *a body in a single being* above is referring to the final product. *Divine water* is a cover-name for butter of antimony and implies its molten state, whereas the *air that governs them* indicates evaporation that follows hydrolysis and fixes the White Stone. This unique chemical reaction fascinated onlookers at this time in history. The astonished philosophers in attendance then ask Cleopatra to explain this marvellous wonder:

> In thee is concealed a *strange and terrible mystery. Enlighten us,* **casting your light upon the elements.** *Tell us how the highest descends to the lowest and how the lowest rises to the highest,* **and how that which is in the midst approaches the highest and is united to it,** *and what is the element which accomplishes these things.*

> And the philosophers, considering what had been revealed to them, rejoiced.

That Cleopatra is allowing uninitiated spectators to witness the process is significant and argues strongly that she is presenting the spectacle as a model or visual aid to elucidate Hermetic principles. The wording

CHAPTER 4 - WHITE STONE OF HERMES

reflects that the philosophers are familiar with Hermetic writings. They are witnessing a visual model of *as above, so below* in their discussion of high and low. The high is understood as the sky god Uranus, and the low the Earth goddess Gaia, whose union produces Saturn as symbolized by the scythe or sickle in the diagram.

From a chemistry perspective, the assembled philosophers are marvelling at the union of a powder and liquid distilled to Divine Water, which is then transfixed to a powder by the addition and evaporation of water. Clearly, the reaction is taking place in a glass vessel – an innovation attributed to Maria Hebrea – suggesting that this recipe post-dates Maria's innovative *Ars Magna*. Cleopatra provides a veiled explanation of the chemical reaction:

> *The waters, when they come, awake the bodies and the spirits which are imprisoned and weak. For they again undergo oppression and are enclosed in Hades, and yet in a little while they grow and rise up and put on divers glorious colours like the flowers in springtime and the spring itself rejoices and is glad at the beauty that they wear ...*
>
> *When they are clothed in the glory from the fire and the shining colour thereof, then rather will appear their hidden glory, their sought-for beauty, being transformed to the divine state of fusion. For they are nourished in the fire and the embryo grows little by little nourished in its mother's womb, and when the appointed month approaches is not restrained from issuing forth. Such is the procedure of this worthy art.*

Cleopatra brilliantly combines poetry and chemistry to describe a chemical reaction that visually illustrates the descent into hell and a rebirth. Her recipe ingeniously protected the secret of "this worthy art" while clothing it in such a way that it could be openly demonstrated or taught to others without breaking Diocletian's ban against *chrysopœia* or revealing the final secret of confecting the Stone.

Sophic Mercury in its liquid form as *divine water* is then converted to a pristine white powder, followed by a fixation step resulting in its stabilized form – the Hermetic White Stone. The title of Zosimos' book on the same subject, *On the Evaporation of the Divine Water that Fixes Mercury*, is revealing enough. Cleopatra could have finished the Stone with little effort and away from prying eyes by secretly adding gold to the process. This technique could have been performed in private and protected as an operative teaching reserved for initiates or closed-door disciples.

It becomes clear that Cleopatra's operative technique involved a two-step process that parallels Zosimos description of the process – 1) a distillation step, and 2) a fixation step via evaporation or open distillation, which fixed the white powder to its most stable form. This process is beautifully illustrated in the *Codex Parsinus Græcus* with an image that includes Cleopatra's concentric circles, a distillation vessel and a fixing vase. This operative procedure is later described in exacting detail by Synesius.

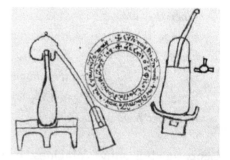

There is no evidence to prove whether Cleopatra's technique was a modification of Maria's *Ars Magna* or that it is an artifact of a separate Hermetic lineage. The commonality it shares with Maria's *Ars Magna* is that the recipe is a chemical reaction in a glass vessel, as opposed to a metallurgical process performed in a crucible. Maria's recipe calls for gold as part of the main reaction, whereas Cleopatra appears to have centered on her serpent of two compositions – dual Sophic Mercury – allowing for gold's inclusion after the fact. Cleopatra's innovation offers important advantages:

CHAPTER 4 – WHITE STONE OF HERMES

1. It was performed without gold and was therefore *technically legal* during the period of Diocletian's edict;
2. It presented the recipe as *a model of the* One at the heart of Hermetic, Gnostic and Neoplatonic philosophies;
3. It is *possible to create large amounts* of the One very inexpensively and store it for later use due to its stability;
4. It is *possible to clandestinely add gold* during or after confection in order to create the Tincture of the Philosophers;
5. It *can be taught openly while still protecting the secret* of how to ferment the Hermetic White Stone with gold to create the Tincture of the Philosophers.

The differences are that the precious metal appears to be omitted or downplayed, and the primary purpose is as a demonstrative model rather than an industrial technology, as evidenced by the *Dialogue of Cleopatra and the Philosophers*. The verbiage presented throughout the dialogue is bursting with overtones of Hermetic philosophy. There is evidence to suggest that Cleopatra's technique was preserved in a later recipe attributed to Hypatia's devoted disciple, Synesius. By the time Cleopatra's work was recorded, her technique for confecting the Stone served as a mystery model that obviously caused the audience of attending philosophers to rejoice at having been introduced to the mystery.

DIVINE WATER AND FIXED MERCURY OF ZOSIMUS'

Go to the waters of the Nile; there you will find a stone which has a spirit;

take it, cut it in two; put your hand in its interior and draw out the heart; because its soul is in its heart. – Zosumus

Zosimus recorded a text with commentary called *On the Evaporation of Divine Water that Fixes Mercury [Hermes]* that clarifies Cleopatra's work. It is a clear and concise recipe to create the White Stone via the traditional two-step process and complements the works recorded by Cleopatra and Synesius. In it he explains that *divine water* is one essence composed of two like natures and that it is neither metal, nor water and that it is destructive, meaning corrosive.

> This is the great and divine mystery object that one seeks. This is the All. It is whole, and it is All. Two natures, one essence, **as one draws one, and one dominates one. This is the silver water, the hermaphrodite …** which is attracted to its own elements. It is the divine water, that everyone ignored, whose nature is difficult to contemplate, for it is neither a metal or water always in motion, or a body, it is not dominated. …
>
> This is the All in all things, **its life and spirit** and it is destructive. One who understands this has the [alchemical] gold and silver.
>
> – *Zosimus*, Authentic Memoirs on Divine Water

He reveals that he found the technique in "the book of the Jews" next to a picture of a Tribikos, which suggests that it originated with Maria, and strengthens the case that she made a union of her two powders, *flowers of antimony* and sal ammoniac via distillation.

> Take the finely pulverised matter and grind it with Gum Elsaron, grind it finely and unite the two powders. – *Maria*

Zosimus begins with an overview then discusses the process of whitening the stone:

> If the flame is too high, the yellow product [occurs], but [it] does not serve you now, because you want to whiten the body [of the black stone].

> If the material is bleached in this way, not only on the surface, the metal will be completely white **and it will lose color through the fire** and will be white in the interior and the surface. … But if the composition [still] contains sulfur… add bitumen, so that thereby the All is desulfurized and become pure and bright.

> The operation took place in a single day.

CHAPTER 4 – WHITE STONE OF HERMES

Zosimus then concisely details the procedure as involving 1) distillation followed by, 2) fixation via drying.

> *I know a class that contains* two operations: one for the flow is generated by the extraction and the second for the moisture having dried leads to exhaustion. Because it [the product] will dry out and fix.

Zosimos clearly states that *divine water* is a distillate and implies that the reaction is quite forgiving with regard to ratios:

> *Do not take care of the weight, or fresh eggs [flowers of antimony], or their yolk [deliquesced sal ammoniac],* only ground together liquids and solids, as has been said before, and put them in the still. ...
>
> *The first water flowing is white. The second dripping has an unpleasant odor ... Then,* when the rise of the water has stopped, you remove the container in which the water has flowed ... and keep it carefully.

He then explains that the first flowing water is white, which he and later Stephanos both refer to by the cover-name *rainwater*. The residue left in the still is wetted again and redistilled twice more. The product is then warmed on sawdust, likely to liquefy the butter of antimony and remove any free ammonia. In the final instructions, Zosimos explains that the Sophic Mercury is transferred to a glass vase without touching it, this being the origin of the encryption *"the water that does not wet the hands"*:

> ... *extract it – the* divine water, you receive, not in your hand, *but in a glass vase.* Then, taking water, put it in the [open or evaporatory] vessel, *as it was written above, and let* heat for two or three days *[until dry]*. After removing [the contents], grind, and sun dry *on a shell*.

In this work, Zosimus is describing the process beginning with calcining stibnite to antimony oxide, followed by distilling divine water and its fixing via digestion / evaporation:

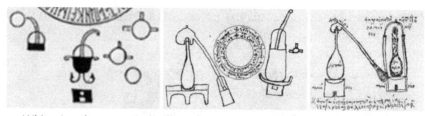

Whitening the stone → distilling divine water → fixing the Hermetic Stone

L   *flat-bottomed evaporation vessel (L), distillation vessel and stove (C), symbols for two distillations and one evaporation (R) – from an original version of* Chrysopœia *of Cleopatra*

M   *distillation vessel (L) with large stove, fixing vessel (R) with small stove underneath – from Zosimus*

R   *distillation vessel (L), fixing vessel (R), with similar stoves underneath each – from later copy of* Chrysopœia *of Cleopatra*

The commonality between Zosimus' work and Cleopatra's is a two-stage process beginning with the distillation of butter of antimony followed by a drying or fixing stage in a separate vessel that appears to remain sealed save for a tube to allow evaporation and reduce a buildup of pressure inside. From an operative perspective, distilling butter of antimony is followed by rectifying the distillate via subsequent distillation. The second step involves the addition of water or alternately allowing the *divine water* to absorb ambient moisture. In either case a reaction occurs followed by a color change in the clear translucent semi-liquid, which then is fixed to a stable dry white powder via evaporation over low heat, a process known as "whitening" yielding a product referred to as the Hermetic White Stone, Stone of the Philosophers or Great-Universal Stone. While the texts do not explicitly say, powdered *gold-antimony glass* (Latten) may be added to *divine water* before being fixed as Maria Hebrea suggests in her Ars Magna recipe. Alternately, powdered gold may be fused with the fixed White Stone in a crucible after the fact as suggested later in alchemical history by Synesius, Stephanos, Khunrath and many others. The addition of a gold product via either of the two methods suggested here was known as "yellowing", which transitions

CHAPTER 4 - WHITE STONE OF HERMES

to "reddening" yielding a product referred to as the Tincture of the Philosophers, al-Iksir (Elixir) or Lesser-Specific Stone. A work attributed to Synesius explains the process in precise detail.

SYNESIUS' HERMETIC WHITE STONE

Much of what is known about Hypatia and the turbulent times in which she lived is preserved in a corpus of texts and letters of one of her most devoted disciples, Synesius of Cyrene. Synesius was a Christian bishop, Neoplatonist, Gnostic and Hermeticist, yet he has never been considered as one of Alexandria's great philosophers, nor an accomplished alchemist. He was, however, a member of Alexandria's intellectual elite at the center of the dramatic clash between Paganism and Christianity. Synesius' main role with regard to alchemy appears to have been one of preservation and commentary.

Synesius was born to a wealthy family of Cyrene who claimed descent from the royal house of Sparta. His parents died early and Synesius inherited his father's library in his youth. His high social status provided Synesius the opportunity to travel to Alexandria, Constantinople and Greece, where he visited the Academy of Athens and received initiation into the Eleusinian mysteries. Sometime after 393 CE, he made his way to Alexandria where his letters reveal that he sought tutelage under Hypatia on the subjects of mathematics, astronomy and philosophy. In his comparison between Hypatia's school and the Academy of Athens, he found Hypatia's to be far superior. Hypatia introduced Synesius to a unique worldview through the lens of Neoplatonic philosophy. Alexandria introduced Synesius to the cosmopolitan lifestyle of residing in the most thriving cultural and commercial city of the period. A fine description is presented by historian W. S. Crawford in his comprehensive biography *Synesius the Hellene*:

> *The mob, among whom were representatives of all three religions, were notoriously hard to keep in order – idle, fickle, treacherous, bloodthirsty, needing to be bribed into good behaviour by gorgeous spectacles, every now and then, in the theatre – a form of amusement to which they were especially devoted. Numbers of monks, ignorant and fanatical, were constantly swarming into the city from the neighbouring deserts. Sailors from all parts of the Empire were ever putting in to the magnificent harbour, recompensing themselves, doubtless, for the many dangers and the long privations incident to their occupation, by a reckless pursuit of pleasure during their sojourn in the luxurious town. And the students attached to the university probably did their part in adding to the surprises of the place.*

Synesius, aged in his late 20s or early 30s, arrived in Alexandria and entered Hypatia's school where he and others were introduced to Hypatia's unique approach to science and philosophy. Hypatia's Neoplatonic philosophy was based upon the system of Plotinus and Porphyry, influenced primarily by north Indian immigrant Vedanta teacher Ammonius Saccas (Śākya Ammonius), known as "the God-taught" teacher residing and teaching in Alexandria during the 3rd century. This particular form of Neoplatonic philosophy aimed to elevate the mind above the drudgery of life through mystic contemplation into union with the *One* as higher truth and aspect of the unchanging eternal source of all things. Synesius would have been encouraged to seek an all-embracing realization or gnosis of the truth of his relationship between himself and the world structure, absolute law and knowing as it relates to the *One* unified reality. Hypatia's philosophy taught to think from a higher perspective – from the mind of God – in order to apprehend the essence of natural phenomena.

Although Hypatia's school and other Alexandrian philosophies were practical in a strange and wonderful way, they appeared too profound and held little appeal for the general populace. The collective public was looking for immediate consolation relevant to his or her personal life's

CHAPTER 4 - WHITE STONE OF HERMES

drama in the present. Christianity offered hope and faith by introducing vitality, and moral and ethical simplicity that answered the needs of the mainstream. Because of this, its influence grew steadily.

Synesius expressed in his letters the desire for a way to "become linked with the spheres, that is to say, be carried up as if to its own natural state of being" by individual effort, rather than by ritual or religion. He studied and admired teachings of mytho-legendary philosophers such as Amun, Zoroaster and Hermes, and was profoundly influenced by Hypatia. His philosophical interests reinforced the idea of a supreme truth and universal harmony. In 'Letter 154' to Hypatia, he derides rambling Christian preachers and corrupt sophist philosophers, finishing with a discussion of the intellect and the light of truth:

> ... only those receive the flashes of the emanations of the intellect, for whom in the full health of the mind's eye God kindles a light akin to his own, that light which is the cause of knowledge to the intellectual, and to knowable things the cause of their being known. In the same way, ordinary light connects sight with colour. But remove this light, and its power to discern is ineffective.

In 'Letter 15', Synesius provides a glimpse into his private life on his estate where he occupied his time:

> ... studying philosophy, mathematics, astronomy, everything; farming, hunting, having many a brush with hordes of pilfering Libyans; and every now and then upholding the cause of someone who had undeservedly fallen into difficulties ...

Synesius' home study in various subjects, and the fact that he kept a library, comes through in his letters. For a land-owning gentleman wielding power and influence in the Byzantine Empire with strong ties to Alexandria's intellectual inner circle, continually adding to his vast library would not only be considered natural, it would be expected. In 'Letter 154', he even complains of books having disappeared from his library. The profile of Synesius reveals an individualistic eclectic philosopher, an educated and intelligent upper class gentleman serving as a protecting influence in his community, who divided his time between his home in Cyrene and his country house near the Siwa Oasis. His letters also reveal that he was athletic and a willing soldier when called to protect his lands and people.

The main regional center of Christianity during Synesius' period was at the port of Ptolemais, just west of his estate. Synesius has been described as an honest man who was very well liked. He appeared to care little for self-aggrandizement, possessing a Christian heart and a philosopher's intellect. Perhaps due to his status and likeability rather than his beliefs, he was offered the position of bishop of Ptolemais, a powerful position and – considering the religio-political climate at the time – an offer he would have been unwise to refuse. Bearing in mind the heated religious climate, it is possible that he was offered the position as a means to protect him from persecution.

Synesius was married to a woman he appeared to greatly care for. As such, he indicated that he would take the position of bishop only if he could remain married. Ultimately, he was converted to the new faith, and baptized and consecrated by Theophilus of Alexandria around 410 CE. Synesius became bishop of Ptolemais without having gone through ordination. He hoped to find a balance between Christian and Neoplatonic thought, yet his teachings remained orthodox – although in his unique approach to Christianity, reason took precedent over scripture. Synesius always remained more fluent in the *Dialogues of Plato* than Christian gospel.

In 'Letter 10', Synesius reveals that his three sons had died, and 'Letter 16' is a desperate plea for Hypatia's reply. Scholars date his death at 53 years old, around 413 CE – a few years prior to Hypatia's tragic murder. There is no record as to fate of Synesius' library; however, scholars study

CHAPTER 4 - WHITE STONE OF HERMES

his personal writings and letters today. Perhaps his library was turned over to the Church, or secreted away by his fellow philosophers. Like Cleopatra, one of the most recognisable images of the *Ouroboros* is attributed to Synesius and found in a Paris codex of Greek alchemy. In the *Codex Marcianus Græcus*, a dialogue between Synesius and Dioscorus is preserved in which Synesius appears to initiate Dioscorus, Alexandrian priest of the Temple of Serapis, into the secret identity of Sophic Mercury. Dioscorus, like so many attempting to penetrate alchemical enigmas, appears to be confounded by the many cover-names used in alchemy. The dialogue begins with Dioscorus questioning Synesius regarding the oath that expressly forbids alchemists from clearly revealing ingredient identities. Synesius confirms that it is true, yet explains that the rule only relates to the uninitiated. Dioscorus continues by asking about the *Key* – the *divine water*, Philosophers' Mercury [Hermes]:

> *Synesius: Confirming it, he [Pseudo-Democritus] has added* the untouched water of sulphur, the [water] produced only from sulphur, the *divine water.*

It is tempting to interpret the above passage as relating to water made from or containing sulfur, but this is exactly the type of red herring and cryptic wordplay that alchemists as early as Pseudo-Democritus were specialized in. Aside from verbal gymnastics employed to conceal the identity of antimony, the cover-name *divine water* was used for more than one substance, historically. Therefore, a proper interpretation requires sensitivity to context. Purified and molten antimony is *divine water* that is separated from its sulfur matrix, and accordingly it was understood as *untouched* (purified) *water of sulfur* (as reduced from stibnite), *produced only from sulfur* (requiring nothing else in the process of reduction to *flowers of antimony* via calcination) resulting in molten *glass of antimony*. This is then converted to liquid *divine water* (antimony trichloride) by distillation.

Whitening the Stone was a necessary preparatory process regardless of whether Alexandrian alchemists were operating according to metallurgical or chemical technique. The process of witnessing a black product being purified to a white state, giving off noxious sulfur vapours until no unpleasant smell remains, and then melting into a translucent liquid that solidifies into a ruddy glass would have been viewed as divine and magical during antiquity.

> *Dioscorus: Your explanation is clear, O philosopher, but back to what he said: if [we are to understand it] in a general sense [lit. without qualification]: [water] made with lime [of stones; metallic white calx].*
>
> *Synesius: O Dioscuros, you do not pay attention:* **the** *lime is white,* **and the** *water made by it is the water made white and astringent by this very substance;* **sulphur, furthermore,** *when it is vaporised, has a whitening power. So, clearly he immediately adds sulphur vapour. He has made these questions plain, you know!*

An interesting aspect of the dialogue is that Synesius continues to use cover-names, never explicitly revealing the identity of the substances. He is pointing out the law of correspondences as it relates to color. Lime is trade-jargon for *lime of stones*, meaning calx of a sulfide ore post-calcination. The word lime also simply indicates a calx similar to the way fine gold powder is sometimes referred to as *gold-lime*. The white lime is *flowers of antimony*; astringent once converted to liquid *divine water*. Synesius hints rather clearly that when sulfur is vaporized the Stone is whitened, yet he incorrectly characterizes sulfur as having a whitening power or being a whitening agent that causes the antimony to become white. Here he is addressing the process of *whitening the Stone* to *flowers of antimony* and the sulfur vapour (sulfur dioxide; SO_2) byproduct of the reaction in accordance with operative technique found in the works of Maria, Cleopatra, Zosimus and others.

CHAPTER 4 - WHITE STONE OF HERMES

Synesius explains that there is more than one kind of mercury, describing it with identifiable cover-names, and finally reveals the identity towards the end of the dialogue when he explains clearly to Dioscorus that *"Here is the body of magnesia, the magnesia only, or that of Roman antimony."* During the dialogue, he tutors Dioscorus in the following manner:

(Synesius' still pictured)

> *... sharpen your mind on the text of his speech, and apply yourself (to understand) in what sense he said ...*
>
> *Sharpen your mind, Dioscorus, and watch out for terms used.*
>
> *... we should exercise our minds and thoughts.*
>
> *Listen to the talk ...*
>
> *O Dioscorus, you're not careful.*
>
> *Note the liberality of the author.*
>
> *Really, you understand, Dioscorus, but observe carefully the direction in which he said ...*

This is the language of a teacher providing instruction, or at the very least, an advanced operator assisting a novice, rather than simple commentary or a discussion between peers of equal standing. The most intriguing aspect of the dialogue is its conclusion where Synesius says:

> *With God's approval, I will begin my interpretation ...*

This implies that his efforts to initiate Dioscorus into the use of cover-names, the identity of Sophic Mercury and the process of *whitening the Stone* – information that would be foundational for a new initiate – were only the beginning. Unfortunately, the text ends and nothing is known of Synesius' complete interpretation or what subject he would go on to interpret. What is important is that most researchers agree that the fragment is an authentic account between Synesius and Dioscorus and date it to around 391 CE or earlier — around the time of the Theodosian decrees, the closing of non-Christian temples in Alexandria, and two years prior to Synesius becoming Hypatia's disciple. This implies that Synesius was familiar with alchemy relatively early in his career.

Synesius has typically been depicted as a speculative alchemist. The more likely scenario given his influences, character and social status, is that Synesius was acquainted with the Art and the Philosophers' Stone as a part of Alexandria's historical technology and as a cosmological / philosophical model; what in modern terms might be called a visual study aid. It would have been typical for Synesius to have alchemical works in his library, either in whole or as fragments. Furthermore, it would only be expected of a character such as Synesius to comment on alchemical matters. Both his *Dialogue with Dioscorus* and *The True Book of Synesius* focus upon white powdered Sophic Mercury as the White Stone of Hermes, thus indicating that Synesius, if not an actual adept, was very well versed in Alexandrian alchemy.

The True Book of Synesius

That Synesius was familiar with substances and processes vital to Alexandria's gold making tradition is evidenced by his commentaries upon specific aspects of the Philosophers' Stone. He detailed the thriving Libyan sal ammoniac trade with Egypt, indicating that he was more than a little acquainted with the key substance that made Alexandrian alchemy possible. His country house was located near the Siwa Oasis, the first stop on the trans-Saharan caravan route where the sal ammoniac trade originated. In a letter, Synesius comments upon Pseudo-Democritus in a dialogue with a priest of Serapis named Dioscorus, in a series of questions and answers. In addition, Synesius is one of 40 alchemists appearing in a Byzantine alchemical corpus of Alexandrian alchemy called the *Codex Marcianus Græcus* compiled sometime between the 7th and 11th centuries in Constantinople, and later in the *Codex Parsinus Græcus* of the 13th century.

The two most intriguing works attributed to Synesius – *The True Book of the Learned Synesius, a Greek Abbot, Taken out of the Emperor's Library* and the *Epilogue According to Hermes* – appeared first in French in 1612, followed by a Latin version published in Amsterdam in 1671 and an English version published in London during 1678. Even a cursory glance at these works reveals that Synesius could not have authored them as they appear in their final forms. Therefore, the primary question of Synesius' authorship is already answered. A more appropriate question would be whether there is any evidence to indicate that the chemistry described in the texts reflects genuine Alexandrian technique, and whether the work may have been known to Synesius and/or in his possession at some point.

As previously discussed, it would be natural for Synesius to possess alchemical works in his library, either in whole or as fragments, and possibly with commentary. As such, researchers must concede the possibility that these texts could ultimately have ended up in a royal library somewhere in Byzantium, only to be rediscovered and adapted by a European editor centuries later. If this hypothesis is correct, it should be possible to distinguish parallel features between the chemistry presented in *The True Book of Synesius* and the chemistry in the recipes of Maria, Cleopatra and Zosimus.

The interpolation theory concerning *The True Book of Synesius* was discussed at length by Israel Regardie, Jewish occultist and author of books on, among numerous subjects, the *Hermetic Order of the Golden Dawn*. The following excerpts from Regardie's introduction to *The True Book of Synesius* reveal that Regardie did not consider Synesius to be a mystic nor a genuine alchemical author:

> *Synesius was not in reality a mystic,* **even although a strain of mysticism peers out here and there from his writing.** *Historically he appears to us rather as a very positive extraverted character.*
>
> *The nature of his very few extant writings ... depict him as of another nature than a writer of alchemy.*

Regardie then presents his interpolation theory for Synesius' works:

> *It is not at all impossible that in Alexandria of Synesius' day were alchemists and alchemical writers.* **It is not even impossible, though this we do not definitely know, that he himself had met them and had been influenced by them.** **In fact,** *it is highly probable that he did.*

Synesius was a powerful public figure with strong ties to both the highest level of Alexandrian Christian clergy and Hypatia's circle of intellectual and political influence. He held social status, authority and a position of privilege higher than most, and would have enjoyed access to information on any form of history, philosophy, science, or technology available at the time. Alchemy would have been known at the time as the Art or historical *chrysopœia* and thus considered a past technology with philosophical or spiritual overtones. Regardie also addresses the probability of European interpolation:

> *It may well have been that some over-zealous owner of the text in, let us say, the fourteenth century, interpolated quotations from alchemical writings of his day* **in order to buttress up his own belief in the art. In so doing, he failed to recognise that** *he had almost completely ruined an authentic text of an earlier age.*

Upon reading the text in the light of Regardie's suggestion that the text has been adapted by a European editor, the following becomes apparent:

- *Geber is mentioned three times;*
- *The word* alembick *is mentioned in the* Epilogue;
- *Water of the Sun is the operative technique for adding gold.*

Synesius lived during the 4th century. Geber is a reference to either Jābir ibn Hayyān, an alchemist of the Islamic tradition from the late 8th century CE, or Pseudo-Geber, who is generally Paul of Taranto, a Franciscan alchemist operating during the 13th century CE. The word *alembic* – spelled *alembick* in the text – has its roots in the Islamic alchemical tradition. *Water of the Sun* is perhaps a reference to the use of aqua regia, a combination acid that dissolves gold via corrosion. Once the gold is fully dissolved into a liquid solution, it is known by a number of cover-names including *water of the Sun*. Since aqua regia was discovered during either the 8th or the 9th century at the earliest, this operative detail cannot be positively connected to Synesius or any Alexandrian recipe unless *Water of the Sun* is used by Synesius to indicate the red *oil of gold* that occurs when fine gold calx is dissolved in butter of antimony.

Despite the above considerations, many compelling details in *The True Book of Synesius* fully accord with Alexandrian methods. Upon extracting the operative details and scrutinising the chemistry, the recipe appears to be a reaction between sal ammoniac and *flowers of antimony*, resulting in a stable dry white powder that is easily identified as antimony oxychloride. This is the chemical identity of the Hermetic White Stone, which came to be known in European alchemy as the *Stone of the Philosophers* by Paracelsus, the *Great* or *Universal Stone* by Khunrath and later as *Algarot Powder*. The use of two powders to create a chemical reaction in a glass digestion vessel without any byproduct or waste decanting required is in accord with Alexandrian technique. The *Epilogue* hints at its use as a philosophical visual aid. Cleopatra's fragments appear to be Hermetic in origin and deal primarily with sal ammoniac and *flowers of antimony*. The recipe in *The True Book of Synesius* is nearly identical in theory to the Cleopatra fragments and Zosimus' texts on *divine water* and fixed Mercury / Hermes as interpreted in this treatment.

Regardie's interpolation / rescension theory is plausible on many accounts. The excerpts below, taken from *The True Book of Synesius*, exemplify operative details that reflect Alexandrian techniques. Analysis of the chemistry reveals a simple and efficient approach to confecting the White Stone of Hermes. Synesius inherited his father's already vast library that included at least one known alchemical text and likely many more. He had access to the most important works available in Alexandria through Hypatia and others. He is known to have commented upon alchemical substances and processes. He is listed as an alchemist in two of the most important alchemical corpuses known to history, the Byzantine *Codex Marcianus Græcus* manuscript and later in the 13th century *Parsinus Græcus* manuscript, which features an illustration known as *figure 43, The Still of Synesius*. It is not a great leap to consider that upon his death, the contents of Synesius' library – including alchemical texts, fragments or commentaries – may have ultimately ended up in a royal library belonging to a Byzantine emperor such as Marcian or Heraclius. The literary history may remain unresolved, but what is convincing is that the chemistry presented in *The True Book of Synesius*, as interpreted here, strongly suggests an Alexandrian origin. A note on the term Mercury that follows: Mercury was Roman for the Greek god Hermes and the two names are synonymous. Synesius lived in Roman Alexandria and understandably, he used the Roman variant of Hermes. When he describes or discusses White Mercury, this can also be understood to mean White Hermes, etc. hence the product described by Cleopatra, Zosimus and Synesius in this treatment is interpreted as the White Stone of Hermes.

Prologue

> *Avoid then that which is mixt, and take the simple,* **for that proceeds from the Quintessence. Note that we have** *two bodies of very great perfection, full of Mercury: Out of these extract thy Mercury,* **and of that thou shalt make the Medicine, called by some the Quintessence ...**

The 'simple' version of an unmixed *prime material* is either antimony regulus or *flowers of antimony*. Since there is nothing in the recipe to indicate antimony regulus, this argues strongly for *flowers of antimony*, which can aptly be considered a body of perfection full of Sophic Mercury, i.e. antimony. The other perfect mercurial body is sal ammoniac. In the correct ratio, these two ingredients will create 'living doubled mercury' in the form of a stable and brilliant white powder.

> *Decoct it gently by little and little, until it have changed the false colour into a perfect,* and *have a great care at the beginning that thou burn not its Flowers and vivacity, and make not too much hast to come to an end of thy work. Shut thy Vessel well,* that what is within may not breathe out, and so thou mayst bring it to some effect. And note, that *to dissolve, to calcine, to tinge, to whiten, to renew, to bath, to wash, to coagulate, to imbibe, to decoct, to fix, to grind, to dry and to distil, are all one, and signify no more than to concoct Nature, until such time as it be perfect.*

The reference to *Flowers* (capitalized in the original text) in the above excerpt is more than coincidental. It is a cover-name for one of the ingredients, that being *flowers of antimony*. The *false color* is the white color of the *flowers of antimony*, which will ultimately become the perfect brilliant white of the White Stone of Hermes.

> *... to extract the soul, or the spirit, or the body, is nothing else than the above said Calcinations, in regard* they signify the operation of Venus.

The *regimen of Venus* is a term later used by European alchemists to denote the process of fixing and perfecting the White Stone.

> **And so** *that which is below, is like that which is above, and consequently there are made therein two luminaries, the one fixt the other not, whereof the fix'd remains below, and the volatile above ...*

This Hermetic passage matches Synesius' personal dual-soul theory and parallels the *Dialogue of Cleopatra and the Philosophers* where the philosophers ask Cleopatra to explain the high and low aspects of the chemical reaction. The cover-name *two luminaries* parallels Cleopatra's moon and star symbolism associated with the lunar, feminine and mercurial aspects. The *fix'd* lunar chemical is *flowers of antimony* and the volatile is sal ammoniac. The words "the volatile above" are in reference to vapors ascending in the vessel during distillation.

> *And as in the beginning, there was only one, so in this Matter, all proceeds from one and returns to one,* which is called a conversion of the Elements, and to convert the Elements, is as much as to *make the humid dry and the volatile fixed ...*

This conveys the same meaning as the Hermetic quote found in the Cleopatra diagram. In this context, the *One* gives birth to crude *prime matter* – antimony ore (stibnite; Sb_2S_3; mśdmt / kahâl / ḳoḥl) – that returns to the *One* as spiritualized *prime matter*. Sal ammoniac is not considered a 'two' but rather a spiritual agent that affects change, causing union between the fixed and volatile substances inside the vessel. The result is philosophical doubled-mercury often likened to the *dual-soul* or soul-spirit unity. Mercury in its 1st (liquid) state is *philosophical* or *sophic mercury* and when fixed to its 2nd (powder) state is *living doubled sophic mercury*. For this reason, *sophic mercury* is sometimes perceived in European alchemy as synonymous with *mercurius duplicatus*; doubled-mercury often symbolized by two eagles, birds of Hermes or dual-serpents. The Prologue is essentially a synopsis of the synthesis to follow.

Practice

> *... work with the Mercury of the Philosophers and the wise, which is not the Vulgar, nor hath anything of the Vulgar, but according to them is the first Matter ...*

CHAPTER 4 - WHITE STONE OF HERMES

Alexandrian alchemy in any form always makes it perfectly clear that *Mercury of the Philosophers* has nothing to do with elemental mercury or quicksilver, which is referred to as *vulgar mercury*. Within the Alexandrian tradition, alchemists never deviated from this operative detail. Antimony in its molten state is identical in appearance and behaviour to quicksilver.

> ... *one only Stone [antimony; Mercury of the Philosophers]* **wherein consists** the whole Magistery; **to which you shall** not add any strange thing, **save that in the preparation thereof you shall** take away from it whatsoever is superfluous, **by reason that** in this matter, all things requisite to this Art are contained ...

This is a very intriguing excerpt with strong implications. The *Key* to the entire art, according to this alchemical tradition, is initially to create the Hermetic White Stone / Philosophical Mercury [Hermes]. This is not hollow or prideful boasting, but rather is presented in fact.

The First Operation: Sublimation [Exaltation]

> *It is not Vulgar but* Philosophical whereby we take away from the Stone whatever is superfluous, **which in effect is nothing else, but the elevation of the** not-fixed part by fume and vapour, **for** the fixed part should remain in the bottom, **nor would we that one should be separated from the other, but** that they remain and be fixed together.

> **Know also that he,** who shall sublime our Philosophical Mercury as it ought to be done, shall perfect the Magistery.

The author is describing an extremely simple and efficient chemical reaction, which he presents in clear and concise language. The primary message is that knowledge of how to perform this reaction is the *Key* to the entire art. Distillation was an operative essential for Synesius as indicated by an illustration of his still in the *Codex Parsinus Græcus*. Following the sublimation process of whitening the stone as a preparatory step, distillation was the primary operative technique in the First Operation. Here the term sublimation is perhaps a copyist's interpretation of the Greek *exýpsosi* (εξύψωση), which means sublimation but also *exaltation*. The term may have served as a double-entendre to indicate the initial *sublimation* and subsequent *distillation*.

> *Hence also is it, that* in our Stone there are but two formal Elements, that is to say, Earth and Water; *but the* Earth hath in its grossness, the virtue and drought of Fire; *and the* Water contains in itself the air **with its humidity.** Thus we have in our Stone visibly but two elements, but effectually there are four.

Fiery Earth = *flowers-of-antimony*, associated with fire due to its ability to dissolve gold so easily. Airy Water = deliquesced sal ammoniac. Neither the exact ratio of ingredients nor the sublimation (i.e. distillation) temperature is given. This is not necessarily a method of hiding a secret. One may assume that an ancient operator reading this recipe would have understood what to do.

> *Flowers (earth-fire) + Eagle (water-air) = Sophic Hermes* (Tetrasomia)

The Second Operation: Whitening

> *It* [this operation] *converts our* [liquid Sophic] *Mercury into the white Stone, and that* by decoction only. When the earth is [being] separated from its water, then must the Vessel be set on the Ashes ...

This is an example of a clear indicator – water is to be evaporated – as a marker showing a change in operations.

> *Then take the earth which though shalt have reserved* in a Vessel of Glass, with its distilled water, and with a soft and gentle fire, *such as was that of Distillation, or purification, or rather one somewhat stronger,* continue it, till such time as the earth be dry and white, *and by reason of its drought, drunk up all its water.*

CHAPTER 4 - WHITE STONE OF HERMES

The above refers to hydrolysis and evaporation. Butter of antimony is hydrolysed by the addition of water. The water is distilled leaving crude antimony oxychloride remaining. The distillate is re-added to the powder and then evaporated. The visual indicator signifying the end of the reaction is a dry white powder with no visible liquid in the vessel.

> *And note that* the said earth will be washed from its blackness [foulness] by the decoction, as I have said, because it is easily putrefied by its own water, and *is cleansed, which is the end of the Magistery,* and then be sure to keep that white earth very carefully. *For that is the White Mercury, White Magnesia, Foliated Earth.*

The key operative detail here is that the stone "is cleansed, which is the end of the Magistery". White Mercury was washed with pure water and then dried via either distillation or evaporation, following which the product is gently dehydrated to fix the powder to its most stable form. In the language of Alexandrian *chrysopœia*, this product was known as the *White Mercury, White Magnesia, Foliated Earth*, or the Hermetic White Stone. Cleopatra called it the *One* and the *serpent of two compositions*. European alchemists would later it *Stone of the Philosophers, Great* or *Universal Stone, Mercurius Duplicatus* (Doubled Mercury), *Mercurius Vitae* (Mercury of Life) or, later in early-modern Europe, *Algarot Powder* (after the Italian who commodified it, Vittorio Algarotti – the chemical identity of which is antimony oxychloride (SbOCl; $Sb_4O_5Cl_2$).

> *White Hermes =* the *One,* the *All,* the *Serpent with its Venom*

The Third Operation: Reddening

> *Take of the white Medicine,* as much as though wilt, and put it in its Glass upon the hot ashes, till it becomes as dry as the ashes. Then *put to it some of the Water of the Sun ... and* continue the fire to the second degree, *until it become dry ...* until the matter be rubified, and fluxible as wax ...

This technique would likely have been a highly guarded secret immediately following Diocletian's criminalization of gold making. It is the final procedure to create a Tincture. The above passage originally called for *water of the Sun*, yet this may either reflect interpolation by the European editor, or refer to Maria's instructions to combine gold and antimony *until they flow like water*, a plausible explanation for the cover-term *water of the sun*. Alexandrians would have used gold calx to perform this operation. This after-the-fact fusion of the Hermetic White Stone with a precious metal would have been performed via fusion in a crucible. During the 6th century, Stephanos described this process as the White Stone "perfecting" gold or other metals. Heinrich Khunrath referred to the Hermetic White Stone as the *Great Stone* or *Universal Stone* because it could be united with any metal – gold, silver, copper, iron, tin or zinc – via fusion. He differentiated the *Great* or *Universal Stone* from the type that requires a specific metal integral to the recipe such as is the case with Maria's Tincture. He referred to the latter as the *Lesser* or *Specific Stone* for this reason. Paracelsus also made the distinction between the two referring to the Hermetic White Stone as *Stone of the Philosophers* and to Maria's Tincture as *Tincture of the Philosophers*.

Two temperature indicators are presented in the excerpt above, yet the details are not provided. The first is identified by the simple description "hot ashes" and the second is fire of "the second degree." Once the contents are dry and the powder begins to change color, the fire is increased. This causes a fusion between the White Stone and the gold calx, resulting in a fermented, finished or refined Philosophers' Stone known throughout history:

Synesius	Tetrasomia + Gold	= Quintessence
Stephanos	White Magnesia + Gold	= Coral Gold
Paracelsus	Stone of Philosophers + Gold	= Tincture of Philosophers
Khunrath	Universal Stone + Gold	= Specific Stone

CHAPTER 4 – WHITE STONE OF HERMES

The European interpolation of the section of the text regarding *projection* contains very little that can be correlated to Alexandrian traditions. The section of the book titled 'Projection' describes the European technique of *multiplication* or *fermentation*, whereby the Philosophers' Stone is increased in potency or volume. Due to its overall European character, discussion of this section of *The True Book of Synesius* has been omitted from this discussion.

Epilogue According to Hermes

> *And when he said,* it ascends from the earth up into Heaven and returns again into the earth, *there is no more to be understood by it then the Sublimation of the Bodies.*

> ... *he says* the Wind carries it in its belly, *that is* ... *where* it first ascends by a wind full of Fume and Vapour, and afterwards returns to the bottom of the Vessel in water again.

Synesius give the last word to Hermes. These descriptions suggest that an earth becomes spiritualized, ascending as air, descends as liquid and is fixed once again as an earth. The cover-term *Wind carries it in its belly* is taken directly from the Emerald Tablet of Hermes in reference to distillation. Its intermediate stage in liquid form alluded to below would later come to be known as *pharmakon* and *eudica* in the late-Alexandrian tradition, *al-koḥl* in Islamic alchemy and *vinegre, sophic mercury* or *butter of antimony* in Europe.

> ... *he says, Its force is absolute, if it be turned into earth, that is to say, be converted by decoction. And to make* a general demonstration of all hath been said, *he says, It shall receive both the inferior and superior force, that is to say, that of* the Elements, *for as much as, if the Medicine receive the force of the lighter parts, that is to say,* air and fire, *it shall also receive that of the more grave and weighty parts, changing itself into* water and earth, *to the end, that* the Matters being thus perpetually

205

joined together, may have permanence, durance, constancy, and stability.

In the passage above, the line "a general demonstration of all that hath been said" can be interpreted as a visually captivating chemical reaction; a demonstrative model of a spoken Hermetic, Gnostic or Neoplatonic philosophy.

Template for **Synesius' White Stone of Hermes**:

White Serpent + Eagle of Hermes = White Stone of Hermes
White Stone + Gold Calx = Tincture

1. **Exaltation**: calcine stibnite (Sb_2S_3) to *flowers of antimony*, and then distil the *flowers* with sal ammoniac (NH_4Cl) to achieve butter of antimony (SbOCl);
2. **Whitening**: hydrolysis of *butter of antimony* to create the *dual-soul*, the *One,* the serpent, the White Stone, *Mercurius Vitæ* ($Sb_4O_5Cl_2$);
3. **Reddening**: a reaction between the White Stone (*dual-soul*; $Sb_4O_5Cl_2$) and the Sun (body; Au), referred to by alchemical jargon as *fermenting*, or joining *dual-soul* with body to create the Tincture of the Philosophers.

Synesius' Philosophical Model

For Synesius, the Hermetic White Stone was far removed from industrial *aurifaction / chrysopœia* of Maria's tradition. Synesius may not have been an original thinker, but he was a genuine philosopher in Hypatia's tradition and that qualified him as a man of heart and mind apart from the mob. As an earnest student of science and philosophy, he naturally would have approached alchemy from this perspective. His commentaries are at once practical yet approaching mystical.

Synesius' work *The Egyptian Tale* 1:1 presents his view that the *dual-soul* is derived from two different sources, both of which "flow from the same fountain" – a demonstrated familiarity with Egyptian Ka-Ba *dual-soul* theory and the Chaldean Oracles. One part, he says, comes from above; the other from below. These two parts must be rejoined and reabsorbed into their primal source. In Synesius' model, the soul and spirit arise independently from an already divided heaven above and Earth below. His reference to "flow from the same fountain" indicates that although they are separated, this was not always the case. Accordingly, the two must be reunited and return to their singular source.

> *Now the relationship of souls and that of bodies is not one thing and the same. It does not belong to souls to be born on earth from the same parents, but rather to flow from the same fountain.* **And** *the nature of the universe furnishes two kinds, one luminous, and the other indistinct, this last gushing forth from the ground, since* its source is somewhere below, *and leaping out of the earth's cavities, if perchance it might so compel the divine law. But* the former is suspended from the back of the Heavens; for it is indeed sent down *that it may order the earthly lot ...*

Synesius later describes reconciliation between the two souls into a unity that returns to "its own source":

> ... should flow back again **by way of the selfsame road,** and be commingled with its own source ...

In Synesius' theory, the *dual-soul* apart from the body, arises from above and below and must rejoin in order to return to the source, the *One*. The entire recipe presented in *The True Book of Synesius* is a very precise model, in the form of a chemical process, embodying Synesius' *dual-soul* views. If *The True Book of Synesius* did not originate from his works, he likely would have been pleasantly surprised to find such a perfect working model for his *dual-soul* theory. The question remains whether Synesius would have viewed the confection of the Stone as a valid philosophical or religious model. W.S. Crawford's biographical characterization of Synesius on such matters points to the affirmative:

> *He [Synesius] is not content to take physical phenomena as illustrations, or even as hints, of theological verities.* He employs them as actual proofs.

The two substances, sal ammoniac and *flowers of antimony* represent energetic rather than material aspects. Together they comprise a unified feminine or lunar *dual-soul* that can later be conjoined with the perfect male or solar body of gold.

> soul (*flowers of antimony*) + spirit (*sal ammoniac*)
> = dual-soul (*doubled living mercury*)
>
> dual-soul (*living mercury*) + body (*gold calx*)
> = perfect triune being (*soul, spirit and body unified*)

An intriguing and unique aspect of *The True Book of Synesius* is the role that gold plays. In this recipe as with Cleopatra's, it is an afterthought. Primacy of place is given to the united ethereal soul-spirit principle as Hermes / Mercury. Gold as the corporeal body is downplayed to such a degree that it appears almost irrelevant and value is placed primarily upon the ability to confect the White Stone of Hermes – *Philosophical Dual Mercury of Life*. Did *The True Book of Synesius* originate from Synesius himself? The answer remains unresolved, but the core operative technique and quasi-religious or philosophical nature of the work is in accordance with Synesius' character and the Hermetic, Gnostic and Neoplatonic philosophical climate to which it is attributed.

It is possible that Synesius' was simply providing commentary upon a copy of Cleopatra or Zosimus' recipes. There is no arguing the fact that a tremendous amount of literature in general, and Alexandrian *chrysopœian* literature in particular, has been lost to history long ago, especially following Diocletian's bonfire and continued Christian destruction of non-Christian property.

CHAPTER 4 – WHITE STONE OF HERMES

Alexandrian *chrysopœia* with its gold-based *Tincture of Philosophers* enjoyed its golden years prior to Diocletian's edict. However, this period came to an abrupt end, which forced Alexandrian alchemy and the Philosophers' Stone to evolve by shifting the focus on its role in order to survive. The period between Diocletian's edict and Hypatia's death represents a short transition period where the Art found new purpose, with the Hermetic White Stone serving as a cosmological / philosophical model shielded beneath a layer of symbolism. Yet shielding the beauty or even the truth of the industrial origins of the Art from unworthy eyes that might be blinded by its radiance is in keeping with Synesius' personal views:

> *A philosophical mind, being a watcher of the True, yields to the necessity of falsehood; for light is like truth, and the eye like the populace. In the same way, then, as an eye would be injured by enjoying light to excess, and in the same way as darkness is more serviceable to those who are afflicted with ophthalmia, so, I consider, falsehood also is of service to the populace, and the truth is hurtful to those who have not strength to gaze upon things as they really are.*

Strong parallels emerge upon comparing the recipe attributed to Synesius' works with Cleopatra's fragments and Zosimus' text.

- **Philosophical model:** In these works, strong parallels are made between Hermetic notions and the reaction witnessed in a glass vessel. Concepts such as heaven and Earth, above and below, purification and *dual-soul* theory are interwoven within these works and are clearly accepted as demonstrating the relationship between psychospiritual notions of the *One* and the material reality of the chemical reaction according to natural order. This correspondence would have been understood to represent truth, in that it relates and accurately describes both the mental and physical reality addressed. Emphasis rests squarely upon the soul-spirit principle leaving the body principle – gold or silver – as an afterthought.

- **Dual nature:** Cleopatra and Zosimus independently identify their product as the *One* and describe it as a serpent with a poison according to two compositions, with the power to accomplish the *All*. Synesius explains that only one Stone – the White Stone – *consists of the whole magistery,* is comprised of *only two formal elements* and is the end of the *end of the magistery*. It's dual nature is soul-spirit representing pure energy devoid of bodily incarnation and it was for this reason that it was valued over the Tincture by Alexandrian philosophers, who would have associated incarnation via the body of gold as a declension away from the purely energetic Light, Goodness and energetic source of material manifestation.

- **Sophic Mercury / Lead** – In the Cleopatra fragments, the symbol for Sophic Mercury / Lead is larger than the other two symbols and occupies the center of the diagram. The serpent is a cover-name for Sophic Mercury / Lead in various stages and is a dual composition. In its beginning stages, Cleopatra's serpent is black, but in its refined stage white. Likewise, *The True Book of Synesius* places emphasis upon the creation of a white Sophic Mercury of a dual nature from black stibnite. Zosimus' work is entirely devoted to the distillation of *divine water* that fixes Mercury (Hermes).

- **Operative details**: These recipes indicate a 2-stage chemical reaction based on distilling a liquid followed by fixing it to a powder. Based upon Maria's precedent and Synesius' clear instructions, it is evident that the reaction is between sal ammoniac and *flowers of antimony*. Three candidates emerge from this description: *butter of antimony* ($SbCl_3$), *liquid antimony menstruum* ($SbCl_5$) or *Algarot Powder* ($SbOCl$). Cleopatra describes her product as having been "given body in a single being," which may indicate a solid, yet she remains unclear on the issue. Synesius makes it clear that his unfixed product is a *dry white earth*, which allows it to be identified with a relatively high degree of certainty as *antimony oxychloride* ($SbOCl$). The fixing instructions suggest that the fixed White Stone is *antimony oxychloride* ($Sb_4O_5Cl_2$). In these recipes, gold and silver remain an afterthought. Cleopatra's images and Zosimus' instructions suggest distillation and fixation, a joint method of operation crucial to Zosimus and Synesius' recipes and a commonality of all three.

The original Alexandrian technique preserved in the works of Cleopatra, Zosimus and those attributed to Synesius, went on to influence Islamic and European alchemical traditions in a number of ways. Paracelsus referred to this product as the *Stone of the Philosophers*, whereas Heinrich Khunrath referred to it as the *Great* or *Universal Stone* as being superior to the Tincture, which he called the *Lesser* or *Specific Stone*, and *Tincture of the Philosophers* by Paracelsus. This particular Hermetic version of the Philosophers' Stone is a strong candidate for the process and product preserved in the Emerald Tablet of Hermes. If this proves to be the case, the implications are that a Græco-Egyptian tradition existed in tandem with or slightly following Maria Hebrea's Judeo-Egyptian school in Alexandria – a hypothesis requiring additional research in order to solidify. Any recipe that calls for omitting the gold in the primary stages, only to be added later – via a fermentation step with either gold calx or a liquid gold solution after achieving the White Stone – is likely a descendant of the archetypal recipe for the Hermetic White Stone. The recipe to confect it is preserved collectively in the *Chrysopœia of Cleopatra*, Zosimus' *Divine Water to Fix Mercury [Hermes]* and *The True Book of Synesius*.

The End of Classical Antiquity
It was not only Hellenistic Pagan religions that were considered Pagan but also Hermetic, Gnostic and Neoplatonic philosophers, their ideologies and literature. As such, they were subject to the same ongoing Christian persecution that Hellenistic religions suffered. The Philosophers' Stone, and what it came to represent was undeniably an artifact of non-Christian

traditions, and as such simply could not survive given the religious persecution of the period. Emperor Theodosius began a severe and cruel campaign to stamp out Paganism in the Roman Empire through a series of decrees known as the Theodosian Decrees. These forced non-Christian ritual observances to be rendered into Christian variations. One of his more damaging decrees, delivered in 391 CE, stated that:

> ... *no one is to go to the sanctuaries, walk through the temples, or raise his eyes to statues created by the labour of man.*
> – *decree 'Nemo se hostiis polluat',* Codex Theodosianus *xvi.10.10*

As a result, non-Christian temples were abandoned and demolished; their foundations utilized to construct Christian churches. The Serapæum was abandoned and later destroyed, while the Cæsareum was converted to the Cathedral of Alexandria. The Musæum, viewed as a university, was spared until the Christian riots of 415 CE finally shut it down. A glimpse at Christian persecution documented during the Roman Empire during Synesius and Hypatia's era puts this into perspective.

- ***291 CE**: Diocletian orders all works addressing chrysopœia destroyed*
- ***300**: Zosimos collects, catalogues and preserves remaining pre-Diocletian chrysopœia literature and adds his own*
- ***350**: Constantius II – non-Christian temples were closed and new laws prescribed the death penalty for performing or attending non-Christian sacrifices and the worship of non-Christian idols, accompanied by Christian laity destroying, pillaging, desecrating, or otherwise vandalising ancient non-Christian temples, monuments and tombs*
- ***361-363**: Julian the Apostate – an attempt at restoration*
- ***363-375**: relative tolerance for non-Christians*
- ***381**: Theodosius I – reiterated ban on non-Christians sacrifice, prohibited divination with the death penalty, prosecuted local magistrates who failed to enforce anti-Pagan laws, dissolved non-Christian associations and destroyed temples*

- **382**: *Gratian – colleges of non-Christian priests lost all privileges and immunities and all temples, shrines and revenues were confiscated and added to the Royal Treasury*
- **389-391**: *Theodosius I – Theodosian Decrees established a practical ban on non-Christians, forbade visitation to non-Christian temples, non-Christian holidays were abolished*
- **391**: *Bishop Theophilus – closed non-Christian temples in Alexandria*
- **392** *onward: Theodosius I – closed the Eleusinian Mysteries, unified the empire and passed a comprehensive law forbidding the practice of any non-Christian ritual within the privacy of one's home, non-Christian ideology in any form was declared an illicit religion*
- **393**: *Synesius became Hypatia's disciple*
- **395**: *Arcadius – reiterated the ban against non-Christian holidays, sacrifices and visitation to a non-Christian temple or sanctuary*
- **396**: *Arcadius – officially revoked the privileges of non-Christian priests and clerics, ordered remaining countryside non-Christian sanctuaries destroyed*
- **408**: *Honorius – renewed laws ordering all statues and altars removed and all non-Christian property appropriated by the government, power to enforce anti-pagan laws was given to bishops, two new laws decreed that all property belonging to non-Christians and heretics be appropriated by the Church, a new law was enacted that prohibited non-Catholics from performing imperial service within the palace*
- **409**: *Honorius – enacted a law that punished judges and officials who do not enforce anti-Pagan laws*
- **410**: *Synesius Consecrated as bishop of Ptolemais*
- **412**: *Death of Synesius*
- **412-415**: *Cyril, Christian Patriarch of Alexandria, tolerates Hypatia's school*
- **415**: *Honorius – enacted a law to appropriate all non-Christian temples throughout the empire to the government, all non-Christian objects were removed from public places*
- **415**: *Death of Hypatia*
- **416**: *Honorius and Theodosius II – edict that non-Christians could no longer be admitted to imperial service or receive rank as administrator or judge*

- *423*: Theodosius II – reiterated previous anti-Pagan laws, declared confiscation of all property and exile for anyone caught performing an ancient rite
- *425*: Theodosius II – ordered the rooting out of all non-Christian superstition, disqualified non-Christians from serving as soldiers or pleading a court case
- *426*: Theodosius II – enacted laws against Christian conversion to non-Christian ideology and against false Pagan conversions to Christianity
- *435*: Theodosius II – reiterated laws against non-Christian rites or sacrifices enforced by death penalty, all existing non-Christian shrines, temples and sanctuaries were ordered destroyed, magistrates faced death penalty for non-compliance
- *438*: Theodosius II – amended previous 423 anti-Pagan law, confiscation of property and death penalty
- *451*: Marcian – decreed death penalty and confiscation of property for performing non-Christian rites, prohibited any attempt to re-open a non-Christian temple, initiated a fine of 50 pounds of gold for non-compliance by any judge, governor or official
- *472*: Leo I – imposed severe penalties for property owners who allowed non-Christian rites on their premises
- *476-491*: Zeno – instituted harsh persecution of non-Christian intellectuals
- *491*: Anastasius – forced to sign a written declaration of orthodoxy prior to his crowning, *Christianization of the Roman Empire completed*

With Diocletian's criminalization of *chrysopœia*, the Art was transformed and adapted from Maria's industrial chemical technology into Zosimos' alchemical literary anthology, with esoteric associations that later became an applied model of cosmological and philosophical ideas by Synesius and others. The tidal wave of Christian reform throughout the 5th century, combined with the deaths of Synesius and Hypatia, marked the end of classical antiquity. *Chrysopœia* as a practical chemical technology and industrial application of the Philosophers' Stone ended a century prior. Zosimos attempted to rescue the Noble Art from its death throes, highlighting its importance as parallel spiritual processes, yet this would not eventuate and the legacy of the Art became ever more

CHAPTER 4 - WHITE STONE OF HERMES

abstract as time went on. From the perspective of operative technique for creating the Tincture, all innovations had been achieved prior to Zosimos. If the recipe attributed to Synesius as interpreted here may be legitimately counted as an early or mid-Alexandrian process, at least three archetypal Philosophers' Stone recipes exist from which countless variations evolved throughout alchemy's long cultural history:

1. **Tincture via Ars Brevis** – the original Bronze Age metallurgical technique for confecting the Stone;
2. **Tincture via Ars Magna** – a slow equilibrium chemical reaction of all ingredients in a sealed glass vessel;
3. **White Stone of Hermes** – a two-step process whereby a double Sophic Mercury, (butter of antimony, liquid antimony menstruum converted to Algarot Powder) is first created, which can then be fermented with gold, silver or other alchemical metal afterwards.

A comprehensive understanding of these archetypal recipes allows, for the first time, the possibility of mapping the evolution of methodology to achieve the Philosophers' Stone. It helps to answer various questions regarding why and how later Islamic and European alchemists evolved techniques and developed new approaches for confecting the Stone. By juxtaposing the archetypal recipes with those that came after, one thing becomes certain: although operative details ultimately became highly sophisticated over time, greater efficiency over Alexandrian techniques was never achieved. Alexandrian archetypal recipes serve as standards of excellence.

The Alexandrian *chrysopœians* worked according to a template; a set of recipes refined to maximum efficiency due at least partially to the work's industrial nature – albeit with spiritual overtones and meaning – which required the ability to reproduce the Stone repeatedly, with qualitative consistency and with the least amount of wasted effort, materials or time. These details characterize a master artisan at their trade. The recipes for the Stone that Alexandrian *chrysopœians* bequeathed future generations are not the work of chemical technology generalists, such as was practiced by other craft traditions in Alexandria and the ancient world at the time. Nor was the Art one of experimental chemistry. Early Alexandrian recipes for the Philosophers' Stone were specific blueprints used by specialists – trade secrets or industrial formulas for creating a substance crucial to their industrial bronze craft; a craft they called *chrysopœia* but would later come to be known as alchemy.

Successive generations of alchemists from all traditions would attempt to recapture the magic and glory of those early Alexandrian *chrysopœians*, but such attempts were doomed to failure because the context of time and culture were just as important as the Art itself. Time moved on, the culture changed and social, religious and political change no longer favoured *chrysopœia* in its original pristine form. In the era of Christian persecution throughout the 5th century, the Art changed with the times. Indeed, it later underwent Christian conversion and acceptance. The philosophical Alexandrian School survived by avoiding politics and focusing on the *Quadrivium* – arithmetic, geometry, music, and astronomy. Theology and Neoplatonic philosophy were ultimately brought in line with Christian doctrine.

The purpose and value of confecting the Stone evolved in various stages. Court astronomers remained important to several emperors of the Eastern Roman Empire, Byzantium, until the 7th century; by which time the Philosophers' Stone evolved to its most abstract form. It developed from its use as proposed in this chapter as a cosmological / philosophical model of the interplay between sky and Earth, the above and below, and as a form of purification. This would ultimately progress in such a way that confecting the Stone became fully incorporated into a highly complex and unified system of astrology, medicine and philosophy taught by Stephanos at the Byzantine Royal Court of King Heraclius in Constantinople. Stephanos prized the White Stone, which he referred

CHAPTER 4 - WHITE STONE OF HERMES

to as *Comprehensive Magnesia* (universal attractor) over the Tincture called *Chrysocorallos* (coral-gold) in his lectures. Greek poet-alchemists following Stephanos' lead, left subtle clues that the act of confecting the Stone was ultimately valued as a form of high initiation into Stephanos' universal mystery tradition and as a memorial or monument symbolic of personal gnosis or redemption. The *chrysopœia* of the late Alexandrian period as presented in the next chapter developed into gold making and transmutation of the human condition as a fundamental component in a system of personal redemption.

217

Chapter 5 – Stephanos' *Chrysopœia*

*Why should we marvel at the species Chrysocorallos?
We should wonder rather at the infinite Beauty. So also I will fulfill your desire,
that you may be made worthy to love such a One [and] with hymnody
to discourse of the more than good goodness of God.*

Stephanos of Alexandria

Western alchemy in general, and the Philosophers' Stone in particular, reached the most abstract and esoteric form via Stephanos of Alexandria. Stephanos was a philosopher and 7th century Byzantine court lecturer and astronomer at the court of Heraclius, 610 to 641 CE. During the period in which Stephanos was active, philosophy, theology and science had become increasingly inseparable. Stephanos was a Christian, fully educated in Neoplatonic philosophy, specialising in astronomy and astrology. An ecumenical approach to reconciling his interests earned him a place as one of the great thinkers to emerge from Alexandria. His goal was *Einsteinian* in that he was searching for a grand unified theory of everything.

Stephanos' work towards this aim earned him the title *Universal Philosopher*. Like other Gnostic philosophers, Stephanos viewed creation as a system interwoven by a deep and indescribable interconnectedness. This web of correlations and interconnections was known to philosophers as the *Law of Correspondence* or by its medical counterpart the *Doctrine of Universal Sympathy*. Stephanos attempted to expand upon this doctrine by creating a synthesis of existing fields of knowledge. Aside from a relatively orthodox Neoplatonic stance, Stephanos developed a fusion between the four subjects of the *quadrivium* (arithmetic, geometry, music, and astronomy) with astrology, medicine and alchemy to demonstrate the interrelationship between these subjects. His lectures, far from being directed towards the novice, are better understood as the 7th century version of post-graduate work presented to an audience of philosophers and royalty already possessing great depth of understanding.

Religious and political conditions in Byzantium allowed for a continuation of astrology and astronomy surviving from ancient times to persist until the 11th century. Astrology, even during times of Christian oppression, was not considered a heretical subject but rather a system of characteristics bestowed by God upon celestial bodies that revealed premonitions to those who could interpret the signs accurately. During the year 617 CE, Stephanos left Alexandria and journeyed to the court of Heraclius in Constantinople to cast a personal horoscope for the Emperor. He cast an additional horoscope that addressed the future of Islam, for which he is well remembered in Islamic sources.

> ... *the considerable number of astrological manuscripts belonging to the private libraries of state and church figures suggests that* many Byzantine scholars and intellectuals had reconciled their Christian faith with astrology. *– Maria Papathanassiou*

Although Stephanos has traditionally been cast as an alchemist, as well as an ecumenical Christian theologian, *chrysopœia* was just one of eight subjects in his unified ideology. It was not alchemy or gold making, but rather Stephanos' unified system of philosophy that became fundamental to Islamic and European alchemy.

> *By the time of Stephanos, then, alchemy had very largely become a theme for rhetorical, poetical, and religious compositions,* **and the mere physical transmutation of base metals into gold was used as** *a symbol of man's regeneration and transformation to a nobler and more spiritual state.* **This convention was strengthened by Stephanos and is characteristic of all late Greek alchemy ...** *– E.J. Holmyard*

CHAPTER 5 - STEPHANOS' *CHRYSOPŒIA*

Holmyard and others have attempted to establish the predominant and longstanding view that Stephanos' alchemy was wholly esoteric. They appear to have discounted any possibility that he was an operative alchemist and did so from an academic frame of reference. This perspective demonstrates both an incomplete grasp of an adept's mindset, and an artificial and premature division between operative and speculative alchemy. What seems to have remained unnoticed is that Stephanos presents precise details regarding the process to confect the Stone, and its application to create alchemical gold from copper, without violating any operative principles of Alexandrian alchemy. Either Stephanos was regurgitating the basic tenets of *chrysopœia* from an unchanged literary alchemical tradition, or he had witnessed or even performed the processes firsthand. In either case, in the sections of his lecture series that address substances or operative technique, foundational operative principles remain intact. In his own words, Stephanos apparently draws attention away from operative alchemy:

> *The most eminent man and counsellor of all virtue turns them around and draws them to the view of truth, that you may not, as I said (take note of) material furnaces and apparatus of glasses, alembics, various flasks, kerotakides and sublimates. And* those who are occupied with such things in vain, the burden of weariness is declared by them.

According to distinguished alchemy researcher F. Sherwood Taylor, it is precisely at this point in alchemical history, that *chrysopœia* had fully transitioned from the realm of matter to one of spirit:

> *To the author's knowledge* this is the earliest passage wherein it is implied that Alchemy is not a quest to be carried on in the laboratory. In this lecture Stephanos indicates clearly that he views Alchemy as a mental process. *– F. Sherwood Taylor*

THE UNIVERSAL PHILOSOPHER

It was Stephanos' vision of alchemy as a component of a much larger ideology that was adopted by Islamic and European traditions. His influence is both unmistakable and extensive. Yet this becomes problematic upon considering just how experimental, operative and applied Islamic and European alchemical traditions actually became. If Stephanos' alchemy had been reduced to *a theme for rhetorical, poetical and religious compositions* as Holmyard proposes, or simply *a mental process* as suggested by Taylor, then the undeniable experimental and operative impulses clearly manifesting in later traditions – whose foundations are based upon Stephanos' alchemy – demand explanation. Stephanos' direct disciple, a hermit-monk named Morienus, proved to be incredibly operative as regards his alchemy.

Academic treatments of Stephanos have downplayed, minimized and even removed any value attributable to operative alchemy in the late Alexandrian period. This misunderstanding came about primarily because of examining Stephanos and later poet-alchemists' literature from the perspective of literary academia, rather than operative alchemy. By analyzing these alchemists' writings – poetic as they may be – with the archetypal process for confecting the Stone in mind, it becomes apparent that the descriptions of operative alchemical processes described by these poet-alchemists were not only detailed, but also accurate. The descriptive quality suggests firsthand observation rather than literary regurgitation.

It is evident that Stephanos' *chrysopœia* was no longer a stand-alone ideology, but this begs the question: what role did alchemy play in his unified system of philosophy? His *chrysopœtic* lectures are filled with esoteric overtones, yet display incredible accuracy concerning anything that might be considered operative. Later traditions would adopt Stephanos' alchemy as foundational, yet apply his fundamentals

CHAPTER 5 – STEPHANOS' *CHRYSOPŒIA*

to intensely experimental and practical forms of alchemy. It therefore seems reasonable to suggest that those following Stephanos' brand of alchemy, Morienus as a case in point, were actually able to confect the material Stone, yet prized its esoteric value to such a degree that it became central to a broader unified philosophy.

Stephanos' works have understandably been interpreted as an allegory for personal redemption, an idea inherent in alchemy since its beginnings and clearly expressed by Zosimus three hundred years earlier.

> *Copper is like a man; it has a soul and a body ... the soul is the most subtile part ... that is to say, the tinctorial spirit. The body is the ponderable, material, terrestrial thing, endowed with a shadow ... After a series of suitable treatments copper [like man] becomes without shadow [purified; without blemish] and better than gold ...*
> – Stephanos of Alexandria, 7th century

This explanation however is insufficient upon considering the increasingly authoritative role that Christianity and Islam claimed on the subject of redemption. *Chrysopœia* and religion coexisted with the other branches of natural philosophy, each with their own unique role to play. What appears to have been overlooked is the possibility that *chrysopœia's* role was not as an alternate or supportive ideology to prevailing religious impulses, but rather served as an expedient initiation into philosophic and religious mysteries. This possibility and the process by which this may have occurred merits further investigation.

> *Generally speaking, the loose structure of Stephanos' lectures* On Making Gold *should not be attributed to his penchant for a personal rhetorical style. Rather, it is the result of his effort to synthesize various ideas originating in a wide array of disciplines into a logical sequence and fashion them into a whole. This, says Stephanos, is exactly the research method of the philosopher; it is clearly his own method, too. His intention to unify various philosophical theories under the umbrella*

of a single theory able to account for all phenomena observed in the universe seems very modern. – Maria Papathanassiou

As an ecumenical Christian theologian versed in the Alexandrian school of philosophy, Stephanos was inclusive concerning philosophy and religion. Integrated into his religious and philosophical system was the inheritance of Hermetic, Gnostic and Neoplatonic wisdom and expressions that have traditionally characterized the speculative aspects of alchemy since its origin. Yet Stephanos did not merely examine and combine different ideologies into an eclectic mix, nor was he simply demonstrating correspondences; he was attempting to reveal the *One* unifying principle underlying all creation that found expression in these separate but interrelated ideologies. Like Stephanos, many other great thinkers shared this stoic vision, finding themselves musing the creative source or fountainhead of existence:

> *... there is both one cosmos of all things, and one God through all, and one Substance, and one Law and one common Reason of intelligent beings, and one Truth. – Marcus Aurelius,* Meditations, VII, 9

By coming into Gnostic experience of the *One* via a *chrysopœtic* model, Stephanos believed that the secret principle governing the processes of creation, the world of humanity and the natural order could be intimately known. Subtle clues in his writings and those of the poet-alchemists that followed in his tradition appear to indicate that *chrysopœia* served as an introduction or initiation into deeper mysteries of reality. Furthermore, their writings suggest that actually confecting the Stone in its material form was part of the initiation process. The Stone and the process to confect it served as a pragmatic analogy of the mysteries, approachable via observation and participation in alchemy. These mysteries involved man's identity and constitution, his role in the natural order and – most importantly – that the apparent multiplicity in the cosmos was actually a unified reality whose surface appearances were interconnected by an infinitely complex system of relationships. Aspirants underwent a series of preparations, ceremonies and various degrees of initiation until gnosis of these mysteries resulted in spiritual redemption. In Stephanos' system, alchemy played a fundamental and important initiatory role at the very heart of his universal philosophy.

> *O unspoken mysteries of a wise God, O rich gifts to those who have loved the Lord, O depth of wealth and wisdom and gnosis of the mysteries. If the present things are such marvels and extraordinary, from what source are everlasting things which no mind is able to explain? If the material work is displayed thus to us by some unspeakable discourse, from what source are thy undefiled good and unfading beauties, which no one is capable of perceiving?*
> — *Stephanos of Alexandria, 7th century*

Legendary Chemistry and Mystical Words

> *Stephanos was certainly not a practical laboratory worker and his treatise embodies no new experimental results,* **but if we view it as a document in the history of Alchemy we must accord it high importance. In the first place the treatises of Stephanos and Olympiodoros are the only long and complete, or nearly complete, works on Greek Alchemy which have come down to us. Stephanos gives us a full exposition of the theory of Alchemy as it was understood in the seventh century A.D.**
>
> **He may not be an original writer, but** *comparatively little of his work is taken from extant alchemical texts.* **He is the first to avow that the art includes mental operations.** — F. Sherwood Taylor

Although Stephanos may have been the first to focus on the speculative aspects of alchemy, these concepts did not originate with him. He crystallized various theoretical impulses and correspondences into a unified system that became part of the intellectual luggage common to all later alchemical traditions. If the Philosophers' Stone and its confection had been fully refined during the course of the previous 500 years, there understandably would be *no new experimental results* as Taylor rightly observed. What makes Stephanos' work so important is the paradigm shift *chrysopœia* underwent with reference to role and meaning as they pertain to mysteries and initiation. Assuming that Stephanos was an adept alchemist – meaning that he could actually confect the material Philosophers' Stone – he naturally would integrate this unique ability into his universal worldview.

Stephanos' writings reveal that he did just that. His work, *De Chrysopœia* (or, *On Gold Making*) is part of two compendiums of alchemical manuscripts; *Venice Codex Marcianus Græcus* 299 and Paris BNF 2327. Sections of his nine-part lecture series were translated by F. Sherwood Taylor and treated by Taylor and Maria Papathanassiou. Most researchers, in agreement with Taylor and Holmyard, maintain that Stephanos was strictly a poet-alchemist. There is enough in Stephanos' writings, however, to indicate otherwise. In part two, in a letter apparently to Emperor Heraclius' brother Theodoros, Stephanos accurately outlines the basic materials and methods of alchemy, which he calls the *field*, veiled by the various cover-names of the day. He begins by explaining that six metals are involved in the making of *Corinthian bronze*:

> *There are all the six brothers* attendant on claudianos [bronze of Maria and Claudianos] and the others together.

The six brothers are the bronze-age metals gold, silver, copper, iron, tin, and lead. Stephanos follows this with a description of antimony – the *serpent* – as being a coagulant by nature, and presents the process of *whitening the Stone*, beginning with greenish-black *stibnite*, followed by saffron and ending with white *flowers of antimony*:

> *For the field has a serpent and he dries up the place* with his breath, *where also men grow feeble. And I saw him and the spotted scales of his*

CHAPTER 5 – STEPHANOS' *CHRYSOPŒIA*

body. The beginning of his tail was white as milk, but his belly and back were saffron-coloured and his head was greenish-black.

Stephanos then advises how best to approach alchemy in language that is more comprehensible to one familiar with operative details of confecting the Stone. Stephanos writes, *"... as I found out"* – clearly indicating, in his own words, that he had learned the operative details of confecting the Stone.

You should divide the field into three, the four brothers one part and the great stone one part, for thus the ancients attempted to do with the field, as I found out.

Dividing the field into three = to identify gold, antimony and fluxing salt as the *tria prima*, i.e. three primary ingredients - △.
The four brothers = preparatory work, creating the *Tetrasomia*, i.e. *divine water*, fiery earth + airy water - □.
The great stone = finishing work, final creation of the Stone called the *coral of gold* (or *chrysocorallos*) - ○.

Stephanos then provides a detailed description of his method of operation. He explains that *moist and dry vapours* are required to confect his White Stone. He does not reveal the composition of his *vapours* but it is clear that they are compound substances requiring preparation. Stephanos mentions *pharmakon* ten times in his lectures. Homer used the term *pharmakon* to mean *poison, beneficial drug* and *spell*. It later came to mean sacrament, remedy, poison, talisman, tincture, cosmetic, perfume and psychoactive substance. Stephanos' alchemical *pharmakon* is synonymous with *divine water* – the secret to his White Stone that he prized over the Tincture, which he referred to by the cover-name *chrysocorallos* (coral-gold). Stephanos' description disqualifies quicksilver, the *white vapour from cinnabar*, from being an ingredient in his recipe, yet demonstrates his familiarity with the technique for isolating mercury.

True is a certain moist vapour and the dry vapour. For the moist is sublimed by the phanoi which have nipples. But the dry vapour by the pot and bronze cover [sublimation], as is the white vapour from cinnabar. Therefore if you imbibe the dry vapour with the moist vapour, you perfect the divine work.

By cross-referencing the final line with similar quotes from Maria, it becomes clear that Stephanos was well acquainted with the archetypal recipe to confect the Philosophers' Stone.

Therefore if you imbibe the dry vapour with the moist vapour, you perfect the divine work. – Stephanos

... two fumes that contain two Lights ... – Maria Hebrea

Join the male and the female and you will find what is sought.
– Maria Hebrea

... marry Gum with Gum together by a true union. – Maria Hebrea

Researchers looking for evidence of experimental chemistry or *new results* will be disappointed with the Alexandrian alchemists, and this is equally true for Stephanos. However, there is a tremendous difference between evidence of experimental chemistry and evidence of Stephanos being fully capable of creating the Philosophers' Stone. He appears to be very orthodox concerning his personal methodology for creating both the White Hermetic Stone and Red Tincture of the Philosophers. His two *vapours* were compounds requiring skill at sublimation or distillation and crucible work. Where Stephanos diverges is the role that alchemy plays in his universal philosophy and the emphasis that he places upon the tremendous value *chrysopœtic* operations hold as an initiatory method for learning greater mysteries. Although alchemy and spirituality initially appear to be separate entities, for Stephanos they both serve a common purpose:

CHAPTER 5 - STEPHANOS' *CHRYSOPŒIA*

And the legendary chemistry is one thing, and the mystical and hidden is another. *For the legendary chemistry is confounded in a multitude of words, but the mystical is operated by the word of the Creator of the world, that* the man who is holy and born of God may learn by the direct operation and by theological and mystical words.

In the passage below, Stephanos describes the Stone by the cover-names *Tetrasomia* (four elements; □), *threefold triad* (the *Tria Prima*; △) and *universal seal* (the *One* and *All*; ○). He also emphasizes the *body of magnesia*, meaning antimony as being the active ingredient. The *whole mystery* alluded to in the following passage is the Stone's confection:

> *O* conjunction of the tetrasomia *adorned upon the surface, O inscription of* the threefold triad *and completion of* the universal seal, body of magnesia by which the whole mystery is brought about ...

In another passage, Stephanos introduces a divine attribute, and then asks the identity of a material counterpart that reflects the same properties. He is speaking of antimony, which is a *great gift* or secret that he will reveal. What is interesting about the impending revelation is that the criteria for receiving the *great gift* includes intelligence and virtue and, more importantly, bearing understanding of both *theoretical practice* and *practical theory*. These final two criteria are only necessary if operative subject matter is being revealed.

> *... O single nature itself and no other nature, rejoicing and rejoiced over, mastering and mastered, saved and saviour,* what have you in common with the multitude of material things, since one thing is natural and is a single nature conquering all? *Of what kind are you, tell me, of what kind? To you who are of good understanding I dedicate this great gift.* To you who are clothed with virtue, who are adorned with respect to theoretical practice and settled in practical theory.

The following passage initially appears to contradict the previous one that requires a solid grounding in both practical and theoretical alchemy.

Yet Stephanos' intention is to redirect one's focus beyond materialistic valuation towards one of deeper meaning and insight. This is not a shift away from practical alchemy, but rather an urging towards greater correspondences between substances, processes and divine nature, approachable through the conceptual framework of practical alchemy. The *single natural thing*, antimony, is still required.

> *Put away the material theory so that* ye may be deemed worthy to see with your intellectual eyes the hidden mystery. For there is need of a single natural [thing] and of one nature conquering the all.

In the passage below, Stephanos succinctly sums up his approach. He begins by paraphrasing Pseudo-Demokritos but adds that the *corporeal* – meaning the physical operation of the process – is, in fact, alchemy's method of initiation into the mysteries. The *operation of the process* mentioned in this remarkable passage can only be the process for confecting the Philosophers' Stone, as only this could engender the potential for profound insight that is alchemy's initiation into the mysteries.

> *For it rejoices on account of the nature being its own, and it masters it because it has kinship with it, and, superior to nature, it conquers the nature when* the corporeal operation of the process shall fulfill the initiation into the mysteries.

Comprehensive magnesia appears to be considered a sacred substance by Stephanos. He describes it as a marvellous spirit that perfects the *coral of gold* (or *chrysocorallos*), a cover-name for the Philosophers' Stone in its industrial form as a gold-containing Tincture for coloring copper. In the passage below, an understanding of antimony as manifesting divine properties brings the mystery to perfection. When read in context, the difference between Stephanos' use of the cover-terms *body of magnesia* and *comprehensive magnesia* appears to be the difference between antimony as an ingredient, such as *flowers of antimony*, and a fully prepared antimony compound, such as *divine water* or the *White Stone*. Perfection in the alchemical sense implies incorruptibility, indestructibility and immortality while also being imbued with properties of growth and regeneration; properties which alchemists of Stephanos' era hoped to emulate spiritually.

> *... it is as a marvellous spirit; then it masters the body moved (by it), then it rejoices as over its own habitation, then it conquers that which in disembodied fashion haunts the whole which is engendered of the whole, that is admirable above nature, which I say to you is the comprehensive magnesia, who will not wonder at the coral of gold perfected from thee? From thee the whole mystery is fully brought to perfection ...*

In the following passage, Stephanos remarks upon teaching the preparation and displaying the work, and waxes poetically on whiteness as a symbol of the spiritual purity and beauty of the substance that is responsible for the *wondrous work*.

> *O wisdom of teaching of such a preparation, displaying the work, O moon clad in white and vehemently shining abroad whiteness,* **let us learn what is the lunar radiance that we may not miss what is doubtful. For the same is the whitening snow, the brilliant eye of whiteness, the bridal procession-robe of the management of the process, the stainless chiton, the mind-constructed beauty of fair form, the whitest composition of the perfection, the coagulated milk of fulfillment, the moon-froth of the sea of dawn;** *the magnesia of Lydia, the Italian stibnite,* **the pyrites of Achæa, that of Albania,** *the many-named matter of the good work,* **that which lulls the All to sleep,** *that which bears the One which is the All, that which fulfills the wondrous work.*

Clearly, Stephanos is not using the technique of literary analogy to describe the White Stone or its confection, but rather is presenting the Stone as being a microcosmic physical analog of greater philosophical and religious concepts. Stephanos stresses the *wisdom of teaching of such a preparation*, but qualifies his meaning with the passage *displaying the work*. The passage appears far more descriptive than imaginative, to anyone familiar with the substance that Stephanos is describing. The act of *displaying the work* should be stressed and merits further examination. He appears to be using the White Stone – by either *displaying the work* as the process of its creation, *displaying the work* as product, or possibly both – as a means of introducing concepts such as shining radiance, management of process, perfection, beauty and goodness, or the *One which is the All*. The tone and emphasis upon a white product bearing *the One which is the All, that which fulfills the wondrous work*, parallels sentiments expressed centuries earlier by Cleopatra, Zosiumus and Synesius. The operative explanation for the passage *that which fulfills the wondrous work* can be explained when we understand that the White Stone, called the *Great* or *Universal Stone* by Heinrich Khunrath, can be compounded with gold, silver or other metal of choice by fusion in a crucible to create a fermented stone, thus its universality in the physical sense. Stephanos often referred to the White Stone as *Universal Seal* or *Comprehensive Magnesia* because it could be fused with other metals in this fashion. For this reason, it was considered superior to the Tincture by some alchemists including Khunrath who referred to the gold or silver-based Tincture as the *Lesser* or *Specific Stone*.

> One is the serpent *which has its poison according to two compositions. One is All and through it is All and by it is All and if you have not All, All is nothing.* – Chrysopœia of Cleopatra, circa late 3rd – early 4th century

> ... one only Stone wherein consists the whole Magistery ... *in this matter, all things requisite to this Art are contained ...*
> – attributed to Synesius, 5th century

> ... *that which bears the One which is the All, that which fulfills the wondrous work.* – Stephanos, 7th century

CHAPTER 5 – STEPHANOS' *CHRYSOPŒIA*

The parallels are unmistakable and suggest Stephanos was discussing the White Stone as being superior to the Red Stone, which was perfected by it. However, in order to make a clear correlation between the Stone and the *One*, an object or model must play a role if it is to be anything other than a thought experiment. The purpose of using an appropriate physical model lies with the fact that juxtaposing an abstract idea with a practical physical representation engenders a deeper, immediate and holistic comprehension beyond surface thought than would be possible without the model. In the passage below, Stephanos all but states that he is visibly displaying the White Stone in order to potentially reveal the mystery hidden within it. He follows this with a list of 16 cover-terms for the finished Red Stone that, according to Stephanos, is the mystery brought to perfection by the addition of gold after-the-fact. The impression that Stephanos is displaying either a physical object, process or both is apparent, as is the notion that a hidden mystery may be *gnostically seen* because of the direct experience of firsthand observation.

> *What is this emanation of the same [moon]? I will not conceal it, but will display visibly the sought-for beauty. For the emanation of it is the mystery hidden in it,* the most worthy pearl, the flame-bearing moonstone, the most gold-besprinkled chiton, *the food of the liquor of gold, the chrysocosmic spark,* the victorious warrior, *the royal covering,* the veritable purple, the most worthy garland, *the sulphur without fire, the ruler of the bodies,* the entire yellow species, the hidden treasure, that which has the moon as couch, *that which in the moon is gnostically seen as in the sense that the moon is perceived ...*

Stephanos, being formally educated and influenced by Alexandrian Neoplatonic philosophy, appears to have employed the White Stone as a physical model, an initiation into the mysteries as a means to more efficiently communicate gnosis of an evolutionary process. He also demonstrates familiarity with the Red Stone through his use of the cover-name *coral of gold* (or *chrysocorallos*) as being a product perfected by the use of the White Stone. In Stephanos' alchemy, the White Stone as a symbol of purity and potentiality perfects the *chrysocorallos*, bringing it to completion as *coral of gold*. Nevertheless, initiation by way of analogy cannot be substituted for personal gnosis. This was achieved by attempting to backtrack from the multiplicity of existence towards a return to the *One* as the source of wisdom and truth of unified reality approached by, yet beyond, human intellect. The *One* represented a quality – a type of *Truth* beyond duality or multiplicity – and this absolute truth was viewed by philosophers such as Stephanos as being an attribute of divinity; the *One* was the fountainhead or ultimate source underlying and interconnecting the *All*, the *Light* of the Gnostics, and imbued with perfect goodness.

Stephanos urged direct psychospiritual experience of *Truth* beyond all distinctions, descriptions or explanations by moving beyond *material theory* in order to *see with your intellectual eyes the hidden mystery*. It was a mystical approach to wisdom as gnosis via the workbench of alchemy. It was not a truth about things or ideas that was sought, but rather an imminent revelation of the relationship between self and landscape as a web of connections. This required a broad worldview and intuitive insight into the natural order of things; a view aided by the use of the Philosophers' Stone as an appropriate model and direct personal insight into correspondences between alchemical, natural and spiritual processes. Only from such a frame of reference could the philosopher move beyond thinking about the *One* to thinking in harmony and partnership with the *One*, thus uniting microcosm and macrocosm.

MYSTERIES AND INITIATION

The act of confecting the Philosophers' Stone is an experience, not an ideology. The ideology is crafted around, and because of, the experience. Immediate undifferentiated experience is the sacred ritual of initiation into the mystery of the *One*, its expression or manifestation as multiplicity and return to singularity commonly summarized by the maxim *solve et coagula*. The material monument or sacred object that serves as a symbolic memorial of the experience is the finished Philosophers' Stone. The internalizing of the sacred object required the success of confecting it oneself; firsthand observation and involvement in the Great Work from the beginning to the end.

> **Mystery** = *mysterion of the philosophers, secret, mystic rite, an object used in mystic rites, talisman, symbol.*
> **Initiation**, *mystic rite = confection of the Stone.*
> **Monument** = *a construction, sacred object or talisman, symbol of achieving the One on a physical, mental and spiritual level.*

Contemporary research in neurology reveals that religious and spiritual ideas and experiences can be stimulated by a visual object, which can result in profound spiritual states. The internal state of the psyche undergoes transformation through the process of witnessing, participating and fully comprehending the alchemical process of confecting the Stone. Thus confecting the Stone served as a vehicle of initiation, yet once the philosopher had arrived at the end of his journey – meaning that he had undergone the transformative process resulting in gnosis and redemption – the vehicle was no longer necessary. To Stephanos, this was the true value of confecting the Stone as a spiritual, rather than physical, act. To ignore this redemptive principle was to miss the point. The following passage has convinced most researchers that Stephanos could not have been an operative alchemist. However, the passage lends itself to an alternate and equally valid interpretation that suggests otherwise:

> *The most eminent man and counsellor of all virtue turns them around and draws them to the view of truth, that you may not, as I said (take note of) material furnaces and apparatus of glasses, alembics, various flasks, kerotakides and sublimates. And* those who are occupied with such things in vain, *the burden of weariness is declared by them.*

What of those who are occupied with such *things NOT in vain*? What would validate or justify operative work? Only the role of *chrysopœia* as an integral component of initiation and spiritual redemption; all else, according to Stephanos, would be in vain. His diatribe is not against operative alchemists as such, but rather against those that fail to comprehend that these processes serve a much greater purpose. From this interpretation, the passage above in no way precludes operative alchemy, but rather reinforces its role as a component of the redemptive process. Stephanos follows this by impressing the importance of grasping a lesson of the *All* via observation, followed by an intellectual understanding of alchemical processes:

> But *see how the All is fulfilled* in the phrase. *"Nothing is left remaining, nothing is lacking save the vapour and the raising of the water." What kind of vapour? Say. What is the vapour and what is the work brought to perfection by it? Show us most clearly the way in which we may recognize the power of the word [of God, the natural order]. And on this matter the philosopher says: "the vapour is the work of the composition of the whole,"* that which shines brightly through the divine water, that which makes the trituration naturally, *that which appears in the course of the method, and is apprehended intellectually.*

His use of the word *vapour* is equivalent to Maria's use of the word *fume* as a cover-name to indicate a purified and reduced alchemical substance. In this case, he uses the term *vapour* synonymously for the finished Philosophers' Stone by describing it as *the work of the composition of the whole. That which appears in the course of the method, and is apprehended intellectually* is an insight into the nature of the *All*, the word of God, the natural order. The Stone and its confection initiate this insight, without which alchemists *are occupied with such things in vain*.

Gnosis and redemption for Alexandrian philosophers was far different from the modern Judeo-Christian variety. The process of redemption in Alexandrian philosophical currents meant initiating a partnership with the divine to such a degree that one shared the same perspective. In order to arrive at gnosis of self and God fully, aspirants were encouraged to match personal perspectives with that of the *All* in order to enter into full communion with the *All* and experience its multiplicity of expression as the *One*; the *One* and *All*, *Alpha* and *Omega* being synonymous with God of Gnostic philosophies.

> *Unless you make yourself equal to God, you cannot understand God: for the like is not intelligible save to the like.* **Make yourself grow to a greatness beyond measure, leap forth and free yourself from the body; transcend time, become Eternity; then you will understand God. Believe that nothing is impossible for you, think yourself immortal and capable of understanding all, all arts, all sciences, the nature of every living being. Become loftier than the highest height; descend lower than the lowest depth. Draw into yourself all sensations of everything created, fire and water, dry and moist, imagining that you are everywhere at the same time, on earth, in the sea, in the sky, that you are not yet born, in the maternal womb, adolescent, old, dead, beyond death.** *If you embrace in your thought all things at once, times, places, substances, qualities, quantities, you may understand God.*
> – Hermetica, 'Mind to Hermes' 126-130

The Gnostic heresy presented man not as having fallen from grace or guilty of original sin, but rather as a great miracle of divine origin, possessing marvelous powers of creation. Gnostic Christian tradition posited very similar notions:

> Gospel of Thomas 3 ... *the Kingdom of the Father is within you and it is outside you. When you know yourselves, then you will be known,* **and you will understand that you are children of the living father. But** *if you do not know yourselves, then you live in poverty, and you are the poverty.*

> Gospel of Thomas 70 *If you bring forth what is within you, what you bring forth will save you. If you do not bring forth what is within you, what you do not bring forth will destroy you.*

> *Gospel of the Nazarenes 6* **Raise the stone, and there you will find me. Cleave the wood, and there am I. For in the fire and in the water even as** *in every living form, God is manifest as its life and its substance.*

> *Acts 7:48* ... *the most high does not dwell in houses made by hands ...*

By Stephanos' time, these notions were preserved, although understood as being encrypted, within orthodox Christianity, with a small sub-culture of Christian philosopher-priests secretly adhering to them. Stephanos' unified philosophical approach – replete with Neoplatonic foundations and discourse on initiation and mysteries hints at Gnostic influence underlying his ecumenical approach to Christianity.

Redemption via initiation in the Gnostic sense meant a rediscovery of one's divine nature; a revealing of the mystery of self, one's immortality and role as God's partner, brother and co-creator in the natural order. The Gnostic heresy was largely that man was once, and can become again through processes of initiation and ever-deeper insights and realizations, the reflection of divinity – a divine being possessed with magical creative power. The objective, therefore, was to reawaken to the truth of one's divinity and participate in the act of creation in harmony with natural law:

> ... *to collaborate in the work of Nature, to help her produce at an ever-increasing tempo, to change the modalities of matter – here, in our view, lies one of the key sources of alchemical ideology.*
> – Mircea Eliade, The Forge and the Crucible

It is from this conceptual framework that Stephanos utilized *chrysopœia* as a means of expounding mysteries, conditioning the imagination of the adept alchemist philosopher to apprehend truths hidden in alchemy and the natural order as a means to self-redemption. Themes such as displaying and observing the work, initiation and mysteries are amalgamated with discussions of alchemical substances and processes throughout Stephanos' work. Nearly the entire content of Stephanos' lectures can be understood as a physico-spiritual double entendre better approached in a universal *both-and*, rather than the divisive and reductionist *either-or*, spirit of understanding. The absurd rift between speculative and operative approaches later in the history of alchemy would likely have been incomprehensible to Stephanos. He taught aspirants to acknowledge the Stone's composition, look at its color, properties and uses, to marvel and wonder at it, and then use this wonder as a springboard initiation into deeper realization of the mysteries of man and creation.

> ... *O inscription of the threefold triad and completion of the universal seal, body of magnesia* by which the whole mystery is brought about ...

> ... *O countenance contemplated by virtuous men*, *O sweetly breathing flower of practical philosophers*, *O perfect preparation of a single species*, *o work of wisdom, having a beauty composed of intellect* ...

> **To you who are of good understanding** I dedicate this great gift, **to you who are clothed with virtue,** who are adorned with respect to theoretical practice and settled in practical theory ... **Put away the material theory so that ye may be deemed worthy to** see with your intellectual eyes the hidden mystery.

> ... the corporeal operation of the process shall fulfill the initiation into the mysteries. **For when the incorruptible body shall be released from death, and when** it shall transform the fulfillment which has become spiritual, then superior to nature it is as a marvellous spirit ...

> Who will not wonder at the coral of gold perfected from thee? From thee the whole mystery is fully brought to perfection ... **(you receive the undefiled mystery of nature).**

> **What is this emanation of the same (luna)?** I will not conceal it, but will display visibly the sought-for beauty. For the emanation of it is the mystery hidden in it ...

BOOK 1 - ORIGINS & EVOLUTION

Macrocosm, Microcosm and the Stone

> *This is the etesian stone. With these it is called by all names.*
> *It is the porphyry which is found in the purple mineral, the purple-coloured substance made from tin, the Macedonian <stone>, and if any other name has been spoken of or written or symbolised in the divine and allegorical writing, it is that; for if the writings would signify anything,*
> *they allegorise about many things.*
>
> *– Stephanos of Alexandria*

To Alexandrian philosophers, everything in Nature, at all levels, was bestowed with hidden meaning that could be revealed via mechanical and spiritual analogues – analogues that were not necessarily seen as mutually exclusive. The skills and abilities of an adept alchemist began with an abstract framework but moved towards specific details of working with substances and processes. The abstract framework served as a model of mechanical operations, and therefore, a form of initiation. With talent, guidance and involved experience the apprentice became the adept and moved beyond the initial abstract framework and mechanics into which he was initiated. Once mastery had been achieved and practical operations had become second nature the alchemist could advance the subject, going on to find patterns of meaning in his work and expressing such through fantastic abstractions of an individual manner based upon personal gnosis. For the most part, alchemical writings were more a form of artistic expression. Impulse following mastery very much in the same manner that modern scientists and academics will follow up academic research publications with books or documentaries directed towards a popular audience (a fundamental difference being that ancient alchemists wrote primarily to convey gnosis to each other and not necessarily for the populace).

Stephanos conceived of a universal language of natural law – the *mysterion*; a language of relationships, which he asserted were the essence of creation, articulated through the dialects of mathematics, music, geometry, astronomy, astrology, and medicine. In Stephanos' conceptual framework, these found symbolic material expression in the model of the Philosophers' Stone and *chrysopœia*. Stephanos revealed that a philosopher initiated into the secrets of the Stone and gold making became the embodiment of the object of his search, a living Philosophers' Stone, and therefore a model of the natural order and fluent in its language; he became the Elixir that could affect transmutation or change via gnosis of his creative, and therefore divine, self. The *All* of creation, as Stephanos envisioned it, was a single organic divine entity from which all forms emanated and thus were manifestations of divinity – an oversoul disguised in innumerable forms as a continuum of divine outpouring and emergence. Stephanos' microcosmic and macrocosmic correspondences are not to be understood as distinct or separate entities, but rather expressions of a single unbroken continuity extending from a primal source permeating all forms of matter. This was the *mysterion*.

Just as concepts of Tao or Zen can be applied to art, calligraphy, archery, or motorcycle maintenance in Eastern traditions, the substratum of Western philosophy at the heart of the Alexandrian school was integral to the *quadrivium*, astrology, medicine, and alchemy as an underlying principle. To Stephanos, this root philosophy was embodied in the Philosophers' Stone. The great processes of the natural order and secrets of humanity's divine nature and role were revealed during the process of confecting the Stone. The finished Philosophers' Stone served as a fractal model of self and creation.

> *Fractals have come to represent a way of describing, calculating, and thinking about forms that are fragmented and irregular (like life itself!). The fractal approach embraces the whole structure — in terms of the branching that produces it — and behaves consistently from large scale to small scale ... Reality, on all levels, displays a regular irregularity! Here, 'fractal' means self-similar. Self-similarity is symmetry across*

scale. Every fracture of the shape replicates a fraction of itself on a smaller or larger scale, continuing infinitely. A fractalled view of reality is an infinite recursion of pattern within pattern.
— *from Rabbi Joel David Bakst (quoting James Gleick)*

Stephanos urged an identification of intellect with alchemical substances and processes, fractal models and their micro-macro correspondences, as a form of Gnostic initiation aimed ultimately toward spiritual redemption.

Model of Divine Self

One aspect of initiation into the mystery of the relationship between self and the natural order as it applies to alchemy is the correlation between the Philosophers' Stone as Elixir and self as Elixir. As the Stone passes through its various stages, the alchemist's psychospiritual self mirrors this process of transformation, resulting in a type of living human talisman embodying the creative powers of the planetary spheres. All planets, metals, colors, and governing influences become embodied into the *One* of the philosopher and his Stone, thus the multiplicity of the *All* coalesces into the *One*: the philosopher becomes the Stone. Initiation into this mystery served to inspire, and to impart a new mindset that enabled the artisan to mature into the philosopher as model of, and mediator between, the celestial and terrestrial realms. The adept alchemist realized himself to be fractal of the macrocosm in the same manner that the Philosophers' Stone is seen as a fractal of the macrocosm.

Stephanos opens his lecture series with what might be considered the standard praise to God, but what is encrypted within is the first introduction into self as a divine *triune being*. Stephanos identifies God as the *King of all*; therefore, *his only begotten son* must possess divine royalty as well. He is not describing the Christian Holy Trinity as one might initially expect; this would be far too blatant for Stephanos. Rather he is

CHAPTER 5 - STEPHANOS' *CHRYSOPŒIA*

presenting the main ingredients of the work and demonstrating the work's potential as a means of *tracking down the truth*.

> *Having praised God the cause of all good things and* the King of all, *and his only begotten Son* resplendent before the ages *together with the* Holy Spirit, *and having earnestly entreated for ourselves the illumination of the knowledge of Him,* we will begin to gather the fairest fruits of the work *in hand, of this very treatise,* and we trust to track down the truth.

Royalty appears as a theme throughout alchemy to encrypt the identity of the *Tria Prima*. 1) Gold, 2) sal ammoniac and butter of antimony (*eagle salt, bird of Hermes* or *eagle's gluten*), and 3) antimony (*flowers* or *regulus*) and are routinely depicted symbolically in animal form as a lion, eagle-*bird of Hermes* and serpent-snake respectively, each with a crown of royalty upon its head.

This is the royal and holy trinity of alchemy and serves as an analogue and means of initiation into the secret composition of man, and the stone, as a *triune being* composed of body, spirit and soul united as the total functioning of being. The holy trinity is not to be understood as merely a mixture of three separate principles, but rather a unicity, a single *triune being*, and microcosm of Plotinus' three divine hypostases – soul, spirit and the *One*. In other words, the holy trinity was not separate from man; it was his divine constitution as microcosm of the *One* undivided reality, which could be rediscovered through initiation. The Philosophers' Stone served as an appropriate physical model of this truth.

> ... *when comparing the* man as a perfect union of body and soul with the Whole (or Nature), which is both one and many, *as seen in the art of making gold, he refers to the well-known mystical relation between microcosm and macrocosm* ... *A basic principle in alchemy is that of the unity of the world, expressed by many passages of similar content referring to the one nature, identified with the whole. The* physical bodies are said to be composed of the four cosmic elements, which are in a dynamic state having births, destructions, changes and reversions from one to another. *– Maria Papathanassiou*

In the passage above, Papathanassiou addresses two very important correspondences, the first of which is man as a model of the unity of the natural order and thus both self and landscape. *The four cosmic elements* is another way of saying *all of creation*. The second correspondence is the principle of constant change or energetic transformation in the natural order – birth, change, destruction, and reversion. The hidden implication in this view is the principle of *rebirth*, as evidenced by a parallel Stephanos makes between burnt medical plant substances and the power still contained in the ashes as being *energy in potentio*:

> On the one hand they [medico-alchemical substances] are active bodies, on the other hand a power, according to another discourse, displaying activity. And *such things as he is able to display to you from matters incinerated and reduced to ashes* pertain to the man skilled in medicine. For *such things as come to rebirth, relate to an easily apprehended art...*

Stephanos is urging the aspirant to grasp the mystery of *rebirth* as a natural power or activity – a divine process – via observing an alchemical representation of this process. If the process of *rebirth* is inherent in the interplay of the four elements, and *rebirth* occurs as an underlying principle of the universe, then this process must also be present in man. By interpreting Stephanos' use of the term *burning and reducing to ashes* as a cover-term for physical death, it becomes clear that he is addressing man's immortal nature as mirroring that of the universe.

CHAPTER 5 – STEPHANOS' *CHRYSOPŒIA*

> *So that* there is no need to be afraid of burning and reducing to ashes all these bodies. For *they come again to a certain power and virtue and re-birth, having a nature imitative of the whole universe* and of *the elements themselves,* whence *also they have re-birth, a communion with a certain spirit,* as of things *coming into existence by a material spirit.*

By understanding Stephanos' use of the term *spirit* to mean *breath of life*, his assertion that the cosmic elements – including man – come into existence and commune with *spirit* makes more sense. The words for *spirit* in Greek (*pneuma*), Latin (*spiritus*) and Hebrew (*ruaḥ*) all mean *breath*. In the passage below, Stephanos indicates that this *breath of life* is sourced from *the air that makes all things*:

> *So copper, like a man, has both soul and spirit.* For these melted and metallic bodies *when they are reduced to ashes, being joined to the fire, are again made spirits,* the fire giving freely to them its spirit. For as they manifestly take it from *the air that makes all things,* just as *it also makes men and all things, thence is given them a vital spirit and a soul.*

Spirit and soul attain union with body through death and rebirth analogous to the calcination of medicinal plants or alchemical metals. This process, known as *killing* in both Eastern and Western alchemical traditions, ultimately results in *rebirth* and greater vitality. *Soul* in Sanskrit (*atman*), Greek (*psyche*), Latin (*anima*) and Hebrew (*neshama*) also means *breath*. It is possible, therefore, to surmise that the soul-spirit unity, also known as dual-soul, is responsible for growth and regeneration in the natural order and is the mystery of vitality, whereas the corporeal body is nothing more than a temporal form, an expression or manifestation, an app allowing the eternal dual-soul to interact within the the temporal. The implications are that one's life essence or animating spirit survives the body's death. This notion, however, dates back to the semi-legendary Pythagoras (circa 580-500 BCE) who is described as having journeyed to Egypt and received instruction there from the priests:

> Whether the doctrine of the immortality of the soul came from Egypt through a Pythagorean medium is not clear, but most of the Greeks who accepted that doctrine were in touch with Pythagoreanism.
> – De Lacy O'Leary, D.D.

In somewhat simplistic terms, Stephanos' *rebirth* – modeled by the alchemical process of calcinations – can be understood as a form of reincarnation, in that the alchemist was causing the transmigration of the substance's soul away from its dead body and into a new Elixir body, resulting in a new and improved substance. The process may also be seen as a form of resurrection where the body of the substance is first killed, then infused with a *breath of life*, resulting in an improved, purified and resurrected form as an Elixir. In either case, the alchemist believed that he was simply replicating a natural, yet divine, process.

The metaphor between copper and man was originally voiced by Maria Hebrea who frames the trinity from the Jewish perspective (Genesis 2:7) that man was originally created by the breath of God (spirit), moving upon *water* (liquids) and *earth* (solids), paralleled by Stephanos' *moist and dry vapours*:

> She says: "Just as man is composed of four elements, likewise is copper; and just as man results [from the association] of liquids, of solids and of the spirit, so does copper." – Maria Hebrea

The *Philosophers' Egg* as womb of creation served as a dual microcosmic model for alchemists. The first is the model of man's original body-soul union where the dual-soul is joined with the *body* of gold:

Gold	Antimony	Menstruum / Flux
Body *(immortality / incorruptibility)*	Soul *(growth)*	Spirit *(regeneration)*

Confecting the Stone was viewed by Judeo-Christian alchemists as modeling the creation and birth of Adam. The dual-body, *divine Adamic-earth* of the alchemical tradition, is purified and ground gold-antimony glass (as a dark reddish-brown powder) which is then infused with the dual-soul present in *divine water*. In the Philosophers' Stone synthesis, the dual-soul can be explained by the fluxing salt or liquid menstruum employed in the synthesis. A purified and reduced form of antimony provided the growth principle, can be viewed as soul, whereas either sal ammoniac or butter of antimony is the regenerative principle, and can be viewed as the spirit due to its volatile nature. This process of creation in which body, soul and spirit become unified mirrored – to some very religious alchemists – the creation process by which Adam received the divine spark.

> *For the conduct of this operation, you must have pairing, production of offspring, pregnancy, birth, and rearing. For union is followed by conception, which initiates pregnancy, whereupon birth follows. Now the performance of this composition is likened to the generation of man, whom the great Creator most high made ...*
> — Morienus al-Rumi, 7th century

The idea that the process occuring in the digestion flask served as a microcosmic model of man receiving a spark of divinity dates back to Gnostic and Jewish schools of Alexandrian alchemy and remains immortalized in alchemical symbolism. The microcosmic model of body-soul-spirit as a *triune being* was encrypted by the cover-name *homunculus* (Latin, *little man*). In the Jewish alchemical tradition this model is known as *golem* as a *triune being* composed of *nephesh* (animal body), *ruaḥ* (spirit) and *neshama* (soul). In the Islamic alchemical tradition, Jābir ibn Hayyān referred to this model as *takwin*. Non-initiates and skeptics often made the classic mistake of interpreting alchemical imagery literally. Based upon misinterpretations of these cover-names, alchemists were scrutinized and scorned for the heresy of actually attempting to create microscopic fully formed humanoids or resurrected animal creatures in their vessels.

Through alchemical initiation, the aspirant witnessed the union of body, soul and spirit as three royal and divine alchemical substances underwent unification into a new *triune being*, and through this insight gained gnosis of not only his own constitution, but the living processes of birth, growth, regeneration, and rebirth. To the alchemist, everything was an organic manifestation – minerals, plants and animals – evidenced by all containing growth and regenerative principles embodied and expressed in form. This mystery was a great secret and a dangerous notion at times amounting to the heresy of animism. Initiation into this mystery via Gnostic awareness brought the concepts of an eternal substrate of life essence and the consciousness of the alchemist into union. As a creative divine being in partnership with God, the great Gnostic heresy of self was that realized man, as an immortal spirit-soul, occupied the position of mediator between the forces of heaven and Earth.

Model of the Natural Order
Hermetic, Gnostic and Neoplatonic philosophies have always offered the possibility of gnosis of the natural order, also known as the *word* or *law* in these ideologies. Philosophers gazed into the heavens, observed the natural world, witnessed dramatic change and evolution of culture, art, science, and technology, and attempted to identify systematic patterns of correspondence. The aim was to comprehend self and the cosmos as a unified whole. A combination of observation and creativity served as a primary method to the Gnostic realization of unity between self and landscape for the philosopher. Alexandrian philosophers understood that

each person possesses a capacity to perceive this total reality – *the mystery* – via proper initiation, which acts to reawaken the ability to experience creation as multilayered, interconnected and pulsating with life and divinity. The conceptual model of a unity being broken down into its smallest parts and reassembled into a united whole once again – *solve et coagula* – finds a small-scale physical parallel in the Philosophers' Stone as a practical representation of this process as it occurs on all levels of reality.

The second application of the *Philosophers' Egg* as womb of creation was the Earth as a fertile cosmic egg incubated by heavenly spheres of influence. As a microcosmic model of the Earth, the digestion vessel contained within it the *Tetrasomia* – earth, water, air, and fire – as a type of cosmic homeostasis. Foundational *Tetrasomia* theories significantly influenced every subject in Stephanos' unified system, making it a prime starting point for exploring correspondences. To comprehend the depth of meaning that the *Tetrasomia* may have held for Stephanos and other Alexandrian philosophers, the origins of the theory must first be explored.

Empedocles, a Greek philosopher living in Corinth and writing during the 5th century BCE, taught that all matter is comprised of earth, air, fire, and water – referred to in Greek as the *Tetrasomia*. Fire and air were elements of a masculine and expansive nature, whereas earth and water were feminine and contractive. In his seminal work *Tetrasomia*, or *Doctrine of the Four Elements*, Empedocles presents the theory in which Love and Strife underlie all interactions and transformations in the cosmos and in man. He explains Four Element Theory not only as a doctrine of the emanation of the cosmos – the world with its objects and qualities – but also as man's composition of body possessing mortal and immortal souls:

> *Empedocles explained that* there are two great living forces in the universe, which he called Love and Strife *and assigned to Aphrodite and Ares. According to Hesiod, the Goddess Love and the God Strife, offspring of Night, were ancient deities, predating the Olympians. The original golden age was the Reign of Aphrodite, when* all things were united and Love permeated the length and breadth of the well-rounded cosmic sphere. But Strife ... broke its Unity, and cleaved the One into Many. It divided the four elements, which ever since combine and separate under the opposing actions of Love and Strife to produce the changing world with its manifold objects and qualities. *As Heraclitus said, "Through Strife all things come into being." Empedocles said that* Strife also divided the one immortal soul of Love into many individual souls, each comprising both Love and Strife in some proportion; these immortal souls are reborn time and again into mortal bodies, which are animated by mortal souls compounded from the four elements.
>
> — Apollonius Sophistes, Exercise for Unity

In Plato's dialogue *Philebus*, Socrates makes a correspondence between man and the universe by explaining that each is composed of the *Tetrasomia*. He asks what holds the *Tetrasomia* together in man's body, the answer being soul. Yet in man, explains Socrates, the *Tetrasomia* is weak and impure, whereas in the universe it is strong and pure. In another lecture, *Timaeus*, Socrates reveals the fractal relationship between humanity, the cosmos and divinity by stating that man and the cosmos are in the image of the *Artisan of the Universe*, which implies shared characteristics of intelligence, divinity and possession of soul. According to Socrates, the soul holds the *Tetrasomia* together in man; hence, the soul of the world — Plotinus' *anima mundi* — must do the same on Earth, as does the cosmic soul in the cosmos. The interplay of the *Tetrasomia*, far from being a mere collection of elements, served as model of the presence of soul that embraces the four elements in their allostatic dance. The Philosophers' Stone was seen as an embodiment of the *Tetrasomia* and therefore *world soul* incarnate.

CHAPTER 5 – STEPHANOS' *CHRYSOPŒIA*

The cosmological framework for Stephanos was primarily medico-astrological as evidenced in his correspondences (presented below in tables largely based upon Stephanos' writings). Earth passes through the signs of the zodiac in reverse – a process known as *precession* – and is marked by the position of the equinox in relation to distant stars. *Precession* was identified by Greek astronomer Hipparchus working between 162 to 127 BCE and later commented on by Plato in Ptolemy's *Almagest*. Plato described *precession* as the *celestial sphere* – called the *eighth sphere* by Ptolemy and later Ficino – rotating in an opposite motion around what was believed to be a stationary Earth. The process to confect the Stone begins theoretically in Aries, the sign of antimony, and works in reverse from the darkness of winter to the life and growth of spring. Thus, the Philosophers' Stone correlates to this cycle, with the stages of its confection demonstrating the *precession of the equinoxes* through the entire zodiac. Cosmologically this process is known as the *Great Year*; therefore, the Stone is an astrological working model of the Earth having completed the cycle of the *Great Year*. The finished Philosophers' Stone equates to the thirteenth zodiacal sign between Sagittarius and Scorpio known as the *Crossing*. The Stone, Earth and the 13th sign were all symbolized by an equilateral cross within a circle (⊕).

⊕ Tetrasomia			
⌓ *Air*	△ *Fire*	▽ *Water*	⍌ *Earth*
Zeus	Hades	Nestis (Persephone)	Hera (Demeter)
Octahedron	Tetrahedron	Icosahedron	Cube
Red Blood	Yellow Bile	White Phlegm	Black Bile
Saffron	Yellow	Translucent	White
Heart to Neck	Navel to Heart	Knees to Navel	Feet to Knees
Hot / Wet	Hot / Dry	Cold / Wet	Dry / Cold
Dross-Bronze	Copper	Gold	Mercury
Vernal Equinox	Summer Solstice	Fall Equinox	Winter Solstice
♈ ♉ ♊	♋ ♌ ♍	♎ ♏ ♐	♑ ♒ ♓

Although the hidden presence of stellar virtues was manifest in matter, the seven celestial bodies or spheres were viewed as ruling the material world; Stephanos therefore viewed them as *governors*. In the digestion vessel, the presence of planetary influences was seen as self-evident based upon matter transitioning through different colors in a precise pattern corresponding to those of the seven planetary bodies. Stephanos therefore invoked planetary essences into the Philosophers' Stone, thus creating a magico-physical substance imbued with the collective governing power of all the planets and their associations.

> *Stephanos explains that the bodies and colours of the seven planets are precisely the seven bodies and colours of this composition, the Tetrasomia. In the same manner that the seven planets pass through the signs of the Zodiac, the seven bodies and colours pass through (i.e. appear in) the composition made up of the four elements.*
>
> – Maria Papathanassiou

Mercury ☿	Saturn ♄ or ☌	Jupiter ♃	Moon ☽	Venus ♀	Mars ♂	Sun ☉
Mercury or Tin	Lead or Antimony	Tin or Asem	Silver	Copper	Iron	Gold
Lustrous	Black	Iridescent or Blue	White	Green-Citrine	Orange-Red	Red-Gold
Preparation	Melanosis		Leukosis	Xanthosis		Iosis
	Nigredo		Albedo	Citrinitas		Rubedo

In this model, the process taking place in the vessel was viewed as a microcosmic model of the interplay between the divine realm (or heavens) and the elements that constitute the Earth. The application of the Philosophers' Stone as a microcosmic model of processes that govern man, Earth and the heavens can be aptly and elegantly summarized by Hermes' maxim of *as above, so below*. Dame Frances Yates, in her work *Giordano Bruno and the Hermetic Tradition*, explains that the *Light* of Gnostic philosophy ultimately emanates from the *Father* in a series of

CHAPTER 5 - STEPHANOS' *CHRYSOPŒIA*

declinations resulting in planetary colors; a philosophical view already ancient by the time Stephanos recorded his work:

> Light descends from the Father to the Son and the Holy Spirit, **thence to the angels,** the celestial bodies, **to fire,** to man in the light of reason and knowledge of divine things, to the fantasy, and it communicates itself to luminous bodies as colour, **after which follows the list of** the colours of the planets. — Frances Yates

According to Yates' explanation, these declinations interconnect to include divine reason and knowledge in man. The colors revealed in the process of confecting the Stone intimately connect man's inner world with the celestial one. Thus, one is able to sense a profound depth of meaning in confecting the Philosophers' Stone during this stage of alchemical history. In order for self-transformation to unfold, the alchemist first needed to realize that the process of confecting the Stone paralleled countless phenomena that make up the natural order at every level. Universal philosophers such as Stephanos divided the natural order into three realms — terrestrial, celestial and super celestial — interconnected by a continuity of influences. The Philosophers' Stone as a model was seen as an embodiment of these. By perceiving these processes as occurring during the course of confecting the Stone, the alchemist gained gnosis of the divine process underlying all change and transformation and thus entered into concurrence with it. Stephanos encapsulates this view in the following excerpts:

> The all is one, through which the all exists. The whole is one ... one nature perfecting that which is sought, which is not composed of many forms but is treated by one art. Hence the all is one, through which the all exists ... The all is one, through which the all exists [and] from it is the all. Opposite natures do not perfect the whole, but one nature [does], the same from itself ... what is sought is one, not this and that but one, coming to be from itself ... the all is one ... the all is one, through which the all exists ... the all is one, through which the all exists ... for nature

delights nature, nature conquers nature, nature rules nature, not this or that nature but the same one from itself through the process.

Monument of Light and Divinity

Archelaos was a late Alexandrian period alchemist in the generation following Stephanos who wrote during the late 7th century. He is mentioned in the Islamic alchemical text, *Fihrist*, and by al-Kindī in his *Fada'il Misr*. In Islamic alchemy, Archelaos is regarded as the author of *The Turba Philosophorum* (or, *Assembly of the Sages*). E.J. Holmyard categorizes him as a poet-alchemist in the tradition of Stephanos and his disciple Morienus. Like Stephanos, Archelaos displays the same double entendre tendencies in his alchemical writings. By the late Alexandrian period, poet-alchemists such as Stephanos and Archelaos had developed poetic expression to such a degree that their writings were valid and accurate on at least three levels of interpretation: 1) specific substances and processes; 2) general substances and processes, and 3) spiritual or redemptive correspondences.

Archelaos was among the last of the Greek alchemists and it is therefore fitting, at the risk of redundancy, to analyze the operative aspects of his work. The following is an interpretation based upon specifics, rather than from a general or redemptive perspective. This is done in an effort to demonstrate that the archetypal recipe first revealed in Maria Hebrea's writings remained constant throughout the period of Alexandrian and Byzantine alchemy as an unbroken tradition from the *Ade* of Maria's seminal work to the *Zethet* of Archelaos' concise poetry.

> *When the spirit of darkness and of foul odour is rejected, so that no stench and no shadow of darkness appear, then* the body is clothed with light and the soul and spirit rejoice because darkness has fled from the body.

In the above passage, Archelaos is addressing the process of *whitening the Stone*. The *spirit of darkness and foul odour* that is rejected from the matrix is sulfur, after which the *flowers of antimony* appear brilliantly white. Darkness of stibnite has literally flown from the body as noxious sulfurous fumes.

> *And the soul, calling to the body that has been filled with light, says: "Awaken from Hades! Arise from the tomb and rouse thyself from darkness! For* thou hast clothed thyself with spirituality and divinity, since the voice of the resurrection has sounded and the medicine of life has entered into thee.*"*

Purified antimony – in this case *flowers of antimony* – is considered the soul of the Philosophers' Stone, and contains growth and regenerative principles. It is therefore understood to be the medicine of life. Gold is the enduring, incorruptible and immortal body principle, the body serving as microcosm of the *One*. The body of gold has reached perfection; lacking in growth and regenerative principles it therefore requires a donor of these. The body of antimony is brittle, easily reduced and therefore very weak, yet contains tremendous growth and regenerative properties due to its crystalline nature. These natures, growth and regeneration, are interpreted as a powerful dual-soul, and thus antimony serves as gold's perfect counterpart, each donating what the other needs. The passage below illustrates this:

> *For the spirit is again made glad in the body, as is also the soul, and runs with joyous haste to embrace it and does embrace it.* Darkness no longer has dominion over the body, since it is a subject of light and they will not suffer separation again for eternity.

Although a little more obscure, a secret hint is embedded below where *divinity* = Zeus-Pitar Amun and therefore the *light of divinity* = *divine water* (i.e. *butter of antimony*), which completes the trinity:

> *And the soul rejoices in her home, because after the body* had been hidden in darkness, *she found it* filled with light. *And she united with it,*

> *since it had become divine towards her, and it is now her home. For it had put on the light of divinity and darkness had departed from it.*

This is evidenced in the passage below where body (gold), soul (antimony) and spirit (divine water) become fully united, the color indicator of which is the white stage. According to both Stephanos and Archelaos, this new triune substance symbolizes the mysteries, and greater spiritual truths knowable through the initiation of actually confecting the Stone. Chemically what is being described is a reaction between the three substances to create a new compound.

> *And the body and the soul and the spirit were all united in love and had become one, in which unity the mystery had been concealed. In their being united together the mystery had been concealed.*

The final passage of this wonderfully profound little alchemical poem indicates the unity of initiation, mystery and monument that contains within it the potential for Gnostic enlightenment through the creation of the Elixir or, as Archelaos calls it, the *monument of light and divinity*.

> *In their being united together the mystery has been accomplished, its dwelling place sealed up and a monument erected full of light and divinity.*

Archelaos brilliantly and accurately reveals the archetypal recipe for confecting the Stone, yet disguises the process in Gnostic alchemical poetry addressing mystery, initiation and monument. Maria's original industrial craft-technology had matured and evolved from a material Art associated with wealth to an initiation into Gnostic mysteries of *light and divinity*; Stephanos' *Chrysopœia*, the alchemical processes as a whole, a means of revelation and personal redemption. With Maria and Archelaos as first and last – the *Ade* and *Zethet* of the great Alexandrian alchemists in an unbroken tradition – the *Ouroboros* of Alexandrian *chrysopœia* finally bites its tail.

Legacy and Transmission

Alchemy is fully understood only through its expressions and the social context in which these are embedded. Alexandrian philosophers viewed the processes of creation as occurring not only at the human or Earthly levels but also on a cosmic scale and in correspondence with each other. The world was not held together by language and culture but by a natural order or system that linked the *All* of creation to the *One* source through the *mystery* from which multiplicity came into being – the manifest from the unmanifest. For Stephanos, alchemy's ultimate value lay in its potential for a type of workbench gnosis initiated via the art of confecting the Philosophers' Stone. To be an adept alchemist was to be initiated into a mindset of reflective awareness in accord with alchemical operations. As the alchemist learned how to learn and refined his approach, he brought forth a gnosis that manifested in a new state of being. To be initiated into the mysteries meant first being introduced to intellect, self-reflection, intuitive awareness, and ultimately gnosis and henosis. The crucial role of abstract expression in alchemy's development as exemplified in Stephanos' lecture series was not simply meant to transmit ideas, but to transmit a way of arriving at knowledge by reading between the lines – in effect becoming unchained from formal discursive reasoning.

Stephanos' perfect fusion of operative and speculative alchemy is fundamental to his greater unified philosophy. As the richness and diversity of abstract alchemical expression increased, alchemy's identity unfolded accordingly. Alchemy became increasingly associated with its abstract manifestations, concepts, symbolism, and imagery. Some have attempted to define alchemy based on its language and abstract sophistications, but these represent only a small part of its totality. Gradually, as alchemy's inner world of meaning became ever-more diverse and complex, it became disassociated with its original character as a holistic artisanal craft technology imbued with divine purpose and meaning; a technological disconnect resulting in near-total fragmentation. Late in alchemy's history a rift between operative versus speculative alchemy developed that was actually more indicative of the tension between wholeness and fragmentation; a tension that lingers as culture, science and technology continue to favour reductionism over integration. Alchemy in its original form was neither operative nor speculative; rather it was a fantastic alloy of both in various ratios throughout its history, revolving around the profound experience of confecting the Philosophers' Stone.

Stephanos' work and personal style of delivery influenced later poet-alchemists Heliodoros, Theophrastos, Hierotheos, and Archelaos. His approach to alchemy was foundational to the development of Islamic alchemy where he is known as *Istafan* or *Adfar of Alexandria*; his influence apparent in the formative alchemical corpus attributed to Jābir ibn Hayyān (known in the West as Geber). *The Turba Philosophorum*, *Rosinus*, *Rosarium Philosophicum* and *Dominicus Pizimentius* all have been influenced, cite or comment upon Stephanos. Even Sir Isaac Newton sourced Stephanos' text for his *De Scriptoribus Chemicis*. Artephius, in chapter twelve of his 12th century landmark alchemical work, *Secret Book*, names only a single alchemist – *Adfar of Alexandria* (Stephanos of Alexandria) – as his source for the proper digestion technique for confecting the Stone.

Jābir ibn Hayyān's brand of Islamic alchemy and Artephius' Euro-Islamic alchemy evolved into an explosion of incredibly experimental operative traditions. Artephius' early work laid foundations for European alchemical development and bridges post-mediaeval Islamic-European and Stephanos' Alexandrian alchemy. It was Stephanos' personal student, a hermit-monk and adept alchemist named Morienus who hand-delivered alchemy to the Islamic Empire, thus ending the exciting chapter on the history of Alexandrian alchemy. Indeed, all forms of Islamic and European alchemy that followed Stephanos owe a great debt of gratitude to him and his bold methodology; one that sought personal redemption and gnosis of the mysteries of the cosmos via alchemical initiation. Stephanos' Stone was many things: the White Stone, *coral of gold*, the embodiment of physical and spiritual vitality, and most importantly the adept alchemist and *Universal Philosopher* himself as mediator between the forces of heaven and Earth. Stephanos' *chrysopœia* was the *All*; a symbol of the processes of Nature, the multifold expressions of form, and the universal medicine he called the *pharmakon* imbued with the power to perfect everything it touched due to its own correspondence with the *All* of creation. His *chrysopœia* was the *One* – an approach, methodology, idea, unification of self and landscape, and all manifestation as the collective face of the *One*.

Chapter 6 – From *Chrysopœia* to *Al-Kīmyā'*

The whole key to accomplishment of this operation is in the fire, with which the minerals are prepared and the bad spirits held back, and with which the spirit and body are joined. Fire is the true test of this entire matter.

Morienus al-Rumi

The year 632 CE marked the death of Muhammad (ﷺ) and the birth of an empire. The first Caliphs sent armies to attack Persia and the Byzantine Empire. These two great former powers, after having grown complacent, were crushed by the aggressive invaders. Despite these victories however, the world had never before witnessed the form of post-conquest tolerance demonstrated by Islamic Caliphs.

> *The hallmark of Muslim conquest was its surprising tolerance.* **Following Muslim conquest, the local populace and political infrastructure was generally allowed to continue on, albeit under Muslim control. … The Muslim people were remarkably tolerant of the Jews and Christians of captured regions.** *Many rose to positions of relative power and affluence in the new cities like Baghdad. This led to a stable and smooth running empire.* – Dr. Gustav LeBon, **Civilization of the Arabs**

The conquering forces turned next towards the treasures of Egypt and the remaining Byzantine provinces, where they met little resistance. Egypt was invaded by Caliph Umar whose forces destroyed Byzantine resistance at Heliopolis, modern day Cairo. This victory, coupled with the taking of Babylon, left Alexandria exposed to attack. Umar laid siege to the city for 14 months before the city was finally turned over by Byzantine officials on 8 November 641 CE. The aftermath was not a scene of slaughter and destruction as one might initially imagine. Islamic armies operated in a manner quite apart from conquerors of the past, as Dr. Gustav LeBon highlights:

> The mercy and tolerance of the conquerors were among the reasons for the spread of their conquests and for the nations' adoptions of their faith and regulations and language, which became deeply rooted, resisted all sorts of attack and remained even after the disappearance of the Arabs' control on the world stage, though historians deny the fact. Egypt is the most evident proof of this. It adopted what the Arabs had brought over, and reserved it. Conquerors before the Arabs—the Persians, Greeks and Byzantines—could not overthrow the ancient Pharaoh civilization and impose what they had brought instead.
>
> – Dr. Gustav LeBon, Civilization of the Arabs

Within 20 years, the new Empire stretched from Spain in the west to Pakistan in the east, and with the new territory came the spoils of all the ancient civilizations with their centers of learning. This marked the beginning of Islamic empire building and a new age characterized by bustling intercontinental commerce and tremendous advances in experimental science, medicine, mathematics and, of course, alchemy. The intellectual heritage of Greeks, Persians, Indians, and Egyptians became available to translators almost immediately. During the early years of the Islamic Empire – a time known as the Umayyad Caliphate – the great serpent of Alexandrian alchemy would shed its skin in order to evolve into a new dynamic form of highly experimental, applied and medicinal alchemy. The early efforts to acquire knowledge by the Umayyads resulted in great cultural and technological achievements. Damascus became the center of activity, drawing physicians, philosophers, scholars, and intellectuals to study logic, astronomy, astrology, mathematics, medicine, and a number of industrial tradecrafts.

A new school of thought appeared in Damascus around the same time that Syriac Christianity was thriving in the region. This school of thought,

CHAPTER 6 - FROM *CHRYSOPŒIA* TO *AL-KĪMYĀ'*

known as the *Qadariyya*, challenged Islam's idea of divine fate, preferring instead the notion that God endowed each man with free will. Syrian Semitic Christians were well established in the monastic style of practice and skilled at theological debate. Islamic scholars studied logic and philosophy in order to refine their expertise at debate, which resulted in constructive dialogue between Islam and Christianity. Christians and Muslims both held civil positions within the Umayyad Caliphate. It is important to grasp the cultural, religious and political climate in Damascus in order to frame the relationship between Morienus al-Rumi (of Byzantium) and Khālid ibn Yazīd properly.

Khālid and the Arabic Translation Movement

With conquest came the need for a centralized language of culture throughout the Islamic Empire. The Umayyad administration initiated a translation movement in which Greek, Persian and Coptic texts were gathered and translated locally into Arabic. Khālid was interested in Greek scientific works – especially alchemy – and assisted in the ongoing translation movement by collecting and translating a great number of works. Some of these works are known, while others are yet to be located.

Khālid is revered by Islamic historians as possessing all the right attributes of nobility. He was an eloquent speaker and poet. He is described as having a strong and courageous yet generous character and an expert on practical technology, medicine and alchemy. His lineage was exactly what one might expect from nobility. His grandfather Mu'awiya was from a cultured merchant class family in Damascus, and eventually became Caliph. Mu'awiya's son, Yazīd, was highly educated in genealogy, astronomy and Arabic literature and remembered for his beautiful poetry. Yazīd was also very involved in engineering irrigation projects. Khālid was Yazīd's second son and although the two were very close, Khālid would never succeed his father as Caliph; developing instead, a strong fascination with alchemy as Dr. al-Hassan, quoting al-Nadim, explains:

> *Khalid was an Umayyad prince and a grandson of Mu'awiya the founder of the dynasty. When his brother Mu'awiya II died in 683 CE he was not elected to be a caliph because of his young age. Having been relieved from the concerns of the caliphate, he turned his attention to the pursuit of high culture.* Alchemy and astrology were pursued by rulers and dignitaries throughout history. *In Europe the fascination of rulers and the upper classes with these pursuits lasted until the eighteenth century.* At Khalid's time alchemy and astrology, beside medicine, had the same importance. *Ibn al-Nadim gave the motives of Khalid in pursuing alchemy as follows:*
>
> "He was a generous man, for when someone said to him, 'You have expended most of your energy in seeking the Art,' Khalid replied, 'In so doing I have sought only to enrich my friends and brothers. I coveted the caliphate, but was unsuccessful.' Now I have no alternative other than attaining the culmination of this Art, so that anyone who one day has known me, or whom I have known, will not be obliged to stand at the gate of the sultan, petitioning or afraid."
>
> – Dr. Ahmad Yousef al-Hassan Gabarin, The Culture and Civilization of the Umayyads and Prince Khālid ibn Yazīd

THE COMPOSITION OF AL-KĪMYĀ'

Morienus' book, *The Composition of Al-Kīmyā'*, is a record of alchemy's transmission to Islamic culture. In its Arabic form, the book was also one of the first – if not *the* first – alchemical text to be translated into Latin, thus qualifying it as a landmark work on the subject. To all European alchemists, the story and alchemy of Morienus and Khālid held an esteemed and prominent place as being an authentic canonical text. Lee Stavenhagen of Rice University is an expert on Morienus who translated, edited and published his work as *A Testament of Alchemy* in 1974.

CHAPTER 6 – FROM *CHRYSOPŒIA* TO *AL-KĪMYĀ'*

Stavenhagen points out that the text as it appears in Latin clearly suffered at the hands of editors:

> *The main fact to emerge from this study is that the Morienus as it stands in the printed edition cannot be treated as a unified, original work which either was or was not translated according to such and such a time, place or authorship as purported.* The printed version is only the end product of at least two hundred years, and perhaps twice that, of textual revision and recombinations.
>
> – Lee Stavenhagen, The Original Text of the Latin Morienus

Stavenhagen's research on the earliest unrevised Latin work inspired his 1974 publication. Although he concluded that there must have been an original Arabic account, he remained dubious about the prologue:

> ... *the inconsistencies that have given scholars cause to regard the Morienus with suspicion, the pompous claim to priority and authorship in the prologue* (which should have been quite good enough to obviate any serious argument on its authenticity in the first place), the inconvenient dating and attribution, the improbable-sounding embellishments of the introductory tale, all are peculiarities of the late version that found its way into print. Stripped of these flourishes, *the original Latin Morienus could very well be one of the earliest alchemical translations from Arabic* ...
>
> – Lee Stavenhagen, The Original Text of the Latin Morienus

Stavenhagen's conclusions were reached without reference to the two extant Arabic sources. The question of authenticity, both of the prologue and the dialogue between Morienus and Khālid, was finally settled by noted scholar in the history of Arabic and Islamic science and technology, Professor Dr. Ahmad Yousef al-Hassan Gabarin, who located two original Arabic copies of *The Composition of Al-Kīmyā'* along with 13 additional Arabic sources that cite the text. The original complete account is the Arabic manuscript *Fatih 3227 (fol. 8b-18b)* and secondary source *Şehit Ali Pasha 1749 (fol. 61a-74b)* in the possession of the Prof. Dr. Fuat Sezgin

Research Foundation for the History of Science in Islam at Istanbul. The prologue turned out to be genuine, although slightly edited; Stavenhagen's translation consistent with the Arabic account.

Dr. al-Hassan was in the process of providing a more literal translation from the *Fatih* prior to his passing in April 2012. His son, Ayman Gabarin, generously granted the author permission to source his father's material. The following Arabic prologue, known as *The Account of Ghalib*, is from Dr. al-Hassan's article *The Arabic Origin of Liber de compositione alchimiae*. *The Account of Ghalib* presented here is the controversial prologue addressed by Stavenhagen, yet is a fresh translation from the original Arabic by Dr. al-Hassan. Unfortunately, the Fuat Sezgin Research Foundation declined to provide a copy of an original Arabic manuscript. Interpretations of *The Composition of Al-Kīmyā'* that follow the prologue are based on Stavenhagen's translation from the early Latin source.

The Account According to Ghalib

Credit for the transmission of Alexandrian alchemy to the Islamic Empire is, according to Arabic sources, given to two men: Morienus al-Rumi and Prince Khālid ibn Yazīd. The tale has been handed down as an historical account by Islamic historians. Latin translators viewed the account as a legendary tale signifying the strength of Christianity's moral influence.

> *Prince Khālid and his court were like a marvel from the tales of Es-Sindibad, but to Latin clerics and scribes familiar with Medieval Christianity, Morienus' ascetic reclusion, his splendid erudition chastened by a hair shirt, his remorse in old age and contemplation of death, seemed good proof of Christianity's prior claim to a whole branch of the vast new knowledge discovered in Arabic possession ...* Here, at the feet of a miserable desert saint, the King of the Arabs finds enlightenment! What Christian could have resisted it?
>
> – Lee Stavenhagen, The Original Text of the Latin Morienus

CHAPTER 6 – FROM *CHRYSOPŒIA* TO *AL-KĪMYĀ'*

Given the sociopolitical climate during the early Umayyad Caliphate, it is easy to accept the possibility of a Christian monk as Prince Khālid ibn Yazīd's teacher. Morienus has always been portrayed as Christian and this would certainly be understandable if he was from Stephanos' lineage. Dr. Gustav LeBon, in an analysis of post-conquest regions affected by Islamic expansion, describes the politics of the early Caliphs as being rather tolerant and thus making the story plausible:

> ... *the early Caliphs, who enjoyed a rare ingenuity which was unavailable to the propagandists of new faiths, realised that laws and religion cannot be imposed by force. Hence* they were remarkably kind in the way they treated the peoples of Syria, Egypt, Spain and every other country they subdued, leaving them to practice their laws and regulations and beliefs *and imposing only a small Jizya in return for their protection and keeping peace among them. In truth, nations have never known merciful and tolerant conquerors like the Arabs.*
>
> – *Dr. Gustav LeBon,* Civilization of the Arabs

A strong precedent is preserved in the story of Sergius, known to Islamic historians as Sarjun. He was a Christian and Byzantine official appointed by Heraclius prior to Islamic conquest, who later acted as a negotiator between the two forces. After the conquest, Sergius served as minister of finance for the entire Islamic state and chief paymaster for the army under Caliph Mu'awiya. He remained a Christian after the conquest and later built a Christian church. Capable Christians had a place in the Umayyad Caliphate.

The Account of Ghalib must now be considered a primary historical eyewitness rendition, recorded four years following the death of Emperor Heraclius by Khālid's non-Arabic Muslim retainer Ghalib, of the momentous occasion when the torch of Alexandrian *chrysopœia* was handed over to the Islamic Empire, where it would evolve into its dynamic new, experimental and medicinal form as Islamic *al-Kīmyā'*.

BOOK 1 – ORIGINS & EVOLUTION

رسالة مريانس الراهب الحكيم للامير خالد بن يزيد

بسم الله الرحمن الرحيم

قال غالب (مولى) خالد بن يزيد بن معاوية كان سبب وصول خالد الى الصنعة الكريمة انه) خرج ذات يوم متنزها الى دير <برّان >بدمشق (وكان مغرما بالصنعة كلفا بها لا يؤثر عليها شيئا وكان لا يفتر عن البحث فيها <والتجربة> بها > والسؤال عما يعرض له فيها رجاء ان يصل اليها

فاتاه رجل في يومه ذلك وطلب الاذن بالدخول عليه فادخل فسلم فاحسن
وقال اني اتيت الامير بفايدة لم يات احد بمثلها

فقال له خالد وما الفايدة التي اتيت بها

قال بلغني انك تطلب الصنعة وتسال عنها قال فاستوى
خالد جالسا وقال نعم فقال ايها الامير اني ساكن في البيت المقدس وقد رايت به رجلا سايحا يقال له مريانس الراهب واصل الى الصنعة ياتي في كل عام الى البيت المقدس >ويهب <مالا عظيما ويعطي الفقراء والمساكين

The Epistle of Morienus the Hermit Philosopher to Prince Khālid ibn Yazīd

In the name of God the merciful and compassionate ...

Ghalib the Mawla [non-Arab Muslim retainer] of Khālid ibn Yazīd ibn Mu'awiya said that the reason behind Khālid's accomplishment of the noble art (al-san'a al-karima) is that he went one day on a picnic to Dayr Barran in Damascus. He was fond of the art (al-san'a), fascinated by it so that he would not give preference to anything else. He was constantly searching and experimenting with it and enquiring about what might come his way, hoping that he may arrive at it.

On that day, a man came to him and asked for permission to be allowed to enter and see him. He was allowed in and he saluted eloquently. He said: I brought to the Amir a benefit that nobody else had matched.

Khālid asked him: and what is the benefit that you have brought with you?

He said: I had learned that you are seeking the art (al-san'a) and are asking about it.

(Ghalib said) that Khālid sat straight and said: Yes (The man) said: O Amir, I live in Jerusalem (al-Bayt al Muqaddas) and I saw in it an ascetic called Morienus al-Rahib (the Hermit). Who has attained the Art (san'a). He comes every year to Jerusalem (al-Bayt al-Muqaddas) and donates a huge amount of money and gives the poor and the needy.

CHAPTER 6 – FROM *CHRYSOPŒIA* TO *AL-KĪMYĀ'*

فقال له خالد ان كنت صادقا لاعطينك نهض مسالتك وان كنت كاذبا لابلغنك ما تستحقه.

Then Khālid said to him: If you are telling the truth I shall give you what you will ask for, but if you have lied then you will receive what you deserve.

فقال الرجل حسبي فلقد انصفتني من نفسك

The man said: I am content with this since you have treated me with justice.

ففرح خالد به واعجبه ما كان منه وامر له بجائزة وكسوة <ووعده> خيرا حسبما قال غالب.

Khālid rejoiced and was pleased with what the man had said. He ordered for him a reward and raiment and he promised him good, according to Ghalib.

ثم وجهني معه وجماعة من الموالي

Then he asked me to accompany him with a group of mawali [non-Arab Muslim freemen].

فسرنا في فيا ترفعنا ارض وتضعنا اخرى فلثنا في لك اياما في طل ذلك الايح حتى ظفرنا به فاذاهو شي كبير ضعيف حسن الصورة بهي المنظر عليه جبة من شعر وكان من الذي سنه بالي

We travelled in realms rising up with some terrains and descending with others. We stayed thus for several days in search of that ascetic until we located him. We found that he was an old man, weak, of good appearance and elegant countenance, wearing a woolen robe and his skin showed as if it was worn.

ففرحنا به ورفقنا به وداريناه حتى قدمنا به خالد فادخلناه عليه ففرح به فرحا شديدا ما رايناه على مثل ذلك الفرح قط

We rejoiced at finding him and we treated him kindly and persuaded him until we arrived and brought him to Khālid who was greatly delighted to see him. We have never seen Khālid so pleased at anything before.

ثم التفت الي وسالني عن مسيرنا في البيداء والرجعة فاخبرته بامرنا من اوله الى اخره.

Then he turned towards me and asked me about our journey in the country and our return and I related to him what happened with us from the beginning until the end.

ثم اقبل على الشيخ فقال له ما اسمك قال مريانس الرومي

Then he [Khālid] turned to the old man and asked him: what is your name? He answered: Morienus of Byzantium (lit.; Rumi = Anatolia, Modern Turkey).

فقال منذ كم صرت في هذه الحال

Then Khālid asked: Since when have you been in this state?

قال بعد موت هرقل باربع سنين

He replied: Four years after the death of Heraclius.

BOOK 1 – ORIGINS & EVOLUTION

فقال خالد اجلس يا مريانس فقعد وشرف مجلسه واعجبه ما راى من سمته وادبه.

ثم قال يا مريانس لو كنت في كنيسة او دير كان ارفق بك.

فقال اصلح الله الامير الخير الى الله وبيده يفعل ما يشاء قد صدقت الراحة في ذلك اكثر والنصب في السياحة اشد واتعب وانما يحصد الزارع ما زرع وارجو ان تكون الخيرة فيما انا فيه ان شاءالله تعالى وانه لايدرك الانسان الراحة الا بكثرة التعب.

فقال خالد لو كان هذا من صدر مؤمن ثم قال يا مريانس بلغني عنك فضل ودين فاحببت ان اراك فارسلت اليك.

فال له ميانس ما انا بعجب وفي الناس مثلي كثير الموت لكل اصد وهو اشد على الاجساد> من ذنوبه حوما بعد الموت اطول واقطع واعظم والله المستعان.

فقال خالد اللهم اعنا عليه فانه داهية على كبر سنه.

ثم امر خالد ان اذهب>به <ناحية بالقصر وان اتيه برجل نصراني من الشيوخ العلماء يونسه ويحدثه ليسكن اليه ففعلت ذلك.

Then Khālid said: Sit down Morienus. He sat down and was given a seat of honour. Khālid was pleased by his noble appearance and politeness.

Then he said: O Morienus, would it not have been kinder to you if you were in a church or a monastery?

He said: May God guide the Prince. Good is from God and it is in his hand to do what he wills. You are right, rest in that is more and wandering causes more fatigue and tiredness; but a farmer reaps what he sows, and I hope that good will result from what I am in, if God wills. Man will not achieve rest except by much toil.

Then Khālid said: Had this been said from the heart of a believer. Then he said: O Morienus I heard that you are a virtuous and a devout person, therefore I desired to see you and I have sent after you.

Morienus said to him: I am not unique. Moreover, among people there are many like me. Death is awaiting each; it is harder on bodies than their sins, and what follows death is longer, harder and greater. And God is our aid.

Khālid said: God help us in dealing with him because he is a cunning man despite his old age.

Then Khālid commanded that I take him to a part of the palace and to bring to him a Christian man from among the elder scientists to entertain him and talk to him, so that he can feel at home with him, which I did.

CHAPTER 6 – FROM *CHRYSOPŒIA* TO *AL-KĪMYĀ'*

وكان خالد ياتيه في كل يوم مرتين فيجلس اليه ويحادثه ويساله عن الامم والزمان وسير الملوك واحاديث اليونانين وهو يخبره بعجايب القوم وحكمهم وامورهم لم يسمع خالد بمثلها فوقع منه موقعا عظيما لم يقع منه احد قط قبل ذلك الى بعض الانام.

فقال له خالد يا مريانس اني طلبت الصنعة حينا وبحثت عن امرها وتعبت فيها فلم اجد احدا يخبرني عنها ولا يدلني عليها فاسالك ان تسبب لي من امرها وعلاجها سببا ولك علي ما تسال مع ردك الي من صنعك الذي كنت فيه ولا باس عليك مني.

فقال له مريانس قد علمت انك لم ترسل الي الا لحاجة منك الي واما قولك ايها الامير لا باس علي منك فقد بلغت مبلغا ليس ينبغي لمثلي بعده ان يخاف الا من الله وقد اوليتني ما انت اهله ورايت من رفقك وشفقتك واحسانك ورافتك ومحبتك ما لا ينبغي لمثلي ان يكتم شيئا مما تطلبه> مع <ما ارى من ذكاء فطنتك وفهمك وجميل مذهبك وطلبك فالله المحمود.

Khālid used to visit him twice a day to sit and chat with him asking him about the various nations, past days, biography of kings and the stories of the Greeks (the Byzantines). He told him about the wonders of the people, their rule and their affairs, things that Khālid had never heard before. This caused Morienus to occupy a high place in the esteem of Khālid, more than anybody has ever occupied before.

Then Khālid said to him: O Morienus, I have pursued the Art (al-san'a) for some time, searched about it, and laboured in it but I did not find anybody who can give me information or guide me to it. And I ask you to enable me to know it and learn its treatment and you will have whatever you request, and in addition you will be returned to your original place; and you do not have to be afraid from me.

Morienus said to him: I knew that you did not send after me unless you had a need for me. As to what you said, O Prince, that I do not have to be afraid of you, I have reached the stage at which no body like me should be afraid except from God. You have bestowed on me what is befitting for you, and I have seen of your kindness, your sympathy, your benevolence, your mercy and your love such that a person like me should not hide anything of what you require. Added to this what I see of your intelligence, your comprehension and the nobility of your faith and your pursuit. Praise God.

فتبسم خالد عند ذلك وقال من لم ينفع به الرفق اضر به الخرق والعجلة من الشيطان

قال مريانس أنا أبين لك ولا حول ولا قوة الا بالله العلي العظيم أصلحك الله انصت للحكمة تعرف المطلوب وتفهمه وتعلمه وتفكر في مداخله ومخارجه لتقف عليه ان شاء الله تعالى ان هذا الامر الذي طلبته ليس يقدر عليه احد بالشدة ولا يظفر به احد> بالعنف <ولا يصل اليه من عالم الا بالرفق والتودد والمحبة الصادقة اول ذلك انه رزق من الله تعالى يسوقه الى من يشاء من خله بالقدرة البالغة تى يسبب له >تعلم <ذلك ويكشف له عن مستوره وهو من اعظم مواهب الله تعالى يعلمه من احب من خلقه وعباده والدالين عليه الخاضعين له.

قال خالد >أجل لا حياة الا بالتوفيق من الله عز وجل.<

ثم قال خالد اجلس يا غالب واكتب ما يدور بيني وبينه.

At this, Khālid smiled and said: A person with whom compassion is not effective will be harmed by crudeness. Haste is an act of Satan.

Morienus said: I shall explain to you, and there is neither might nor power but in God, the most high the supreme. May God guide you to the better. Listen to this science (hikma) and you will know what is needed and understand it and learn it and will contemplate its inside and outside traits so that you will become acquainted to it if God, glory to him, wills.

This matter that you have requested cannot be attained by anyone by force and cannot be gained by violence and can only be acquired from a scientist by kindness, affection and true love. First, it is a fortune from God, glory to him. He delivers it to whom he choose from among hi creature by supreme power. He causes him to learn it and discloses to him its secrets; and this is one of the gifts of God, the high. He teaches it to whom he loves from among his creatures and his subjects and to those who are his guides and are submitting to him.

Khālid then said there is no course of action except by guidance from God, the most powerful and dignified.

Then Khālid said: Sit down, O Ghalib, and write what will take place between him and me.

CHAPTER 6 – FROM *CHRYSOPŒIA* TO *AL-KĪMYĀ'*

Morienus said: Know, O Khālid, that God, the high, has created his servants weak from weakness. So They cannot hold back what he had advanced and they cannot advance what he had held back. They cannot know anything except what God reveals to them, and they cannot understand except what God gives them, and they cannot get except what he has opened a way to it by his power. He made those whom he has chosen from among his creatures seek the knowledge of this mental gift that takes out its possessor from the hard toil of this world and lead him to the riches of future life and its delights. They continued transmitting its knowledge by inheritance from one to the other until the science was eradicated and its people had gone and the teachers could no longer be found. From those genuine books that had remained there are the books of the holy men and philosophers that were written by our predecessors and were left as inheritance to those successors whom God has willed to attain this Art that was described to be too elaborate and to be full of falsehoods. If they have said too much and called things by other than their true names and described them by symbols yet without doubt they have explained them, clarified them, and informed about them by the art and by examples and allusions. They tried to keep away the fools and to prevent darkness by intelligent minds and true sayings and so they perplexed men of comprehension and reduced to nothing those who have no belief. They signaled to men of science and comprehension, clarified, and explained. The wise should seek science and should not fall short of it. Let him put his hope in God and desire from him that he enthuses in him true guidance in all his affairs, and to bestow on him critical understanding, good handling, correct interpretation and excellent compiling without deviation.

خلق العباد ضعفا من ضعف لا يؤخرون ما قدم ولا يقدمون ما اخر ولا يعلمون الا ما علمهم الله ولا يدركون الا ما اعطاهم الله ولا ينالون الا ما جعلهم اليه السبيل بقدرته وجعل من اختص من خلقه يطلبون علم هذه الموهبة التي تخرج صاحبها من نصب الدنيا وتوصله الى ملك الاخرة ونعيمها فلم يزالوا يتوارثون علمها واحد فواحد حتى درس العلم وذهب اهله وعدم المعلمون فكان مما وجد في الكتب الصادقة والباقية كتب الاولياء والحكماء التي كتبها من كان قبلنا واورثوها من اعقابهم من اراد الله ان يبلغ هذه الصنعة التي وصفوها بالاكثار والاباطيل وان كانوا اكثروا وسموا الاشياء بغير اسمائها ووصفوها بالرموز وانهم لا محالة قد بينوها واوضحوها واخبروا عنها بالصنعة والامثال والتعريض وحاولوا دفع السفهاء عنها ومنع الظلمة منها بعقول زكية واقاويل صادقة فحيروا ذوي الفهم واهلكوا من لا رأي له وأشاروا لاهل العلم والفهم واوضحوا وبينوا وعلى العاقل طلب العلم فلا يقصر عنه ولكن يكون رجاؤه في الله واليه رغبته ان يلهمه مراشد اموره كلها وأن يرزقه الفهم الناقد والتدبير الجميل والتأويل الصحيح وحسن التاليف من غير زيغ.

The Composition of Al-Kīmyā' is fascinating for a number of reasons. Morienus offers methods to create the Stone in various places throughout his book, the clearest of which is presented here. His general methodology follows Maria's archetypal *Ars Magna* once the cover-names have been decoded. Minor editing and translation errors make decoding the chemistry presented in Latin-based versions of *The Composition of Al-Kīmyā'* a challenge, but not entirely insurmountable.

Quicksilver, or is it Sophic Mercury?

Morienus' first innovation appears during a discussion on working with quicksilver. If the use of the term *quicksilver* is accurate and not a translator / copyist error, it is the first clear indication that elemental mercury – derided as the *vulgar* type of mercury in most alchemical texts – served role in purifying and reducing gold and antimony.

> *According to the authorities,* quicksilver and fire wash latten, or earth, and cleanse it of darkness.

Morienus quotes Maria as his primary source for the technical use of mercury as a purifying agent. Alchemical texts throughout the centuries have always warned against the use of *mercury of the vulgar*. Morienus boldly proposes that it is not only useful, but also that its use began with Maria Hebrea.

> *Maria also said that there is nothing which can* remove from latten, or earth, its darkness, or proper colour of blackness, **but** quicksilver covers it at first, turning it white; **when the** latten overcomes the quicksilver, and reddens it. **And a philosopher said that** quicksilver cannot remove from latten the substance of its colour, or change it except in appearance.

Archeological discoveries reveal that quicksilver had been used in Egypt during antiquity. Greeks and Romans of the ancient world were also very aware of quicksilver and its uses, writing in detail about it long before

CHAPTER 6 - FROM *CHRYSOPŒIA* TO *AL-KĪMYĀ'*

Maria's time. Aristotle discussed quicksilver's role in religious ceremonies. Theophrastus listed Spain as a primary source, addressed its industrial value to Greece and Rome, and provided recipes for creating vermilion pigment and isolating quicksilver. Vitruvius detailed extraction techniques and properties of quicksilver amalgams. Pliny wrote a complete treatise on natural cinnabar and synthetic vermilion, listed mining sources, and even finished with a recipe to create artificial quicksilver. The technique for purifying and reducing gold with quicksilver was already ancient and widely employed by Maria's time. It is plausible that quicksilver's alchemical use began with Maria Hebrea.

> *The ancient Egyptians knew and used tin and copper amalgams. Some mercury flasks have been found in tombs dating back to 1500 or 1600 BCE. From the sixth century BCE their literature makes increasing mention of mercury, its preparations and uses.*
> *– Jiame Wisniak,* The History of Mercury

If quicksilver was perceived to be an agent of purification for gold, it is not only possible but perfectly natural that it would be employed to purify stibnite as well. Zosimus employed bitumen to *whiten the stone.*

From an operative perspective, this technique is efficient in that quicksilver does not amalgamate well with antimony, but bonds easily to sulfur. Stibnite and an excess measure of quicksilver will result in mercury bonding with the sulfur to create black mercuric sulphide (HgS). When mercuric sulphide is formed, it is black, yet when heated it becomes vermilion. This may explain the first blackening followed by reddening in the above passage attributed to Maria. Mercury begins to vaporize at around 357°C. Some of the excess mercury will oxidize and then decompose at 500°C. The remaining mercuric sulphide will decompose at 580°C, thus effectively removing any remaining sulfur from the matrix. Raising the temperature slightly will leave quite pure antimony remaining, which will then oxidize to *flowers of antimony*. This process is well within the technical capabilities of a skilled operator and can be written as follows:

whitening the Stone via mercury reduction

$$3\ Hg + Sb_2S_3 = 2\ Sb + 3\ HgS \quad \rightarrow \quad Sb + \triangle = Sb_2O_3$$

Morienus calls his *whitening the Stone* process and subsequent *latten* creation the *first process*, with the caveat that although it is a composition, it is not the *true composition*. Quoting another unnamed philosopher, he explains how the *whiteness* that occurs during the *whitening the Stone* process can be attributed to the *marvellous power* of quicksilver; yet when the quicksilver technique is applied to gold the color indicator is red rather than white. Morienus is apparently describing two separate processes.

> But *latten, or earth, can take on the substance of whiteness from quicksilver,* which has a marvellous power to cause all the colours to appear after washing, *removing blackness and impurity and rendering white, except in the case of latten, which reddens it.*

In a passage further into the dialogue, Morienus recaps the process, providing some additional details. His use of the term *elixir* below is in a generic sense, referring to gold calx prior to being joined with molten *flowers of antimony*, which must be *fine and pure*.

> In preparation of the *body, or earth,* for purification, you *make it very hard and dry in nature, and then whiten it by purging and washing,* and *invest it with the spirit,* aid the *quest for the white or milky vapour,* for *the red or fiery or virgin's milk,* before the elixir is put onto it, for *the elixir is absorbed only by a fine and pure body* having no foulness in such a way that *it is beautifully tinted after the elixir has passed through it. This is the composition of the authorities, and is the first composition.*

It is impossible to prove conclusively whether Maria used quicksilver to *whiten the Stone*, yet she may have hinted that this was the case by

CHAPTER 6 – FROM *CHRYSOPŒIA* TO *AL-KĪMYĀ'*

discussing a special water and associating Greek Hermes (Latin: Mercury) with *whitening the Stone* in the following passages at the beginning of her dialogue:

> Mary said: O Aros, oftentimes Nations have died about this part. Know you not, O Aros; that *there is a water or a thing which whitens Hendragem*?

And:

> Mary answered: *Hermes in all his Books has mentioned that the Philosophers whiten the Stone in one hour of the day.*

If, however, Morienus' use of the cover-terms *latten*, *body* or *earth* indicates a single compound created by fusing gold calx with molten *flowers of antimony* – known as *glass of antimony* – his reference to *mercury* can only mean *Philosophical* or *Sophic Mercury*; this likely being the case. In this scenario, *latten* represents the *sophic* gold calx, known later by the cover-term *Sophic Sulfur*, which is then amalgamated with *Sophic Mercury* to confect the Philosophers' Stone. This scenario explains why Morienus describes color indicators of *latten* in terms of *darkness* and *proper colour of blackness*. If *latten, body* or *earth* are actually cover-names to indicate a gold-antimony fusion, then he is addressing color indicators according to the traditional color regimen, as *Sophic Mercury* causes a blackening, followed by whitening and finally reddening during the Stone's confection.

Summary of the Secret

Morienus began discussing operative details in the dialogue with Khālid by first presenting an overview of the entire process. This template must serve as Morienus' general method of operation if the chemistry hidden within it is to be accurately decoded and analyzed. In the passage below, he establishes that he is an adept alchemist and that a certain level of expertise and patience is required:

BOOK 1 – ORIGINS & EVOLUTION

> *I have attained to the full knowledge of these things* **by the will of the great Creator most high. And the very root of this knowledge is to** *act with care and perception* **at the time of composition, avoiding all haste and error, and watching patiently day and night for fixation.**

Morienus uses the cover-terms *blood, or virgin's milk* to indicate *butter of antimony*. *Blood* is a cover-name to indicate a liquid nature and a life-giving medium containing spirit-soul that will later be added to the body of *latten*. The term *virgin* refers to *flowers of antimony* as being both a female principle and purified to whiteness. *Whitening the Stone*, followed by creating *blood, or virgin's milk*, serves as one of the first steps in Morienus' process.

> **But** *that which prepares this body is blood, or virgin's milk,* **for it unites and joins all the various substances and properties into one body, it being only necessary to apply to them a gentle heat that long continues at the same degree, neither increasing nor decreasing.**

In the passage below, Morienus refers to *blood* by an additional cover-name: *eudica*. He characterizes *eudica* as reacting easily with other substances to create salts that are fire-resistant; an accurate description of the various antimony chlorides used industrially as modern fire retardants applied to surfaces and clothing. His intriguing hint to *look for it in vitreous minerals* is in reference to *glass of antimony* made from molten *flowers of antimony*, also called the *Green Lion* – a substance with which he was thoroughly familiar.

> **But** *the substance eudica should be used to agitate the bodies you have converted to earth so that they are not burned,* **for if they were quickly consumed, they would not long contain their spirits. Eudica consists well with all bodies,** *enlivening and preparing them* **without confusing them, but** *converting certain of them into others and resisting the heat of the fire. If you wish to obtain eudica, look for it in vitreous minerals. But having found it, you must take care not to use it* **if its spirit has escaped,**

CHAPTER 6 – FROM *CHRYSOPŒIA* TO *AL-KĪMYĀ'*

for you would spoil your operation. Therefore preserve it when you find it.

The last line in the above passage is likely a mistranslation of instructions to prevent vapours from escaping. The *foul earth*, in the following passage, is described as foul only in the sense that it has yet to undergo transformation and final purification:

> *Now* the foul earth readily receives the white sparks and prevents the destruction of the blood, *or air, or virgin's milk, during decoction.*

Morienus' description of *blood* being broken in the passage below is in reference to decomposition that occurs during the beginning stages of the reaction. Like Cleopatra, Synesius and Stephanos before him, Morienus considers the White Stone to be the fruition of the work, yet he acknowledged the option of a Red Stone.

> *But* such is the blood's strength that it must be broken in order to promote rather than impede, and this is done after whatever remaining still dark of the confused minerals has been whitened, thus accomplishing the full fruit of this magistery, *the truth of which you may well not at first have seen. That is in sum the secret of your operation, as I have condensed it and set it forth for you.*

Morienus' summary, as interpreted above, is an accurate and efficient method for confecting the Philosophers' Stone and appears to be a combination of Maria's two recipes.

Morienus' text is the first to so clearly describe the liquid antimony menstruum that he referred to by the terms *eudica, blood, virgin's milk, blood water* or *foul water* and *fire*. Yet this product is identical to *divine water* of the early period. He is consistent with the thematic meanings of his cover-names to indicate a thick liquid containing life essence. He believed that this fluid revivified the body of powdered *latten*. Later in European alchemy, Basil Valentine clearly adopted Eudica as Eurydice in his 11th Key, yet Morienus' *eudica* would also come to be known by the appropriate cover-names *Secret Fire* or *water that burns*.

Morienus' second innovation is that he created *latten* in a sealed crucible first, mirroring Maria's *Ars Brevis* method of operation. However, he followed this with a chemical reaction between the *latten* and *eudica*, in a sealed digestion vessel in the manner of her *Ars Magna*. By doing so, he essentially unified Maria's two methods. Maria's and Morienus' methodologies are efficient ways to create the Philosophers' Stone, yet Morienus' technique – based upon earlier Alexandrian predecessors – would go on to serve as the archetypal recipe foundational to European alchemy.

One Root, Many Names

> ... *as to whether this operation has one root or many, know that* it has but one, and but one matter and one substance of which and with which alone it is done, *nor is anything added to it or subtracted to it. When certain of his disciples asked Herakleios what you have asked me, he told them how* a single root grows into many things which return again to one ...

> Its names may vary, *yet I say it is but one.*

The root of alchemy is antimony, which is found in stibnite ore. A reminder at this point is warranted regarding stibnite's role and availability in antiquity. Ground stibnite for cosmetic use was called *koḥl* during the period in which Morienus and Khālid discussed alchemy. It was easily sourced, found in every home and used by males and females of all ages and all walks of society. It could be purchased in any souk or bazaar and was truly a household item of very little expense. Even the extremely poor could afford it. Alchemists knew that it was possible to extract the *essence of koḥl* from it. The term for extracted *essence of koḥl* is *al-koḥl*, from which we derive our term *alcohol*. The prime material was humble

CHAPTER 6 – FROM *CHRYSOPŒIA* TO *AL-KĪMYĀ'*

ḳohl. This was the first riddle that an alchemist needed to solve. The second key to deciphering the encrypted lexicon of alchemy was to comprehend the function of cover-names as the scientific jargon of the day to express substances and processes in code.

> *It is cast in the streets and trampled in the dung, but let none take pains to extract it. Fools have often wasted much zeal on the dung in hopes of extracting it, but they have acted in ignorance.* The wise know that this unique thing is hidden and that it is what contains the four elements within itself, *having power over them.*

> *The wise used many terms in the books, one being 'sperm', which, when converted, becomes blood, and then is consolidated with the flesh as though part of it. Thus the process of generation proceeds by a succession of forms,* until man is made. Another *simile is that of the palm, on account of the colour and natures of its fruits before they ripen. Others are such as the tree of bad seed, or wheat, or milk,* or many other names. *Though all have but one root,* there is an operation that alters each one, giving many new colours and natures, hence many names.

Antimony is what Morienus refers to as *sperm* or *seed*. It is converted to *blood* (*butter of antimony*) and consolidated with the *flesh* (gold-antimony glass). This is a fair summation of Morienus' method of operation. He is attempting to reveal to Khālid the fundamental lesson in working with cover-terms: that the truth lies in the actual substances or processes and not in the metaphors that conceal or reveal them, of which there can be countless variations.

> *... the conduct of this operation, you must have pairing, production of offspring, pregnancy, birth, and rearing. For union is followed by conception, which initiates pregnancy, whereupon birth follows. Now the performance of this composition is likened to the generation of man ...*

It is clear that Morienus is attempting to impress upon Khālid that a thorough grasp of how cover-terms are developed and used is essential to their deciphering. This lesson was vital to comprehending alchemical substances and processes and remains a challenge even in today's alchemical research. Alchemists of every generation, in every tradition, repeatedly warn against literal interpretations and stress the fact that alchemical language served to encrypt real materials and methods. Indeed *The Composition of Al-Kīmyā'* sheds light upon a great many cover-names that later made their way into Islamic and European alchemy.

> [Morienus quoting Zosimus to Theosebia] *"I will show you that the wise varied their maxims and compositions only because they wished them to be understood by men of wisdom and prudence, while the ignorant should remain blind to them; clearly it was for this reason that the wise wrote variously in their books of the stages of the operation. There is but one stage and one path necessary for its mastery. Although all the authorities used different names and maxims, they meant to refer to but one thing, one path and one stage."*

Morienus then confirms Zosimus' statement about the prime material and its manner of manipulation. In another passage, he stresses the value of practical experience over theory derived from books:

> *It is true that this is but one matter, having a single path, for its composition is the same. The one is just like the other ...*

> *... one who has seen this operation performed is not as one who has sought it only through books, for there are books which mislead those in quest of this knowledge. And the greater part of those books are so obscure and disorganised that only those who wrote them can understand them.*

CHAPTER 6 – FROM *CHRYSOPŒIA* TO *AL-KĪMYĀ'*

First Composition – Preparatory Work

The first operation is to purify, reduce and fuse *flowers of antimony* with gold calx. In doing so, the antimony will begin as black stibnite and be manipulated to white and yellow *flowers of antimony*, ending in a yellowish-red or green *glass of antimony* ultimately fused with gold. This fusion is repeatedly referred to as a *body* or *earth* and is the first operation in Morienus' preparatory work. The second operation is to create *blood* or *water* in the form of *butter of antimony*.

> ... this operation must be done twice, and two compositions must be completed, one after the other, and when the second [composition] has been completed, the whole operation is ended ...

The passage below occurs in a section prior to that on decomposition. That it is not a true composition indicates that Morienus is discussing a compound reagent and not the actual Philosophers' Stone. He describes this composition as being one third of the whole operation.

Composition and operation #1 – latten, body or earth
(Composition of the authorities, not the true composition)

> A composition does precede putrefaction, but this is not a true composition ... It consists in extracting the water from earth and dispersing it, so that the earth begins to putrefy. When it has thus putrefied and been cleansed, the whole operation is done, with the aid of great God most high. Then you will attain that which you have sought, which is the composition of the seers, being a third part of the whole operation. And know that you have accomplished nothing of your operation until you have cleansed, dried, and whitened the impure body, or earth, and so infused it with spirit, or fire, that the tincture descends into it and enters it with spirit, or fire, that the tincture descends into it and enters it after it is so cleansed and improved, there being no further impurity or foulness in it.

In the passage above, Morienus clearly asserts that *whitening the Stone* is the initial process. By the cover-terms *impure body* or *earth* in the above passage, he is specifically referring to stibnite. Alexandrian alchemists believed that this black stone became cleansed and infused with spirit during the process of *whitening the Stone*. The *foul odour* is a reference to the noxious sulfur dioxide fumes given off during the process:

> ... when the white spirit or milk has been made and raised [exalted], and the residual body well prepared, so that *its darkness or blackness is removed and that foul odour withdrawn from it*, the extraction of the spirit from the body has been accomplished. And at the moment of conjunction of that spirit with that body, great wonders will be seen.

The passage below suggests that the whitened Stone – *flowers of antimony* – is a foundational stage towards achieving *red or fiery virgin's milk*, better understood as molten *glass of antimony*:

> In preparation of the body, or earth, for purification, you *make it very hard and dry in nature, and then whiten it by purging and washing*, and invest it with the spirit, *aid the quest for the white or milky vapour, for the red or fiery or virgin's milk, before the elixir is put onto it*, for the elixir is absorbed only by a fine and pure body having no foulness in such a way that it is beautifully tinted after the elixir has passed through it. This is the composition of the authorities, and is the first composition.

Cover-names for working with antimony present an obstacle only to those unfamiliar with antimony in its various forms. Sir Isaac Newton, commenting upon Morienus, provided an equation that might appear nonsensical to anyone other than an alchemist fully familiar with the chemistry of antimony:

> White fume = living mercury, lemony fume = yellow fume = yellow, red fume = red orpiment
> – Sir Isaac Newton, Keynes MS 12 *(from Morienus)*

CHAPTER 6 – FROM *CHRYSOPŒIA* TO *AL-KĪMYĀ'*

Sir Isaac Newton's equation begs the question how a single substance derived from a greyish-black ore could become white, lemony, yellow, red, or ultimately a *Green Lion*. Basil Valentine made the same observation:

> *From Saturn [antimony] proceed many colours, that are made by preparation and art, as black, ash-colour, white, yellow and red ...*

The answer lies in the chemistry of antimony as follows:

- **White fume** = crude antimony trioxide (Sb_2O_3). Forms when stibnite is roasted at 350°C. The sulfur vapourizes leaving a white powdery volatile solid known as *flowers of antimony*, which can be further fixed.

$$2\ Sb_2S_3 + 9\ O_2 = 2\ Sb_2O_3 + 6\ SO_2$$

- **Yellow fume** = antimony tetroxide and pentoxide (Sb_2O_4; Sb_2O_5). Forms when *flowers of antimony* are heated in air and takes the form of a fixed white or yellow powdery solid. The yellow coloration can revert to white upon cooling.

$$Sb_2O_3 + 0.5\ O_2 = Sb_2O_4$$

- **Red fume** = antimony glass begins to form at 656°C as *flowers of antimony* reach their melting temperature. Upon cooling, the glass will appear lustrous, translucent and tending from dark yellowish to deep red, or alternately a green color depending on the sulfur content and other impurities present. The glass is then ground to a fine powder.

Preparatory work that begins with black stibnite converted to white and then yellow *flowers of antimony*, followed by a ruddy *glass of antimony*, proceeds according to the same color regimen for confecting the Stone; Newton's black-white-yellow-red fumes.

- **Green Lion** = this term was first used by Morienus as a cover-name to indicate *glass of antimony*. Green *glass of antimony* occurs due to the amount of sulfur that remains following calcination. Green *glass of antimony* contains more sulfur than does red *glass of antimony*. The cover-term *Green Lion* has also been interpreted to mean *glass of antimony* by notables such as Newton, as indicated by the following equation:

Green Lion = glass of Spanish ochre, also = Diana (Latona) or also = red earth, unclean body or stinking water = tin, lead, Adrop, uncleanness of death
    – Isaac Newton, Keynes MS 12 *(from Morienus)*

Both Morienus and Newton understood the *Green Lion* and *glass* to mean a *glass of antimony* product, ready to be fused with gold calx. In a number of sections from Latin translations, the terms *glass dregs* or *glass impurities* have been misinterpreted from the Latin *faex vitris*. In the context Morienus and Newton use the term, the interpretation should read *grinds of glass, ground glass* or *glass grindings*. This is simply a practical description. Once *glass of antimony* has been achieved, it is then ground to a fine powder before being fused with gold calx. After fusion, it is cooled and ground again. The term *Green Lion* is also used synonymously with *latten, body* or *earth* by Newton. *Green Lion* can be better understood as signifying powdered *glass of antimony*, both before and after fusion with gold calx. The lion (gold), was green, unripe or immature as this signified only the first composition.

Morienus refers to his first composition – gold calx fused with *glass of antimony* – by the cover-names *latten, body* or *earth* synonymously. Later Islamic alchemists would understand this *first compound* as Sophic Sulfur, whereas European alchemists would refer to it synonymously as Sophic Gold Calx and the Stone of the First Order. He seems unsatisfied with these, however, and adds an additional series of earth-based cover-terms: *foul earth, sulphur, foul sulphur, ochre, red ochre, Almagra*

CHAPTER 6 - FROM *CHRYSOPŒIA* TO *AL-KĪMYĀ'*

(Spanish purple-red ochre) and *orpiment*. Anyone unfamiliar with the compound might justifiably interpret these in a literal sense as separate and individual substances, but such an interpretation would be a mistake. This related family of cover-names is intended to suggest an earthy, reddish mineral compound containing gold.

Latten was tradecraft jargon used by Romans to indicate a copper alloy or, alternately, a substance to be alloyed with copper. *Latten* is also associated with *Latona*, a Roman Earth goddess and mother of the Moon goddess *Diana*. *Diana* literally means *heavenly* or *divine*. Sir Isaac Newton appears to have interpreted the following string of cover-names as being synonymous for *latten* derived from a reading of Morienus:

> *Almagra*, or *Latona (Diana)*, or *red earth*
> — *Isaac Newton*, Keynes MS 30/5 *(from Morienus)*

Almagra, vel Laton, vel terra rubea Morien p. 34, 35.

> *Adam*, *red body*, *red earth*, *Summer Ochre*, *Soul*, **Living Gold**, **our gold**, *Summer Corsufle*, *grounds*, *red gum*, *Youth*, *red Lythargyrium*, *Marthek*, *Red Magnesia*, *Olive*, *King Ruby*, *redness*, *red Salt*, *seed*, **Sericon**, **Sun**, *Red Brimstone*, *living Brimstone*, **Medicine**, *red Sulphide*, *glass*, *Kibrith* are the same.
> — *Sir Isaac Newton*, Keynes 30/5 *(from Geber's Three Words)*

Adam, corpus rubeum, terra rubea, Aestas Almagra, Anima, Aurum vivum, aurum nostrum, Corsufle, Aestas, Faex, Gumma rubea, Juventus, Lythargyrium rubeum, Marthek, Magnesia rubea, Oliva, Rex, Rubinus, Rubedo, Sal, rubeum, semen, Sericon, Sol, Sulphur rubeum, Sulphur vivum, Theriaca, Vitriolum rubeum, vitrum, Kibrith, idem sunt.
> Enarratio trium Gebri verborum p. 57, 58.

Morienus describes a simple procedure for creating *latten* by fusing an equal weight of gold calx and powdered *glass of antimony* in a sealed crucible:

> **Fusing latten** – Then begin in the Creator's name and with his vapour takes the whiteness from the white vapour. Now pour out these things and set them about, taking an equal weight of each. Then they are mixed together. Pack the mixture in a vessel that will just hold the full charge. This is the best way to handle these things, for they contain vapours which escape unless sealed in a container. The sublimation [fusion] should be done after sunset and the vessel allowed to stand until the cool of the evening. Then unstop the vessel and shatter it, examining that which is thus extracted, and if you find it well consolidated, like a stone, grind it thoroughly.

Operation #2 – eudica, blood or water

Morienus provides very few operative details for how he creates his fiery liquid menstruum, but a careful reading reveals a short recipe for how he may have accomplished *butter of antimony* via distillation of powdered *glass of antimony* with sal ammoniac. Distillation had been established for more than five hundred years prior to Morienus, along with the use of *glass of antimony* and sal ammoniac as standard alchemical reagents. It is impossible to know whether Morienus carefully weighed out his ingredients or estimated the correct volumetric measures by sight. However, the reaction he is describing is very forgiving as long as the ratio of sal ammoniac is slightly in excess of the antimony, which appears to be the case in the following recipe:

> **Distilling eudica** – Take the white vapour, or virgin's milk, and the green lion, or fire, and red ocher, or fire [all synonymous for glass of antimony], and the impurity of the dead, or earth [sal-ammoniac]; dissolve them and cause them to ascend and unite, using for each part of the green lion three parts of the impurity.

The final line reveals the truth that he is only working with two substances, despite the shopping list of synonymous cover-names presented. Here, *white vapor or virgin's milk* refers to dry and aqueous sal ammoniac respectively. The use of the phrase *ascend and unite* is an

CHAPTER 6 - FROM *CHRYSOPŒIA* TO *AL-KĪMYĀ'*

indication that he is creating a product via sublimation or distillation. Assuming that Morienus had become relatively efficient at this distillation, the chemistry for this reaction would look something like the following:

$$2 Sb_2O_3 + 12 NH_4Cl = 4 SbCl_3 + 3 O_2 + 12 NH_4$$

The reason Morienus did not write more regarding the operative details may have been to maintain secrecy, but more likely, because there is not much more to the operation than simple distillation. Instead, he goes on to characterize the substance and the path to achieve it, which can be understood as:

whiten the Stone (white vapour) → **fuse glass of antimony** (Green Lion) → **distill to butter of antimony** (eudica)

[Ghalib speaking] Now here is an explanation of the various substances according to Morienus, who said that ...

[White vapour] ... the philosophers referred to the impure body [stibnite] as lead. The purified body [flowers of antimony] is tin.

The green lion is glass and almagra is latten, although it may have been called red earth earlier. And blood is orpiment, and foul earth is foul sulphur.

Eudica is apart from all these and is called glaze, or the dregs and impurities of glass. The red vapour is red orpiment, the white vapour quicksilver and the yellow vapour sulphur [white, yellow and red antimony compounds prior to creating eudica].

This is the nature of the white vapour, the green lion, and the foul water.

Although the distillation of *eudica* (a.k.a. *foul* or *stinking water*) is a genuine chemical reaction, Morienus appears to have considered it a stage in the refinement or *exaltation* of antimony, rather than a composition of elements. This synopsis of the three-step path to create *butter of antimony* was studied and commented upon by Sir Isaac Newton in a section entitled 'The Entire Preparation of the Foliated Earth', in which he describes *eudica* – meaning *butter of antimony* – as the greatest of secrets. Newton followed Morienus' synopsis very closely, yet used his own equation format to demonstrate the synonymous nature of the cover-terms involved.

The Entire Preparation of the Foliated Earth
– Sir Isaac Newton, Keynes MS 12

White fume = living mercury, lemony fume = yellow fume = yellow, red fume = red orpiment

Green Lion = glass of Spanish ochre, also = Diana (Latona) or also = red earth, unclean body or stinking water = tin, lead, Adrop, uncleanness of death

Eudica = glass grinds or glass impurities, the greatest of secrets. In these consists efficacious compounds but these three are sufficient, the white fume, the green lion and stinking water

Newton's use of the term *stinking water* is an echo of Morienus' *foul water*. The process of creating *eudica* by distilling powdered *glass of antimony* with sal ammoniac releases strong ammonia vapors and the substance has a characteristic pungent odour. Newton's use of the term indicates that he was not only familiar with the substance, but that he had actually experimented with it himself, as evidenced by the following passage:

Dec 10 1678. *Crude unmelted & finely poudered* ☿ *[stibnite] 240 grains* ✻ *[sal-ammoniac] as much well mixed, by sublimation* left *130 grains* below. The sublimate looked very red.
　　　　　　– Sir Isaac Newton, Portsmouth Collection Add. 3793

It is clear that Newton had deciphered Morienus' cover-names concerning *eudica,* but the question remains whether Newton derived his

CHAPTER 6 - FROM *CHRYSOPŒIA* TO *AL-KĪMYĀ'*

inspiration directly from Morienus or he was simply providing commentary. Later Newton would employ *corrosive sublimate* distilled with regulus of antimony to achieve his *butter of antimony*. The white precipitate mentioned in the passage below is *Algarot powder*.

> Antimony 1lb, ☿ [corrosive-sublimate] 1℔ gives *butter of* ♁ [antimony] 1/2℔, which precipitated with water gives 1/4℔ or ♃4 1/2 of *white precipitate*. But the ☿ dissolves not all the metalline part of the ♁ for by addition of fresh ☿ more butter may be got out of it.
> – Sir Isaac Newton, Portsmouth Collection Add. 3795

In another manuscript, Newton characterized *eudica*, incorporating a number of additional cover-terms for the substance. For Newton, the *Green Lion* was an unclean dead body having been essentially killed by fire. *Eudica* is a saviour, life essence as *blood* – albeit smelly during its production. Perhaps most importantly *eudica* was *morsa-chimeia* [μορσα χημεια], chemistry's mortise and tenon groove joint – a term meant to imply that it was the binder of alchemy, and the greatest of alchemical secrets.

> **Eudica**, the mortise of chemistry [μορσα χημεια], secret of all secrets, venom from glass, (otherwise brine of glass or glass impurities), Pyrrhus, red hair, a golden robe, the collared horseman who snatched the white maiden from the jaws of the dragon, winged Perseus who rescued Andromeda, red Oil, life-giving blood for which bodies are adapted, conjoins all into one and makes fire-resistant. For bodies [whitened] without souls are quickly burned, but if Eudica is added to them the bodies revive when prepared and heated over a slow fire, to an earth cured of all combustibility.
> – Sir Isaac Newton, Keynes 30/1 (from Morienus)

Eudica, moszachumia, omnium secretorum secretum, fel vitri (alias faex vitri vel vitri immunditia) Pyrrhus, rubro capillitio, aureo vestimento, eques torquatus qui virginem Albificam Beiam seu Blancam ex faucibus draconis eripuit, Perseus alatus qui servavit Andromedam, Oleum

> rubrum, sanguis qui corpora coaptat vivificat, conjungit et ex omnibus unum efficit a combustion defendit. Nam corpora [dealbata] absque animis cito comburuntur, sed si Eudica eis apponitur ipsa corpora vivificat aptat et leni igne decocta et in terram mutata ab omni combustione curabit. Morien p. 30, 31, 35.

The etymology of the word *eudica* is perhaps related to *eudice* or *eudit*, meaning *of the tribe of Judah* – possibly an association between the substance and Maria. An alternate possible etymology is that it is an error in transliterating *Eurydice* first to Arabic and then to Latin, as daughter of the god of light and wife of the god Orpheus. Basil Valentine clearly adopted this interpretation for his 11th Key. If this were the case, it may have been an effort to associate the substance with extremely romantic, ecstatic and enslaving love between non-equals.

> Eurydice and Orpheus were young and in love. So deep was their love that they were practically inseparable. So dependent was their love that each felt they could not live without the other. These young lovers were very happy and spent their time frolicking through the meadows. One day Eurydice was gaily running through a meadow with Orpheus when she was bitten by a serpent. The poison of the sting killed her and she descended to Hades immediately.
> – Juliana Podd, Encyclopedia Mythica (originally from Thomas Bulfinch)

Orpheus was ultimately murdered and his soul reunited with Eurydice in the underworld. The above legend includes uniting and reuniting principles, a serpent, poison, descent into and fire of Hades – all themes typically associated with alchemical portrayals of *butter of antimony*.

The Second and True Composition – Uniting Body and Soul-Spirit

Once *latten* and *eudica* had been achieved, the next step was to unite the two philosophically, the final product of which was known to Islamic alchemists as *Sophic Sulfur* and to European alchemists as *Sophic Gold-Calx* and the *Stone of the Second Order*. From this point onward, the final composition was a rather simple affair. The following passages represent

CHAPTER 6 – FROM *CHRYSOPŒIA* TO *AL-KĪMYĀ'*

the most intelligible instructions for this operation found in the Morienus text:

> *Sift it and put it into another vessel, the bottom of which is rounded.*
>
> *Join one part of this of the unpurified [latten], proceeding in this order one after the other until all [latten and eudica] are mixed together, from which the elixir then may be made.*
>
> *Construct for it a philosophers' furnace and let a philosophers' fire be started in it and kept going for a space of twenty-four days. After this time, you should withdraw the vessel from the furnace and dry that which you find in it.*

The fire being discussed by Morienus in the passage below would come to be known in European alchemy as the *Secret Fire* of the philosophers, the identity of which is of course corrosive antimony trichloride ($SbCl_3$), known later as *butter of antimony*:

> *The whole key to accomplishment of this operation is in the fire, with which the minerals are prepared and the bad spirits held back, and with which the spirit and body are joined. Fire is the true test of this entire matter. And if the truth of anything pertaining to this magistery is not at once apparent, that will come to nothing, so when it does not spring up nor mingle with the impure substance to form one body, you will look in vain for something of that which you seek to come forth. For when you do as I have told you, you will find what you seek, God willing. Now know this well, and understand and remember it, and know the composition of this matter and give it much study, which will reveal to you the straight path. Nor can anything more then be hidden from you.*

Morienus rightfully asserts that knowledge of the identity and properties of *butter of antimony* is a key that unlocks the gateway to becoming an adept:

> ... in answer to your question about the white vapour, or virgin's milk, you may know that it is a tincture and spirit of those bodies already dissolved and dead, from which the spirits have been withdrawn. Then the spirits [dual-soul; soul + spirit] are again restored to them. But any body lacking spirit is dark. It is the white vapour that flows in the body and removes its darkness, or earthiness, and impurity, uniting the bodies into one and augmenting their waters.

The cover-term *white vapour* in the passage below is rather ambiguous. It can refer to the original *white vapour* as *flowers of antimony* or it could be referring to *eudica*. In the case of *flowers of antimony*, the oxides present in antimony trioxide provide the potential for chloride-oxide ion exchange that makes the reaction possible. The statement is equally correct if Morienus is referring to *eudica* because without the corrosive nature and chloride ions present in antimony trichloride, there will be no reaction and thus no resulting Philosophers' Stone from which to make alchemical gold. In either case, the following passage holds true.

> You should know this well and understand it, and know that when you have thus accomplished the tincture in it, you are on the right track toward that finest red gold, than which nothing better or purer is to be found. For this reason it is called Roman gold. Without the white vapour, there could have been no pure gold nor any profit in it.

Final Fermentation

Then Morienus did something unexpected: he provided instructions for what European alchemists later termed *fermentation*, the product of which is known in European alchemy as the *Stone of the Third Order*. His instructions are as follows:

> Then take such a quantity of the elixir as to form eleven parts for every ten parts of the white body. Mix these, and for every ounce of this mixture, add one-fourth of a dram of eudica. Then put the jar in a large furnace and build a fire over it, keeping it going for two days without being extinguished at any hour of the day or night. When this is done,

CHAPTER 6 – FROM *CHRYSOPŒIA* TO *AL-KĪMYĀ'*

take out what you find in the jar and praise the great Creator most high for what He has given you.

The difficulty in working with exact measures from ancient texts lies in translation and copyist errors, and non-standardized measurements (i.e. variations in weights and measures between cultures, or different systems in use in a geographic location). This is especially true for measurements recorded prior to metrication and Prussian weight reform. In some texts, measures can be very vague or even non-existent. Some measures are meant to indicate volume by sight, whereas others indicate exact weight, yet in an obsolete unit of measure. The above translation is from Latin to English according to Stavenhagen. If the Arabic to Latin translator was familiar with Arabic units of measure, and accurately converted these measures into Latin counterparts, the question then is to what European system of measurement does the Latin unit refer? The region corresponding to modern Italy alone had eight different variations as late as 1800. To make matters worse, Stavenhagen's Latin translation contains numerous variations of the recipe repeated in different sections of the text, each with variations on the measurements.

Fortunately, principle is a reliable factor in these matters and it must be assumed that Morienus, having directly inherited the centuries-old Alexandrian tradition, had already achieved a high operative level of efficiency. His ratio of *elixir* to *white vapour* is nearly equal and the amount of *eudica* required to catalyse the reaction would allow for a forgiving margin of error.

Morienus uses the cover-terms *ferment* and *fermentation* in a general sense to indicate the medium that he terms *milk,* which is to be fermented by gold as the catalyst. Maria used the term *ferment* to indicate sal ammoniac, whereas European alchemists would adopt the term *fermentation* yet apply it to a specific finishing process.

> When you have treated the body, or earth, in the way we have described, put it into a fourth part of the ferment, or milk. For the fermentation of gold is like that of bread. ... the substance eudica should be used to agitate the bodies ...

In the process above, Morienus combines approximately equal amounts of his just finished *elixir* and white *flowers of antimony* along with a small proportion of *eudica*. The predominant color of the finished product is white. In the following passage, he also indicates that the product is subjected to *the heat of the fire* and *much heating* as part of the heat regimen:

> **Then** examine its interior, and if you find the whiteness retained, well fixed and calcined, not driven off by the heat of the fire even after much heating, you have accomplished both natures of this operation. But know that if you had given your entire kingdom and brought all your subjects together at once to project the whiteness into the purified matter, they would have been powerless to do so by such means.

The product of this reaction certainly meets the criteria as being a fire-resistance substance due to the thermal decomposition properties of antimony oxychloride; its final decomposition temperature being just above 600°C.

Template for **Morienus' al-Kīmyā'**:

1. **First Composition** – purify and reduce gold, create *flowers* (white vapour) then *glass of antimony* (green lion), and fuse both;
2. **Second Operation** – create *butter of antimony* (foul water) via sal ammoniac and powdered *glass of antimony* ($SbCl_3$);
3. **True Composition** – combine the *latten, body* or *earth* ($Au \cdot Sb_2O_2SO_4$) with *eudica, blood* or *water* ($SbCl_3$) and digest in a sealed round-bottomed glass vessel over low continuous heat;
4. **Fermentation** – combine the *elixir, white vapour* and *eudica*, and heat in a furnace for two days.

Legacy and transmission

Morienus' work argues strongly (by his own admission as an operative adept-student of Stephanos' tradition at the court of Heraclius), that Stephanos did indeed incorporate operative alchemy into his unified initiatic mystery tradition. The connection between Stephanos and Morienus is further strengthened considering that, in the manner of Stephanos, Morienus prized a White Stone over a Red one. His operative technique however, either sheds light on the White Stone's confection or reveals unique operative innovations that must be attributed directly to Morienus. In either case, it becomes clear that late-Alexandrian alchemists were skilled at operative procedures and must be considered, despite extremely speculative or poetic tendencies, to embody a holistic approach to alchemy with regard to speculative and operative aspects. Morienus' transmission leaves little doubt that he was an adept of the highest order.

Ghalib's first person account of alchemy's transmission from Morienus to Khālid offers tremendous insights into this momentous event in the history of the Philosophers' Stone. *The Composition of Al-Kīmyā'* was the first alchemical text recorded in Arabic and served as a sourcebook for Islamic alchemists such as Jābir ibn Hayyān. It was also one of the first authentic alchemical texts translated from Arabic to Latin, which went on to color and define European alchemy. One of the finest surviving examples of Morienus' entire method of operation is preserved in a pictorial manuscript known as *Cabala Mineralis*. Morienus' influence can also be seen in Artephius' method and Paracelsus' *Tincture of the Philosophers*, among others in the form of a joining of sulfur (*latten*) and mercury (*eudica*). Paracelsus summarized the confection as a union of lion's blood (*latten*) and eagle's gluten (*eudica*). These later cover-terms simply echo those used in the early-Alexandrian period such as *the two fumes that contain two lights* (sol and luna), *red gum* and *white gum*, *red man* and *white wife*, *kibric* and *zibeth*, etc. Many of the cover-names employed by Morienus would stay in use throughout European alchemy's long history. Morienus combined Maria's *Ars Brevis* technique of fusing *latten* in a crucible as an initial step, with her *Ars Magna* technique of a chemical reaction via digesting in a glass vessel at ascending heat. Like Synesius and Stephanos, the goal was to achieve a White Stone, pure and impervious to the flames.

His major innovations – *butter of antimony* and the *fermentation* technique, in combination with a new *modus operandi* – established a new archetypal recipe that served as a template for European methods of confecting the Stone. His correlating *latten* and *eudica* to *Sophic Sulfur* and *Sophic Mercury* may have inspired Jābir ibn Hayyān's sulfur-mercury theory. Morienus' *liquid fire* inspired Artephius to praise this *vinegre* or *Secret Fire* in his *Secret Book*. Paracelsus' *Blood of the Lion* and *Eagle's Glute*n are cover-terms synonymous with Morienus' *latten* and *eudica*. Basil Valentine's work with *glass of antimony* and his recipe to create *butter of antimony* by distilling antimony with *sal Armoniak* (Valentine's spelling) is based upon recipes first introduced by Morienus. Yet most impressively, Morienus inspired one of history's greatest scientific minds obsessed with alchemystical pursuits – Sir Isaac Newton – to codify and attempt to decipher his cover-terms and possibly even reproduce some of Morienus' recipes. Morienus was among the last of the Alexandrian-style *chrysopœians*, but he was also the source of Islamic *al-Kīmyā'* and the earliest forms of European alchemy that later developed into early-modern Chymistry as practiced by Newton.

Islamic alchemy developed a remarkable culture of theoretical and experimental alchemy that led to great advances in physics, mathematics, astronomy, pharmacy, and applied chemistry. Practical industries developed from these advances, including perfumery, cosmetic and paper production, metallurgy, military technology, and the making of dyes, pigments, glass, and ceramics. Islamic practical alchemy laid firm foundations for a collective cultural value system that embraced

CHAPTER 6 - FROM *CHRYSOPŒIA* TO *AL-KĪMYĀ'*

scientific and technological innovation and free exploration. This incredible work ethic entered Europe via Spain and Sicily during the 12th century. By the 13th century, a tremendous amount of Islamic alchemical knowledge was the subject of a large translation movement that took place in Andalusia, Spain. Islamic scientists, originally inspired by alchemy, contributed greatly to the scientific and chemical revolutions of the 17th and 18th centuries.

The evolution of the recipe to confect the Philosophers' Stone would change very little after Morienus. In some cases, Islamic and European processes to confect the Stone would never again reflect the elegance and efficiency of Alexandrian methodology. Alchemy, as "[an] art, purporting to relate to the transmutation of metals, and described in a terminology at once Physical and Mystical," was another matter altogether evolving far beyond the transmutation of metals. Islamic alchemical innovations included new acids and ways to reduce stibnite using various salts or urine menstruums, yet the fundamental method for confecting the Stone changed very little. Morienus' alchemy was passed on to the Abbasid Caliphate where it inspired a new, highly experimental form of industrial and medical alchemy that served as the forerunner to modern chemistry and pharmacy. European alchemists, upon translating Greek and Arabic alchemical texts, developed their own brand of reproductive and experimental alchemy that evolved into the iconic and recognisable style most are familiar with today.

Paracelsus discovered a new *liquid antimony menstruum* even more efficient than *butter of antimony*, which he called the *Alkahest*. The *Alkahest* was the last great innovation for confecting the Philosophers' Stone, yet European alchemy, losing its fascination with the Stone, eventually made the shift towards modern chemistry. The origins of uniquely medical alchemy and systematic experimental chemistry can be attributed to the Islamic tradition, with European alchemists building upon and fully developing alchemy in many new directions. Since the

299

BOOK 1 – ORIGINS & EVOLUTION

beginning, the Philosophers' Stone has been at the very core of every western alchemical tradition and this is what differentiates alchemy from chemistry. The credit for the origins, evolution and chemistry for confecting the Philosophers' Stone must be attributed to Alexandrian *chrysopœians*, and the Stone's introduction to Islamic and European traditions primarily from Morienus to Khālid ibn Yazīd.

Chapter 7 – Paracelsus and the Alkahest

God has made everything; something from nothing.
This something is a seed; the seed gives the end of its
predestination and of its office.

Paracelsus

He was a robust man, red-faced with a character as large as his overweight body and possessed of a self-righteousness that is available only to those with deeply rooted experiential wisdom. As Europe's premier alchemist-physician and theoretician, he was a universal genius, successful healer, adept-alchemist and theosophical visionary steeped in Neoplatonic and gnostic mysticism. As a person, his genius was misunderstood; he was quite often drunk, rejected by his peers and on many occasions home was the tavern-floor that he happened to be sleeping on that particular night. His name was Phillippus Theophrastus Bombastus von Hohenheim, self-titled Paracelsus belying his sense of self-worth as being greater than Roman physician and writer Celsus. He was born in 1493, died in 1541 a few months before his 48th birthday and there has never been another alchemist like him …

His education began with his father, William of Hohenheim, who has been described as being a Grand Master of the Teutonic Knights. As a young man, Paracelsus was introduced to mining, mineralogy, botany and natural philosophy alongside medicine. He was exposed to alchemy during childhood via the writings of Johann Hollandus and later by famous exponent of the occult, Johannes Trithemius, abbot of Sponheim.

> At twenty years of age he started on his travels through Germany, Hungary, Italy, France, the Netherlands, Denmark, Sweden, and Russia. In Muscovy [Grand Principality of Moscow] he is said to have been taken

> *prisoner by the Tartars, who brought him before "the great Cham." His knowledge of medicine and chemistry made him a favourite at the court of this potentate, who sent him in company with his son on an embassy to* Constantinople. *It was here, according to Helmont, that he was taught the supreme secret of alchemistry by a generous Arabian, who gave him the universal dissolvent, the Azoth of western adepts, the alcahest or sophic fire. Thus initiated, he is said to have proceeded to India. – A.E. Waite,* Lives of Alchemystical Philosophers

Paracelsus received his Doctorate from the University of Ferrara in northern Italy; the same university where Nicolaus Copernicus received his Doctorate of Canon Law. He later served as both physician and military surgeon, which led to his being known as *Doctor of Both Medicines*. Yet to label Paracelsus an alchemist or doctor is far too reductionist for a man of such breadth of character; rather he was a naturalist physician, spiritualist and symbolist intellectual, zealously theosophical and defender against what he felt were injustices – personal, professional or social. He was a prolific writer employing a triple-entendre style that occasionally reflects alchemical, medical and religious models of understanding in a single passage. More importantly, Paracelsus was a fervent empiricist believing that the only valid data worthy of acknowledgment came from direct experience, experimentation and observation and he rejected authoritarianism outright on these grounds.

> *I have not patched up these books, after the fashion of others, from Hippocrates, Galen, or anyone else, but by* experience, *the greatest teacher, and by labour,* have I composed them. *Accordingly, if I wish to prove anything, experiment and reason for me take the place of authorities. – Paracelsus,* Conc. Alchemical Degrees and Compositions

Paracelsus' incredible life and teachings have been recorded by professional biographers and historians in more detail than is possible in this treatise. The following treatment must necessarily remain limited to his unique approach to mineral/metal alchemy as it relates to his confection of the Philosophers' Stone and the Alkahest.

Paracelsus is most known for his exhaustive system of plant medico-alchemy called Spagyrics, which drew on classical knowledge of herbal medicines.

> *Contrary to the lingering misconception that his medical findings derive from new experience or travels, the material sources documented here indicate that most of his healing herbs and stones were traditional remedies found for the most part in Pliny, Dioscorides, and medieval medicine. – Andrew Weeks,* Paracelsus – Essential Theoretical Writings

While Paracelsus may have drawn from any source available, it was his alchemical method of preparation that revolutionized alchemy and pharmacy. His approach to confecting the Philosophers' Stone can be described as relatively orthodox based largely on the use of *butter of antimony* in parallel with Morienus, Artephius, Arnold de Villa Nova and the Hollandus'. His Alkahest however appears to be a manifestation of his own innovative spirit derived from an impulse to bring substances to their highest perfection. For Paracelsus, to be a skilled artisan or an alchemist meant playing the same role in different fields of endeavor.

> *For [nature] brings nothing to light that is complete as it stands. Rather, the human being must perfect [its substances]. This completion is called alchimia. For the alchemist is the baker in baking the bread, the vintner in making the wine, the weaver in weaving cloth. Thus, whatever arises out of nature for human use is brought to that condition ordained by nature by an alchemist. – Paracelsus,* Paragranum, Alchimia H 2:61

Aside from the alchemical teachings of Hollandus, Paracelsus was also familiar with not only works of medieval alchemy but also those of contemporary herbalist-distillers, Hieronymous Brunschwig and Philip Ulstadt. Manley Palmer Hall recounts that upon his return to Europe from Constantinople, Paracelsus declared that among Islamic physicians he had at last discovered men of scientific stature and the scientific attitude of natural inquiry. While Paracelsus makes it clear that he knew the *fourfold arcanum* to confect the Philosophers' Stone since his youth, A.E. Waite suggests that he acquired the secret of the *alkahest* in Constantinople as an adult.

During the period in which Paracelsus was in Constantinople, Suleiman the Great was instituting social change and ushering in the Ottoman Empire's golden age of artistic, literary and architectural development. Suleiman's personal interests included poetry and gold technology, yet he patronized a variety of arts and craft-technologies. Workshops at Suleiman's residence, the Topkapı Palace, accommodated hundreds of artisans from across the Mediterranean world and beyond; from painters, engravers, weavers and tile makers, to bookbinders, goldsmiths, ivory artisans, manuscript illuminators and musical instrument makers. The resulting artistic style, which came to be known as *new international*, was a mix of Byzantine, Italian, French, Central Asian, Persian and Arabic influences. In short, the cultural climate in Constantinople as Paracelsus would have experienced it fostered a fantastic blend of artistic expression that accompanied alchemy, medicine and scientific inquiry.

As Paracelsus' medico-alchemy reached maturity, he attempted to establish a universal method of operation for creating both plant and mineral-based alchemical pharmacy congruent with his theosophical views towards self-redemption. To achieve this aim, he developed a theoretical model or structural progression scheme that he called the four pillars:

> **Accordingly,** *the first pillar [of medicine] is a complete philosophy of earth and water;* **and** *the second pillar is astronomy or astrology, incorporating a full understanding of air and fire. The third pillar is alchemy* **without flaw and encompassing all preparations, properties, and adept [art with its power]** *over the four aforesaid elements. And the*

CHAPTER 7 – PARACELSUS AND THE ALKAHEST

> *fourth pillar is virtue; and it remains with the physician unto death [as a support] that encompasses and sustains the other three pillars. And take note of what I am saying, for you must enter here and know these four pillars ... – Paracelsus,* Paragranum, Preface H 2:10

Paracelsus saw life as strewn with seeds by the hand of God that were predestined to reach maturity in the mineral, plant and animal kingdoms. Yet nature could bring them just so far, an example of which might be a fruit seed. It will sprout and grow to a sapling, reach maturity and achieve its predestination as a ripe fruiting body. It requires human assistance however for the fruit to be transformed into jam, a pie, wine or brandy. Paracelsus cites bread as an example:

> *Bread is created and given to us by God, but not in that shape which the baker confers upon it. Those three Vulcans, the farmer, the miller, and the baker, produce from that first matter [prima materia] a second, namely, bread [ultimae materiae substantiae].*
> *– Paracelsus,* The Book of Alchemy

To Paracelsus, this human intervention to further nature was the very essence of alchemy, and in his case, the result was always a medicine in the form of an organic or inorganic medicinal ingredient, exalted to its highest state, free from impurities and prescribed at non-toxic dosage. He applied this model to human transcendental endeavor as well. His later writings reveal a passionate determination to revolutionize established medicine and pharmacy and bring them into alignment with his four-pillar approach.

> *All of them [my findings] are composed in accordance with the four that follow here, philosophia, astronomia, alchimia, and [the] virtues. It is my intention to bring you to the conviction of accepting nothing that is not founded on these four corner-stones. It is on them that I base what follows in order that you should understand the rationale and cause of my writing and think accordingly of me ...*
> *– Paracelsus,* Paragranum, Preface H 2:21

He began his universal approach with the standard alchemical assertion that in order for a substance to be exalted it must first rot or be killed, which is to say crucified in order to be reborn into an exalted state. He paralleled this process with human digestion pointing out that in order for the nutrition to be bioavailable food must first digest.

> *The rose is magnificent in its first life; and it is adorned by its taste. But as long as it maintains the latter, it is not [yet] a medication.* It must rot, and die as such, and be reborn: Only then can you speak of administering its medicinal powers. *For just as the stomach leaves nothing unrotted that should become the human being, so* nothing may remain unrotted that is to become medication.
>
> *– Paracelsus,* Paramirum I:VI, H 1:92

In the passage above, Paracelsus is paraphrasing biblical scripture from the Book of John:

> John 12:24 *Truly, truly, I say to you, unless a grain of wheat falls into the earth and dies, it remains alone; but* if it dies, it bears much fruit.

He applied this fundamental principle to organic and inorganic chemical pharmacy. His plant alchemy involved isolating water and alcohol soluble essential oil from the plant, *killing* the organic solids by incinerating the remaining plant matter to ashes, which were calcined, leeched and crystallized to a refined salt. He then permanently recombined the oil and salt via reflux distillation; the product understood as the primordial form of the organic, which he termed the *primum ens*. His approach to metals was slightly different. He would first *kill* the metal to remove the metallic body via acid reaction. This process results in a soluble metal-salt he termed the metal's *quintessence*, which in aqueous form was called the *oil of the metal*. Once plant, mineral or metal salts were stabilized either dry or in solution, the substance was considered to have reached its most exalted form. This procedure can be understood as removing the outer shell or body of a substance resulting in its purest ethereal or primordial form.

Another way of stating the above, consistent with Paracelsus' religious ideology, is that the substance had reached its full maturity afforded it by nature, after which it must undergo crucifixion like Christ in order to reach its most exalted and useful form as resurrected medicine. Paracelsus believed that everything in nature, metals, sulfides, stones, gums, resins and herbs could be manipulated alchemically and were therefore potentially both therapeutic and / or toxic. Dosage alone, according to Paracelsus, determined the difference between toxic and therapeutic action:

> *All substances are poisons;* **there is none which is not a poison.** *The right dose differentiates a poison from a remedy.*
> – *Paracelsus,* Von der Besucht

While Paracelsus remains silent with regard to his recipe for the Alkahest, he preserved his formula for the archetypal white and red Philosophers' Stone variants, which he referred to as the *Stone of the Philosophers* and *Tincture of the Philosophers* respectively, in various works as an exposition of the *four arcana*. Researcher T.P. Sherlock, in his article *The Chemical Work of Paracelsus*, suggests familiarity with the process since childhood:

> *There are only four arcana, as Paracelsus has known since his childhood.*
> – *T.P. Sherlock,* The Chemical Work of Paracelsus

If Paracelsus had been introduced to his four-step process for confecting the stone during childhood, this work would have been an integral part of his life-story and character in a manner quite unlike others drawn to alchemy in mature adulthood. It may also explain his undeniable sense of self-righteousness. Aside from his many well-documented personality quirks, he was a true universal genius and an unparalleled master of his art.

> *Now at this time, I, Theophrastus Paracelsus Bombast, Monarch of the Arcana, am endowed by God with special gifts for this end,* **that** *every searcher after this supreme philosophic work may be forced to imitate and to follow me,* **be he Italian, Pole, Gaul, German, or whatsoever or whosoever he be. Come hither after me, all you philosophers, astronomers, and spagyrists, of however lofty a name ye may be,** *I will show and open to you, Alchemists and Doctors, who are exalted by me with the most consummate labours, this corporeal regeneration. I will teach you the tincture, the arcanum, the quintessence, wherein lie hid the foundations of all mysteries and of all works.*
> – Paracelsus, The Tincture of the Philosophers

ARCANA & ESSENCES

> *That alone is an arcanum* **which is** *incorporeal and immortal,* **possessing eternal life,** *above the understanding of nature and the knowledge of men. ... They have power to alter, to change, and to renew, to restore like the arcana of God, according to their judging.*
> – Paracelsus, Archidoxis Book IV

The Four Arcana

The Four Arcana occur in the *Paragranum* and *Archidoxis* in varying levels of detail. Together they constitute a gradational progression or *path* for confecting the Hermetic White Stone and the Tincture. His incredible skill with alchemical processes allows for a tremendous amount of wiggle-room. For example, it is clear that he could create *butter of antimony* by several processes, which allows for leeway when performing the 3rd arcanum. He presents his instructions for confecting the *Tincture of the Philosophers* flippantly as if he deems the process rudimentary common knowledge. Recipes are given in a rambling sentence or two and rarely with supporting details, giving the impression that he feels the reader should already be expert enough to read between the lines and

CHAPTER 7 - PARACELSUS AND THE ALKAHEST

fill in the blanks. In the following passage, one of the most concise summations of the art in alchemical history, he clearly expects his audience to comprehend what he means by *lion's blood, eagle's gluten* and the *old process* that completes the work:

> Take only the *rose-colored* blood from the Lion and the gluten from the Eagle. When you have mixed these, coagulate them according to the old process, and you will have the Tincture of the Philosophers, *which an infinite number have sought after and very few have found.*
> – Paracelsus, The Tincture of the Philosophers

The four mysteries below can be considered *essences*, in the sense that each possesses the potential to effect change and must be brought to the state of an essence, *essensificatum*, by one initiated into the specific *arcanum* or mystery of the process. The lowest essence is *Lion's Blood* with the highest being the *Tincture of the Philosophers*.

The passage below is out of context and spoken in a general voice, yet it applies to the four arcana that follow.

> ... there are *not four but rather one* **arcanum,** *but arrayed in a fourfold way ...* – Paracelsus, Paragranum H 2:30

The 1st Arcanum – or first mystery is the standard process of purifying and reducing stibnite to molten glass of antimony and then fusing it with gold-calx, a product referred to as *flores auri* by Artephius and *latten* by Morienus before him. Basil Valentine detailed the process, as it was known in Paracelsus' time:

> *Take Hungarian or other Antimony [stibnite], the best you can get, grind it, if possible, to an Impalpable Powder; this Powder spread Thin all over the bottom of a Calcining Pan,* **round or square, which hath a Rim round about, the height of two Fingers thickness; set this Pan into a Calcining Furnace, and** *administer to it at first a very moderate Fire of Coals, which afterward increase gradually:* **when you see a Fume beginning to arise**

> *from the Antimony,* stir it continually with an Iron Spatula, without ceasing, as long as it will give forth from itself any Fume.
>
> *... calcine it, continually stirring as we said, until no more Fume will ascend. If need be* repeat this Operation so often and so long, as until that Antimony put into the Fire, will neither fume, nor concrete into Clots, but in Colour resemble White and pure Ashes: *Then is the calcination of Antimony rightly made.*
>
> *Put this Antimony [ash] thus calcined into a Goldsmiths Crucible set in a Furnace...* Antimony may flow, like clear and pure Water. ... the Glass made thereof be sufficiently cocted, and hath acquired a transparent Colour... see whether it be clear, clean and transparent.
>
> – Basil Valentine, 15th century

1st Arcanum – The First Matter (Lion's Blood)

1. *Of the prime material in flux [flowers of antimony],*
2. *gently digest in molten-flux for a month [night] thereafter [molten glass of antimony] ...*
3. *to which afterwards add the Monarch [gold] in equal weight [resulting in the Blood of the Lion].*

$$Au + Sb_2O_3 \cdot Sb_2S_3 = Au \cdot Sb_2O_2SO_4 \text{ gold-antimony glass}$$

The *2nd Arcanum* – is described as the *Stone of the Philosophers* to indicate the Hermetic White Stone. This product fully accords with Hollandus' accurate assertion that antimony alone, referred to below by the cover-name *Saturn*, is all that is required to create alchemical gold:

> *And know, my Child, for a Truth, that in the whole vegetable work* there is no higher nor greater Secret than in Saturn; for we do not find that perfection in Gold which is in Saturn; for internally it is good Gold, herein all Philosophers agree, and it wants nothing else, *but that first you remove what is superfluous in it, that is, it's impurity, and make it clean, and then that* you turn it's inside outwards, which is it's redness, then will it be good Gold; for Gold cannot be made so easily, as you can of Saturn ... – Johann Isaac Hollandus, A Work of Saturn

The process above appears in a concise abridged form in the fifth book of *Archidoxis* entitled *A Very Brief Process for Confecting the Stone*. A.E. Waite suggests that it was an editor's insert:

> Take, then, mercury, *otherwise the element of mercury, and* separate the pure from the impure. **Afterwards** let it be reverberated even to whiteness, **and** sublimate this with sal ammoniac *until it is resolved.* Let it be calcined and dissolved again, and digested in a pelican for one month, being afterwards coagulated into a body. **This is no longer burnt, or in any way consumed, but remains in the same condition.**

One of the most respected physicians and hermetic alchemists of the 17th century, Dr. Heinrich Khunrath, hinted that butter of antimony, encrypted by the cover-terms *Azoth* and *Catholic (Universal) Mercury of the Sages* is all that is required for preparing the Stone. The stone created by this method was called by Khunrath the *Universal* or *Great Stone of the Philosophers*:

> *That which I describe is not a myth: you shall handle it with your hands, see it with your eyes,—that Azoth, or Catholic Mercury of the Sages, which, together with inward and outward fire, in sympathic harmony, through an unavoidable necessity, physico-magically united,* is alone sufficient for the preparation of our Stone.
> – Heinrich Khunrath, Amphitheatrum Sapientiae Aeternae, Fol. 202

The recipe for the 2nd *arcanum* lends itself readily to interpretation as *butter of antimony* via *flowers of antimony* + sal ammoniac, and then subjected to digestion to achieve Hollandus' antimonial *Stone of Saturn*, Khunrath's *Great* or *Universal Stone* and parallels the Alexandrian Hermetic *White Stone* of Cleopatra, Zosimus, Synesius and Stephanos. In this scenario, the *Stone of the Philosophers* is the *White Stone* sans-gold, namely *Algarot powder*. Its relationship to Saturn is alluded to by Morienus. The metal associated with Saturn is lead, yet *philosophers' lead* is antimony as Morienus explains:

> ... *the philosophers referred to* the impure body [stibnite] as lead. The purified body [flowers of antimony] is tin.

This is echoed by Isaac Newton in an equation series for the *green lion*:

> *Green lion = glass of Spanish ochre, also = Diana (Latona) or also = red earth, unclean body or stinking water =* tin, lead, Adrop, *uncleanness of death – Sir Isaac Newton,* Keynes MS 12 (from Morienus)

Artephius described *saturnine antimonial mercury* as *lead of the wise*. Basil Valentine described antimony as the *bastard of lead*, meaning Saturn's illegitimate son. Antimony in a refined form is likened to tin based on the phenomenon that occurs when antimony substitutes for tin in the creation of bronzes. Copper-antimony alloy in the correct ratio results in an alloy that is non-corrosive and with a surface appearance virtually indistinguishable from elemental gold, whereas this is not the case for lead. It is precisely this phenomenon that Hollandus was alluding to in describing Saturn, meaning *secret lead of the wise*, as *good Gold; for [alchemical] Gold cannot be made so easily, as you can of Saturn*. Some alchemical recipes for the Philosophers' Stone avow that the Stone is very inexpensive to create and does not require even the least amount of elemental gold. Alexandrian alchemists recognized this phenomenon during at least the Roman period derived from technical bronze-craft expertise likely predating Alexandria by centuries, if not millennia.

> *It is necessary for a proper comprehension of the nature of this correspondence to consider the position of the planets, and to* pay attention to Saturn, which is the highest of all.
> – A Short Catechism of Alchemy
> Founded on the Manual of Paracelsus preserved in the Vatican Library

While Paracelsus may have indeed viewed the *Stone of the Philosophers* in its original context as the archetypal *White Stone of Hermes*, an argument can be made that he adopted a novel substance as the chemical identity for his new and improved *Stone of the Philosophers*. This new variation entered the European tradition from Islamic alchemy, was an extremely useful substance that alchemically qualifies as a Stone discovered by al-Razi, and adopted by Paracelsus as a reagent and medicine; hence the cover-term. Paracelsus' explains that his unique *Stone* of the *Philosophers'* is variant is created according to his own unique process:

> ... *I will leave that [archetypal] process and pursue* my own, as being that which has been found out by me *through use and practical experiment. And* I call it the Philosophers' Stone, because it affects the bodies of men just as their's does, *that is to say, just as they write of their own.* Mine, however, is not prepared according to their process; *for* that is not what we mean in this place, *nor do we even understand it.*
> – *Paracelsus*, Archidoxis, Book V

The 2nd *arcanum's* purpose as explained by Paracelsus – *it cleanses the human body as fire cleanses the spotted skin of the salamander* – is better understood in the context of the subsequent process. Paracelsus worked with two mercury salts – *corrosive-sublimate* (mercuric chloride; $HgCl_2$) and *horn-quicksilver* (mercurous chloride; Hg_2Cl_2), each of which were alchemical substances originating from medieval Islamic alchemy's cultural intercourse with Indian alchemy. *Corrosive sublimate* was discovered and used by al-Razi as a wound disinfectant and treatment for syphilis. *Horn-mercury*, known later as *Calomel*, was a disinfectant, ingested as a laxative and served as a treatment for syphilis until the early 20th century. Both could be prepared via distilling quicksilver with sal ammoniac.

This *stone* is the reagent that chemically perfects the body of *stibnite*, purifies it (cleanses the body) and brings it to a fire-resistant state (salamander). Both the previously mentioned substances will work as reagents in the 3rd *arcanum* reaction that follows in the series, yet *corrosive sublimate* is the more efficient reagent and Paracelsus was certainly well aware of this. In his work the *Treasure of Treasures for Alchemists*, Paracelsus provides a recipe based on the use of *aqua-regia* to create mercury salts. While the instructions below for creating the *Stone of the Philosophers* can indeed be interpreted in the classical sense as the creation of Algarot Powder as preserved in *Archidoxis*, it also lends itself to an alternate interpretation as a chemical reaction of *quicksilver* with sal ammoniac:

2nd Arcanum – Stone of the Philosophers (Corrosive Sublimate)

1. Take the element of mercury [cinnabar] and separate the pure from the impure [to achieve purified quicksilver].
2. Then reverberate it to whiteness, [by] sublimate[-ing] it with salmiax [sal-ammoniac] till it is resolved;
3. then calcine it and resolve it again, then set it in a pelican and allow it to digest for a month;
4. coagulate it into a body, which will not burn nor will it distort, it remains also uncorrupted …

Corrosive-sublimate	$2\ NH_4Cl + Hg = HgCl_2 + 2\ NH_4$
Horn-quicksilver	$2\ NH_4Cl + 2\ Hg = Hg_2Cl_2 + 2\ NH_4$

Whether the 2nd *arcanum* refers to *butter of antimony* converted to *Algarot powder*, or the creation of a mercury-salt remains open to interpretation. In either case, each of these was employed by Paracelsus in his system of alchemical pharmacy. If he were creating *Algarot powder* via this method, he is in accordance with the archetypal *White Stone of Hermes*, Hollandus' *Stone of Saturn* and Khunrath's *Great* or *Universal Stone*. If the recipe refers to a mercury salt, he is in accord with al-Razi's use of mercury salts applied alchemically and medicinally. Both *Algarot powder* and mercury salts share the commonality that they were prescribed as emetic or laxative medications.

The 3rd *Arcanum* – is described by Paracelsus as *living mercury* with the power to cause hair and nails to grow anew; implying that it is a source of growth and regeneration. It must be remembered that the body he is referring to is not necessarily a human body *per se*, but rather the body of the Philosophers' Stone in its various stages of formulation. Although al-Razi employed *corrosive sublimate* in his system of alchemy, the following reaction can be considered Paracelsus' innovation resulting in an increase in efficiency.

This process combined the ease of reducing *stibnite* chemically, as opposed to metallurgically, with the added bonus of concurrently creating *butter of antimony*. The byproduct, *cinnabar of antimony*, had other alchemical uses apart from creating the Philosophers' Stone. In *De Tinctura Physicorum*, Paracelsus refers to the primary product, *butter of antimony*, by the cover-name *eagle's gluten*. The cover-name is appropriate in that *gluten* as a binder is used in the same manner that Isaac Newton referred to *eudica* (butter of antimony) as the *groove-joint of chemistry* to suggest binding and coagulating properties.

Philosophy realized on Earth what astronomy revealed in the heavens; the alchemist as mediator united the two. Paracelsus associated the salt to create *sophic mercury* and *sophic mercury* itself, with the Aquila constellation:

> *He is a philosophus who has knowledge of the Lower Sphere; he an astronomus who knows of the Upper Sphere. And yet the two together embody a single sense and art. ...*
>
> *For one aspect coincides with knowing about mercurius; the other with knowing about Aquilatus.*
> — *Paracelsus*, Paragranum, Philosophia H 2:29

The cover-name *eagle* is in reference to sal ammoniac, yet the term also came to indicate the process or resulting product of distillation, and was specifically related to *butter of antimony*, called *gluten from the eagle* by Paracelsus.

The reason for this association might derive from classical literature. Aquila belongs to the Hercules family of constellations and is the eagle that carried Zeus' (Jupiter-Ammon's) thunderbolts. The alchemical cover-term *labors of Hercules* indicate materials preparation, which relied on the *eagle* (sal ammoniac) to create *gluten from the eagle* (butter of antimony). The brightest star in the constellation, Altair, is derived from Arabic *al-nasr al-ta'ir* (الطائر النسر) meaning *flying eagle*. Yet this name can be traced back to Sumerians and Babylonians who also named this star Eagle in their own languages.

Morienus and Newton viewed step two of the process below as uniting the *body* with the *spirit*. Paracelsus begins his instructions with a recap of the previous process followed by very vague instructions for coagulation on a marble slab. This is a standard alchemical procedure where powdered *lion's blood* is placed on the marble-topped work-surface and the corrosive *eagle's gluten* is added to it ¼ of the amount at a time. This is better understood as a drink-and-dry process where powdered *lion's blood* is moistened with molten butter of antimony. This is repeated, *resolved and coagulated*, until the powder will no longer dry out; a traditional method of arriving at an efficient ratio between the reagents. The paste is then placed into a sealed digestion flask and digested at low continuous heat.

The two cover-names used by Paracelsus to indicate butter of antimony are *mercurius vitae* and *gluten of the eagle*. These indicate the reagent in each recipe by which *butter of antimony* is achieved. *Mercurius vitae* is achieved by distilling stibnite with a mercury salt, whereas *eagle's gluten* is achieved by distilling *flowers of antimony* with Eagle salt. Paracelsus may have seen these as two different substances or possibly felt that *mercurius vitae* contained some form or essence of mercury derived from mercury sublimate and likewise *eagle's gluten* containing an essence of sal ammoniac, which is to say he named the reactions rather than the substances. Regardless of how he arrived at the names or the conceptual

CHAPTER 7 - PARACELSUS AND THE ALKAHEST

framework underlying them, the yield in question is antimony trichloride or in the case of *mercurius vitae*, its hydrolysed form as antimony oxychloride.

Mercurius Vitae and Cinnabar of Antimony

$$Sb_2S_3 + 3\ HgCl_2 = 2\ SbCl_3 + 3\ HgS$$

Eagle's Gluten

$$12\ NH_4Cl + 2\ Sb_2O_3 = 4\ SbCl_3 + 3\ O_2 + 12\ NH_4$$

3rd Arcanum – Mercurius Vitae (Eagle's Gluten)

1. The Mercurius vitae is made from mercurium essensificatum [corrosive sublimate], which is first purified and then sublimed with antimony [stibnite] so that they both ascend together and become one.
2. The product is resolved on a marble and coagulated [with lion's blood] four times. This gives Mercurius vitae, which comforts us in old age.

Mercurius Vitae (*Lion's Blood + Eagle's Gluten; Lili*)

$$Au \cdot Sb_2O_2SO_4 + SbCl_3$$

Step 1 above indicates a distillation between *essence of mercury* and stibnite, thus strengthening the hypothesis that the product of the 2nd *arcanum* is indeed a mercury-salt intended to be applied to the 3rd *arcanum* in logical progression. Step two of the third *arcanum* has understandably been interpreted by some researchers as the coagulation of *butter of antimony* to *Algarot powder*, which quite possibly was the original version. If *Algarot powder* is the product of the 2nd *arcanum*, then it is a stand-alone product and thus breaks the logical progression of the *four-in-one arcanum*. However, Paracelsus did not explain exactly how coagulation of the 3rd *arcanum* is performed and it can be interpreted as *Algarot powder*. The argument against this interpretation is that coagulation to *Algarot Powder* is typically performed in a vessel with water. Following precipitation of the powder, the supernatant liquid, known as *philosophical spirit of vitriol* or the *first lotion*, is set aside for later use as a medicinal substance. Coagulating on a marble slab suggests a combination of powdered *lion's blood* with *butter of antimony*. This strengthens the notion that the four-in-one arcanum follows a logical progression; the first step in this arcanum being the combination of *lion's blood* with *eagle's gluten* subjected to the traditional heat regimen. The possibilities for the *four-in-one arcanum* flow in order of logical probability is as follows:

1. First Matter – Lion's Blood ($Au \cdot Sb_2O_2SO_4$) →
2. Stone of the Philosophers ($HgCl_2$) →
3. Mercurius Vitae – Eagle's Gluten ($Au \cdot Sb_2O_2SO_4 + SbCl_3$) →
4. Tincture of the Philosophers

The 4th Arcanum – is the heat regimen to confect the Philosophers' Stone, which Paracelsus refers to as the *Tincture of the Philosophers*. It is a continuation of the previous three steps and a summation of the entire four-stage process. Doctor of philosophy and medicine, Alexander von Suchten, wrote an excellent treatment detailing Paracelsus' 4th *arcanum* in *An Explanation of the Natural Philosopher's Tincture, of Theophrastus Paracelsus* that fully accords with the recipe that follows. Paracelsus only glosses the *Tincture of the Philosophers* in *Paragranum*. In a separate tract found in the *Archidoxis, De Tinctura Physicorum*, he devotes the entire text to the subject. His delivery belies his attitude that confecting the stone was in truth nothing more than a simple and straightforward union of *lion's blood* and *eagle's gluten*, concocted until finished.

Paracelsus provided clues to the identity of *lion's blood* and *eagle's gluten* by suggesting:

> *That you may rightly understand me,* seek your Lion in the East, and your Eagle in the South.

Alexander von Suchten interprets Lion as *Green Lion,* meaning antimony, and advises to search for it to the east in Hungary. If the cover-name lion is interpreted to mean gold, Paracelsus' reference to the *East* as corresponding to *Lion* is perhaps an association between gold and ancient Phrygia in what is now modern Turkey. Phrygia in central Anatolia (Greek Ἀνατολή Anatolē; "east" or "dawn") was the home of King Midas and an ancient center of extractive gold metallurgy at Alacahöyük dating back to the 3rd millennium BCE. Paracelsus was intimately familiar with Turkey after spending time in Constantinople, present-day Istanbul, studying with Islamic alchemist-physicians. Alternately, Paracelsus may be paralleling the book of Daniel where Daniel is taken to Babylon in the land of Shinar, the land of God in the East, and educated in Chaldean mysteries culminating in such depth of understanding that he earned a permanent place at the court of King Nebuchadnezzar by interpreting the King's dream via alchemical allegory.

Alexander von Suchten states plainly that the *Eagle* is located in Istria to the south, a peninsula in the Adriatic Sea. Istria is Latin for Danube, which suggests that von Suchten was denoting the liquid nature of *Eagle's Gluten*. Paracelsus' correlation between *South* and *Eagle* is perhaps in reference to the original source of *Eagle* salt in Egypt to the south, home of the ancient mysteries. On the other hand, it could have been a tongue-in-cheek reference to ammonium chloride isolated from urine. In any case, his point remains clear that *Lion's Blood* is combined with *Eagle's Gluten* to achieve the *Tincture of the Philosophers*.

The 4th *arcanum* below lays out in general terms Paracelsus' summation or template for confecting his *Tincture of the Philosophers*. He ends by explaining clearly that the *Tincture's* primary application was as a medicine of a marvelously therapeutic nature.

BOOK 1 – ORIGINS & EVOLUTION

4th Arcanum – Tincture of the Philosophers (The Stone as a Medicine)

1. Take only the rose-colored blood from the Lion and the gluten from the Eagle. When you have mixed these, coagulate them according to the old process, and you will have the Tincture of the Philosophers, which an infinite number have sought after and very few have found. Whether you will or not, sophist, this Magistery is in Nature itself, a wonderful thing of God above Nature, and a most precious treasure in this Valley of Sorrows. That you may rightly understand me, seek your Lion in the East [Anatolia (Greek Ἀνατολή Anatolē; "east" or "(sun)rise" home of extractive metallurgy and King Midas], and your Eagle in the South [Egypt; Siwa, home of Eagle salt] ...

2. Lastly, the ancient Spagyrists having placed Lili [product of the 3rd arcanum] in a pelican and dried it, fixed it by means of a regulated increase of the fire, continued so long until from blackness, by permutation into all the colours, it became red as blood, and therewith assumed the condition of a salamander. Rightly, indeed, did they proceed with such labour, and in the same way it is right and becoming that everyone should proceed who seeks this pearl ...

3. For then at length you will see that soon after your Lili shall have become heated in the Philosophic Egg, it becomes, with wonderful appearances, blacker than the crow; afterwards, in succession of time, whiter than the swan; and at last, passing through a yellow color, it turns out more red than any blood. Seek, seek, says the first Spagyrists, and you shall find; knock, and it shall be opened unto you.

Paracelsus then goes on to characterize the *Tincture's* legendary therapeutic profile, thereby bridging the gap between metallurgical and medicinal alchemy. Descriptions such as these incited a sweeping obsession in Europe for the quest to obtain the Philosophers' Stone as a universal rejuvenative and tonifying medicine aimed towards perfect health and extraordinary longevity.

> This, therefore, is the most excellent foundation of a true physician, the regeneration of the nature, and the restoration of youth. After this, the new essence itself drives out all that is opposed to it. To effect this

> regeneration, the powers and virtues of the Tincture of the Philosophers were miraculously discovered, and up to this time have been used in secret and kept concealed by true Spagyrists.
> — *Paracelsus*, Archidoxis, De Tinctura Physicorum, Chapter VII

> ... many ways have been sought to the Tincture of the Philosophers, which finally all came to the same scope and end.
> — *Paracelsus*, Archidoxis, De Tinctura Physicorum, Chapter I

The Fifth-Essence

> The quinta essentia is a 'materia' which is extracted bodily **from all things, and from all things in which there is life,** separated from all impurity **and all that is mortal,** made subtle and purified **from all, separated from all elements. Now it is to be understood that** the quinta essentia alone is the nature, power, goodness, and medicine, which is shut up in things ... — *Paracelsus*, Archidoxis Book III

For Paracelsus, an essence was a substance's state of being one progressive step away from the bodily shell or corporeal form. The aim therefore was to isolate the metal's *quintessence* via alchemical technique.

> With regard to metals**, then, it must be understood that** they are divided into two parts, into their quintessence and into their body.
> — *Paracelsus*, Archidoxis, Concerning the Quintessence

The metal's *body* was considered as being a composite of the *Tetrasomia* (earth, air, water and fire) in varying proportions. Therefore, the *quintessence*, meaning fifth essence, was a substance's ethereal state of being furthest removed from the solidity of the four elements. It was the substance's primordial state, as it existed in the æther beyond the worldly realm of the four elements. Concerning the chemistry of metals, Paracelsus achieved a metal's *quintessence* by converting it into a fusible or soluble salt via *aqua-regia*. This method of operation may have been derived from the work of Paracelsus' childhood influence, Johann and Isaac Hollandus who operated in a parallel manner:

> The ancients worked long before they produced the Stone. With subtlety, they shortened the work, just as it is being done today. Understand, our parents required three or four years before they could perfect the Stone. This was because at that time, they knew no strong water, only distilled vinegar. Now, their descendants have invented aquafort [nitric-acid and aqua-regia]; which has greatly shortened the work. You should know that the work can be shortened even more through the first labor, **in as much as one must make the metals subtle and mingled, so that it turns into a dough-like matter.**
> – *Johann Isaac Hollandus,* De Lapide Philosophorum

Metals that had been dissolved in *aquafort* or *aqua-regia* were converted to salts in ionic solution. The excess liquid was removed to the point where the metal-salts achieved a thick liquid consistency, at which point they were understood to be *oils* of the particular metal being dissolved. In the case of gold for example, *oil of gold* referred to a gold-salt in an aqueous state. Metal-salts, as dry elixirs, were then derived from their *oils* via evaporation. The salt form of a metal showed no signs of having a metal shell and was soluble; therefore, alchemists considered the metal as having been released from its earthly four-elements (or *tetrasomia*). Today we would say it was converted to ionic form. In a passage where Paracelsus addressed the difference between *Potable Gold*, *Oil of Gold* and *Quintessence of Gold*, his definition of these terms becomes clear:

> Aurum Potabile, that is, Potable Gold, Oil of Gold, and Quintessence of Gold, are distinguished thus. Aurum Potabile is gold rendered potable by intermixture with other substances, **and with liquids [such as wine or honey]**. Oil of Gold is an oil extracted from the precious metal **without the addition of anything**. The Quintessence of Gold is the redness of gold extracted therefrom [from its oil] and separated from the body of the metal. – *Paracelsus,* De Membris Contractis, Tract II, c. 2.

CHAPTER 7 – PARACELSUS AND THE ALKAHEST

The passage above lends itself to the following interpretation…

Aurum Potabile = *bioavailable gold as salt or nanoparticles, typically combined with organics in syrup or wine as a medicine*

Oil of Gold = *aqueous yellow to red gold-salt also known as sulfur of gold [pictured left and center]; tetrachloroauric acid;* $HAuCl_4 \cdot H_2O$

Quintessence of Gold = *bioactive gold-salts or nanoparticles; cover-term for the Philosophers' Stone by some, soluble anhydrous purple* $AuCl_3$ *crystals [pictured right], also known as starry or celestial sulfur.*

Paracelsus is quite clear and unyielding with regard to this scheme, which remains consistent throughout his writings. It therefore seems reasonable to apply the scheme to antimony as well. The *quintessence of antimony* in this case would be a stabilized salt of antimony; *butter* or *oil of antimony*, also called *glacial oil of antimony* by Christophe Glaser in *Cours de Chymie 1663*, referred to antimony trichloride. The challenge that would have faced Paracelsus in this regard is that the stable salt of antimony, *Algarot powder* (antimony-oxychloride; SbOCl) is not water-soluble. *Butter of antimony* does not stay in liquid form but solidifies as it cools and then fumes via a reaction with humidity in ambient air ultimately resulting in *Algarot powder*. Achieving *quintessence of antimony* that could be stabilized as a fluid was an alchemical riddle.

Chaos and Stars

Hence many have looked for the quintum esse [fifth being] in alchemy which is nothing else but that thus the four elements are taken away from the arcana, and what remains afterwards is the arcanum. This arcanum furthermore is a 'chaos' [volatile compound; spiritus], and it is possible to *carry it to the stars [distil] like a feather before the wind. Now*

> *the preparation of the medicine should be done in this way.* The four elements [corporeal manifestations] are taken away *from the arcana, and then* it should be known what astrum [star, constellation or æther] *is in this particular arcanum, and what is the astrum of this particular disease, and what astrum is in the medicine against the disease.*
> — *Paracelsus*, Paragranum

The passage above provides a number of suggestive clues as to the nature of *the quintum esse* as it must apply to antimony. It is a *spiritus*, meaning either volatile compound that is a product of distillation, easily vaporizes or both. Its corporeal (metallic) form has been removed as discussed above and three *astrums* are associated with the *quintessence*; *astrum* of the arcanum, disease and medicine. The term *astrum* typically means star, constellation or æther but can also mean glory, nobility or even immortality and is associated with the heavens. Paracelsus was likely insinuating a correspondence between all of these. The *astrum* residing in the body of a substance was the *soul-spirit* principle responsible for its nature and hidden power, which could be isolated via alchemical preparation and applied.

Paracelsus held that each substance contains an *astrum* as its vital principle in correspondence with a celestial body. The passage below discusses the astrum of mercury, yet it remains unclear as to whether he is addressing *quicksilver* or *sophic mercury*. Both substances are mercurial as they pertain to the salt-sulfur-mercury scheme and thus the following applies equally to both:

> *And let it be understood* regarding Mercurius: *that it is not male unless it is sublimated by the astrum of the sun: otherwise it will not ascend.* It *has many [methods of] preparations, but only one corpus. The corpus, however,* is nothing but the sulphur or sal, *which has many sorts of corpora, for which reason* they yield many sorts of salia and sulphura. Here *it is but a single corpus, but the astrum prepares it variously in many natures ...* – *Paracelsus*, Paramirum, I:III H 1:80

CHAPTER 7 – PARACELSUS AND THE ALKAHEST

To comprehend *sophic mercury* as Paracelsus meant for it to be understood, a reading of the passage below taken from *A Short Catechism of Alchemy* definitively clarifies the issue:

> **Q:** *How is the generation of seed comprised in the metallic kingdom?*
> **A:** *By the artifice of Archeus [separator and organizer] the four elements, in the first generation of Nature, distil a ponderous vapour of water into the center of the earth; this is the seed of metals, and it is called Mercury, not on account of its essence, but because of its fluidity, and the facility to which it will adhere to each and every thing.*
> **Q:** *Why is this vapor compared to sulphur?*
> **A:** *Because of its internal heat.*
> **Q:** *From what species of Mercury are we to conclude that the metals are composed?*
> **A:** *The reference is exclusively to the Mercury of the Philosophers, and in no sense to the common or vulgar substance, which cannot become a seed, seeing that, like other metals, it already contains its own seed.*
> **Q:** *What, therefore, must actually be accepted as the subject of our matter?*
> **A:** *The seed alone, otherwise the fixed grain, and not the whole body, which is differentiated into sulfur, or living male, and into mercury, or living female. ... [and further along...]*
> **Q:** *What kind of mercury, therefore, must we make use of in performing the work?*
> **A:** *Of a mercury which, as such, is not found on the earth, but is extracted from bodies, yet not from vulgar mercury, as has been falsely said.*
>
> – A Short Catechism of Alchemy Founded on the Manual of Paracelsus preserved in the Vatican Library

In the *Catechism* excerpt above, the point is made abundantly clear that the seed of a sulfide-ore is the identity of sophic mercury, which is then separated and distilled to a *ponderous vapour of water*. Like most alchemists, the author states conclusively that he is not referring to quicksilver.

The *astrum of the sun* in combination with antimony results in a variety of antimony-salt products. The *astrum of the sun* is sal ammoniac as understood from Hollandus' text *Hand of the Philosophers with its Secret Signs*. Sal ammoniac is designated by the alchemical symbol of a star – an asterisk or astrum (✱):

> The third sign [✱] of the philosopher's Hand is the Sun, **standing above the third finger.** By it, Sal Ammoniacum is designated ...
> – *Johann Isaac Hollandus*, Hand of the Philosophers

Paracelsus is alluding to the fact that *sophic mercury* in the form of *regulus, flowers* or *glass of antimony*, when sublimated with *sal-ammoniac*, yields a variety of different salt products such as *butter of antimony, antimony-pentachloride* and *Algarot powder*, which he describes as *many [kinds of] preparations* or *many sorts of salia and sulphura*. What remains unsaid is that the varying products that result from distilling *antimony* with sal ammoniac are dependent upon ratio and technique. Just as *sophic mercury* can result in the three forms mentioned above, Paracelsus cryptically introduces three parallel mercury products, called the *three modes* as follows:

> *First of all,* let us speak about mercurius. **As we have already said, the mercurius is the liquor in the human being** and this is of many kinds, so that many different things proceed from it. It must be kept in mind in all of these that *there are three modes of disintegration.* The one **in which the m[ercurius] rises up** is distillatio. The *second is sublimatio.* The third *is praecipitatio.* – *Paracelsus*, Paramirum II:IV

From the perspective of alchemical substances and processes, he is referring to manipulating antimony via distillation, sublimation and precipitation. The problem as Paracelsus may have viewed it was that *butter of antimony* still retained a *corpus* or body, meaning it still had a form of solidity to it. The proof of its inherent power was that this liquid *quintessence of antimony* could dissolve gold and convert it to a salt in

CHAPTER 7 - PARACELSUS AND THE ALKAHEST

solution without losing its power to dissolve the way acids such as *aqua-regia* do.

A simple experiment by adding the body of gold, in the form of gold leaf, to antimony pentachloride provides a visual model of what possibly inspired Paracelsus' narrative below:

> *The mercurius is rendered so subtle by this preparation that no one may withstand it in its inherent, natural force.* **And this is the reason why: the other two substances [sulfur and salt] cannot check it because of the excessive heat** *which pushes them back. In consequence, it becomes so subtle that it penetrates the bones and the flesh [dissolving the substance];* **[it does so] not only through the poros, but indeed even outside of these,** *sweating through and penetrating.* **You should be aware that** *there then arise pustulae, Morbus Gallicus, lepra, and other things of this kind;* **and** *from this they receive their primitiva materia and their causa;* **and much more could be said about things of this kind ...**
> — *Paracelsus*, Paramirum II:IV

He is describing the *natural force* or powerfully corrosive nature of the substance, which he appropriately terms *excessive heat*. His visual portrayal that *it penetrates the bones and the flesh* refers to dissolution of the body of gold as it reacts to its chloride salt form; the *sweating through and penetrating* refer to the fuming and corrosive nature witnessed during the reaction. His description of open sores is in reference to the crystallization of gold during the reaction. Once it becomes a gold-salt, described by Paracelsus as receiving its *primitive materia*, it dissolves into solution or as Paracelsus would have understood – its oil. Further along in the same chapter he addresses the fuming phenomenon that must have left quite an impression. He likens the ensuing crystallization to a *chill* or *frost*:

> *As for the form and way in which it rises in this sort of heat, take note with respect to this that it causes many kinds of chill (frost), heats, shuddering, [and] trembling, when its paroxysmus, or something similar*

> to it, is initiated. *For when it happens that such a sharp poison and subtlety is initiated [in] nature; it descends then into an opposite condition, that is, a fright: this taking fright is a physical trembling which results from fear:* the chill [or] the heat accompanies it: *for there results an obstruction and an excess of vapors,* as with a closed container which is boiling and lifting itself up; and the chill is the materia and the nature of any fright: it [always] causes chilling. But *when the heat increases so powerfully, the chill decreases and lets the heat prevail.* So you can see the peculiar nature of the mercurius.
>
> — *Paracelsus*, Paramirum II:IV H 1:26

It is natural, but somewhat incorrect to interpret the above as simply an observation of illness, and equally incorrect to assert that what is being described is a mere chemical reaction. Paracelsus was a master chemist who undoubtedly correlated chemical observations in the lab, in this case *sophic mercury* in its most exalted form, with his overall theory of *in-vivo* activity of mercury by using gold as an analog for the body in an *in-vitro* experiment. The above experiment applies equally to *butter of antimony* or antimony pentachloride. His primary challenge when following the path of exalting antimony to its logical conclusion, its most exalted liquid saline form, would have been to stabilize such a volatile fuming corrosive antimony salt.

> *And although these may by the Lord God be made manifest to anyone, still, the rumor of this Art does not on that account at once break forth, but* the Almighty gives therewith the understanding how to conceal these and other like arts even to the coming of Elias the Artist, *at which time there shall be nothing so occult that it shall not be revealed.* — *Paracelsus*, Archidoxis, De Tinctura Physicorum, Chapter IV

CHAPTER 7 – PARACELSUS AND THE ALKAHEST

LIQUOR ALKAHEST – IGNIS AQUA

Spirit Divine, blest be thy state,
That art in Salt incorporate
And in the Worlds true virgin wombe
A pure Quintessence art becomme.
Lord have mercy upon us.

– Heinrich Khunrath, Incorporating the Spirit of the Lord in Salt, 1597

In the poem above, Khunrath has encrypted a number of telling clues towards the identity of the *Spirit of the Lord [embodied] in Salt*. It is characterized as a divine spirit (distillate), a salt that embodies the entire art (power to dissolve and coagulate gold) and a *quintessence* (corporeality removed). This characterization of the *alkahest* as a salt distillate and a *quintessence* runs throughout Paracelsian and Helmontian literature. The following section will approach the *alkahest* via a study of descriptions and cover-names employed to both encrypt and reveal the identity of the substance.

Characterization
Paracelsus mentions the *alkahest* by name only once in his writings:

> There is, also, *the liquor alkahest*, which possesses great force and efficacy in preserving and fortifying the liver, as well as in preserving against all forms of dropsy that come from the vices of the liver ... And even if the liver was already ruined and destroyed, [the alkahest] *itself plays the role of the liver*, as if this had never been ruined nor destroyed. Thus, whosoever of you labors in medicine ought to strive with the greatest zeal to learn how to prepare the alkahest in order to turn away the many diseases that arise from the liver.

Paracelsus was speaking euphemistically when referring to the liver. He states that the *alkahest itself plays the role of the liver*. The liver was considered the fountain of blood, seat of the soul and the center of vitality next to the heart. As an analog to the function of the liver, it is important to understand how Paracelsus viewed its function:

> *As soon as it happens that the drink and food are cleansed of their waste and [that which has been cleansed] is sent on to the liver, it is to be noted first of all that the urine is generated outside of the region of the stomach. What happens is that first of all the* nutrient is attracted by the liver and in the process of this attraction the urine is separated from the nutrient... It is just like a rain that falls in drops as these are generated, and not in the form of an entire body of water. It is a generation in the form of drops which falls down in this way (as is described in terms of its mechanici). And so it is too that the materia of *the nutrient that belongs to the liver is mixed in with the urine, and [now] it is extracted from the urine; so that the latter remains to itself... It is a noble organ which serves many other parts of the body, indeed nearly all of them.* – Paracelsus, On the Origin and Cause of
> Diseases, Tractatus Tertius H 1:159

To Paracelsus, the liver receives pre-digested and separated nutrient from the stomach and then separates that from urine. His clue that the *alkahest* operates in similar manner, can also be understood as playing the role of dissolving and purifying, followed by producing a new life-giving substance the way the liver cleans and produces blood and nutrient according to medical views at the time:

> *... tis plain by experience; that* this Liquor will by greater length of time, dissolve all mixt Beings *by its Active, Thin, Spirituous Penetrative, Dissolving and Homegeneous Nature,* in a Natural degree of Heat equal to that of the Liver, *and separate them into their distinct Substances ...*
> – Cleidophorus Mystagogus, Trifertes Sagani or Immortal Solvent

Paracelsus stated that the *circulated salt* had the ability to extract the *prima entia* (primordial body) of metals, meaning it could be employed to separate and convert these metals to their salts analogous to the way the liver separates *materia nutrimenti* and *urina*.

His liver analogy may also allude to the Greek legend of Prometheus, a titan who stole the secret of fire, a ray of sun, from heaven. He smuggled it down to Earth and presented it to man thus catalysing humanity's process towards technological progress and civilization. Prometheus is also credited for creating man from clay. Furious that Prometheus acted without permission, Zeus sentenced him to be bound to a stone and stripped of his garments. Each day Aquila [Eagle] would come to feed on his liver, which regenerated only to be eaten again the following day. Hercules, agreeing with Prometheus' selfless service to humanity, eventually saved him by killing Aquila with his poisoned arrows. In this imagery, we have a stone, sacred fire, regeneration, an eagle and divine creation all of which may be interpreted as alchemical themes. Prometheus eventually became a cultural symbol of humanity's striving towards scientific and technological ends, and the inadvertent dangers and inherent consequences that this entails.

German Paracelsians, Gerard Dorn and Martin Ruland claimed that the *alkahest* was *prepared mercury*. This was an alchemical manner of speaking, which can be better understood as a preparation of *sophic mercury*. Joan Baptista van Helmont was a Belgian nobleman and alchemist-physician in Paracelsus' tradition who lived during the late 16th to early 17th century. He was educated in Kabbalah and western mysticism before taking up medicine, earning his M.D. after ten years of study. Shortly thereafter, he devoted his energies to *pyrotechny*, the term for chemistry at the time, and proudly titled himself *Philosophus per Ignem* (Philosopher by Fire). He carefully researched Paracelsus' works, replicating many experiments and went on to develop his own brand of medical alchemy for which he was well remembered. Van Helmont expanded on Paracelsus' *alkahest* and was largely responsible for the ensuing widespread fascination for it. He characterized the *alkahest* as follows:

BOOK 1 – ORIGINS & EVOLUTION

> *There is a certain* universal solvent that dissolves, changes, separates, and reduces all bodies... *To such a point that an herb so dissolved, may be completely distilled, and will leave neither coal nor residual ashes in the bottom... In the aforementioned solvent there is a very* powerful destruction *of every thing; in relation to it, the elements are impotent, and even fire is of no importance or force.*

... and elsewhere:

> *His [Paracelsus'] Liquor alkahest is* more eminent, being an immortal, unchangeable, and loosening or *solving water, and his* circulated Salt, *which* reduceth every tangible body into the liquor of its concrete or composed body. *– Joan Baptista van Helmont*

In the passages above, van Helmont characterized the *alkahest* as a powerful liquid solvent and a *circulated salt*. In his writings on the subject, he further expands on these fundamental properties. The salt could purify a substance, dissolve it to its primordial liquid state of being yet preserve the potency and medicinal properties of the substance in solution. It had the power to dissolve plant matter, minerals, stones, metals and even charcoal without leaving any residue and without suffering any reaction. Furthermore, the substance being dissolved, including gold would simply melt as if snow placed in hot water. In one experiment mentioned in the *Ortus Medicinae*, van Helmont describes placing charcoal into the *alkahest* where it dissolved into an oil at gentle heat:

> *I have put* equall parts of an Oaken Coal, and of a certain water, in a glasse *Hermetically shut:* in the space of three dayes, *the whole Coal was turned* by the luke-warmth of a Bath, *into two transparent Liquors, divers in their ground and colour ... But the dissolving Liquor, remains in the bottom, being of equall weight and virtues with it self.*

This notion that the *alkahest* was a *circulated salt* was expanded upon by Dutch physician and chemist William Yarworth, active during the early

332

CHAPTER 7 - PARACELSUS AND THE ALKAHEST

18th century under the patronage of Isaac Newton. The excerpts below characterizing the *alkahest* were published in *Trifertes Sagani or Immortal Solvent* in 1705 under the pseudonym Cleidophorus Mystagogus:

> *Tis a Noble Circulated Salt* prepared with wonderful Art, till it answers the desires of an Ingenious Artist; yet 'tis *not any Corporeal Salt made liquid by a bare Solution, but is a Saline Spirit,* which Heat cannot Coagulate by evaporation of the Moisture, but is of a *Spiritual Uniform Substance, Volatile;* which in *a gentle Heat will Distill over, leaving nothing behind;* that is to be understood in a Requisite Heat of Sand; so is there an Exaltation made far above what Nature was able ever to perform. ...the Principles being Centrally contained in the Original Chaos, which being separated and brought again to an Indissoluble Union, is, *the Serpent devouring his own Tail and so renovating into that,* upon which Death can have no Power...
>
> ... the *Alkahest or Sal Circulatum is a Saline Liquor,* and therefore by Paracelsus sometimes called the *Liquor of Salts* and doth Dissolve Bodies, but remains not with them, being as easy separable from them ... the Liquor Alkahest *dissolves not only Gold, but also all the other Metals,* by way of Destruction, so that the Generative Virtue is defaced and wholly obliterated, and in this Reduction into their First Matter... Tis *a Ponderous Liquor, being indeed all Salt,* without and Watery Phlegm; it is *all Volatile being wholly a Spirit, without and Corporeity* left in it ...
>
> – Cleidophorus Mystagogus, Trifertes Sagani or Immortal Solvent

The *alkahest* then is a highly corrosive heavy nonaqueous liquid saline solvent, a volatile *circulated salt* containing no water, which can be distilled and is associated with the *Serpent* or *Ouroboros* biting its tail, implying an antimonial product. Modern chemistry historians have stoically declared that no such substance exists, but this is a failure to factor human nature into the equation. The more likely scenario is that Paracelsians were working with a specific compound solvent;

descriptions of whose properties suffered over-embellishment to the point where the genuine substance, regardless how fascinating its properties may have initially appeared, could never live up to the spin that developed around it by those unfamiliar with its true identity and properties.

The *alkahest* became the object of an obsessive quest for rediscovery during the 17th and 18th centuries. Yet since Paracelsus' time, a small number of German, Dutch and Belgian alchemist-physicians in his tradition appear to have been fully familiar with a very real substance called the *alkahest* or *circulated salt*. A study of additional cover-names employed for the substance may reveal commonalities and offer deeper insight into the identity of the *alkahest*.

Cover-Names

Paracelsus encrypted the identity of white *Sal Artis Mirificum* (or, remarkable salt of the art) by a few different cover-names and associations via the combined use of anagrams and linguistic word play. *Salt* and *essence of antimony* figure strongly in two of his cover-terms. The *alkahest* and *Elias Artista* are terms that have become firmly established in alchemical jargon. Deciphering these two terms reveals much about Paracelsus and the secrets he encoded in them.

> *Every congelation, [every]* coagulation is from the salt.
> – *Paracelsus,* Paramirum II:I H 1:107

Elias Artista is encrypted in the form of an anagram as commented on by Glauber and can be arranged in various ways to express a similar notion:

    et artis salia = art and salts (from Glauber)
    sale a artisti = salt (coagulation) by the artist
  arti ista e sal = this art (or technique) from the salt (coagulation)
  artist i a sale = Latin: the artist of the salt, Italian: the artist in salt
   artisti a sale = (Italian: artists' salt, Italian: artist from the salt)

The term *alkahest* is a little more difficult and requires the researcher to assume the same sense of religious inclination and symbolic meaning that would have been second nature to Paracelsus. He encrypted the identity of the seed of the *alkahest* into its name. This is significant because Paracelsus believed in predestination, meaning that seeds of life were created by God and they were predestined to achieve their highest essence, which Paracelsus termed *ultimae materiae substantiae* (substances of the final matter) via the aid of a human artisan; an echo of Plotinus' Logos:

> *The Logos within a seed contains all the parts and qualities concentrated in identity ... – Plotinus*

In the case of the *alkahest*, he is referring to its seed as being the seed found in *stibnite* that must be brought to ever-higher essences or exaltations. This is part of Paracelsus' personal theory of matter and reflects his unique theosophical views. According to Paracelsus, a substance in a specific form will demonstrate the property of the predominant element, the other three imperfect due to the potency of the predominant finishing one. The *quintessence* arrives at the seed of an element in its most exalted state beyond the corporeality of the four elements. His essence schema for metals also applies to antimony, the four essences or gradations loosely described in the following passage. The outline in the passage below can be understood as an expression of the *Tetrasomia* of antimony...

> *This spirit in its fiery form is called a Sandarac, in the aerial a kybrick, in the watery an Azoth, in the earthly Alcohoph and Aliocosoph.*
>
> *1ˢᵗ esse* = **stibnite** – *fiery Sandarac*
> *2ⁿᵈ esse* = **regulus** or **flowers** – *earthly Alcohoph or Aliocosoph*
> *3ʳᵈ esse* = **glass** of antimony – *aerial Kybrick*
> *4ᵗʰ esse* = **butter** of antimony – *watery Azoth*

In ascending order, each essence is slightly more purified, volatile and somewhat less solid that its former state of being, culminating in a highly corrosive antimony salt in an ethereal liquid form. The cover-terms *doubled-mercury* or *circulated-salt* are appropriate because the *alkahest* can be made via reflux distillation (re-extraction or re-essencification) of *butter of antimony* to achieve the *quintessence* (5th esse), the *mercurius-duplicatus* (doubled-mercury) or *sal-circulatum* (circulated salt); all of which in this context are synonymous.

> 5th *esse* = *alkahest* (spiritus urine + spiritus vini) – *circulated salt or quintessence of antimony*

The theory and method of creating a *circulated salt* via *reverberatio*, known today as reflux distillation, is explained in the following passage:

> *But that which is reverberated is a different sal and is a liquidum humidum: the same thing is distilled back and forth in its [own] anatomy and is called reverberatio. The reason is that no heat nor foreign richness can enter into its substance. Rather, just as water and oil cannot be mixed, other, foreign things cannot enter into it. For this reason, on account of this salt, the spiritus go round and round and up and down, until there arises a mucilago or viscositas; whereupon it has more of its acerbity than it should have ...* – Paracelsus, Paramirum, II:V H 1:129

Paracelsus never explicitly states that the *quintessence of antimony* is the *alkahest*. This must be inferred by association and careful examination of the cover-names and characterization of this highly secret salt. His cover-name for the *quintessence* or *circulated salt* is the anagram *alkahest*:

kahâl es † = antimony is the savior
kahâl es † = antimony is the initiation
es † kahâl = it is the renewal/rebirth of antimony
est kahâl = it is antimony (Italian: Levante stibnite)
or (Latin: ...is of stibnite)

... or as Paracelsus once wrote in his letter to *Theodore* in 1536 in a treatise concerning the *tincture of antimony*:

Dear Theodore;

With great truth the ancient Spagyrists have said "Est in ☿ io quicquid quaerunt sapientes!" *but they have not told us what kind of ☿ they meant. Although I dare not mention openly what they intended, let it suffice that, from ocular demonstrations. I am convinced they intended such a ☿ as I treat of in the following work. That with this tincture or ☿ of ♁ ...*

"Est in ☿ (of ♁) io quicquid quaerunt sapientes!"
"It is in mercury (of antimony) what seek the wise!"

Savior – The first interpretation is based on the Tau symbol that rests atop the circle of antimony - ♁ - and known in Latin as the *salvator mundi* symbol, meaning the *savior of the world*. This symbol held great importance to Semitic populations in antiquity. It was understood as symbolizing Moses with his outstretched arms in Exodus 17:11. In Judaism, Tav is the last letter of the Hebrew alphabet, which means mark or seal and symbolizes redemption or the Messiah. In Ezekiel 9:4, the Tav sign was marked on the forehead of the saved ones. It may have been the mark of Cain, or more historically, the insignia of the Kenite metallurgico-religious priesthood encountered by Moses in the Sinai. In the Christian tradition, Tav symbolized the crucifiction, was adopted by St. Anthony of Egypt, St. Francis of Assisi, and was discussed by Pope Innocent III.

Initiation – Manly P. Hall, in his book *The Secret Teachings of All Ages*, explained that when a King had been fully initiated into the mysteries, a Tau cross was placed upon his lips. This may have been meant to seal the lips from ever speaking of what was experienced. Tav has an occult relationship with Saturn, thus the correlation with antimony, earth and divinity.

Rebirth – The Tau cross can also signify rebirth or renewal and is the final letter of Azoth when spelled in Greek, AZΩT, implying renewal and regeneration. In a work entitled *De renovatione et restaruatione* (Of renovation and restoration), Paracelsus describes his salt as *dissolved salt, circulated salt* and *the prima ens* (primary entity) *of salts*.

Kahâl – The last interpretation is a straightforward declaration that the *alkahest* is nothing more than *stibnite* in its highest essence, the *quintessence of antimony*. In a general sense, Paracelsus created *quintessences* of metals by reacting metals in aqua-regia to create metal-salts. According to this standard method of operation, the *quintessence* of antimony must also be a chloride-salt of antimony to qualify as a *quintessence*. Paracelsus was biblically fluent and the term kahâl would certainly be familiar to him as was the association between stibnite and the Hebrew word for it. If any of these is an accurate interpretation of the message Paracelsus was attempting to convey, there is no need to choose one over the other. Paracelsus' gift for linguistic word play suggests he would have considered all the above and then some.

Mercury – Michael Toxites, a 16th century Tyrolean poet, teacher, Paracelsian physician and alchemical manuscript publisher, in a commentary on the *alkahest*, described it as a *liquor of mercury*. This is in accordance with Gerard Dorn and Martin Ruland's assertion that the alkahest was *prepared mercury*. In true alchemical fashion, this description was a cover-name for a prepared liquor of *sophic mercury*, meaning a *liquor of antimony*. According to this interpretation, the word alkahest may simply be a derivative of German *geist*, meaning mercury where:

Alkahest = al-geist = of [sophic] mercury = of butter of antimony

Many of the *alkahest's* synonyms originate with van Helmont who coined the following cover-terms; *universale solvens, dissolvens immutabile,*

CHAPTER 7 - PARACELSUS AND THE ALKAHEST

ignis aqua, ignis gehennae, summus atque felicimmus salium, liquor unicus, liquor exiguus, small liquor, chiefest and most successful of salts and tincture of the Lile of Antimony.

> It is a salt, most blessed and perfect of all salts; the secret of its preparation is beyond human comprehension and God alone can reveal it to the chosen. – Johannes Baptista van Helmont

During the period in which these great alchemists were practicing medicine and chemical pharmacy, and up until the modern period, antimony was considered a powerful medicine due to its purgative powers and therefore was a revered therapeutic substance. According to Dr. R. Ian McCallum's book *Antimony in Medical History*, by the 18th century *"there were more compounds of antimony than any other modern element in the medical chemists' repertoire"*.

> Of all minerals antimony contains the highest and strongest Arcanum. It purifies itself and at the same time everything else that is impure. Furthermore, if there is nothing healthy at all inside the body, it transforms the impure body into a pure one ...
> – Paracelsus, The Complete Works

The characterizations and cover-names all suggest that the *alkahest* was a thick volatile liquid salt compound derived from antimony, and that it was corrosive enough to dissolve gold with the added bonus of converting it to a chloride salt. It purified substances in the same manner that the Liver purified food. These great founders of pharmaceutical chemistry may have been referring to antimony pentachloride ($SbCl_5$) and its unique ability to generate chloride salts from nearly anything that were exposed to it effortlessly. Alchemists and physicians saw tremendous potential to create new medicines based on plant and mineral salts via the *alkahest*. The quest to discover the recipe to create it was the holy grail of alchemical pursuit during the 17th and 18th centuries.

DISTILLING THE IMMORTAL LIQUOR

The unfortunate fact of the matter is that Paracelsus did not leave exact details for how he produced his *alkahest*. The best that can be hoped for is an exploration of his influences and world-view coupled with an examination of canonical recipes in his lineage weighed against probable chemical candidates. What follows is a hypothesis for the identity of Paracelsus' *alkahest*. The legacy of the *alkahest* may have its origins in a unique compound solvent detailed in the work of Johann and Isaac Hollandus.

Paracelsus was tutored in chemical medicine by his father and was exposed to the writings another alchemist-physician father and son team, Johann and Isaac Hollandus. Isaac Hollandus wrote an excellent tract on the separation, crystallization, recrystallization and combination of urine salts and rectified distillate. He used this menstruum as the main ingredient in a compound solvent that included vinegar, ethanol, sal ammoniac, sea-salt and quicklime. Paracelsus would certainly have been familiar with Hollandus' urine product. The method of operation for Hollandus' work with urine is an exact model for the separate-combine tone of Spagyrics attributed to Paracelsus. Furthermore, Hollandus' instructions for using his solvent to create *Oleum Solis* (Oil of Gold) echo nearly verbatim Paracelsus in his instructions for creating Aurum Potabile (Potable Gold) via *circulated salt*.

One of the primary ingredients found in traditional recipes for the *alkahest* is a urine product often described as *spirit of urine, sophic sal ammoniac* or sometimes simply *urine*. *Spirit of urine* is generally listed as an alkahest ingredient in combination with *spiritus vini*, which some have interpreted to mean alcohol. These two ingredients make up two of Hollandus' six ingredients for his solvent. Paracelsus would have understood Hollandus' *spirit of urine* as a substance to be used in combination with others. It seems reasonable to suggest that Paracelsus too would have considered *spirit of urine* as an ingredient rather than a complete product in and of itself. Writings in later alchemical texts indicate as much due to the two-ingredient common denominator found in treatises that address the *alkahest*.

$$\text{\tiny sp}\square + \overset{\text{\tiny 87}}{V}$$

Spiritus Urine

Paracelsus, in a comment railing against sophist physicians employed knowledge of urine, or lack thereof, in the same manner a fencer uses his weapon to reveal the perceived ineptitude of his opponent:

> *If you lack all knowledge of urine, what should I think of you* **but that you are extracting dues and interest for your Mrs. Doctor to do her humble bidding, just like a pimp would do his business.**
> – Paracelsus, Paragranum H 2:14

Paracelsus considered urine to be the *central* or *circulated salt* of the body, isolated by the liver and a derivative of blood. Moreover he viewed it as a sort of microcosm of the entire human organism, thus he prized uroscopy as a valid form of diagnosis:

> *Urine is reliable and it is worthy of an important evaluation; [for] it contains the entire physiognomy [of the patient], the whole anatomy [of the patient] with all properties [thereof].* **Since all of that is contained in the urine, it must be evaluated reliably by the physician and formulated in its entirety, since these things are indeed present in it.**
> – Paracelsus, Paragranum H 2:21

Urine is very rich in sodium, potassium and ammonium chlorides. In addition, urine contains considerable amounts of urea, iodine and oxalic, uric and hippuric acids along with other components in trace amounts. The recipe below has been rendered into contemporary language in order to highlight its nature as demonstrating sound chemistry. Stale urine is used as a starting point because ammonium and nitrogen reach their highest concentrations, phosphorus falls to its lowest and these levels begin to stabilize at almost exactly the 40-day mark. Early alchemists must have discovered this phenomenon through trial, error and careful observation.

Spirit of Urine – Johann Isaac Hollandus

Separate
1. Distil 40-day+ old stale urine to dryness and retain the urine-distillate.
2. Calcine the remaining salts to dryness for three hours.

Crystallize
3. Dissolve the salts in water and digest for two hours.
4. Separate the liquid from the solids via filtration while warm.
5. Reduce the liquid to saturation point and refrigerate. Collect the resulting salt crystals.
6. Repeat step 5 with the filtrate.

Recrystallize
7. Dissolve the collected salts in water and distil.
8. Digest the distillate and separate the liquid from the solids via filtration while warm.
9. Reduce the liquid to saturation point and refrigerate. Collect the resulting salt crystals.
10. Repeat step 9 with the filtrate.

Combine
11. Dry all collected salts.
12. Rectify the retained urine-distillate from step 1 via distillation 10-12 times until no solids remain.
13. Dissolve the purified salts into the rectified urine distillate.

This rectified urine distillate served as a starting point to create Hollandus' compound solvent. The ratios below have been converted

CHAPTER 7 – PARACELSUS AND THE ALKAHEST

into metric measurements and adjusted to yield 500 ml of solvent, based on Hollandus' original instructions in *Tractatus de Urina*. The ingredients below are first dissolved and then distilled 10-12 times and rectified. Hollandus explains that the application of this distillate was primarily to turn calxes (powdered materials) into their *quintessences*, suggesting that it converted powdered substances to chloride-salt products.

Dissolve, Distil and Rectify

250 ml	urine distillate above
125 ml	distilled vinegar
125 ml	ethanol [spiritus vini]
10 g	sea-salt
10 g	sal-ammoniac
10 g	quicklime

A simplified variation of the above recipe appears as a comment written by Dutch physician Theodore Kerckring as a notation in Basil Valentine's *Triumphal Chariot of Antimony*. Kerckring's *menstruum* follows the two-ingredient formula for the *alkahest* composed of *spiritus urine* + *spiritus vini* if these ingredients are interpreted to be ammonium chloride + ethanol:

> Not common spirit of wine, which would be useless for this operation, but that of the Sages, which is prepared as follows for the extraction of this tincture:
>
> 1. Take four ounces of thrice sublimed [purified] *salt of ammonia*;
> 2. of *spirit of wine [spiritus vini] distilled over salt of tartar, so that it is quite clear – ten ounces;
> 3. place ill phial over digestive fire till the spirit of wine is filled with the fire or sulphur of the salt of ammonia;
> 4. distil thrice in the alembic, and you have *our true menstruum* ...

Apart from the Philosophers' Stone, the *alkahest* was a great object of re-discovery sought by would-be adepts. During the mid-17th century alchemist-physician and member of the Hartlib circle, George Starkey, wrote treatises devoted to the subject of *philosophic sal ammoniac* and the *alkahest*. Starkey had determined that *sophic sal ammoniac* was ammonium carbonate, also called *spirit of hartshorn*, which he believed to be one of the two ingredients that comprise the *alkahest*. In *Gehennical Fire*, Professor William R. Newman detailed Starkey's experimentation in an effort to reconstruct the recipe, which Starkey described as follows:

> *That most acute subtile penetrative* Spirit of Man's Urine, by the help of another medium, *not of a diverse ferment from itself, but centrally one I say with it,* must be united to an Acidum, not Corrosive, *sed naturae suae grattissimum.* This Acidum must be equally volatile with the Salt of Urine, *before it can be Married or United intimately with it.* Then by other Circulations [reflux distillation] it attains the height of purity *to be entitled Ens salium, summum salium & foelicissimum.*
>
> – George Starkey, Liquor Alkahest

Starkey wrote another short work, *The Unheard of Unorthodox Doctrine of Sal-Ammoniac both Vulgar and Philosophical*. Professor William R. Newman demonstrates that Starkey, convinced that the cover-name *spirit of urine* indicated ammonium-carbonate, experimented with acetic, sulfuric, nitric and hydrochloric acids in an effort to determine just exactly which acid was the identity of the non-corrosive, volatile *acidum* that served as the second ingredient in the recipe. To Starkey, this mysterious *acidum* was the *spirit of wine* that, when reacted with *spirit of urine*, produced the *alkahest*. What Starkey and other researchers in the Hartlib Circle and the Royal Society were attempting to identify was an acid or *central salt* that dissolves gold without violence and is a product of (sophic) mercury. They had completely overlooked *butter of antimony* as the chemical identity of philosophers' mercury, which is to say they failed to realize *al-koḥl* (butter of antimony) as being synonymous with *spiritus vini*.

Ultimately, Starkey's version of the *alkahest* was rejected and the search continued. His *spirit of urine*, ammonium-carbonate, degrades to urea + water and is indeed a useful substance but falls far short of the *urine menstruum* of Hollandus unless used in combination with a metal-salt such as *butter of antimony* or *antimony-pentachloride*. Hollandus' chloride-salt saturated *urine menstruum* can be understood in contemporary terms as a low-temperature *ionic liquid* (IL) in a general sense, or more accurately as an *ionic solution* due to the concentration of chloride-salts in solution; these being absent in Starkey's version.

Spiritus Vini
Spiritus vini or *spirit of wine* is a Latin derivative of Arabic *al-koḥl* (essence of stibnite) originally meaning *butter of antimony*. Butter of antimony ☿ was the *central* or *circulated salt* of the world or Earth; the alchemical symbol for the world ☿ (Latin: mundi) being related to the symbols for stibnite ☿ or antimony ☿. *Butter of antimony* is sometimes referred to by the cover-term *spiritus mundi* meaning *spirit of the earth/world*, which can be further understood as *spirit of antimony* where *spirit = distillate*. Arnold de Villa Nova, favorably mentioned by Paracelsus for his leprosy cure and similar attempt to reform medicine, explained that *spirit of wine* or *aqua vitae*, along with three additional cover-names, indicate a single substance interpreted here as *butter of antimony*. The cover-term "*juices of Lunaria*" simply means *liquid of the moon*, the lunar principle being refined antimony.

> *The juices of Lunaria, Aqua Vitae, Fifth Essence, Spirit of wine, [living] mercury vegetable, are all one. The juices of Lunaria is made of our wine, which thing is known but to few of our children, and with it is our solution made, and our potable gold is made, that being the mean thereof and cannot be without it.*
> — Arnold de Villa Nova, Rosarium Philosophorum

Both Morienus and Newton employed the cover-names *eudica* or *blood* to indicate *butter of antimony*. Newton shared a close relationship with a distillation chemist from Rotterdam named William Yarworth (a.k.a. Y-Worth), for whom Newton acted as both patron and editor assisting Yarworth with authoring alchemical texts under the pen name Cleidophorus Mystagogus. Of interest is Yarworth's two-ingredient recipe for the *alkahest, urine + blood*, presented in his work *Trifertes Sagani or Immortal Solvent*, which is wholly devoted to subject. In the passage below, he suggests that the *Body* (stibnite), as universal or prime material, is the source of *Blood* (butter of antimony).

> *Now that which is centrally one with this Philosophical Urine is Blood; for the Blood is the Universal Form, as the Body is the Universal matter* ...

Cleidophorus suggests that *spirit of urine* must be united with *blood*, described elsewhere by him as a volatile acid and a salt. This characterization accurately applies to *butter of antimony* and agrees with Isaac Newton's equating *eudica* with *blood*:

> *Eudica, the mortise [and tenon joint] of chemistry [μορσα χημεια], secret of all secrets ...life-giving blood for which bodies are adapted...*
> – Isaac Newton, Keynes 30/1 (from Morienus)

William Yarworth serves as a direct link between alchemist-physicians of Paracelsian tradition and English chymists in the Hartlib circle through the intermediary of Isaac Newton. Newton was fully familiar with creating *butter of antimony* via several processes as evidenced in the Sloane manuscripts, he was familiar with Morienus' cover-names *eudica* and *blood* to indicate *butter of antimony* and this understanding may have been a factor in his working relationship with Yarworth. Yarworth, writing under the pseudonym Cleidophorus Mystagogus, was able to expand on the Starkey/Philalethes writings concerning *sophic sal ammoniac, urine* and the *alkahest*. He describes the *alkahest* as:

CHAPTER 7 – PARACELSUS AND THE ALKAHEST

> *... one Saline Liquor, Homogenous **and Immortal. Here this** Body of two faces**, or that** of old Saturn's Urine and the Blood o the Great World **are reduced to one, ad that you may know it when so reduced,** it is a Fire, yet in the form of Water; 'tis an Air, yet Condensed; 'tis no Corrosive, yet the most sharp and perpetual Corrosive; 'tis not Medicinal yet the Crown of all true Medicine, being a Cleanser and Purifier in nature, **a Destroyer and Conqueror of Bodies**; 'tis **called the Fire of Hell, because the Spirit that comes from the Center is united to the Blood** ...*
> – *Cleidophorus Mystagogus,* Trifertes Sagani or Immortal Solvent

In the conceptual framework of alchemy, alchemists believed that joining the very essence of the *central circulated salt* of mankind ☉ (spiritus urine) with the essence of the *central circulated salt* of the earth ♁ (al-koḥl; spiritus vini) resulted in a divine creation artificed by man. These *central salts* were considered natural organic products. Urine as a *central circulated salt* of the body is relatively self-explanatory. *Butter of antimony* was understood as being an equivalent to *gur, central salt of the Earth* and the substance found in mines believed responsible for keeping metals in liquid solution deep underground. Paracelsus spent time working with miners in both metal and salt mines and was thoroughly familiar with their theory of *gur* as the earth's natural saline solvent. His work on the generation of metals falls under the heading of the element water in a treatise named *Philosophia de Generationibus et Fructibus Quatuor Elementorum*. A more concise passage from the *Paragranum* is sufficient to make the point clear:

> *It is not from earth but from water **that the rocks of the earth grow. They are like the ore of water. With regard to the birth of metals, they forget about that** which becomes coagulated.*
> – *Paracelsus,* Paragranum, Philosophia H 2:31

Both *urine salts* and *gur* occurred naturally and their alchemical analogs were considered synthetic versions of naturally occurring substances, urine being derived from the organic domain and *gur* from the inorganic.

When combined however, the resulting product was understood to be artificial, in the sense that it could only be produced by the aid of a human operator.

> ... *they do both [sophic mercury and the alkahest] proceed from the first Chaos [stibnite], before Art hath undertaken to work upon it: But here the difference comes,* one [sophic mercury] *is prepared in a way agreeable to Nature, the other* [the alkahest] *Artificial; and consequently really divested from the Generative Power, being drawn beyond the Predestination of its Natural Seed ...*
> – *Cleidophorus Mystagogus,* Trifertes Sagani or Immortal Solvent

Paracelsus' *alkahest*, as interpreted here, is grounded in sound scientific principles. In contemporary terminology, *butter of antimony* is classified as an *inorganic nonaqueous aprotic solvent*, but it is also an acid salt and a metal chloride. When a metal chloride is combined with urea, the urea serves as a hydrogen bond donor, which qualifies it as a *protic solvent*. In a letter written by John Locke to Robert Boyle, Locke anticipated such a combination when he wrote the following:

> *I thinke a principall remedy may be made by* Armoniacke salts satiated with acid salts and volatilized *which I beleeve may be by* a short way effected and farther advanced to allmost the Alkahest.
> – *John Locke to Robert Boyle,* Correspondence, 24 March 1677

$$\text{Armoniacke salts + acid salts volatized}$$
$$2\ NH_4Cl + SbCl_3 = 2\ NH_4 + SbCl_5$$

Alchemical *spirit of urine* such as the one prepared by Hollandus', rich in chlorides, ammonia and urea when interacted with *spiritus vini* as antimony trichloride results in what is known today as a *deep eutectic solvent* or *DES*. The features of a *deep eutectic solvent*, a volatile compound saline solvent that remains liquid at low temperature, accurately matches descriptions of the *alkahest* of the 16th century, yet the deep eutectic phenomenon was only first described recently by Abbott et al. in 2003.

Alkahest as Deep Eutectic Solvent (DES)

Spiritus Urine + *Spiritus Vini*

(Urine Salts; $(NH_4)_2CO_3$ + NH_4Cl) + (al-Koḥl / Butter of Antimony; $SbCl_3$)

Uniting the Spirits

Having hypothesized a plausible chemical identity for both cover-terms *spiritus urine* and *spiritus vini*, the question remains; how would alchemists have philosophically joined the two spirits? Paracelsus cryptically describes the processes as:

> [So] that from the coagulation *it is re-dissolved, and then coagulated the form is changed,* as the process of coagulating and of re-dissolving teaches. – Paracelsus, De viribus Membrorum Cap. de Hepate

This is Paracelsus' way of alluding to *reverberatio* or in modern parlance, reflux distillation in a pelican vessel. This very well may have been the original method for creating the *circulated-salt, the alkahest*.

Common denominators can be identified between Hollandus' work with urine, Van Helmont and Paracelsus' *alkahest*. They each describe saline liquors implying salt-content, they each involve *spiritus urine* and in each recipe the urine distillate was combined with one or more additional substances. Hollandus added sea-salt and sal ammoniac thus fortifying the preexisting sodium, potassium and ammonium chloride content of his distilled product. He then combined his urine product with quicklime, vinegar and ethanol if his recipe is to be taken literally. If however, any of these ingredients are cover-names, the identity of his solvent will be quite different from what is presented. Paracelsus likely worked with a *spirit of urine* more in the style of Hollandus' than the ammonium carbonate product proposed later by Starkey. Whether Hollandus' use of the terms *acetum* and *aqua vitae* were intended to indicate vinegar and ethanol, or were meant to serve as cover-terms for two other substances is debatable. What appears to have influenced Paracelsus was that Hollandus' urine distillate was rich in chloride salts and it was meant as an ingredient in a compound product.

Just as Paracelsus left no recipe for the *alkahest*, neither did he reveal its origins. Paracelsus may have experimented with attempting to exalt *butter of antimony* by redistilling it with *sal-ammoniac* thus achieving antimony pentachloride. He would have observed that antimony pentachloride is extremely volatile, fumes by drawing moisture from the air, solidifies almost instantly if left exposed to even the driest ambient air, but more interestingly he would have observed gold readily dissolving in it, crystallising and then melting to an *oil of gold*. He may have attempted to stabilize antimony-pentachloride via reflux distillation with *spiritus-urine* or ammonium carbonate.

He may have simply attempted to optimize Hollandus' process to a more elegant solvent that involved only two key ingredients, *spiritus urine* and *spiritus vini*, rather than Hollandus' unwieldy six-ingredient formula. Alternately, he may have been taught the solvent by the Islamic alchemist-physicians that he admired so much during his time in Constantinople according to legend suggested in A. E. Waite's biographical account. What remains clear is that Hollandus, Paracelsus and van Helmont were each familiar with a universal solvent, employed it in their work, described its properties, mechanism of action and left clues as to its composition and identity.

Regardless of recipe, the product was clearly a corrosive fluid salt solvent. As a compound fluid salt, it would have matched Paracelsus' personal view of the corrosive nature of a *heightened* or exalted salt product.

> *For if the sal should be heightened and separate itself, what is it then other than a corrosive thing? Where its upstart pride lies, there it devours and eats away.* – *Paracelsus,* Paramirum, II:II H 1:113

CHAPTER 7 - PARACELSUS AND THE ALKAHEST

SOLVENT MAGISTERIES
Sophic Mercury or the Alkahest?

> *With this generation* many have made attempts [to confect the Stone] but [have] failed: *nothing wants to come of it when a sow is [at work] in the carrot patch. Accordingly,* there is yet another transmutation after this one which yields all the genera sulphurea, as well as the mercuralia and salia, *its demonstration is of value to the microcosmic world; for from it many applications are to be sought* in the human being: his health, *his aqua vitae, his lapis philosophorum, his arcanum, his balsamum, his aurum potabile,* and other things of this kind.
> – *Paracelsus,* Paramirum II:II H 1:115

It can be difficult to distinguish between *sophic mercury* and the *alkahest* from descriptions in alchemical texts. They both are characterized as being a thick translucent liquid (butter of antimony melts easily into a clear liquid), both are salts, both are corrosive, reactive and will confect the stone. However, the *alkahest* is applied in an entirely different manner to *sophic mercury* according to Paracelsus, van Helmont and other German, Dutch and Belgian Spagyrists. The difference between *sophic mercury* and the *alkahest* is analogous to the difference between a coagulant and a solvent. Sophic mercury dissolves matter, but only temporarily before becoming permanently united, i.e. coagulated with it. The *alkahest* on the other hand, dissolves matter to a fine state of reduction before being removed from it by decantation, an "attractorium" or "feather filtration". Though they share similar properties, are crafted in an analogous manner with parallel materials and are each characterized as fiery liquids, they remain quite different in their intended applications ... or do they?

Oil of Gold via Antimonial Vinegre – Artephius

One of the earliest European templates for a process analogous to Paracelsus' application of the *alkahest* occurs in chapter three of the

Secret Book of Artephius, in which Artephius details a unique use for his *antimonial vinegre* of the philosophers to create *Oil of Gold*.

1. *Take (saith he) crude leaf gold, or calcined with mercury,*
2. *and put it into our vinegre, made of saturnine antimony (mercurial) and sal ammoniac, in a broad glass vessel, and four inches high or more;*
3. *put it into a gentle heat, and in a short time you will see elevated a liquor, as it were oil swimming atop, much like a scum.*
4. *Gather this with a spoon or feather dipping it in; and in doing so often times a day until nothing more arises; evaporate the water with a gentle heat, i.e., the superfluous humidity of the vinegre...*

... and there will remain the quintessence, potestates or powers of gold in the form of a white oil incombustible.

In this oil the philosophers have placed their greatest secrets; it is exceeding sweet, and of great virtue for easing the pains of wounds.
— Artephius, The Secret Book, Chapter III:6-8

See Chapter 8 for a detailed explanation of this reaction. Artephius' *quintessence* is in perfect accordance with Paracelsus' definition of the term. Although the solvent is *sophic mercury*, it is applied in the exact manner described by Paracelsus and van Helmont for the *alkahest* centuries later. This may have been the prototype for Paracelsus' *alkahest*. It is a salt composed of two substances, sal ammoniac and *antimony*, and is a heavy viscous fluid that dissolves all bodies, meaning the metallic bodies employed in alchemy.

Oleum Solis via the Urine Stone – Isaac Hollandus

Hollandus' application for the *Urine Stone* works according to a similar principle. *Urea* helps break down gold and convert it to a soluble gold salt, largely aided by the concentration of chlorides in solution. He uses the cover-term *our Stone in its coarseness* to indicate his urine-salt distillate prior to its combination with additional ingredients. The only ingredients in the distillate are those that were originally present in the raw urine to begin with, in other words *such as it comes out of man's minera*.

1. *Take our Stone in its coarseness (or: in its raw state), such as it comes out of the minera of man. Understand well what I am saying!*
2. *Put it into a wide, glass vessel and* add the powdered gold leaves.
3. *Pour on this some of our Stone, which must be old and well settled and purified.* Pour of the Stone, two fingers' width over the gold. Set the vessel with gold and the Stone of summer into the heat of the sun. A white-golden skin or oil will form on top.
4. *Remove it carefully with a feather,* in such a way that you move the matter as little as possible. Put it into a glass. Proceed in this way several times a day, removing the oil till no more oil forms on top.

Thus you can obtain oleum solis with our Stone in its coarseness, as it comes out of man's minera. ... Understand well what I have hinted at here, because there has never been a greater secret in nature concerning our Stone which also, in spite of its coarseness, transforms ☉ [gold] into oil. – Johann Isaac Hollandus, Tractatus de Urina

The product is a gold-salt that forms a surface layer described as a white-golden skin, which is then removed for use. In Hollandus' case, the white-golden skin or surface layer consists of gold-salts in solution, described as *Oleum solis* (oil of gold).

Aurum Potabile via the *Magistery* – Paracelsus

Paracelsus also provides a recipe in *Archidoxis, Book VI*, to create *Aurum Potabile* via a process that echoes Artephius and Hollandus called the *Magistery*. In order for this process to create a gold salt in solution efficiently, the solvent must first be able to break the valence bonds of gold and contain a chloride ion donor, thus his *circulatum* must be a corrosive salt compound as implied.

1. *Take circulatum (circulated salt) well purified and in the highest essence.*
2. *In it place thin beaten or filed metal which has been wrought and cleansed to the most pure and subtle condition.*
3. *Place the two together and let them circulate for four weeks so that the sheets of metals are reduced by the temperate medium to an oil which floats on top like fat, colored according to the kind of metal.*
4. *Skim this off by a silver 'drawer' (attractorium) to separate it from the circulation.*

Thus you make potable gold, potable silver, and the same for the other metals, which are fit to be drunk and consumed without any harm ...
– Paracelsus, Archidoxis Book V

Of all the Elixirs, the highest and foremost is gold.
– Paracelsus, Da Vita Longa, 3:273

It has been well established by researchers and practitioners in his lineage long ago that Paracelsus knew several methods to achieve butter of antimony, and that he included it in his alchemical repertoire. The chemical identity of his *alkahest* could simply have been butter of antimony, as it was the archetypal universal solvent originating with Alexandrian alchemy according to interpretations set forth in this treatise. Yet the *alkahest* just as easily could have been an exalted butter of antimony product in the form of antimony pentachloride or a deep eutectic solvent. Just as *sal ammoniac + antimony* results in *sophic mercury* (butter of antimony), *sophic sal ammoniac + sophic mercury* may have been the chemical identity of the *alkahest* as the supreme exaltation of antimony. A product containing ethanol and refined urine salts or ammonia salts certainly makes sense as regards plant alchemy, yet is inadequate as regards the dissolution of metals. The notion that *spiritus vini / blood of the earth* (butter of antimony) combined with *spiritus urine / sophic sal ammoniac* (refined urine salts) via reflux distillation to achieve *circulated salt* is certainly plausible and accords with Paracelsian materials, methodology and ideology. On the other hand, he may have developed a range of solvents specifically to extract plant oils and another to extract oils of metals. Paracelsus was an original and innovative genius as to alchemical products and their production. One thing remains certain; he did indeed create a unique and powerful saline solvent that he called the *Alkahest*, the identity of which was known to some in his tradition and passionately pursued by outsiders. Whether later alchemists were able to reproduce it accurately or not remains open to investigation. Today the word *alkahest* is sometimes used as a generic term to describe various solvents as "the alkahest of tartar" or "the alkahest of urine", which amounts to little more than the admission; "we still haven't identified the chemical identity of the original".

BOOK 2

Chemistry
&
Confection

It is both customary and right ...

that those who have accomplished anything worth mentioning

in any art or science should make known their discoveries to the world,

in order that mankind at large may be benefited by them.

Petrus Bonus, 14th century

Chapter 8 – Labors of Hercules

I know that some self-constituted "Sages" will take exception to this ...

But one question will suffice to silence their objections: Have they ever actually prepared our Tincture? I have prepared it more than once, and daily have it in my power; hence I may perhaps be permitted to speak as one having authority.

Go on babbling about your rain water collected in May, your Salts, your sperm which is more potent than the foul fiend himself, ye self-styled philosophers; rail at me, if you like; all you say is conclusively refuted by this one fact – you cannot make the Stone. ... of our Stone I know what I am writing about.

Eirenaeus Philalethes

REPRODUCIBILITY

The quest by uninitiated independent researchers for reproducibility of the Philosophers' Stone is, and always has been at the very heart of alchemy. Reproducibility is the crux that differentiated an adept from an aspirant. The focus of this book is essentially to provide researchers with a history and evolution of reproducible methods for creating the Philosophers' Stone. As Paracelsus so eloquently points out in the passage below, the main hindrance to reproducibility is human error.

> *All the fault and cause of difficulty in Alchemy,* **whereby very many persons are reduced to poverty, and others labour in vain,** *is wholly and solely lack of skill in the operator, and the defect or excess of materials,* whether in quantity or quality, whence it ensues that, in the course of operation, things are wasted or reduced to nothing. If the true process shall have been found, the substance itself while transmuting

> *approaches daily more and more towards perfection.* The straight road is easy, but it is found by very few. – Paracelsus, 16th century

The lions-share of the workload in confecting the Stone rests in the preparation of the primary materials; a process known to alchemists as *The Labors of Hercules*. Once a substance has been prepared, it is known alchemically as a *philosophic* or *sophic* substance. Today these labors are easily circumvented by beginning with prepared materials sourced from chemical-supply companies. Admittedly, much of the tradition of alchemy revolves around artisanally preparing one's own materials from scratch, and simply the suggestion of purchasing them might be considered an outrage by those adhering to tradition. There are however ethical considerations in some cases for beginning with purchased chemicals rather than creating them in the traditional manner. These considerations including safety hazards, chemical pollution and others will be addressed later where relevant.

Reproducibility in alchemy when following canonical methods unfortunately often includes replicating operative inefficiencies or in some cases outright errors. Prior to the advent of chemistry, reagent proportions or ratios were determined by whole numbers, divine balance, philosophical or religious notions, etc. Consider the example of Mr. Ford's technique for creating butter of antimony as recounted by Bacstrom:

> *Mr. Ford takes equil parts of the martial regulus of antimony and corrosive-sublimate and distils over a butter [of antimony] in the usual manner. A part of the regulus remains behind with the revived mercury; to this he adds a fresh portion of corrosive-sublimate and distils again.*

Some researchers will accept instructions such as these as *canonical* simply on the basis that the instructions were written in the past and then attempt to reproduce the experiment *canonically* using a process-based approach. Such a researcher would be practicing and ultimately perfecting a mistake; a practice entirely unnecessary in light of current understanding. Clearly, Mr. Ford is attempting to achieve butter of antimony (antimony trichloride; $SbCl_3$) via a reaction between purified metallic antimony (Sb) and corrosive sublimate (mercuric chloride; $HgCl_2$):

> Mr. Hand's Butter $2\ Sb + 3\ HgCl_2 = 2\ SbCl_3 + 3\ Hg$

It is a simple and elegant reaction, in which the reaction and product may be considered *canonical*, yet Mr. Hand's ratios and technique show room for improvement. It is a perfect example of Paracelsus' warning against "...*the defect or excess of materials...*" mentioned in the quote above. Today's chemistry allows for greater efficiency concerning exact ingredient ratios based on molecular weights. The following section will address canonical reactions and products while disregarding inefficiency and inexact ingredient ratios. Using a principle and product-based methodology in this way will allow researchers to focus on what alchemists' got right while avoiding *canonical* errors and pitfalls. Bacstrom appears to suggest a similar approach based on balanced mathematics:

> *Alchemy appears to me as a study in so single a point of view that it strikes with astonishment that any person acquainted with Mathematical philosophy should not instantly discover it, for in the study of Mathematics, before you can give a solution or answer to any question proposed you must bring your numbers to the same denomination.* – Sigismund Bacstrom

Once a working knowledge of the materials and methods to confect the Stone is fully understood, it can be created by almost anyone with little or no knowledge of chemistry or even alchemy. As unromantic as this may seem initially, it is the primary reason for the incredible secrecy throughout the ages. It is truly as the ancients say; women's-work and child's-play.

> *Warning*: *Some materials and methods in the following section are of a hazardous nature and should only be handled under strict supervision in a controlled environment with the use of protective clothing, safety glasses and breathing apparatus.*

Philosophical Materials-Prep

Confecting the Philosophers Stone takes place in two fundamental stages – 1) preparation and 2) confection. Once all materials have been sourced, they must be prepared for use as reagents. Reproducing the Philosophers' Stone today will prove much easier than at any other time in history due to the ease of sourcing highly pure materials, quality laboratory glassware and equipment and the benefits these afford. With just a few starting materials and some basic equipment, the modern aspirant can begin the process of reproduction. The first consideration is to determine whether to begin with a product-based approach or a more traditional process-based methodology. This simply translates as whether to use commercially sourced materials or to learn to create them artisanally from scratch.

While materials preparation became ever more complex as alchemy transitioned to medical Islamic and and experimental European forms, Alexandrian methods in Maria's tradition were limited to just a few simple procedures. Ultimately, the aim is create a pair of chemical compounds that will be united to confect the Philosophers' Stone. During the early-Alexandrian period, Maria referred to these two chemical compounds by pairs of cover-terms *white gum* ↔ *red gum, kibric* ↔ *zibeth* or *two fumes that contain two lights* (sun and moon) among others. Cover-terms for these two compounds found fullest expression in the 7th century writings of Morienus and commentary by Sir Isaac Newton where they are known by the cover-terms *latten* ↔ *eudica, body* ↔ *blood* or *earth* ↔ *water*. When alchemy was transmitted to the Islamic Empire, these two compounds became known as *(sophic) sulfur* ↔ *(sophic) mercury*. Centuries later in the European tradition, Paracelsus referred to this pair of compounds by the cover-terms *blood of the lion* ↔ *gluten from the eagle*.

Prior to creating the two compounds, the *prime material* must first be calcined to a fine greyish-white powder by the process known as *whitening the stone*. Powdered *flowers of antimony* were then divided into two portions due to it being an ingredient in each of the compounds. Following calcination, *gold-antimony glass* was created by fusing the *flowers of antimony* with gold powder. This glass was then cooled, ground to a fine powder, re-melted, cooled and ground a second time before being fit for use. This compound was known by the cover-names *red gum*, *latten*, *body* or *earth* by Alexandrian alchemists. The second portion of *flowers of antimony* was then reacted or distilled with sal ammoniac, which resulted in the *white gum, eudica, blood* or *water* – known principally as *divine* or *sulfur water* throughout the history of Alexandrian alchemy. In the European tradition, it became generally known by the cover-terms *sophic mercury* or *philosophers' mercury* in its viscous liquid form. Achieving this unique pair of compounds meant preparatory work was complete and confection of the Philosophers' Stone could begin.

A general summation of materials preparation based on original Alexandrian / Byzantine *chrysopœia* is as follows:

1. **Whiten the Stone (White Vapor)** – calcine stibnite according to proper technique to achieve *flowers of antimony*. Once the stone had been whitened, the product was divided into two portions and used for the following processes. Whitening the stone taught fundamental calcination; the mastery of fire being man's first and most ancient technological achievement. Creating oxides by roasting was a standard material technology used in bronze and glass making from antiquity until today and introduced the fundamental technique for creating the primary mineral and metal oxide precursors.

2. **Fuse Latten (Green Lion)** — smelt *flowers of antimony* to a molten state, add the appropriate measure of gold-calx or purified gold-leaf to the molten *glass of antimony* and keep in flux until fusion is complete. Cool, grind to a fine calx, flux and grind once again. During the early Alexandrian period, this process was chiefly the only preparatory process required. The product was then combined with sand, *natron* or *sal ammoniac* via projection or detonation to create the Stone via Ars-Brevis, or digested with *divine water* at low heat in a sealed vessel to create the Stone via Ars-Magna technique. Latten fusion taught fundamental metallurgical and glassmaking crucible technique.

3. **Distil Eudica (Divine/Sulfur/Foul Water)** — distil *flowers of antimony* with the correct measure of sal ammoniac to achieve *butter of antimony*. During the middle-Alexandrian period, this process was employed via digestion to create *Algarot powder* as the White Stone of Hermes. Zosimus, Synesius and Morienus clearly demonstrate familiarity with *butter of antimony*, known from the early-Alexandrian period as *divine-water* and during the Byzantine period as *eudica*. Morienus transmitted the technology to Khālid ibn Yazīd at the fall of the late-Alexandrian period. Distilling eudica introduced the fundamental chemistry technique of distillation.

Today all but *latten* may be bought ready-prepared and at high quality, yet if one were to attempt to prepare all materials artisanally from scratch a few more processes must be added to the above three preparations. The first is the purification and reduction of gold and the second is the creation of pure ammonium chloride.

The above three processes represent a collective walk through the history of science and technology — fire, glassmaking and metallurgy, followed by chemistry. By mastering the three preparatory processes, an apprentice acquired the skills of calcination, vitrification, fusion, trituration, projection, detonation, digestion, sublimation and distillation. Although the preparatory processes initially seem simple and straightforward, they afforded an apprentice the opportunity to refine and perfect fundamental alchemical techniques. Work on materials-prep introduced the apprentice to basic theory and practice from which point a growing appreciation and depth of understanding for the elegance and intelligence of the Art began to manifest. In the early Alexandrian period, alchemy was not experimental chemistry but rather a tradecraft similar to trade-guilds that developed in medieval Europe. The mysteries of gold-making from the earlier period were trade-secrets revealed to the aspirant, usually amalgamated with a religious or philosophical ideology, as the aspirant progressed through the ranks. Alchemical and philosophical mysteries revealed themselves as the apprentice slowly worked towards adept-hood and ultimately mastery. Materials preparation was the first step of the journey.

Gold Refinement and Reduction
When confecting the *White Stone* of the Hermetic tradition, gold, silver or any other metal is an afterthought and not part of the main process. The case is much different with regard to the *Tincture of the Philosophers* of Maria Hebrea's tradition, in which gold or silver serves as a primary ingredient from the outset. In the Dialogue of Maria and Aros, Maria states that she purchased some materials and collected sal ammoniac.

> *These are the principles of this Art,* part is bought and part is found on small hills [hillocks].

For most of Egypt's long history gold mining, refining and reduction was monopolized by the ruling elite and priestly class who carefully controlled gold technology. In the great Alexandrian centers of commerce during the Roman period in which Maria lived, highly refined gold and gold powders were widely available at very high purity and at a fine state of division. For this reason gold purification and reduction in Alexandrian alchemy was practically a non-issue. A look at Egypt's gold-mining history makes it clear that very high quality gold-dust was the stock in trade. Diodorus Siculus, in his 40-volume history written 1st century BCE details the labor-intensive process of Egyptian gold mining as originally recorded by Agatharcides a century earlier:

1. The *gold-bearing earth which is hardest they first burn with a hot fire, and when they have crumbled it... they continue the working of it by hand*; *and the soft rock which can yield to moderate effort is crushed with a sledge by myriads of unfortunate wretches.*
2. *And the entire operations are in charge of a skilled worker who distinguishes the stone and points it out to the labourers; ...* the *physically strongest break the quartz-rock with iron hammers, applying no skill to the task, but only force.*
3. *The boys there who have not yet come to maturity, entering through the tunnels into the galleries formed by the removal of the rock, laboriously gather up the rock... piece by piece and carry it out into the open to the place outside the entrance.*
4. *Then those who are above thirty years of age take this quarried stone from them* and *with iron pestles pound a specified amount of it in stone mortars, until they have worked it down to the size of a vetch [slightly smaller than a pea].*
5. *Thereupon the* women and older men receive from them the rock of this size and cast it into mills *of which a number stand there in a row, and taking their places in groups of two or three* at the spoke or handle of each mill they grind it until they have worked [the ore] down... to the consistency of the finest flour.
6. *In the last steps the skilled workmen receive the stone which has been ground to powder; ...they rub [it] upon a broad board which is slightly inclined, pouring water over it all the while; whereupon the earthy matter in it, melted away by the action of the water, runs down the board, while that which contains the gold remains* on the wood because of its weight.
7. *And* repeating this a number of times, they first of all rub it gently with their hands, *and then lightly pressing it with sponges of loose*

texture they remove... whatever is porous and earthy, until there remains only the pure gold-dust.
– Diodorus of Sicily, *Translation by C.H. Oldfather*, Cambridge, MA: Harvard U. Press, 1967

Most Egyptian gold was mined in this fashion resulting in either very fine gold-dust of high purity or electrum which is a naturally-occurring gold-alloy containing small impurities of 5% silver or less. This may explain why Maria never mentions her process for gold refining and reduction; 23-24 karat gold-dust was readily available and silver would certainly have been seen as an acceptable impurity. Like gold-dust, stibnite as mśdmt/kahâl/koḥl was also a trade good, easily sourced in Alexandrian markets and a common household item. The need for gold or silver refining in alchemy comes to the forefront in Islamic and European traditions, in which alchemists developed their own reduction and purification techniques.

Metallurgical Purification – The art of refining gold stretches back into great antiquity. The earliest forms of gold purification and reduction employed metallurgical techniques involving stibnite and much later in history quicksilver, lead or cementation techniques such as those recorded by Pliny the Elder. Stibnite is an excellent medium for purifying gold and one of the most ancient. Cinnabar and mercury were also used in antiquity but awareness of environmental mercury toxicity makes the use of mercury unethical today. Early Alexandrian alchemists displayed a fascination for working with stibnite, which they termed *the serpent*. The first step is to heat stibnite until it liquefies, then add gold until it fully dissolves into the molten stibnite. Antimony will bind to the base metal impurities present in unrefined gold. The temperature range is between the melting point of antimony and gold; between 630.63 and 1,064.18 °C. This process, described in detail below, was known to alchemists as *The Grey Wolf Devours the Sun*; the wolf being a cover-name for stibnite and the sun a representation of gold.

BOOK 2 – CHEMSTRY & CONFECTION

1. The gold is first placed in a red hot earthen crucible, and when melted it swells, and a little stibium is added to it **lest it run over; in a short space of time, when this has melted, it likewise again swells, and when this occurs it is advisable to** put in all the remainder of the stibium, and to cover the crucible **with a lid, and then to** heat the mixture for the time required to walk thirty-five paces.
2. Then it is at once poured out into an iron pot, **wide at the top and narrow at the bottom, which was first heated and smeared over with tallow or wax, and set on an iron or wooden block.** It is shaken violently, and by this agitation the gold lump settles to the bottom, and when the pot has cooled it is tapped loose,
3. **and is again melted** four times in the same way. But each time a less weight of stibium is added to the gold, **until finally only twice as much stibium is added as there is gold, or a little more;**
4. then the gold lump is melted in a cupel.
5. The stibium is melted again three or four times in an earthen crucible, and each time a gold lump settles, so that there are three or four gold lumps,
6. and these are all melted together in a cupel.

– Georgius Agricola, 16th century

The sulfur present in stibnite combines with silver impurities, whereas the antimony combines with all other impurities present. Cupellation is the process whereby the gold-antimony matrix is heated to vaporize the antimony. As the antimony vaporizes, so do the impurities along with it leaving a high-purity gold remaining. Lead was used for this process as early as 1,500 BCE and mercury for the same purpose estimated at circa 400 BCE. The process can be repeated until a pure product is achieved. This process via quicksilver is still in use by small-scale industrial and hobby miners worldwide. Ethically these practices are incredibly environmentally harmful when performed without the proper protocols, and are best avoided in favor of today's more eco-friendly methods.

Metallurgical Reduction — The aim of gold preparation was first to achieve gold as close to 24 karat as possible, then to reduce the particle size of gold as small as technology would allow. One of the earliest techniques to achieve this aim was to create a gold-antimony alloy by combining purified gold with a small amount of *flowers of antimony* or *antimony regulus* just enough to cause the gold to become brittle. The small amount of antimony creates an antimony-glass. For this reason, the process is known as *vitrification* or *glass making*. The result is a very brittle mass, which could then be ground to a fine powder known as *gold-calx* or *bezoar of gold*.

Another metallurgical technique for achieving *gold-calx* is to create a gold-mercury amalgam and vaporize the mercury. This process can be repeated until a fluffy yellow-brown to dark-brown powder is achieved known as *gold-calx, gold-sponge or gold-lime*. Today bench-top mercury distillation units fired by a small handheld brazing torch or jewelers' torch are inexpensive and easy to purchase, making this once-dangerous process safe, economical and ethical.

> *Considerations*: For an amateur experimenter, high-quality gold-leaf sold for food-use or gilding is commercially available in most cities and is suitable for most applications. Gold-leaf used for religious purposes comes in various qualities. If sourcing gold-leaf originally intended for religious purposes, always be sure to check the karat or percentage of gold-content prior to purchase.

Chemical Purification and Reduction — Islamic alchemists were familiar with gold purification and reduction via the metallurgical technique, but later developed mineral acids from experimenting with salt distillation. Jābir ibn Hayyān's tradition is credited with the discovery of the first crude compound acid or *aqua-regia* that could dissolve gold. This was achieved by a reaction between sal ammoniac (ammonium chloride) and *saltpeter* (potassium nitrate). When the two salts are mixed in the proper proportions and heated, a violent reaction occurs that can be explosive.

Jābir's aqua-regia resulted in a dirty olive-green powder. It's chemical identity is gold-hydrazide ($[Au(NH_3)_4]^{3+}$), also known as *gold-fulminate* or *explosive gold*. *Gold-fulminate* was not only one of the first gold compounds ever recorded, it was also the first high explosive. The fluffy green, yellow or brown colored powder is one form of *gold-calx* found in Islamic and European traditions. Europeans eventually evolved the technique with the discovery of nitro-hydrochloric acid, known first in alchemy and later in chemistry as *aqua-regia* (king's water).

The King's Bath (European) $Au + 3 NO_3 + 6 H = Au^3 + 3 NO_2 + 3 H_2O$

and $Au^3 + 4 Cl^- = HAuCl_4$

Aqua-regia greatly simplified the process of purifying and reducing gold; a discovery so valuable that the alchemical cover-name is still used in chemistry labs today. When gold is dissolved in *aqua-regia*, the result is *chlorauric-acid* ($HAuCl_4$). *Chlorauric acid* takes the form of golden-yellow or sometimes orangeish-red crystals. Heating *chlorauric-acid* crystals releases hydrogen chloride resulting in *gold-trichloride* salts.

$$2 HAuCl_4 = 2 HCl + AuCl_3$$

Gold-trichloride was one of the first forms of medicinal gold discovered by alchemists but was found to be too harsh for direct ingestion. It was combined with sea-salt (sodium chloride; NaCl) to create a more user-friendly medicinal *gold-sodium chloride* elixir that remained in use well into the 20th century, prescribed by Dr. Leslie E. Keeley and later by Edgar Cayce who buffered it with baking soda. *Gold-sodium chloride* is still used as a rejuvenative or tonifying elixir by some today. When *gold-trichloride* is heated to around 160 °C, it decomposes to *gold-chloride*. When *gold-chloride* is heated to 298 °C, it decomposes to metallic gold.

$$AuCl_3 = AuCl + Cl_2$$

$$AuCl + \Delta = Au + HCl$$

Today high purity gold calx can be easily precipitated from gold chloride solution using sodium or potassium metabisulfite or a number of other substances to yield gold powders that can be easily further purified, a technique pioneered by European alchemists. Alchemists of old would be surprised at the degree to which modern methods routinely achieve gold-reduction to nanoparticle size with relatively little effort and at extremely high purity.

> *Considerations*: From a cultural or artistic perspective, ancient methods can be fun experiments. From an operative perspective however, desired purity and particle-size using modern methodology is far more efficient to achieve these aims resulting in extremely high purity gold within nanoparticle size range.

Sal Ammoniac and Urine-Menstruums

Fluxing Salt (Seal of Hermes) – The Philosophers' Stone was originally created via metallurgy by creating a gold-antimony alloy. During the early Alexandrian period, Maria's Ars-Brevis recipe indicates a fusion between a fluxing salt and gold-antimony alloy. Natron, a naturally occurring matrix of sodium chloride, carbonate and bicarbonate, was likely the earliest fluxing salt used for this purpose. It was routinely employed in Egyptian chemical technologies such as metallurgy and glassmaking. As a household product during antiquity, it was employed as toothpaste and appeared in the Bible as Hebrew *néter*, meaning soap. The Ancient Egyptian word for natron was *ntry* or *netjeri*, which evolved to Greek nitron (νίτρον) and Latin natrium from which the chemical identifier Na is derived to indicate sodium. It was the original *nitre* of the alchemists but the word *nitre* came to denote saltpeter (potassium nitrate; KNO_3) much later in alchemy's history.

Sal ammoniac – Natron fusion technique was improved upon via the use of sal ammoniac. Sal ammoniac reacts with gold-antimony glass or alloy to create the Philosophers' Stone. Various tricky metallurgical techniques were employed such as layering in a sealed crucible or repeated projections or detonations, meaning casting sal ammoniac or a sal ammoniac matrix into the crucible to initiate an instant reaction. Maria evolved metallurgical technology by affecting a chemical reaction reliant upon sal ammoniac, gold-calx and *flowers of antimony* inside a glass sublimation/digestion vessel. One of the keys to success is the proper ratio of ingredients, a trade secret likely achieved through empirical observation during the developmental period. This operative technique laid the foundation for creating chemical reactions between sal ammoniac and *flowers of antimony*, resulting in *butter of antimony* and Algarot powder.

Salt-saturated Urine – Salt-saturated urine is one of humanity's most ancient chemical technologies evolving from leather tanning and fabric dying. A *urine-menstruum* can be understood as a low-cost, readily available alternative to sal ammoniac. The earliest mention of a urine-menstruum as regards alchemy is by Pseudo-Democritus in his 2nd century work, *Physika kai Mystika*. He utilized urine and a menstruum of unslaked lime and urine in a number of operations, although these may also indicate cover-terminology. Islamic alchemist, Al-Razi, incorporated urine and urine salts in his alchemical repertoire. The use of urine-based menstruums for reducing stibnite is an innovation that should properly be attributed to the Islamic alchemical tradition, which was later adopted by some European alchemists to reduce stibnite and achieve *butter of antimony*. Isaac Hollandus and others employed urine menstruums alchemically and, although they played an important role in experimental Islamic and European alchemy, they are not an historical component of archetypal recipes typified by Maria's style of Alexandrian *chrysopœia*.

CHAPTER 8 – LABORS OF HERCULES

CALCINING THE WHITE VAPOR

In my opinion antimony is a composition made by Nature to create a metallic mineral that is overflowing with an undue proportion of hot and dry material and with its moisture poorly mixed, with an effect wholly contrary to the composition of metals. Therefore, it comes to be, like quicksilver, a mineral deformity and monstrosity among metals. Or it might be a material that is about to reach metallic perfection... Obviously it has in it much earthiness, as is attested by the odor of its sulphurous burning, its imperfect blending, its defective mixture, its difficult fusion, and finally the character of its regulus, for although this is very white and almost more shining than silver, it is much more brittle than glass. – Vannoccio Biringuccio, 16th century

Whitening the Stone – Archetypal Alexandrian methods begin with a single all-important material fundamental to glassmaking and bronze-craft, which dates back to bronze-age antiquity. Stibnite was well known to Alexandrian alchemists yet its identity remained encrypted by cover-names such as *serpent, bull, sword, the One, the black stone, our lead, our mercury, body of magnesia, single root, sperm or seed, body or earth*. Each tradition added cover-names for this substance but all traditions understood it to be the *Prime Material* required to confect the Stone. This mystery was the first riddle that an aspiring adept needed to solve.

In Alexandrian and Islamic traditions of alchemy, a pure form of antimony was achieved by first calcining *mśdmt/kahâl/ḳoḥl* to a fine white powder. This process was originally called *whitening the stone* by Maria Hebrea who attributed the technique to Hermes. Stibnite (Sb_2S_3) is composed of antimony (Sb) and sulfur (S). When finely powdered stibnite is pan-roasted at an optimal 350 °C, the sulfur first melts then vaporizes as sulfur dioxide, a highly toxic gas with a pungent irritating smell.

What remains is pure antimony that oxidizes resulting in crude antimony trioxide (Sb$_2$O$_3 \cdot$ Sb$_2$O$_3$), known in alchemy as *flowers of antimony*. In its molten or vitrified state it converts to antimony oxysulfate (Sb$_2$O$_2$SO$_4$), known as *glass of antimony*. The process was likened to removing the wings of the *black dragon* to yield the *white serpent*, a technology derived from copper smelting originating during the Bronze Age.

Copper sulfides were first calcined to copper oxide, which served as a primary source of copper in Egypt and the ancient Near East; a process adopted by alchemists such as Maria and applied to stibnite, galena and other sulfide ores.

Whitening the Stone by Fire $2\ Sb_2S_3 + 9\ O_2 =$ crude $2\ Sb_2O_3 + 6\ SO2$

In the European alchemical tradition, the process of *whitening the stone* was perfected by Basil Valentine:

1. *Take Hungarian or other Antimony [stibnite], the best you can get, grind it, if possible, to an Impalpable Powder; this Powder spread Thin all over the bottom of a Calcining Pan*, round or square, which hath a Rim round about, the height of two Fingers thickness; set this Pan into a Calcining Furnace, and *administer to it at first a very moderate Fire of Coals, which afterward increase gradually [stay as close to 350 °C as possible]:* when you see a Fume beginning to arise from the Antimony, *stir it continually with an Iron Spatula, without ceasing, as long as it will give forth from itself any Fume.*

2. *... calcine it, continually stirring as we said, until no more Fume will ascend.* If need be *repeat this Operation so often and so long, as until that Antimony put into the Fire, will neither fume, nor concrete into Clots, but in Colour resemble White and pure Ashes:* Then is the calcination of Antimony rightly made.

> 3. *Put this Antimony thus calcined into a Goldsmiths Crucible set in a Furnace ... Antimony may flow, like clear and pure Water. ... the Glass made thereof be sufficiently cocted, and hath acquired a transparent Colour ... see whether it be clear, clean and transparent.*

Contemporary researchers have discovered that iron, silica and sulfur were often present in European *flowers of antimony.*

> *Considerations*: While this process is rather simple, it is unethical to release sulfur dioxide into the environment. It is a toxic gas and is responsible for generating acid rain in the atmosphere. *Whitening the stone* should only be attempted within proper laboratory conditions according to protocols for working with sulfur-dioxide vapors. Commercially prepared *flowers of antimony* is inexpensive and readily available at varying degrees of purity. Basil Valentine gave a recipe that combines hot sulfuric acid with antimony oxychloride that can be interpreted as a recipe for synthetic glass of antimony or kermes mineral depending on reagent ratios.

Desulfurization of Stibnite via Iron – Pliny the Elder wrote of male and female forms of antimony demonstrating a distinction between crude *stibnite* and refined *antimony*. An artifact, thought to be part of a vase, made of metallic antimony dating to apporoximately 3,000 BCE was discovered at Telloh, Chaldea (in present-day Iraq), and a copper object plated with antimony dating between 2,500 and 2,200 BCE was dicovered in Egypt. *Regulus of antimony* may have been known to Islamic alchemists such as Jābir ibn Hayyān, however due to the tremendous volume of Islamic alchemical literature yet to be translated this remains unverified. Biringuccio is typically credited with the first description of *regulus of antimony,* as it appears in the quote at the beginning of this section, yet further into his book he clearly credits alchemists with knowledge of reducing stibnite to *regulus* and how to work with it. Basil Valentine thoroughly understood the subject in detail thus vindicating Biringuccio's kind words towards alchemists. Fourteen years after Biringuccio published his work *De La Pirotechnia*, German scholar and scientist Georges Agricola, "Father of Metallurgy", published an account of antimony.

Stibnite is composed of approximately 72% antimony + 28% sulfur. Biringuccio explained that when stibnite is heated slowly, a red liquid (molten sulfur) forms, which was then decanted from the antimony, yet he did not elaborate any further. The temperature range to achieve separation is between the melting point of sulfur and antimony; between 115.21 – 630.63 °C. This is the original process known as *Cutting the Crow's Head*.

European metallurgists developed techniques for reducing stibnite to metallic antimony *regulus* by creating a reaction between stibnite's sulfur content and iron filings. The reaction results in *green vitriol* (iron-sulfate; FeS) and *star-regulus*. The *regulus* then underwent further purification with fluxing salts. This technique was highly valued as an operative secret and remains so in some surviving alchemical traditions today. Precise metallurgical technique and equipment such as high-temperature furnaces are required, making this form of preparation beyond the reach of most casual enthusiasts. This process is known as *desulfurization (reduction) of antimony by iron*.

It too dates back to Bronze Age copper refining where copper sulfide ores were mixed with charcoal and iron or magnesium additives and heated slowly, increasing the temperature gradually to not more than 800 °C to dehydrate and ignite the sulfur matrix, which was removed prior to oxidation. This technique was an ancient borrowed Bronze Age technology applied to stibnite by alchemists.

Stibnite Reduction via Iron Precipitation $\quad 2\,Fe + Sb_2S_3 = Fe_2S_3 + 2\,Sb$

The resulting product was known as the *martial star-regulus of antimony*. Antimony is commercially available in a number of different forms at levels of purity that alchemists could only have dreamt of in the past.

Considerations: Metallurgical alchemical procedures require specialized equipment and skill-set. These processes should only be attempted under supervision or by someone specifically trained in metallurgical technique. Dangers include high-heat and toxic fumes and vapors.

VITRIFYING THE GREEN LION

Latten, Rebis – The crux of confecting the archetypal Philosophers' Stone is to create glass of antimony as a precursor to *latten* (gold-antimony glass). *Latten,* or *Rebis* are cover-terms for what Maria called *Red-Man* or *Red-Gum* and Morienus synonymously termed *Latten, Earth, Body* or *Almagra* and Newton referred to as the *Green-Lion*; basically a *gold-antimony glass* or alloy of varying proportions symbolized by a *green lion* (glass of antimony) devouring the sun (gold calx). It is based on the most ancient metallurgical technique for creating the Philosophers' Stone. *Latten* satisfies the gold-antimony content while serving  to divide gold's particle size even further. *Latten* is a matter of choice rather than necessity, but does speed up the process of confecting the Stone.

Latten has been traditionally described by color-indicators that suggest either a red or a green coloration, which demands further explanation. This discrepancy arises based on how carefully stibnite has been calcined to flowers of antimony. The remaining antimony trisulfide proportion to antimony trioxide determines whether *glass of antimony* ends up red or green, or various other colors. Carefully calcined stibnite will yield a golden to ruddy or red glass, whereas a higher sulfur content will yield green glass, higher still and it becomes dark red to brown. This is why descriptions of *glass of antimony* vary as regards color. Maria describes her gold-antimony glass as a *"Red Man"* or *"Red Gum"* suggesting a ruddy or red coloration, whereas Morienus simply states, *"The green lion is glass..."* References to the color green in Islamic alchemy suggest a higher sulfur content in Islamic glass of antimony. Reproducibility experiments to achieve these color variations can be performed as follows:

1. **Red glass of antimony** – mix 1 measure of antimony trisulfide with 8 measures antimony trioxide and heat to 1,000-1,050 °C for 15-30 minutes.
2. **Green glass of antimony** – mix 1 measure of antimony trisulfide with 3 measures antimony trioxide and heat to 1,000-1,050 °C for 15-30 minutes.

Glass of antimony may have been produced synthetically by Basil Valentine. His recipe calls for adding a small amount of hot *spirit of iron vitriol* (sulfuric acid) to *living mercury of antimony* (antimony oxychloride).

$$H_2SO_4 + 2\ SbOCl = Sb_2O_2SO_4 + 2\ HCl$$
$$2\ H_2SO_4 + Sb_4O_5Cl_2 = 2\ Sb_2O_2SO_4 + 2\ HCl + H_2O$$

All Ars-Brevis via Sicca (Brief-Art Dry Path) recipes are reliant upon creating *Latten*. Creating *latten* shortens the reaction time to create the Stone.

Fusing Latten

We realized early in our research that finding exact ratios in Alexandrian texts would be problematic. This is primarily because different sources provide different ratios. The gold to antimony ratio is typically presented in European texts as 1:1, 1:2, 1:3, 2:1 and 3:1, typically in a veiled and cryptic rendering. It became clear upon further experimentation that the conflicting gold to antimony ratios could be explained in a very general sense in the following manner:

1. *gold in excess of antimony – suggests the use of antimony regulus measured by volume*
2. *antimony in excess of gold – suggests the use of flowers of antimony measured by volume*
3. *an equal ratio of the two – suggests equal measure by mass (weight)*

The ratios are forgiving. For those aspirants who cannot afford to experiment with gold, silver is a far more affordable medium. For those who prefer to practice for a while before adding precious metal, copper or iron may be substituted for gold. Experimentation and observation has shown that fusing *latten*, while being the preferred method by Alexandrian alchemists, is not necessary when using quality gold leaf or gold powder at a very fine particle size.

DISTILLING DIVINE WATER

Behold, now I have doubled mercury in my possession: Now I own it – white lily, powder of adamantine, chief central poison of the dragon, spirit of arsenic, green lion, incombustible spirit of the moon, life and death of all metals, moist radical, universal dissolving nutrimen, true menstruum of the philosophers, which without doing any harm reduces metal to first matter. This is the true water for sprinkling, in which the living seeds of metal inhere, and from which other metals can be produced.

– Solinus Saltzal, Fountain of Philosophical Salts

Divine/Sulfur/Foul Water – *Butter of antimony* was first unmistakably described by Morienus, which he called *eudica*. When confecting the Stone, the general rule is to avoid introducing any undesirable substance into the recipe. *Butter of antimony* can be seen as an evolution or innovation in technique over the early Ars Brevis method because it contains two of the three ingredients desired in the finished product, antimony and chloride ions, with no waste or byproducts. Of course, ancient alchemists would not have understood it this way.

Butter of antimony is known today as antimony trichloride and is an inorganic nonaqueous solvent. It takes the form of a translucent oily mass when warmed, crystals or fatty waxy paste when cool, and thus the term *butter of antimony*. Sir Isaac Newton knew several recipes for it and called it the *greatest of secrets*. Other alchemists referred to it as the *great Key*:

> ... *if we can find something which will serve as a key, the house of mystery will give up its treasures ...* – A.E. Waite

Maria's Divine Water – Maria and other alchemists discuss a very special *sulfur-water* (theío ýdǫr; θεῖο ὕδωρ) or synonymously *divine-water* (theía ýdǫr; θεία ὕδωρ) that has always mystified researchers. In consideration of the possibility that divine-water was actually *butter of antimony*, inspiration for the synonymous cover-names *sulfur-water* and *divine-water* begin to make sense and seem very appropriate for this amazing alchemical reaction and resulting solvent. A telling clue that *divine-water* was indeed a product of antimony is the common synonymous cover-name *serpent's-bile* that appears is some Alexandrian alchemical texts – the word *serpent* corresponding to antimony. Synesius wrote of *divine-water*, specifically stating that it was a product of distillation, could dissolve the (alchemical) metals during the decomposition or solve stage, and could produce a change in physical properties – a fitting description of *butter of antimony*:

> *Take the water which escapes from the end of the eduction tube and keep it for the decomposition. It is what is called Sulphur Water [or Divine Water]. This produces "transformation," i.e., the operation which brings out the hidden nature, which operation is called "solution of the metals."*

Most researchers have noted the unusual respect Zosimus displayed for Maria and how he dedicated his writings to his enigmatic sister *Theosebia* (Theoseveia; θεοσεβεια). *Theosebia's* name means *divine reverence*, but in an interesting turn of wordplay can also be interpreted as *reverence for sulfur* and may actually be an encrypted expression of awe and respect for this alchemical solvent. Maria's *white sulphur* or *Kibric*, Arabic *kibrīt* (كبريت) meaning *sulfur*, indicated *flowers of antimony* – a primary ingredient in *sulfur-water*. Alchemists such as Stephanos and Paracelsus have written puzzling personal letters to a mysterious and unidentified *Theodore*, which can now be understood as an encryption praising this particular *sulfur / divine-water*. This distinctive alchemical solvent hypothesized here is encoded by the cover-names *Maria, Theosebia* and *Theodore*; a standard practice in antiquity for protecting valuable or sacred knowledge:

Maria = Mariah = מאריה = lion; Miriam = מרים = corrosive sea; Hamarim = המרים = water of bitter or poisonous things

Theosebia = sulfur / divine reverence (theío sévas; θεῖο σέβας) = reverence (theoseveia; θεοσεβεια)

Theodore = divine / holy water (theía ýdor; θεία ύδωρ) = sulfur-water (theío ýdor; θείο ύδωρ)

Theodore = divine gift (theó-dóro θεό-δώρο) = sulfur gift (theío-dóro θείο-δώρο)

Theodosius = sulfur dose or sulfur giving (theío dosios θείο δοσιος)

Theodosius = divine dose or divine giving (theo dosios θεο δοσιος)

While no early-Alexandrian records explicitly reveal a recipe for *butter of antimony*, it is possible to create *butter of antimony* via an extremely low-tech method detailed later in this section that would have been easily achievable during the period in which Maria operated, which makes the notion of early Alexandrian *butter of antimony* at least plausible. The reaction is achieved by combining finely ground *stibnite* with an aqueous solution of *sal ammoniac* and heating to around 60 °C, which immediately catalyzes the reaction. Once the reaction is complete, it is simply exposed to evaporation. Although perhaps not as efficient or elegant as distillation, it is an effective method of achieving crude *butter of antimony*. The sulfurous vapor given off is ammonium sulfide, a salt that decomposes at ambient temperature and the primary ingredient in modern stink bombs. The odor is truly hideous, nauseating and it is certainly understandable that the ancients perceived it to be a *sulfurous-water*. After evaporation, the waxy substance that remains is crude *butter of antimony*. Now free from its noxious odor and possessing the amazing ability to dissolve minerals and metals including prepared gold and silver, it is easy to see how this process and its resulting product may have been considered by early alchemists as sulfurous and simultaneously sacred or divine. The process can also be applied to *flowers* or *glass of antimony* with less nauseating results.

It is plausible and actually quite possible that Maria was in possession of this alchemical solvent, or was the identity of the secret solvent personified in name. F. Sherwood Taylor references *the school typified by Maria and Comarius* when describing Maria's explicit brand of alchemy. In Greek, *Comarius* essentially means *Mr. Maria* or its Latin equivalent *of the sea*, of which may also be interpreted as encrypting *butter of antimony* as a secret solvent. It is precisely this solvent that differentiates *the school typified by Maria and Comarius* from the Pseudo-Democritan type and that has survived to become the core of recognisable Islamic and European alchemical traditions. Zosimus' praise of his sister "Theosebia" or a "Dear Theodosius, or Dear Theodore..." letter such as were written by Stephanos and Paracelsus may have actually been a covert way in which alchemists declared their initial milestone success at achieving butter of antimony. The cover-term *vinegar / vinegre* used later by Artephius and others to indicate *butter of antimony* may have originated with 5th century Alexandrian alchemist John the Archpriest who makes mention of *the harsh white vinegar* or alternately *the white vinegar under fire* (to lefkon oxos drimytaton; το λευκον οξος δριμυτατον) in reference to his solvent. During the 6th century, Stephanos gave butter of antimony a new cover-name, *pharmakon* (farmákọn; φαρμάκων), meaning *medicine* of which he converted to his White Stone known as Comprehensive Magnesia. This incredible *sulfur / divine water* was viewed as the all-important *Key* to the art, remaining a carefully guarded alchemical secret for several hundred years. One of *divine-water's* unique characteristics is that it could liquefy any of the known alchemical metals, then mysteriously solidify or crystallize ultimately to a fine powder.

Damascus Eudica – There are no written accounts by early Alexandrian alchemists that specifically address creating *butter of antimony* aside from encrypted implications addressed above. However, Morienus may be credited with delivering late Alexandrian / Byzantine style alchemy to Khālid ibn Yazīd in Damascus, and in doing so, revealed a simple recipe for its creation. He used the cover-names *eudica* and *foul water* among others to indicate antimony trichloride. The recipe preserved in the writings of Morienus suggests that he created *eudica* by distilling ground *glass of antimony* with sal ammoniac:

Morienus' Eudica, Blood or Foul Water
$Sb_2O_4 + 6\ NH_4Cl = 2\ SbCl_3 + 2\ O_2 + 6\ NH_4$

Islamic Vinegre – The Islamic tradition worked with both sal ammoniac and *urine-menstruums* according to the writings of Al-Razi and others. Late Islamic alchemy also appeared to create reactions beginning with both raw *stibnite* ore and *antimony regulus*. One of the clearest descriptions to create butter of antimony in the late Islamic tradition is preserved in *The Secret Book of Artephius* (Ibn Ar-Tafiz) from the 12th century. Artephius refers to butter of antimony as *"our vinegre"*, hints that he begins with mercurial (refined) antimony as opposed to *stibnite* and clearly states that *sal ammoniac* is used accordingly:

> ... *you may draw a great arcanum, viz. a* water of saturnine antimony, mercurial and white... *not burning, but dissolving, and afterwards congealing to the consistence or likeness of white cream.*
>
> ... *our* vinegre, *made of saturnine antimony (mercurial) and sal ammoniac ...*
>
> *The whole, then, of this antimonial secret is, that we know how by it to extract or draw forth argent vive, out of the body of Magnesia,* **not burning,** *and this is antimony,* **and a mercurial sublimate.**
>
> — Ibn Ar-Tafiz, 12th century

Refined *antimony* is the *Magnesia* mentioned in the passage above, yet there is no indication as to whether Artephius employs *flowers of antimony* or *regulus*. He declares that *argent vive* is antimony, which is also a *mercurial sublimate.* This has led some researchers to conclude that mercuric or mercurous chloride, known as *corrosive sublimate*, was employed. From a product-based perspective, even if this were true, the result would still be *butter of antimony* with quicksilver as a byproduct of the reaction. An alternate and equally valid interpretation of mercurial sublimate is that the mercurial sublimate being discussed is *butter of antimony* itself; it is a product of sublimation (distillation) and is therefore a sublimate that is mercurial by nature. His description of *saturnine antimonial water* that dissolves gold and congeals to the consistency of white cream is a very appropriate description of *butter of antimony*:

> ... *a water of saturnine antimony, mercurial and white;* **to the end that it may whiten sol,** *not burning, but dissolving, and afterwards congealing to the consistence or likeness of white cream.*

Artephius remains unclear as to whether he employed *flowers* or *regulus*, but his mention of Stephanos and a reference to *"flores auri"* can be interpreted as *gold + flowers of antimony fusion* according to Stephanos' (Adfar of Alexandria's) methodology:

> *For if this fire of the lamp be not measured, or duly proportioned or fitted to the furnace, it will be, that either for the want of heat you will not see the expected signs, in their limited times, whereby you will lose your hopes and expectation by a too long delay; or else,* by reason of too much heat, you will burn the "flores auri", the golden flowers, *and so foolishly bewail your lost expense.*

The chemistry of Artephius' *antimonial secret fire*, his *vinegre*, presented below demonstrates that Artephius could have employed *flowers of antimony* or *regulus*, as either will produce his *vinegre*:

Artephius Vinegre via Flowers $Sb_2O_4 + 6\ NH_4Cl = 2\ SbCl_3 + 2\ O_2 + 6\ NH_4$
Artephius Vinegre via Regulus $2\ Sb + 6\ NH_4Cl = 2\ SbCl_3 + 6\ NH_3 + 3\ H_2$

His recipe results in pure *butter of antimony* (antimony-trichloride; $SbCl_3$) with ammonia and hydrogen byproducts or quicksilver as a byproduct if one interprets the passage to include mercuric or mercurous chloride. Artephius' technique for creating *butter of antimony* perfectly accords with Morienus' methodology, which profoundly influenced European alchemy. Butter of antimony appears in the work of Basil Valentine and a great many others in the veiled manner of alchemical tradition. There is no clear indication whether Artephius' process is derived directly from Morienus or not, but he does cite Morienus' teacher *Adfar*, also known as Stephanos of Alexandria, in a section that addresses digestion temperatures. He gives very clear descriptions of his water, which he clearly identifies as a salt in one passage, and lauds its potential:

> *In it also is a power of liquefying or melting all things that can be melted or dissolved; it is a water ponderous, viscous, precious, and worthy to be esteemed, resolving all crude bodies into their prima materia, or first matter ...*
>
> *If therefore you put into this water, leaves, filings, or calx of any metal, and set it in a gentle heat for a time, the whole will be dissolved, and converted into a viscous water, or white oil as aforesaid.*

> *Work therefore with it, and you shall obtain from it what you desire, for it is the spirit and soul of sol [gold] and luna [antimony]; it is the oil, the dissolving water, the fountain, the Balneum Mariae, the praeternatural fire, the moist fire, the secret, hidden and invisible fire.*
>
> *It is also the most acrid vinegar,* concerning which an ancient philosopher saith, I besought the Lord, and he showed me *a pure clear water,* which I knew to be *the pure vinegar, altering, penetrating, and digesting.*
>
> I say *a penetrating vinegar,* and the moving instrument for putrefying, resolving and reducing gold or silver into their prima materia or first matter.
>
> And therefore *our ultimate, or highest secret is, by this our water,* to make bodies volatile, spiritual, and a tincture, or tinging water, which may have ingress or entrance into bodies.
>
> And *this argent vive is called our esteemed and valuable salt,* being animated and pregnant, and our fire ... – Ibn Ar-Tafiz, 12th century

For those tempted to interpret elemental mercury as having anything to do with Artephius' mercurial water, Artephius speaks strongly against such an interpretation:

> And in this our philosophical sublimation, *not in the impure, corrupt, vulgar mercury, which has no qualities or properties like* to those, with which *our mercury, drawn from its vitriolic caverns is adorned.*

European Eagle's Gluten – Paracelsus provided the next innovation concerning *butter of antimony*, which he called *Gluten of the White Eagle* or *The Eagle's Gluten*. That Paracelsus referred to *butter of antimony* by the term *Eagle's Gluten* indicates that he knew it was originally crafted via sal ammoniac, which was encrypted by the cover-name *eagle* by alchemists of his day. He evolved the process to create *butter of antimony* by initiating a chemical reaction between *stibnite* and *corrosive sublimate* (mercuric chloride; HgCl) to achieve *butter of antimony* with *cinnabar of antimony* (mercuric sulfide; HgS) as a byproduct according to the following equation...

> Paracelsus' Eagle's Gluten and Cinnabar of Antimony
> $Sb_2S_3 + 3\ HgCl_2 = 2\ SbCl_3 + 3\ HgS$

This reaction is Paracelsus' own innovation and offered increased efficiency because it did not require *antimony regulus* as starting material. Initially creating *flowers or glass of antimony* or *antimony regulus* from *stibnite* can be a time consuming and labor-intensive process. His process combined the ease of chemically reducing *stibnite*, as opposed to metallurgically, with the added bonus of creating *butter of antimony* at the same time. The byproduct, *cinnabar of antimony*, was reserved for other alchemical uses apart from creating the Philosophers' Stone.

In a letter to Theodore, Paracelsus reveals that the secret to his alchemy, indeed his greatest treasure, the identity of *mercury of the philosophers* (sophic mercury) is *butter of antimony*, called the *mercury of antimony* in this particular communication:

> *Dear Theodore;*
> *With great truth the ancient Spagyrists have said* "Est in mercurio quicquid quærunt sapientes" *["Whatsoever the wise seek after, is in mercury"] but they have not told us what kind of mercury they meant ...*
>
> *Therefore I boast of and praise this* my Tincture made out of mercury of antimony, and can say, with great truth, that whatever you wish to obtain in medicine or Alchemy, you may find it in this mercury of antimony, for it is the specified mercury of the Philosophers, *wherewith I have done everything myself that I am about to communicate to you and for which reason* I esteem this as one of my greatest treasures and secrets in medicine and Alchemy ...
>
> – *Paracelsus,* The Tincture of Antimony

BOOK 2 – CHEMSTRY & CONFECTION

Basil Valentine knew a number of methods for distilling *butter of antimony*; the recipe below demonstrates familiarity with either Morienus' or Artephius' method and identifies the product as a type of salt. What remains unclear is the exact ratio of ingredients, yet this distillation is very forgiving if sal ammoniac is used slightly in excess. The final clue clinches the identity as butter of antimony, which can then be easily precipitated to *Algarot powder*.

> *Take of Antimony* **most subtlely pulverized One Part, of** *Sal Armoniack pulverized; mix these, and putting them into a Retort* **distill them together.**

> *… in one certain way of Preparation, from Antimony by distillations is drawn forth* **an Humour acid and sharp, like true perfect Vinegar. … Of it is made a true and natural Salt …**

> *Preparation of Antimony consists in the Key of Alchemy … in* **extracting its Essence, and in vivifying its Mercury;** *which* **Mercury must afterward be precipitated into a fixed powder.** *Likewise by Arts and due Method, of it may be made an Oyl … – Basil Valentine*

Basil Valentine preserved his recipe in a famous image known as the *Second Key*, in which *sal ammoniac* is depicted as a swordsman with a hidden face in reference to Egypt's Ammon, meaning *hidden*.

The *eagle* or *bird of Hermes* (Latin; Mercury) is perched upon the sword. The caduceus associated with *sal ammoniac* is topped with an opened fleur-de-lys implying that it is the volatile of the two. A second swordsman wields a sword with a crowned-serpent coiled around it. The serpent originally and throughout alchemical history is in reference to antimony; the wing-less serpent symbolizing refined antimony as opposed to the winged serpent of stibnite. The caduceus associated with antimony is topped with an unopened fleur-de-lys implying that it is the fixed of the two. The crown worn by serpent and by Sophic Mercury itself demonstrates that Sophic Mercury is a form of antimony; antimony being a "little king" and related to gold. At the bottom of the picture is a large pair of wings implying volatization, meaning distillation to achieve a stable substance. Sophic mercury is an angelic winged figure wielding a caduceus in each hand; he is *mercurious duplicatus*, the heavenly *Key* to confecting the Philosophers' Stone.

Valentine's 2nd Key (Antimony + Sal-Armoniack)

$2 Sb + 6 NH_4Cl = 2 SbCl_3 + 6 NH_3 + 3 H_2$

Therefore, in the preparation of Antimony consists the Key of Alchemy, by which it is dissolved, divided, and separated; as in calcination, reverberation, sublimation, etc.; also in extracting its essence, and in vivifying its mercury which mercury must afterwards be precipitated in a fixed powder: likewise, by Art and a due method, of it may be made an oil for the cure of diseases. – Basil Valentine

Distilling Eudica

Calcination and fusion are established chemical technologies derived from metallurgy and glassmaking and therefore conventional operative procedures. Most historians of science and technology would have no reservation in accepting these procedures as being prevalent in Maria's Alexandria. The notion that Maria was not only capable of distillation, but that she also created antimony trichloride so early in history however strains credibility just a bit. To alchemists, it makes perfect sense, but a historian or scientist demands evidence or at the very least, a well-tested hypothesis.

We decided to explore the possibility that Maria, Cleopatra, Zosimus and others may have created *divine / sulfur water* via chemical reaction and that the product of the reaction was indeed antimony trichloride. We began with two products known to Maria, stibnite and sal ammoniac. So much for the materials list, plausible equipment and methodology was another matter altogether.

Cleopatra's iconic distillation and evaporation setup has always intrigued and haunted researchers. It was depicted with an exhaust tube but without any form of receiving vessel. In fact, the distillation vessel appeared to have the same exhaust tube as the evaporation unit in the earliest depictions. We surmised that perhaps the reaction was not necessarily distillation but rather a chemical reaction that releases noxious vapors to achieve a crude form of butter of antimony. Maria remained silent concerning her method to *unite the two powders*:

> *And Mary said to him: Lift up your head Aros, because I will teach you to proceed by the shortest way, with the clear fixed body, found on small hills:* **this matter cannot be conquered by putrefaction.** *Take the finely pulverised matter [sal ammoniac] and grind it with Gum Elsaron [flowers of antimony], grind it finely and unite the two powders.*

Her powders were sal ammoniac and *flowers of antimony*, which explains the distillation technique Zosimus sourced from Maria to distil *divine water*. Cleopatra's method however was not so straightforward. The drawings of her reaction and evaporation vessels appear rather enigmatic. Each vessel displays an exhaust tube yet neither is connected to a receiver of any sort. I reasoned that perhaps *divine water* could be achieved via a reaction between stibnite and aqueous sal ammoniac without the need for distillation. We decided to test our hypothesis. Private unrevised communication between the authors was as follows:

Experiment – hypothetical Alexandrian SbCl₃

Just a brief summary what we talked about! It works!
1. I used 0.1 M solution of $NH_4Cl_{(aq)}$ and heated it up to 60 °C together with Sb_2S_3. Almost instantly the reaction started to smell like sh*t and I let it run for one hour.
2. After extraction, I collected a waxy substance that was identified as $SbCl_3$. This was only the first attempt, with care so the yield and optimized temperature, solvent mixtures and concentrations I need to calculate, but the yield from this attempt was around 40%, really smelly (almost puked)!!

$$6\ NH_4Cl_{(aq)} + Sb_2S_3 = 2\ SbCl_3 + 3\ (NH_4)_2S$$

Essentially, we added a sal ammoniac solution to powdered stibnite, catalysed the reaction and made a stink bomb! ... In addition, we also had achieved crude *butter of antimony* at a somewhat inefficient yield. Analysis confirmed that we had achieved *butter of antimony* by this method. We had intentionally neglected to calculate the measures in a "let's just see if it works" approach. Nevertheless, we were later able to optimize the synthesis easily. The experiment proved that *divine/sulfur water* was indeed within the material and technological capabilities of early-Alexandrian period alchemists. It also explained Cleopatra's odd vessel depictions with exhaust tubes unconnected to any sort of receiver. Late-period Alexandrian alchemist, Morienus and commenter Sir Isaac Newton, associated butter of antimony with *foul* or *stinking water*:

> *This is the nature of the white vapour, the green lion, and the foul water.*
> – Morienus

> *Eudica = glass grinds or glass impurities, the greatest of secrets. In these consists efficacious compounds but these three are sufficient, the white fume, the green lion and* stinking water. *– Sir Isaac Newton*

This also may explain why early-period Alexandrian alchemists used a double entendre pun in Greek that identified *butter of antimony* as both *divine water* and *sulfur water*, made by a process always associated with foul, stinky odor sometimes associated with death and rotting corpses. Passages that address the noxious odor that arises when creating *divine water* may simply refer to the release of gaseous ammonia byproduct as well.

Following this experiment, we began to wonder if a hydrolysis reaction between sal ammoniac and *flowers of antimony* would yield butter of antimony efficiently. The combination results in *butter of antimony* and liquid ammonium hydroxide, better known as household ammonia. Since ammonium hydroxide boils at 37.7 °C and butter of antimony at 220.3 °C, it seemed a logical hypothesis that we could easily separate the two via low temperature evaporation. Private unrevised communication between the authors based on the following chemical reaction was as follows:

> We want to try this with $6\ NH_4Cl_{(aq)} + Sb_2O_3 = 2\ SbCl_3 + 6\ NH_4OH$ (hydrolysis product that will be evaporated during heating as water and ammonia).

$$6\ NH_4Cl + Sb_2O_3 + 3\ H_2O = 2\ SbCl_3 + 6\ NH_4OH$$

The above reaction suggests that Cleopatra's two vessels may have represented two variations for creating *divine water*, one by heat catalyst and the other by evaporation in the hot Egyptian sun. The earliest version of the image does show one vessel with a heat-source under it and the other flat-bottomed vessel without a heat source. It is most likely that the evaporatory vessel is as Zosimus stated – the vessel that *evaporates divine water to fix Hermes*.

In typical alchemical fashion, Zosimus encrypted his two-ingredient list with the cover-names *sulfur* (flowers of antimony) and *quicklime* (sal ammoniac). Had Zosimus meant sulfur and quicklime in the literal sense, the product would be calcium polysulfide. This substance would serve well for surface-treating metals in Pseudo-Democritan style applications. Aside from the fact that it would be highly naïve to interpret Zosimus literally, he clearly states that he sourced his methodology from the texts of the Jews near a picture of a Tribikos, which indicates he was following a Judeo-Egyptian methodology typified by Maria. Furthermore, ingredients are almost never presented so openly in authentic alchemical texts. In addition to Zosimus' use of the term, *flowers of antimony* are also referred to by the cover-name *sulfur/divine* in Maria's texts irrespective whether translated from Greek (theío / theía; θεῖο / θεία) or Arabic (kibrīt كبريت).

> **Over the** *Kibric [sulfur; God] and Zibeth [offering],* **over** *the two fumes that contain two Lights ... – Maria*

According to Zosimus, *divine water* was the *Key* to the entire art:

> *Without the divine water, nothing exists.* For the entire compound is taken up through it, is roasted through it, burned through it, fixed through it, made yellow through it, decomposed through it, tinctured through it, subjected to iosis through it, refined through it, cooked through it.

Zosimus' practical instructions for distilling *divine water* clearly indicate the use of a digestion / distillation vessel and receiver. He advises a 2-3 week digestion period prior to distillation. This step parallels the hydrolysis stage suggested by Dr. Gabrielsson in the communication above. The odor discussed in the passages below most likely indicates a combination of gaseous ammonia and excess hydrogen chloride byproducts escaping upon opening the digestion vessel:

> *Allow digestion for 14 or 21 days until constantly rising vapors. Carefully attach, lute and maintain the device to retain the smell, because if she escapes, all the work is lost.* Indeed, the smell is quite unpleasant, and it is in this work [distillation step] that the smell resides.

> *The first water flowing (distillation) is white. The second flows drop by drop and has an unpleasant odor,* all uniform. Then, when the rise of the water has stopped, you remove the container in which the water has flowed, you cultivate, and you preserve it carefully. Approaching the still, you plug your nose because of the smell, and you will find in the vase female slag (caput mortuum) [skouriés: σκουριές].

> *Do not refuse the death to attain the resurrection, but* look for the resurrection of the (dead) *which was hopeless.*

Zosimus is describing the process of collecting two distillates, 1) crude hydrochloric acid, and 2) aqueous ammonia, which is retained for some other purpose. The most important part of the instruction is the *female skouriés*, the dead matter remaining in the digestion / distillation flask. This is the desired product of the reaction that will be resurrected via evaporation. Its chemical identity is antimony trichloride. The product must be kept in a sealed vessel and is called *divine* in either Hermetic reference to the vapour that ascends from below to above, or Gnostic Christian reference to its *divine* resurrection. If Zosimus' instructions to *cultivate and preserve it carefully* were in reference to *the first water flowing*, meaning the hydrochloric acid distillate, which could be added to the remaining *skouriés* and redistilled to achieve a purer *divine water* ($SbCl_3$) followed by its hydrolyzation to Hermes (SbOCl).

$$Sb_2O_3 + 6\ HCl = 2\ SbCl_3 + 3\ H_2O$$

CHAPTER 8 – LABORS OF HERCULES

Hypothesis: The chemical identity of *divine / sulfur water* in Maria's Judeo-Egyptian and the Hermetic Greco-Egyptian schools of Alexandrian alchemy is antimony trichloride. The archetypal recipe for its preparation was preserved in its clearest forms by Cleopatra, Zosimus and Synesius. We are in full agreement with Zosimus that it was central to Alexandrian alchemy, which is to say it was the *Key*. This particular form of *divine water* should be differentiated from the product of the same name used by the Babylonian-Egyptian school of alchemy typified by Pseudo-Democritus, the Leyden and Stockholm papyri. Credit for the discovery of hydrochloric acid and antimony trichloride should rightly be attributed to early Alexandrian *chrysopœians*.

Chapter 9 - Chemical Wedding

Let me assure you that in our whole work there is nothing hidden but the regimen, of which it was truly said by the Sage that whoever knows it perfectly will be honoured by princes and potentates. I tell you plainly that if this one point were clearly set forth, our art would become mere women's work: there would be nothing in it but a simple process of cooking.

Eirenaeus Philalethes

According to Philalethes above, and indeed the hypothesis presented throughout this work, the color indicators and heating temperatures that collectively comprise the *regimen* mentioned in the passage above remain a common factor throughout the history of alchemy. The color regimen was originally divided into three or four stages – blackening, whitening, yellowing and reddening. These color changes occur during both the preparatory and confection stages and intimately correspond to the temperature regimen.

Blackening is associated with decomposition. Stibnite ore naturally occurs as a silver-grey crystal or sulfide ore, which must be reduced to a fine amorphous black powder. This stage was performed with pestle and mortar and without heat. *Whitening* of the black powder occurs when stibnite is sulfur-reduced using heat to achieve white *flowers of antimony* or lustrous *antimony regulus* – a process known as the *first operation*. At elevated heat, *flowers of antimony* transition through a brief yellowing stage before they melt to red molten glass of antimony, hence the yellowing and reddening stages of antimony preparation. This preparatory procedure precisely matches the standard alchemical color and ramped heat regimen. The final step in this process involves vitrifying purified gold calx with *glass of antimony* – a process known as the *first composition*. The product mentioned above is known by a number of different cover-terms such as *red gum*, *latten*, *body* or *earth* in Alexandrian alchemy, and *lion's blood*, *sophic gold calx*, *stone of the 1st order* and many others in European alchemy. According to Maria's instructions, it was powdered, re-melted and then powdered once more.

Following preparatory work, the basic color and temperature regimen repeats itself a second time. Powdered *gold-antimony glass* was then combined with a chloride donor to finish the stone. In the case of the Quick-Dry method, this was achieved with crude sal ammoniac in a crucible and by the Great-Wet method in a glass digestion vessel. The secret story of the Philosophers' Stone is affecting this union with a suitable medium. The archetypal medium was known as *divine water* throughout most of the Alexandrian period, but also *white gum*, *blood*, *water* and *eudica* until the Byzantine period. Later traditions referred to it by the cover-terms *bird of Hermes*, *living sophic mercury*, *eagle's gluten*, *butter of antimony* and many others.

Powdered *gold-antimony glass* or alloy was then combined with *divine water* at usually unspecified ratios, placed into a vessel and gently heated. The product will initially melt and assume a pitch-black coloration. After a period, the product begins to dry and takes on a dark greenish-brown coloration around the edges, slowly changing to greyish-white. If *divine water* has been added in excess, a deep red liquid layer will form over the white mass, which can be removed and converted to a white oil by the addition of water. If the ratios are well balanced, the product will proceed through all colors and phases described in alchemical texts.

> Therefore Simplicio, come either with arguments and demonstrations and bring us no more Texts and authorities, for our disputes are about the Sensible World, and not one of Paper. – Galileo Galilei, Dialogo

CHAPTER 9 – CHEMICAL WEDDING

CONFECTING THE STONE

Rather than relying on highly conflicting and divergent canonical sources, we opted for an experimental approach instead. Our initial experiments consisted of little more than equal volumetric measures of gold-calx, flowers of antimony and *butter of antimony* simply tossed into a digestion vessel together to see what happens. Private unrevised communication between the authors is as follows:

Experiment #1 – fast run to force color-indicators

1. I mixed 5 Au + 2 Sb_2O_3 and grounded them to a fine powder;
2. and then added $SbCl_3$ at room temperature in a vessel under argon gas and sealed it;
3. then I slowly raised the temperature, 1 degree for every ten minutes.

Around 42 °C the reaction started to go black [melanosis; nigredo] and then greenish around 50 °C. I then let it go to 100 °C and obtained a white / yellow cloudy solution [leukosis; albedo / xanthosis; citrinitas], which crystallyses upon cooling [pictured right]. I know that this is not the way they did it according to the [traditional heat / time regimen] recipe, but I just wanted to try a fast run and see if I could observe a color change. It seems to work dudes and I will analyze the different steps and then try the real recipe. I think it is important to see what happens on mixing Sb_2O_4 and $SbCl_3$ and then add gold to that mixture...

Our initial experiment indicated that the color regimen is based on temperature control rather than time duration. Controlling the temperature at each successive stage allowed the reaction to stabilize. We performed a second experiment to verify this and to establish a temperature-based control as follows:

BOOK 2 – CHEMSTRY & CONFECTION

Experiment #2 – slow-run to identify correlation between temperature and color-indicators

1. Ok dude, I mixed Sb_2O_3 + $SbCl_3$ and heated it to 80 °C. This created a grey mixture in molten state;
2. cooled that to room temperature and added ground [powdered] gold leaf, sealed the vessel;
3. ...then I raised the temperature to 45 °C with addition of one degree every hour. First the matter transitions through (you can call it) greenish stage as before, followed by yellow-grey to white [leukosis; albedo], all within five degrees increment.
4. Then around 70-80 °C (happened at night) the mixture became liquid.
5. From 100-110 °C the liquid dried out and the "salt" crystallized (I would say amorphous powder).
6. This salt (unsealed vessel) melted between 120-130 °C, let it stand for one hour and then cooled it by just taking it off heating.
7. Around 60-70 °C the color started to become reddish-blue or purple [iosis / rubedo] and at room temperature, the mixture was darker with reddish-brown character, but not brown!

This was our initial product-based approach to proof of concept, which yielded successful results matching traditional temperature controls and color indicators. It also demonstrated that the manner in which gold is introduced is not as crucial to success as we had originally thought, providing a solid foundation from which to attempt reproductions via more canonical methodology in the future. Most interesting was the realization that despite countless traditional sources describing a timeframe of months to over a year to confect the stone, progress is actually based on temperature regulation according to color-phase indicators. We should have known this in advance because Maria clearly explained this to Aros nearly two millennia ago:

> The *Philosophers have always indicated a long regimen, and have concealed the work,* that no man should easily undertake it, and *they*

CHAPTER 9 – CHEMICAL WEDDING

pretend to take a whole year to performing the Magistery; but only for hiding it from the ignorant. – Maria Hebrea

Temperature and Color Regimen

We made the following observations concerning temperature control, color indicators and heating durations:

1. **Blackening** occurs only when gold is united with antimony as *gold-antimony glass* or alloy (latten). The black stage, known in European alchemy as **sophic decomposition**, is catalysed at 40-42 °C. Butter of antimony should be molten before being added to powdered latten/rebis. Depending on the amount of butter of antimony used, the product at this stage can be liquid, semi-solid or even a dried-resin texture. The product becomes pitch-black.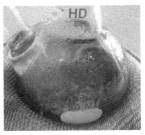

2. A greenish transition stage around the edge of matter in the vessel begins to spread before transitioning to grey. The green-yellow-grey transition stage, known traditionally as **vegetative germination**, is catalysed at 47-50 °C. Some alchemists include green and grey coloration in the color regimen. The color indicators are very subtle and unspectacular.

3. **Whitening** occurs as a liquefying stage that begins like a frost or snow that spreads throughout the matter as it transitions from grey to white. The white stage, known as **volatization by evaporation**, is catalysed at 70-80 °C or higher. Some alchemists include a light greenish-blue stage that can sometimes occur before yellowing begins.

4. **Yellowing** transitions to **Reddening** as the matter dries in the visual way a sunset does by shifting from yellow to red. The yellow-red stage, known as **volatization by sublimation**, is catalysed at 100-110 °C.

5. Fixation begins at 120-130 °C and finishes at around 150 °C. Upon cooling and drying, the matter appears as a heavy reddish-purple-brown colored powder [analysis samples of the beginning of yellowing and reddening stages pictured below].

Heat control is crucial to success. We ultimately settled on a digital hotplate-stirrer setup with an oil-bath of centrifuge-extracted coconut oil (green chemistry) in a clear translucent vessel. The oil is circulated continuously to ensure even heating and the coconut oil and glass bowl allow for viewing into the submerged digestion vessel to check for color-indicators. It is helpful to use an accurate thermometer, but not entirely necessary. All one needs to do is begin with a low temperature and look for color indicators, gently raise the temperature and be patient. Confecting the Philosophers' Stone is truly is a matter of observation and temperature regulation. I decided to check a few alchemical sources to corroborate this observation:

> *...note, that to dissolve, to calcine, to tinge, to whiten, to renew, to bath, to wash, to coagulate, to imbibe, to decoct, to fix, to grind, to dry, and to distil, are all one, and signify no more than to concoct Nature, until such time as it be perfect.* – Synesius, The True Book

> *Know you that in one thing, to wit, the stone, by one way, to wit, decoction, and in one vessel the whole mastery is performed.*
> – Roger Bacon, Mirror of Alchemy

CHAPTER 9 – CHEMICAL WEDDING

> *The white and the red spring from one root without any intermediary. It is dissolved by itself, it copulates by itself... It is therefore to be decocted, to be baked, to be fused; it ascends, and it descends.* All these operations are a single operation and produced by the fire alone.
> — Paracelsus, Aurora of the Philosophers

> *I know that the Sages describe this simple process under a great number of misleading names. But this puzzling variety of nomenclature is only intended to veil* the fact that nothing is required but simple coction.
> — Anonymous, A Very Brief Tract Concerning the Philosophical Stone

> *He coagulates himself and dissolves himself, and passes all the color ... which wants nothing but excitation, or,* to speak plainly, a simple, natural coction. *–Thomas Vaughan,* House of Light

One major concern arose due to our observation that the colors are not necessarily as dramatic as some alchemical texts would have one believe. The three main colors – black, white and red – indicate stabilization and the need to increase the temperature. The other four, dark green, yellow-grey, light greenish-blue and yellow-orange indicate transitions and occur only very subtly. One might even say that black, white and red were really the only predominant colors. We checked numerous traditional alchemical sources considered by most to be authentic in order to corroborate our observations:

> *Venerate the king and his wife, and do not burn them, since you know not when you may have need of these things, which improve the king and his wife.* Cook them, therefore, until they become black, then white, afterwards red. *– Diamedes,* Turba Philosophorum, 29th Dictum

> *And as heat working upon that which is moist, causeth or generates* blackness, which is the prime or first color, *so always by decoction more and more heat working upon that which is dry begets* whiteness, which is the second color; *and* then working upon that which is purely and perfectly dry, it produces citrinity and redness, thus much for colors.
> — Ibn ar-Tafiz (Artephius), Secret Book

> *... it dissolves itself, it coagulates itself, it putrifies itself, it colors itself, it mortifies itself, it quickens itself,* it makes itself black, it makes itself white, it makes itself red. – Roger Bacon, Mirror of Alchemy

> *The constant and essential Colors, that appear in the Digestion of the Matter, and before it comes to a Perfection, are three,* **viz.** *Black, which signifies the Putrefaction and Conjunction* **of the Elements;** *White, which demonstrates its Purification;* **and** *Red, which demonstrates its Maturation. The rest of the Colors,* **that appear and disappear in the Progress of the Work,** *are only accidental, and unconstant.*
> – Urbigerus, Aphorisms of Urbigerus

> *Combine two things, decompose them,* **let them become black. Digest them and** *change them to white* **by your skill; at last** *let the compound change to a deep red,* **let it be coagulated, and fix it ...**
> – Siebmacher, Sophic Hydrolith

> **It is to be noted also that** *our Stone in digestion is moved to all the colours in the World, but three are principal,* **of which good care and notice are to be taken, to wit,** *Black colour, which is first* **and it is the key of the Beginning of the Work; of the Second kind or degree,** *the* **White** *colour is the Second,* **and** *the Red is the third.*
> – Count Bernard Trevisan, Verbum Dismissum

Based on our suspicion that *butter of antimony* is the universal solvent and *Key* to archetypal Alexandrian alchemy, we then decided to perform an experiment described first by Artephius and four centuries later by Paracelsus to see if this solvent stood the test of medieval late-Islamic and renaissance European alchemical applications.

Oil of Gold via Antimonial Vinegre	**Aurum Potabile via the Magistery**
Ibn ar-Tafiz – Secret Book, Chapter 3	*Paracelsus* – Archidoxis, Book VI
1. Take (saith he) crude leaf gold, or calcined with mercury,	1. Take *circulatum* (circulated salt) well purified and in the highest essence.

CHAPTER 9 – CHEMICAL WEDDING

2. *and* put it into our vinegre, *made of saturnine antimony (mercurial) and sal ammoniac,* in a broad glass vessel, and four inches high or more;

3. put it into a gentle heat, *and in a short time* you will see elevated a liquor, as it were oil swimming atop, *much like a scum.*

4. Gather this with a spoon or feather *dipping it in; and in doing so often times a day* until nothing more arises; evaporate the water with a gentle heat, i.e., the superfluous humidity of the vinegre ...

2. *In it place thin beaten or filed metal* which has been wrought and cleansed to the most pure and subtle condition.

3. Place the two together and *let them circulate for four weeks* so that the sheets of metals are reduced by the temperate medium to *an oil which floats on top like fat,* colored according to the kind of metal.

4. Skim this off *by a silver 'drawer' (attractorium) to separate it from the circulation.*

Artephius' recipe is for *Oil of Gold via Antimonial Vinegre* and Paracelsus' *Aurum Potabile via the Magistery*, both detailed at the end of chapter seven of this treatise. A line-by-line comparison revealed that they were possibly describing the same chemical reaction.

Our intention was to test the dual hypothesis that 1) antimony trichloride was the chemical identity for Artephius' *saturnine antimonial vinegre* and for Paracelsus' *sal circulatum*, and 2) that both recipes were describing the same chemical reaction. If our experiment results matched the descriptions from these recipes, our hypothesis could then be considered plausible.

Experiment #3 and #4 – creating white and red via Artephius and Paracelsus' methodology

1. We heated an unmeasured amount of $SbCl_3$ to 80 °C until it liquefied in an open vessel;
2. added 99.99% purity gold powder finely divided to a small particle size. (Artephius advised gold powder or leaf and Paracelsus gold leaf);
3. digested the mixture without heat. The mixture turned black and solidified somewhat. Greening occurred after a few days in this phase.
4. The temperature was then raised to 40 °C and the matter liquefied and remained molten.
5. We turned off the heat to see if the matter would solidify but it did not. We attributed this to butter of antimony's very hygroscopic nature in which it attracts moisture from ambient air.
6. A phase transition occurred in the form of semi-solid matter at the bottom and a liquid surface layer. The solid turned white and the liquid a dark blood red.
7. We separated the liquid layer from the solid, applying traditional "feather filtration" to hold back any remaining solid particulates and stored the red oil in a sealed 50ml glass vessel. The semi-solid layer took on a greyish-white coloration.

Step 5 suggests slow and natural hydrolysis of butter of antimony, yielding hydrogen chloride and soluble gold salt. The red oil initially appears to be a crude gold-salt solution whereas the white "oil" remaining appears to be a crude antimony oxychloride. Artephius describes a white oil with the power of gold. He also instructs to "evaporate the water with a gentle heat" but remains unspecific as to which oil, white or red, he is referring. Paracelsus circulated his mixture in a sealed pelican then removed the oil with a shallow silver spoon. His desired product was medicinal gold-laden red oil.

This experiment demonstrates plausibility that Paracelsus' Alkahest may have been *butter of antimony*. Paracelsus' use of the word *magistery* can be read here as synonymous with Artephius' *saturnine antimonial vinegre*, the chemical identity of which is antimony trichloride. This experiment series also demonstrates that gold leaf will work just as well as gold calx, sponge or powder at varying particle sizes. An enigmatic, yet highly descriptive passage from Artephius still left us puzzled:

> *... a water saturnine antimonial should be mercurial and white, so that it whitens the gold without burning, but only dissolving it and later coagulating it as white cream ...*
>
> *If you place in this water any metal that is, filed or attenuated and if you leave it for some time in sweet and slow heat, it dissolves everything and it will becomes viscous water and in white oil just as it was already said.*

The gold dissolved without violence just as Artephius described. We had a greyish-white sludge (pictured left) and a red oil (pictured right) but no white cream of gold. This is the operative crux of the reaction, the details of which must be rediscovered by direct experimentation. The mystery eluded us until we washed a glass beaker that temporarily contained the red oil. Upon adding water, the traces of red oil turned to a white cream instantaneously, suggesting butter of antimony converting to antimony oxychloride by hydrolysis. We decided to convert the remainder of the red oil to a white cream and coagulate it via gentle evaporation.

The question as to whether Paracelsus' Alkahest may have been a deep eutectic solvent (DES) derived from *butter of antimony* combined with refined urine salts (urea and / or ammonium and sodium chlorides)

remained unanswered. The next experiment was to explore gold in combination with antimony pentachloride. I had performed the experiment successfully, observed gold dissolving and forming red crystals before yielding an oil layer, and communicated the results to my colleagues. Roger wanted to re-run the experiment himself to verify the results. Private unrevised communication between the authors is as follows:

> *I also confirmed the reaction Au + SbCl$_5$ as you did Erik but I have to analyze the product before I can say anything about it, but it is cool anyway ...*

This confirmed the possibility of dissolving gold leaf or calx in antimony trichloride and pentachloride to yield a product Artephius and Paracelsus referred to as *oil of gold* to indicate gold salt in solution.

The primary difference between this approach and Maria's *Ars Magna* lies largely with the selection of digestion vessel type. For this type of reaction, Artephius used a four-inch deep glass open digestion bowl for this experiment whereas Maria and Paracelsus preferred a sealed glass digestion flask. An open digestion experiment appeared the most intriguing to us so we opted for that. Both style of vessel will work and there is evidence that Artephius employed traditional methodology as well. The experiment series also provided a plausible explanation for Alexandrian *water of the sun* mentioned in the *True Book of Synesius*. In this scenario, Synesius could have created *water of the sun* using a similar methodology to the one described above.

Armed with this new understanding of *butter of antimony* (the white oil) and aqueous gold salts (the red oil), other alchemical recipes that describe converting the white product to a stable dry powder followed by reintroducing the red oil to the fixed white powder began to make sense. A number of these recipes can be found scattered throughout alchemical literature. It also allowed us to formulate a hypothesis as to how the Ingalese couple may have confected their stone via *oil of antimony* and *oil of gold* (*butter of antimony* and aqueous gold salt). An additional insight was that perhaps gold dissolved in antimony trichloride or pentachloride, hydrolysed and then evaporated to a dry powder might be a shortcut (or optimized synthesis if one prefers) to confect the Philosophers' Stone. Silver can replace gold to create a silver-based Tincture of the Philosophers.

> *... he who has the material will also find a furnace in which to prepare it, just as he who has flour will not be at a loss for an oven in which it may be baked. ... You cannot go wrong, so long as you observe the proper degree of heat,* **which holds a middle place between hot and cold.** *If you discover this, you are in possession of the secret, and can practise the Art,* **for which the CREATOR of all nature be praised world without end. AMEN.** *– Basil Valentine,* Twelve Keys (Postscript)

ARCHETYPAL RECIPES

This treatment presents several hypotheses based upon confecting the Philosophers' Stone, taking into consideration a materials list known to each alchemist discussed below and a skill-set that far surpassed the basic requirements of Stone's confection in their respective traditions. Much of the evidence is circumstantial, yet a strong argument is made here for sal ammoniac and metal oxides and powders as being both prevalent throughout 1st to 7th century Alexandria and foundational to each of the hypothesized recipes. Our early experiments establish plausibility for the basic formulations, products and their applications presented below. The objective of this book remains an attempt to broaden popular familiarity with alchemy's family tree, advance understanding of the origins and evolution of archetypal Philosophers' Stone recipes prevalent in Alexandrian alchemy, and to supplement research efforts in this area.

Moses' Parvaim Gold

Hypothesis: Moses made use of stibnite to dissolve elemental gold and convert it to *gold-antimony glass*. In its friable state is was then ground to a powder, suspended in liquid and ingested as biblical sources claim. Furthermore, he may have used his product in combination with copper found in the region to create alchemical gold, or combined his product with sand and alkali ash to create the *Urim* and *Thummim* mentioned in biblical sources. Evidence of copper mining and tons of alkali ash found at Serabít el-Khâdim support this hypothesis. Moses' technique served as the proto-archetypal recipe for creating *gold-antimony glass* used in Judeo-Egyptian alchemy typified by Maria and her school and known to European alchemists as the Stone of the 1st Order.

Maria's Ars-Brevis – The Brief Art (Dry-Path)

Hypothesis: Maria's Ars Brevis is a method of creating a chemical compound described as Ios (Tincture) via metallurgical technique. The materials she used were flowers of antimony, gold powder, natron and/or sal ammoniac. Her technique can be linked to antimonial bronzes of Ghassulian culture that flourished in southern Israel at Beersheba near the Dead Sea approximately 3,800-3,500 BCE. Ghassulian antimonial bronze primarily served religious ceremonial and decorative purposes. This distinguished it from industrial tin bronze that later defined the bronze age in Mesopotamia and the Levant. Maria's Tincture was also employed to create colored glass via a technique derived from ancient Assyrian or Egyptian methods. These two products would come to be known as alchemical gold and alchemical gemstones. Evidence of copper, antimony, zinc, tin and lead calcination to create metals oxides common to metallurgy and glassmaking supply an ancient precedent that supports this hypothesis. Maria's methodology is the archetypal recipe for the Ars Brevis via Sicca recipe to create *Ios* (Tincture; Elixir).

CHAPTER 9 - CHEMICAL WEDDING

Maria's Ars-Magna – The Great-Art (Wet-Path)

Hypothesis: Maria's Ars Magna is her own innovative technique for creating *Ios* via a chemical reaction between flowers of antimony and aqueous sal ammoniac. She distilled these together to create a solvent. *Gold-antimony glass* was ground to a powder and reacted with the solvent in a glass digestion vessel. Distillation and the use of glass vessels were chemical technologies common to perfumery, cosmetics, and essential and anointing oils, unguents, pigment and dye manufacture. Distillation arts originating from Sumerian / Akkadian perfumery were fully developed in Babylon. Perfumery, incense and anointing oil manufacture were trade crafts often carried out by female artisans. Maria appears to have combined distillation / digestion technologies with artisanal bronze-craft, made possible by the use of sal ammoniac. Her foundational branch of alchemy represents a Judeo-Egyptian school of industrial alchemy with an emphasis on bronze-craft via her elixir, artisanal alloying and patination during 1st century Alexandria, Jerusalem and likely Syria. Alloying methodology differentiated Judeo-Egyptian alchemy from Pseudo-Democritus' Babylonian style school, which focused on gilding and surface-treating technology and medicinal products. A number of historical sources support this hypothesis. Maria's recipe is the archetypal Ars Magna via Humida recipe to confect *Ios* (Tincture or Elixir). This version of the Philosophers' Stone is the identity of Paracelsus' *Tincture of the Philosophers* and Khunrath's *Lesser* or *Specific Stone,* a.k.a. *Sophic Sulfur* or the *Stone of the 2nd Order.*

White Stone of fixed Hermes – Doubled Living Sophic-Mercury

Hypothesis: A separate branch Greco-Egyptian alchemy was popular at Panopolis, present day Akhmim or possibly elsewhere in upper Egypt in tandem with or slightly later than the Judeo-Egyptian and Babylonian-style schools located in Alexandria. This school employed a two-stage (doubled) methodology that involved creating *butter of antimony*, known then as *Divine Water* via chemical reaction, followed by hydrolysis and evaporation to yield a white powder compound known originally in Greek as *Hermes*, and later by the Latin *Mercury*. No evidence exists that would enable an exact dating for the origins of this technology and it may derive from a tradition far more ancient than that of Maria. The emerald tablet of Hermes is an early encryption of the southern Egyptian school's recipe for this product. Unearthing an original tablet would be a great step towards determining the origins and age of the tradition that produced the Hermetic White Stone. The product can undergo fusion with any of the metals known to alchemy to create a powder of projection for use in bronze-craft. It may have been employed as a precursor to artisanal antimonial bronze manufacture as-is without being combined with any metal other than copper via carbon and crucible technique. The recipe encrypted in Cleopatra's *Chrysopœia*, Zosimus' *On the Evaporation of Divine Water that Fixes Mercury* and Synesius' dialogue with Dioscorus in which he identifies the Prime Material as Roman antimony supports this hypothesis. The recipe attributed to Synesius in the *True Book* accords with the three examples above. This version of the Philosophers' Stone is the identity of Stephanos' *White Stone*, Paracelsus' *Stone of the Philosophers* and Khunrath's *Great* or *Universal Stone*.

Morienus' al-Kīmyā' – Latten & Eudica

Hypothesis: Morienus' work accords with Maria's Ars Magna methodology to create a Philosophers' Stone that contains gold, yet his was the first to indicate a fermentation process by adding powdered Philosophers' Stone to an additional measure of the primary ingredients and concocting it throughout its temperature and color regimen a second time. The chemical identity for the Stone of the 1st Order was *gold-antimony glass*, the Stone of the 2nd Order was the product typified by Maria, and the Stone of the 3rd Order was the product resulting from Morienus' innovative fermentation finishing technique. His methodology was foundational to early Islamic and European alchemical traditions. His operative methods were derived from, yet slightly different to those of his teacher, Stephanos. The Cabala Mineralis is an accurate representation of the methodology typified by Morienus, as are many other alchemical texts that appeared in the European tradition. Morienus combined *gold antimony glass* (latten, body, earth) with *butter of antimony* (eudica, blood, water). These came to be known in the Islamic alchemical tradition as (sophic) *Sulfur* and (sophic) *Mercury*. Paracelsus referred to these two substances as *Blood of the Lion* and *Gluten of the Eagle*. Morienus then subjected the Stone to a process he called *fermentation* to achieve a refined or exalted version of the Philosophers' Stone, referred to later in the European tradition as the Stone of the 3rd Order.

Ingalese Stone – Oils of Gold & Antimony

Hypothesis: The Ingalese couple arrived at a modified recipe to confect the Philosophers' Stone based on an interpretation of Paracelsus, who suggested combining oils to confect the stone. According to Paracelsus, the definition of an *oil of metal* is an *aqueous metal salt*, thus *oil of gold* is aqueous gold trichloride and *oil of silver* is aqueous silver nitrate, etc. This is supported by texts in which he details dissolving gold in aqua-regia and evaporating it to its *oil*, or specifically to *oil of gold*. By this definition, the chemical identity of *oil of antimony*, called *glacial oil of antimony* by Glaser and others, is antimony trichloride. Other versions of *oil of antimony* exist, but these are largely based on the Acetate-Ethanol Tincture branch of alchemy typified by Roger Bacon and Basil Valentine and have nothing to do with the type of *oil of antimony* central to this hypothesis. The combination and coction of aqueous *oil of gold* and *oil of antimony* will result in a product similar if not identical to one made by more traditional methods.

The One/All of Alexandrian alchemy

Hypothesis: The chemical identity of *The One / All* mentioned in the context of Alexandrian *chrysopœia*, as differentiated from the use of the term in religious or philosophical contexts, was antimony in general and stibnite as Prima Materia (or first material) in specific usage. This alchemical use of the term/s was employed additionally as an umbrella term to cover all of the antimony compounds known to Alexandrian *chrysopœians*. It was also known as the Serpent and was symbolized in alchemical texts by the ouroboros serpent or viper biting its tail. The terms *ios* and *iosis*, meaning *venom* and *to envenomate* in the alchemical context as cover-terms synonymous with *tincture* or *to tinge*, is derived from the serpent or viper representation of antimony.

Administration – Honey-Vinegar Tonics and Sweet White Wine

Hypothesis: The soluble gold salts, trace iron and silica impurities from the stibnite along with sub-trace levels of antimony were brought into solution and the insoluble separated by filtration. In the Islamic tradition, this would have been achieved by water or vinegar, whereas European alchemists used wine. In Islamic elixirs, the honey acted as reducing agent, whereas in wine the acid content and ethanol achieved the same.

A Final Few Lines

> We have arrived at the point in our investigations where we are satisfied that the art is true, and that it is all and possibly more than is claimed for it in the books of the Sages. We would not wish anyone to believe that it is work quickly performed, as many have held out. ... But the whole work can only be proven by trial.
> – Delmar de Forest Bryant, 20th century

Our experiments informed us that two (flowers of antimony and aqueous sal ammoniac) or three ingredients (gold, flowers of antimony, butter of antimony) depending on whether we were confecting the Hermetic White Stone or the Tincture, placed in a vessel and then gently cooked, yielded the Philosophers Stone. We did not set out to become alchemists, yet pursuing and enjoying alchemical reproducibility experimentation has been so rewarding that we plan to continue our efforts for years to come. Our aim here is simply to present hypotheses for the chemical

CHAPTER 9 – CHEMICAL WEDDING

identity of the various archetypal products of Alexandrian alchemy and to place them in historical context so that credit for their discovery may be attributed properly. The recipe to confect the Philosophers' Stone is relatively easy and anyone with an interest to try it should do so with proper guidance and protective measures. Alchemy is addictive, especially when practical, theoretical and ideological streams merge and begin to carry one along down a widening course. We have included the above accounts for the primary purpose of sharing our sense of discovery with the reader, whereas experimentation and balanced chemical equations support our hypotheses for the most part.

We have since performed subsequent alchemical experiments and reproductions in a manner far more controlled, molecular weights carefully calculated, controls in place, etc. That is the process of optimization and due to space constraints in this work, detailed recipes including systematic instructions, measurements, photographs, etc. will be presented in a companion instructional guidebook entitled *Téchni – The Art of Confecting the Philosophers' Stone* by the authors. The aspiring adept should now have all the tools necessary to step foot on the path of discovery towards becoming adept at confecting an archetypal Alexandrian-style Philosophers' Stone. The following section details the applications for the Philosophers' Stone in historical context. The final chapter is a discussion of the theoretical and ideological bedrock upon which the edifice of the alchemical process was built, both speculative and operative.

> *It has been set forth by the Sages in the most perplexing and misleading manner, in order to baffle foolish and idly curious persons, who look rather at the sound than at the meaning of what is said. Yet, in spite of foolish and ignorant people,* the Art is one, and it is true. Were it stripped of all figures and parables, it would be possible to compress it into the space of eight or twelve lines.
>
> – *Petrus Bonus*, The New Pearl of Great Price

BOOK 3

Applications
&
Observations

This natural process by help of craft thus consummate

Dissolveth the Elixir spiritual in our unctuous humidity;

Then in the balneo of Mary together let them be circulate,

Like new honey or oil till they perfectly thickened be,

Then will that medicine heal all manner infirmity,

And thus turn all metals to sun and moon most perfectly;

Thus shall ye have both great Elixir and Aurum Potabile,

By the grace and will of God, to whom be laud eternally.

George Ripley

Chapter 10 – Powder of Projection

If men say that gold is the most precious stone
they do not mean common gold extracted from gold ores,
because common gold does not produce a tincture by which
other [metals] may be colored, since it contains only just
sufficient color for its own body. It contains no surplus tincture.
But our gold which has the desired quality can make gold and tint into gold.

Aydamir al-Jaldaki

Approximately 200,000 years ago, during the Middle-Pleistocene, man encountered the first ternary alchemical reaction in the form of fire. The achievement and mastery of fire was profoundly significant and left such a deep and lasting impression on the collective psyche of our species, that it became the first object of veneration as fire-worship. Fire-worship appeared predominantly in the zone of ancient Anatolia, Syria, Egypt, Mesopotamia and Iran and lasted many thousands of years.

Around 26,000 years ago, this new chemical technology gave rise to humanity's first great alchemical transmutation; with the aid of fire man learned that he could magically transmute soft wet clay into dry hard stone and thus developed earthenware pottery. Earthenware was never worshiped as such, yet gods of the furnace certainly were. Not only could pottery be fired in an oven, around the time of the agricultural revolution it was discovered that bread could also be prepared by this magic. Sacred breads appeared and kitchen and hearth gods earned a place of veneration in many cultures. While these pyro technologies were magical and revered they were not alchemical as the term is currently understood.

It was not bakers' nor potters', but rather glass-makers' ovens, in which sand was transmuted into gemstones, that influenced early metallurgy. Alchemy's foundations were built upon the chemical technology of glassmaking and metallurgy, the advent of which was a great industrial leap forward in the history of humanity. In Alexandria, the origins of alchemy are intimately linked to bronze technology, specifically the production of gold and silver tinged and liver-purple patinated decorative bronzes. Yet in early Alexandria's embryonic stage of alchemical development, bronze technology was already considered ancient, even by Maria. True origins of alchemy date further back into remote antiquity to an event for which no written record exists. At some point in our ancient past, likely in Anatolia or Sumeria, a curious fire-magician discovered that by placing special stones in kilns originally used for glassmaking, he could transmute a stone into metal. These magic stones were naturally occurring copper oxide and copper carbonate minerals. Copper sulfides followed soon thereafter and were first mined and smelted at Kültepe in Turkey, which led to specialized metallurgy in Cyprus and Greece.

The earliest evidence of *native copper* exploited by humans is found in southern Anatolia, modern Turkey, at the Neolithic settlement of Çayönü Tepesi, situated at the foot of the Taurus Mountains near the upper Tigris River. The settlement dates from approximately the 8th millennium BCE. By this time, copper metallurgy had already reached technological breakthroughs in basic cold and heat hammering, annealing, melting and casting. Sumerian coppersmiths in Mesopotamia were able to extract copper from ores and achieve levels of purity higher than 98%.

Semi-precious stones originally used as pigments, malachite and azurite, later became central to the copper and bronze ages in Mesopotamia and the Levant. Turquoise and malachite in the Sinai indicated copper deposits, which were mined by the Egyptians assisted by Kenite smiths and Midianite and Amalekite labor.

By 4,500 BCE, vast copper smelting and foundry sites began to crop up in Jordan and around the Dead Sea ushering in complex crucible technologies. According to a growing body of research, crucible smelting during the 5th millennium BCE developed elsewhere in Iran, Northern Mesopotamia, Central Europe, the Balkans and Greece. Later, advanced copper sulfide smelting, refining and furnace technologies developed and spread northward from the Dead Sea area towards the northern Euphrates and Caucasus.

The word for copper is from the Latin *cyprium*, meaning a malleable copper-based metal known to the Romans alternately as *aes* (cupreous; brass/bronze). The contraction of *aes-cyprium* is *cuprum* from which the chemical symbol Cu is derived. The word *aes* passed into Germanic and ultimately became the word *ore*. However, *cyprium* is derived from Greek *Kýpros* (Κύπρος) in connection with the island of Cyprus, which was named after Sumerian *zubar* or *kubar* (cupreous; copper-alloy). The seashore at Paphos on Cyprus is the mythological birthplace of the goddess Aphrodite, then known as *Kýpria*, and the island signifies a connection between copper-based metals, the sea, trade and the predominant local earth-goddess at the time.

The correspondence between copper and earth-goddesses such as Venus, Aphrodite or even earlier variants Astarte, Ishtar, Isis or Hathor originate in remote antiquity. This is a result of diverse cultural and economic intercourse between eastern Mediterranean cultures, Egypt, Mesopotamia and the Levant at the beginning of the Bronze Age. Early Akkadian texts speak of *male* and *female* stones (ores) and metals. A great many earth-goddesses were associated with delicate iridescent green or blue copper-containing ores. Some of the earliest symbols for life, woman or female deities were also symbolic of distinctly feminine copper:

BOOK 3 – APPLICATIONS & OBSERVATIONS

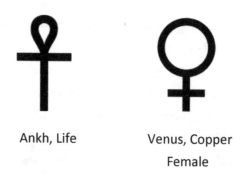

Ankh, Life   Venus, Copper Female

As commonplace as metallurgy is today, it was sacred magic during the evolutionary technology interval between the Stone Age and Bronze Age. Copper in its developmental phase was used primarily for prestige and ceremonial items rather than utilitarian purposes and consequently great metallurgico-religious traditions developed around this revered interplay of fire and earth. In many mythological traditions from antiquity, the copper and bronze smelter was a "civilizing hero". From the furnaces of the religious copper-smelters and refiners, the next stage in the evolution of alchemical fire-magic emerged in the form of bronze-craft; the catalyst for technological and economic intensification that occurred during the Bronze Age.

Gavin Menzies, Royal Navy Commander and Historian in his book *The Lost Empire of Atlantis*, presents a compelling hypothesis for an extensive Minoan maritime trade-network centered upon a thriving bronze-trade and its impact on the region. He explains the importance of bronze during the period of civilization building in the Mediterranean, Levant, Egypt and Mesopotamia:

> The Bronze Age is named after a copper ally. It was a miraculous material. Suddenly, metalworkers could change rock to metal, using the magic of fire. Today, world politics is dominated by areas rich in oil, uranium and knowledge. Many thousands of years ago, metal was as important for the development of wealth and power as energy supply and information are today.

The discovery revolutionized the technology of the world. Bronze was sharp-edged, strong and durable. The texture and strength of this material makes it ideal for creating effective weapons and astoundingly flexible tools. This precious metal – for then, it truly was precious – totally transformed the world. It was essential to advancing technology and crucial to the development of civilisation. It gave man a modern material – an alloy – that could be moulded and beaten into any shape.

Suddenly, Bronze Age man possessed hard, corrosion-resistant metallic tools – shovels, axes, chisels and hammers. For the first time in the history of the world, with the free time bought by this helpful technology man was able to create products of pure luxury in large amounts, such as sumptuous jewelry.

Copper thoroughly dominated metallurgy from 7,000 to 1,500 BCE. Artisanal crucible smelting gave way to the more industrial bowl-furnace around 3,500 BCE, while arsenical and *antimonial bronze* as it pertains to alchemy diverged or branched off from the utilitarian tin-bronze that came to dominate much of the Bronze Age from around 1,500 BCE onwards.

Alchemical gold's ancient origins lie in the earliest *arsenical / antimonial bronze* metallurgy developed in the Mesopotamian area and southern Levant, covering areas of modern Iran, Iraq, Israel, Jordan, Syria and Turkey. Originally, *arsenical* and *antimonial copper alloys* were developed by reducing *fahlore*, a copper-arsenate ore rich in antimony, or possibly other silver-sulfosalts such as *pyragyrite* (Ag_3SbS_3), *miargyrite* ($AgSbS_2$) or *stephanite* (Ag_5SbS_4). The Sumerians were among the first in the region to identify and name both *antimony* and *brass/bronze*. Yet *antimonial bronze* technology was widespread and the ancients soon discovered that antimony-silver combinations alloyed with copper improves hardness, malleability, tensile-strength, corrosion-resistance and other properties of copper. Antimony ore was mined at Ghebi in the southern Caucasus Mountains and transported as far as Beersheba in the Negev, southern Israel to make religious antimonial bronze.

In a study of nineteen metal-combinations of *fahlore*-metallurgy of the Bell-Beaker culture during pre-Bronze Age Europe, antimony featured in eight of the nineteen metal combinations. *Silver-antimonial bronzes* were some of the recurrent prehistoric bronzes and were likely precursors to *gold-antimonial bronze*. These *fahlore*-based *silver-antimonial bronzes* from the Bell-Beaker culture of the Carpathian region have been dated as early as the 5th millennium BCE. Around the same time, *antimonial bronze* was in production in southern Israel, with *arsenical bronze* developing on the Iranian plateau a thousand years later. These centers of *antimonial bronze* production serve as veritable missing links connecting prehistoric metallurgy to alchemical gold-making in early Alexandria. During this developmental period of the copper age, it becomes clear that a number of ores were experimented with for their alloy potential. Prehistoric origins for alchemical gold and silver likely began with *fahlore*, yet the ancients may have also experimented with unique ores such as *pyragyrite*.

The Egyptians began mining malachite copper ore by exploiting Midianite and Amalekite labor, which malachite is named for, at Timna in modern Israel. This site was so rich that it is considered a plausible candidate for King Solomon's mines. Near the high place of offering at Timna, traces of metallurgy can be found, the earliest furnaces dating to 4,200 BCE or earlier. Another workshop at Timna is located in the courtyard of the shrine devoted to the goddess Hathor. In addition, the mining site Serabît el-Khâdim in the Sinai is an example of a sprawling temple complex also devoted to Hathor. Near the temple, remains of centuries of copper and turquoise mining can be found. Although signs of copper smelting and slag were found near the mines, great workshops were also located in the temple complex proper. These are just two examples of religion and ritual associated with deity, copper, metallurgy and commodity. The sacred magic to transmute stone into metal was reserved for initiated priests or members of particular metallurgico-religious Canaanite tribes, such as Kenites, Midianites, and Amalekites in addition to Egyptian priests. Connection between the goddess and copper-ores is prevalent in the Egyptian tradition where Hathor is the principle goddess of copper. Old and Middle Kingdom Egyptian texts indicate that the most precious metals, and specifically gold, were considered the bodies of the Gods. The Ugaritic Canaanite word for gold is *ḫrṣ* (Biblical Hebrew; hārûs הרס). Hathor images are often accompanied or in the presence of the Golden Sparrowhawk of Horus or the sun-disk to represent gold accompanied by antimony symbolized by a bull / cow, serpent or both. The result of this symbolic combination was alchemical gold.

ALCHEMICAL GOLD

Alchemy is nothing else but the set purpose, intention, and subtle endeavour to transmute the kinds of metals from one to another.
– Paracelsus, Coelum Philosophorum

Alexandrian Chrysopœia

In 1992, the *American Journal of Archeology* published an article written by Jacobson and Weitzman entitled *What Was Corinthian Bronze?* This was followed by an article in the *Gold Bulletin* in 1999, written by David Jacobson entitled *Corinthian Bronze and Gold of the Alchemists*. Both articles highlight parallels between surviving accounts from antiquity regarding *Corinthian bronze* and chemical technologies for surface-treating bronzes sourced from the *Leyden Papyrus X* and the *Papyrus Græcus Holmiensis*, also known as the *Stockholm Papyrus*, found together at Thebes and likely written by the same scribe sometime around the 3rd century CE.

> *Alloys that went under the name of Corinthian Bronze were highly prized in the Roman Empire at the beginning of the Christian era, when Corinthian Bronze was used to embellish the great gate of Herod's*

> Temple in Jerusalem. From the ancient texts it emerges that Corinthian Bronze was the name given to a family of copper alloys with gold and silver which were depletion gilded to give them a golden or silver lustre. An important centre of production appears to have been Egypt where, by tradition, alchemy had its origins. From an analysis of the earliest alchemical texts, it is suggested that the concept of transmutation of base metals into gold arose from the depletion gilding process.
>
> – David M. Jacobson

The last eleven recipes of the *Leyden codex* are short medical tracts taken directly from *De Materia Medica*, written by Greek physician, pharmacologist and botanist Pedanius Dioscorides in the 1st century CE, thus loosely connecting medicine to Pseudo-Democritan chemical technology. Democritus is mentioned by name in the *Stockholm Papyrus* and together these texts represent an incomplete yet accurate record of the style of alchemy practiced by Pseudo-Democritus and his school. Babylonian style gilding technologies preferred by Democritus can be dated from the 3rd millennium BC, based on archeological evidence recovered from the Royal Tombs at Ur in Mesopotamia.

These articles serve as valuable tools demonstrating to researchers the connection between alchemical gold as a valuable Alexandrian commodity with the *Corinthian bronze* products of the time. They make no mention however of *Ios, Xēre/xêrion, Powder of Projection, Chrysocorallos,* or *Pharmakon* – all synonymous for the Philosophers' Stone – nor do these articles or papyri make any mention of antimony; the Philosophers' Stone and antimony both being central to recognizable alchemy of most alchemical traditions. Many researchers have focused on these papyri attempting to answer lingering alchemical questions and unlock mysteries such as the Philosophers' Stone and its application towards gold-making. The challenge with this approach is that Maria Hebrea's school and that of Pseudo-Democritus were extremely dissimilar in their choice of materials and overall methods of operation.

Pseudo-Democritus operated according to a Persian-Babylonian methodology in contrast with Maria's Judeo-Egyptian processes. Synesius, in a letter to Dioscorus, explains that Pseudo-Democritus did not operate in the manner of the Egyptians, and while he did employ alloys, he did not follow typical Egyptian-alloying methodology. Pseudo-Democritus' method of operation involved surface-treating metals in order to alter or enrich the metal's surface coloration.

> *Speaking to the great Ostanes, he [Democritus] attested that he did not make use of the projections of the Egyptians, nor of their processes of heating and melting [alloying]; but that he operated upon substances by application from outside, effecting the chemical result by means of the fire [fire-gilding]. And he said that it was the custom of the Persians to operate in this manner.* – Letter of Synesius to Dioscorus

From this it is possible to surmise that at least two alchemical movements existed in early Alexandria, 1) a Babylonian-style guild concerned with chemical dying and gilding technologies that incorporated medicinal plant and herbs, and 2) a Judeo-Egyptian guild of talented bronze-artisans concerned with very fine gold and silver *antimonial bronzes* along with a bronze featuring liver-purple colored patina and possibly another complex alloy composed of copper and additional metals. This division of approaches was discussed by F. Sherwood Taylor in *A Survey of Greek Alchemy*:

> *The earliest alchemical authors are sharply divided into two schools...*
>
> 1. **The followers of Democritus** – *These carry out their alchemical work by superficial colourings of metals and by the preparation of alloys by fusion ...*
> 2. **The school typified by Maria and Comarius** – *These employ complex apparatus for distillation and sublimation [to initially create a powder of projection].*

Maria's *elixir* was alloyed or brought into fusion with copper but only after the production of a *powder of projection*. The alloy underwent fusion by the Egyptian technique of *projection*, giving rise to the Philosophers' Stone being termed *powder of projection* when applied to gold-making. The difference between these two approaches has caused longstanding confusion as to the nature of early Alexandrian *chrysopœia*; a clearer understanding of which changes the entire landscape of alchemy research. It is worth repeating Zosimus' views about the Judeo-Egyptian approach to alchemy:

> *There are two sciences and two wisdoms: that of the Egyptians and that of the Hebrews, which latter is rendered more sound by divine justice. The science and wisdom of the best dominate both: they come from ancient centuries.*
>
> *The Jews, having been initiated, have transmitted these procedures which had been confided to them.*

Even Pseudo-Democritus appears to defer to Jewish understanding when he wrote the following:

> *It was the law of the Egyptians that nobody must divulge these things in writing ... The Jews alone have attained a knowledge of its practice, and also have described and exposed these things in a secret language.*

Joseph Needham differentiated between the two traditions by describing the Pseudo-Democritan approach as *aurifiction*, and that typified by Maria's tradition, and the legacy of alchemy that followed, as *aurifaction*:

> *Thus looking backward we can see clearly two great traditions of metallurgy related to* **aurifiction** *and* **aurifaction,** *the making of uniform-substrate alloys on the one hand,* **and the** *'tingeing' or 'dyeing' of metal surfaces* **by altering the composition of surface layers,** *or by depositing microscopically thin coloured films, on the other. There can be no sharp distinction of course between them, since what can be done*

CHAPTER 10 - POWDER OF PROJECTION

with the surface layers or the films depends very often on the exact composition of the mass of metal beneath them. Some students of Hellenistic proto-chemistry have felt that they could descry a difference here between two cultural traditions, the Egyptian and the Persian respectively, and this is of some importance for us if Persian should be taken to imply influences still further Eastern in character. – Joseph Needham, Science and Civilization in China

In November of 1950, M.L. and J. Dufrenoy of University of California San Francisco published an article in the *Journal of Chemical Education* entitled *The Significance of Antimony in the History of Chemistry*. Their position was far more accurate concerning alchemical gold; antimony was the key ingredient. They associated antimony with Moses and to Jewish biblical sources where it is mentioned in a number of passages, one of which describes antimony as an item included in the Temple:

Indeed, in Egypt, stem (the neutral sulfide of antimony) must have enjoyed much consideration among the priests and learned, since Moses brought along from Egypt the stone Puch which was to be embodied in the Temple, and of which we read in I Chronicles ...

The term *stem* in the passage above is the Egyptian verb form that means *to apply stibnite eye makeup*. There is also a noun form and a number of hieroglyphs that are all loosely related to stibnite.

ṣtem = to apply stibnite eye makeup; to assist a priest; to hear; to bring to completion; to join
ṣtem-t = stibnite, antimony, eye makeup

Egyptian *ṣtem-t* can be written *sdm*, *mśdmt*, or *mesdemet*. The hieroglyphic form always includes an owl in profile for the phonetic sound of M but also possibly because stibnite was used to widen the eyes. At some point during Hellenistic Egypt, a Greek-Egyptian compound word was developed that combined Greek *vafí* or *baphē* (βαφη), meaning *tincture* and Egyptian *met* derived from *mesdemet* to form the compound

429

word *baphēmet*. It was symbolized by the ram head of the horned Ethiopian and Libyan god *Amun / Ammun*, Greek *Zeus-patèr Hammon* (Ζεύς- πατὴρ Ἄμμων) or Roman *Jupiter* (Iuppiter) *Ammon / Jove* (Giove) *Ammone*; all referring to the same deity.

Early Greek *stimmi* (στίμμι) and later *stibi* (στίβι) were both derived from the Egyptian *stem-t*, which later became Latin *stibium*. Pliny, aside from discerning between male and female forms of stibnite / antimony, also calls antimony *larbaris* (Latin; stibinos) meaning *alabaster* thus suggesting familiarity with the white female calcined form as *flowers of antimony*. Moses has always been connected, albeit figuratively, with *antimonial bronze*; the stone of *pûkh* referenced by the Dufrenoys above being a Hebrew variant for stibnite. In Islamic tradition, when Allah revealed himself to Moses on Sinai, Moses fainted and the mountain caught fire after which Allah said - *Henceforth shalt thou and thy seed grind the earth of this mountain and apply it to your eyes!*

One of the earliest symbolic expressions for antimony has been the dragon or serpent, and later the *ouroboros* or serpentine stone of Naassene and Ophitic Gnosticism. With this in mind, the story of Moses and the *brazen serpent* takes on new meaning. One of Moses' first tasks was to create a *brazen* (cupreous; brass/bronze) *serpent* to heal the Israelites. By interpreting *serpent* as synonymous with *antimony*, the notion of Moses performing *antimonial bronze* metallurgy becomes plausible:

> Brazen (cupreous; brass/bronze) Serpent (stibnite/antimony)
> = antimonial bronze

The *brazen serpent* was kept in Jerusalem's Temple sanctuary, where the Israelites began to worship it as a representation of God by offering sacrifices and burning incense to it. It was later destroyed by King Hezekiah.

CHAPTER 10 – POWDER OF PROJECTION

The reference to I Chronicles above is followed in the article by a mention of Isaiah, both of which read:

לבית הכינותי כחי וככל ב
והכסף לזהב הזהב אלהי
הברזל לנחשת והנחשת לכסף
אבני לעצים והעצים לברזל
פוך אבני ומלואים שהם
ואבני יקרה אבן וכל ורקמה
לרב--שיש

I Chronicles 29:2 So I have provided for the house of my God, so far as I was able, the gold for the things of gold, the silver for the things of silver, and the bronze for the things of bronze, the iron for the things of iron, and wood for the things of wood, besides great quantities of onyx and stones for setting, antimony, colored stones, all sorts of precious stones and marble.

הנה נחמה לא סערה עניה יא
אבניך בפוך מרביץ אנכי
בספירים ויסדתיך

Isaiah 54:11 O afflicted one, storm-tossed and not comforted, behold, I will set your stones in antimony, and lay your foundations with sapphires.

The article excerpt below links antimony to the symbol for Saturn and identifies it as alchemical *lead of the wise*. Lead is traditionally the metal associated with Saturn, yet alchemists have always maintained that the planet shared a primary occult relationship with antimony rather than lead. Article excerpts and source quotes are as follows:

> *Although antimony has been repeatedly identified as the metal wherefrom were made many ornaments or implements unearthed along with other prehistoric relics of man, antimony as a metal had received no specific name until a few centuries ago, and* **the symbol for Saturn was applied indifferently to lead and to antimony ...**
>
> *Dioscorides described as "lead-like" the metal obtained from the roasting of stibnite,* **and** *for many centuries thereafter no distinction was drawn between lead and the other "imperfect metals."* **Fallopian, lecturing in Venice in 1569, charged his former teacher Brasavola with having confused antimony with lead. To Basilius Valentinus, antimony was the "bastard of lead" and** *as late as 1652, René Chartier referred to antimony as being the "Sacred Lead of the Wise."*

Primary sources strengthen the Dufrenoy couple's case considerably:

> One roasts this ore [stibnite] by placing it on charcoal and heating to incandescence; *if one continues the roasting, it changes into lead ...*
>
> *— Dioscorides*

> *But the main thing of all is to* observe such a degree of nicety in heating it [stibnite], as not to let it become lead ... *— Pliny*

> *But that we may also say something of the* Lead of the philosophers, **let the curious searchers of nature know, that** between Antimony and common Lead there is a certain near affinity, **and they hold a strict friendship, the one with the other.** *— Basil Valentine*

> *Antimony is an imperfect metal, and in philosophy is called Saturn,* **of which Rhasis [al-Razi] writeth, saying, that** *in Saturn, Sol and Luna are contained in potentia, not visibly.* **Also Pythagoras saith,** *all secrets are in Saturn; and this Saturn is our Lead,* **which we physicians must know and understand; for it is created for us; it is ours and not another's ...**
>
> *— Alexander von Suchten*

> *It has been named Proteus by some,* **because it takes on diverse forms and colours,** *by others the* Sacred Lead, *the* Lead *of* Philosophers, *the* Lead *of the Wise, because it was believed that it was akin to* Saturn, **who devoureth his Children as this doth Metals. Many other names have been given to it, which need not be mentioned here.** *They have laboured with great application to find out the Philosophers stone by this Mineral.*
>
> *— Nicolas Lémery*

This acknowledgement by the Dufrenoy couple is simply a rediscovery of what alchemists have always alluded to in their writings. The important question to be answered is; just why was antimony so important to Alexandrian *chrysopœia* in Maria's tradition? Was it simply a matter of religious and cultural correlations, or could it be something more tangible? This question too is answered by the Dufrenoys:

> *Antimony is recorded by Geilmann to be a constituent of Roman bronzes from the first century on. Corrosion-resistant CuSb [copper-antimony] alloys have been prepared for several thousand years,* **and in our days,**

CHAPTER 10 - POWDER OF PROJECTION

> *patents are still granted for such alloys. Berthelot emphasizes that a patent was granted to Dingler in 1891 for alloying* 6 per cent of Sb [antimony] into a copper alloy having most of the characteristic features of gold. This alloy would be difficult to distinguish from gold without density determination, or exposure to high temperatures; *any such alloy may have represented the "gold" of the alchemists,* which *Albertus Magnus knew could not stand the test of fire.*

This revealing article highlights the possibility that the story of alchemical gold is in fact the story of antimony as an ancient chemical technology. Another fascinating revelation from the article clears up an additional long-standing mystery connected with the confection of the Philosophers' Stone. Some alchemical texts clearly state that not a single iota of metallic gold is required to confect the Stone and we find surviving examples of these recipes in Synesius Hermetic *White Stone* and Hollandus' *Stone of Saturn*. Other traditions unequivocally employ gold as a component such as Maria and Morienus' recipes and most other orthodox variants based on the archetypal template originating with her tradition. Still others present a two-fold perspective; Stephanos' white *Comprehensive Magnesia* and red *Chrysocorallos*, Khunrath's *Great Universal* and *Lesser Specific Stones* or Paracelsus' *Stone of the Philosophers* and *Tincture of the Philosophers*.

Antimonial bronze dates back to prehistoric times when copper and bronze-craft was still embryonic. It was a time when religious impulses permeated daily life and human endeavor. These unique antimonial-bronzes survived as sacred metallurgical traditions in Egyptian and Jewish cultures, while the rest of the ancient world favored commercial tin-bronze as a valuable commodity, hypothesized as an impetus second only to grain behind empire-building and conquest. Researcher Nissim Amzallag in his article *From Metallurgy to Bronze Age Civilizations* reveals that copper and antimony trade routes led to and converged in the Southern Levant. *Antimonial bronze* prestige and ceremonial artifacts from the *Nahal Mishmar horde*, possibly linked to the Ghassulian *Temple of Ein Gedi*, date from around 3,750 BCE and contain up to 12% antimony. During the Copper Age in the Holy Land, the settlement Tulaylat al-Ghassul overlooking the Dead Sea was home to the dominant culture in the region, which lasted about a thousand years. Ghassulian antimony was sourced from Ghebi in the southern Caucasus Mountains, intentionally imported and specifically employed to create the 416 cupreous religious artifacts for use in Ghassulian rituals. During the 4th millennium BCE the area from the Dead Sea in the East, to Gaza in the west, and south into the Sinai as far as Timna, appears to be the homeland of proto-industrial bronze metallurgy. By the time of Roman-period Alexandria when Maria inherited the art, *antimonial bronze* technology, known later to the Romans as *aes-stimmi*, was already a relic of great antiquity originating from a proto-*chrysopœia* that thrived for nearly four thousand years prior.

High quality alchemical gold could be produced by alloying 94% copper with 6% antimony, was corrosion-resistant, lustrous, hard yet malleable, and displayed the overall look of elemental gold. The inclusion of gold via the *powder of projection* resulted in a higher quality and likely much more valuable product with the added benefit that it could pass the touchstone test for gold. Nevertheless, this begs the question; why bother creating a powder of projection in the first place? Why not simply create an alloy by mixing metals or metal ores? F. Sherwood Taylor in *A Survey of Greek Alchemy* addressed the difficulties that confronted the ancients when creating alloys in former times:

> *These alloys are prepared by methods which seem unnecessarily complicated to us.* **The complication is due in part, at least, to the fact that** *the alchemist had no means of judging the purity of his materials or of finding out the composition of a satisfactory product.* **Small** *differences of composition often profoundly modify the colour and other properties of an alloy,* **and a chance success has often been attributed**

CHAPTER 10 - POWDER OF PROJECTION

to the use of some inert ingredient. The retention of such ingredients leads to the adoption of these complicated mixtures. The making of alloys is not easy even to-day, for, during fusion, volatilisation or oxidation removes such metals as zinc, arsenic, lead and mercury to an extent which cannot be certainly predicted. Thus slight variations in the conditions of fusion often alter considerably the appearance of the product.

The challenges identified by Taylor could largely be minimized by creating a *powder of projection* as an initial step in the overall process by processing stibnite by the exact same preparatory technique as was done for copper sulfide ores in bronze making. This involved roasting a metal sulfide in order to create a metal oxide.

1. Antimony must be transformed to an oxide form as an initial step. By creating *flowers of antimony* first, the antimony undergoes further purification as a necessary preparatory stage.
2. By creating a *powder of projection* prior to alloying with copper, antimony or antimonial-bronze, the gold-antimony alloy is reduced to a very small particle-size, possibly aiding gold particle dispersion.

Without a *powder of projection,* it is far more difficult to achieve consistent quality control. The making of *powder of projection* as a fundamental initial step results in a much higher quality alloy of controllable consistency and thus a finer product. In light of this, it becomes readily apparent that Maria and those in her lineage were deeply concerned with purity, reproducibility and unfailing quality. Far from being an experimental chemist, Maria can now be better understood as the consummate master-artisan, a *chrysopœian* bronze-smith in pursuit of nothing less than an output of excellence and consistency.

Such a technology appears to arrive on the scene in Alexandria fully developed, yet *antimonial bronze* held a prominent place in the culture, religion and politics of Egypt, Mesopotamia and the Levant for countless generations prior. *Antimonial bronze* pre-dated Alexandrian alchemy by millennia, and was encoded in myth, symbolism and religious expression associated with Ugaritic and Canaanite (Kenite) Semitic cultures throughout the region. Although scholars have long-since identified correlations between constellations, planetary bodies, their corresponding colors and metals, tremendous opportunities remain for exploring the connections between metallurgy, socio-political economic structures in antiquity and their key roles in formative religion. Nissim Amzallag touches on this in the article *From Metallurgy to Bronze Age Civilizations*:

> ... the production of copper in a furnace was probably interpreted as a process of creation of a matter previously unknown, **since the oxide ore from the southern Levant is totally devoid of native copper. In such a context, it is not surprising that the** smelters became invested with demiurgic (magical) powers, encouraging them to transcend the boundaries of "existing matter," the universe given by the gods. **As many ancient mythologies bear witness,** this "Promethean attitude" is probably the source of the civilizing impetus accompanying, from its origin, the spread of furnace metallurgy.

To date, relatively little is known of metallurgy's central importance and influence as it pertains to mythico-religious, artistic and literary expression as a crucial adjunct to the rise of complex societies and social elites. Greeks called the Canaanites the Phoenix (Φοίνιξ) people, later known as the Phoenicians. In Canaanite literature from Ugarit, modern Syria, a poem probably originating sometime between 1,450 and 1,200 BCE provides insight into ritualized gold and silver making with references to the Canaanite god of metallurgists and their supreme deity El-Elyon:

CHAPTER 10 – POWDER OF PROJECTION

> *Heyan [Kothar-wa-Khasis] goes to the bellows*
> *In the hands of the craftsman are the tongs*
> *He casts silver*
> *He casts gold*
> *He casts silver by thousands*
> *Gold he casts by myriads*
> *He casts day and night*
> *El's crown of 2 myriads*
> *El's crown studded with silver*
> *Adorned with red gold*

Written accounts describe bronzes as fine as gold being equally valuable or even more so than genuine gold by weight. Anyone in possession of this commodity or its trade secret could literally manufacture fortunes. There are practical and ideological reasons for unique bronzes being this valuable. One of the clearest descriptions of *Corinthian bronze* in antiquity comes from accounts of massive doors and pillar-caps at Herod's Temple in Jerusalem. Solid elemental gold would be an incredibly unsuitable medium for the manufacture of temple pillars or large gates; gold being far too soft and heavy for such applications. Cicero and Plutarch address the fact that, unlike other copper alloys, *Corinthian Bronze* was corrosion-resistant. The advantage that corrosion-resistant *gold-antimonial bronze*, superficially identical to elemental gold, offers is that it is much lighter and harder than gold and could hold a sharp edge thus protecting it from surface damage. Temple engineers in Alexandria, the Levant and around the Mediterranean needed a substance that was durable, yet displayed all the characteristic features of gold to adorn temples and shrines, create pillar-caps, statuary or mechanical idols that could move and talk such as those crafted by Heron designed for Alexandrian temples.

From a contemporary perspective, such alchemical gold is artificial and it might seem strange that it could demand a price near to or greater than elemental gold by weight. To understand this phenomenon better, one must adopt the hermetic or gnostic mindset of Alexandria's educated or religious class fully initiated into the mystery tradition. Alchemical gold was indeed artificed by a master artisan and therefore artificial, but the notion of artificial carried entirely different connotations in antiquity than it does today. The hermetic-gnostic tradition takes the position that man, as a creative being, is co-partner in unity with the divine. They posited that God created the natural order, the earth and the immutable laws that govern creation. Yet, he left opportunities for humanity who, upon coming to a deep understanding of the natural order, might work with all that has been created in order to exalt it to a higher expression via resourceful human output. To put it another way, God did 90% of the work, and it was left to those philosophers initiated into the mysteries, to finish the remaining 10% with the caveat that man's productivity be in harmony with the natural order. By first creating a powder of projection and then creating alchemical gold, silver, and royal purple patinated bronzes or diamonds and rubies with it, was the highest exaltation of naturally occurring source materials. Art and artifice applied to reproducing or improving the most valuable elements in nature such as gold or gems, when viewed through this ideological framework, was among the highest manifestations of man's creative and divine nature.

The question then becomes ... Where are the remaining examples of such treasures? An excellent question indeed, yet one might also ask where the magnificent gold riches of Croesus of Lydia, Nebuchadnezzar II of Babylon or Akhenaten of Egypt lie; only fractions of which have ever been recovered. However, the question must be posed in consideration of the darker aspects of humanity's animal nature. The scarcity of copper and bronze artifacts from the Copper and early Bronze Ages is mainly due to continuous recycling of prestige artifacts and utilitarian tools over many centuries. Gold in any form is subject to recycling for a number of reasons. From the time of city-state civilization-building in the ancient

CHAPTER 10 - POWDER OF PROJECTION

near-east to the age of empire and conquest culminating in Roman and later medieval Islamic Empires, it was common practice for conquering invaders to confiscate all valuables, especially anything resembling gold. All metals were smelted either on-site or alternately transported home where the bounty entered the local economy or treasure store thus making it susceptible to further invasion. Edward Gibbons, in *The Decline and Fall of the Roman Empire*, recounts just such an act perpetrated by Emperor Maximinus Thrax sometime between 235 – 238 CE for little other reason than to support his decadent lifestyle:

> *By a single act of authority [by Emperor Maximinus Thrax], the whole mass of wealth was at once confiscated for the use of the Imperial treasury. The temples were stripped of their most valuable offerings of gold and silver, and the statues of gods, heroes, and emperors, were melted down and coined into money.* These impious orders could not be executed without tumults and massacres, as in many places the people chose rather to die in the defence of their altars, than to behold in the midst of peace their cities exposed to the rapine and cruelty of war.

Another poignant example will suffice to illustrate this practice in antiquity; an event that occurred during the early Alexandrian period. On July 1, 69 CE Vespasian became Emperor of Rome and his son Titus, serving as military commander under his father in Judea, was charged with ending the *Great Jewish Revolt*.

Herod had completely rebuilt and enlarged the *Second Temple of Jerusalem* around 20 BCE, employing 10,000 workers and 1,000 wagons in the process. The temple complex doubled in size and featured various courts, gates and gleaming expansive cloisters. The Temple itself stood surrounded by extensive walls with massive pillars featuring pillar-caps and gates of *Corinthian bronze*. The pillaging and total destruction of the temple yielded Roman plunder worth tremendous untold fortunes. The various plundering of Jerusalem's temples throughout history according to tradition and biblical accounts has always been depicted as being exhaustive and complete.

Titus sacked Jerusalem in 70 CE, destroying the city and Herod's Temple along with it. He was awarded the Arch of Titus as commemoration for his feat. The Arch pictographically displays the treasures looted from Herod's Temple being paraded in triumph through the streets of Rome as was custom, before being placed in the Temple of Peace. There they remained until Rome was sacked by Genseric the Vandal in 455 CE, who according to Procopius:

> ... *placing* an exceedingly great amount of gold and other imperial treasure *in his ships sailed to Carthage,* having spared neither bronze nor anything else whatsoever *in the palace [including golden-bronze roof tiles from the Temple of Jupiter Capitolinus] ... and* among these were the treasures of the Jews.
>
> – *Procopius,* The Vandalic War, III.5.3, IV.9.5

In 614 CE, during the final phase of the Byzantine-Sassanid war, Jerusalem was besieged and captured by the Persians who, according to Antiochus, slaughtered all the Christians and burnt any remaining church. Unfortunately, this sort of post-conquest destruction, along with widespread looting, is the rule rather than the exception and it should come as no surprise that very few *Corinthian bronze* or alchemical gold artifacts have surfaced from these volatile areas; likely succumbing to the smelting-flames of the victors. During these turbulent times, a nation's treasures as well as entire libraries disappeared, save for a few scrolls found secreted away in desert caves or treasure buried in tombs unknown and untouched by the invading forces or subsequent looters. None of the treasures from the first or second Temples of Solomon, Herod's Temple and relatively little of the magnificent metallic statuary from Greek or Alexandrian temples has ever been recovered. Looting also took its toll with relatively little of the great wealth of ancient Egypt, in terms of precious metals and gems being recovered and this is just a small sampling of what has been lost to history. Alchemical gold during turbulent ages of conquest and empire building did not stand a chance of survival except in legends, literary accounts of Alexandrian alchemists or secret hoards yet to be discovered.

Maria's Body of Magnesia

> *Mari, the first, says: Copper burnt with sulfur [white sulfur, i.e. flowers of antimony],* **treated with the oil of natron [as a fluxing agent], and** *recovered after having undergone the same treatment several times, becomes excellent gold, and without shadow. ...according to experience, by burning the copper [first], the sulphur produces no effect. But* **if you burn the [white] sulphur [first], then it not only renders the copper without blemish, but also makes it approximate gold.**

In the passage above, the term *melt* has been translated as *burn, burning* or *burnt*; these terms are also synonymous with *fuse, fusion* or *alloy*. Her method was to *melt* the entire fusion multiple times with *natron* as a fluxing agent. The term *sulfur* is in reference to Assyrian *kibritu* and Arabic *kibrīt* (كبريت), meaning sulfur but used by Maria in the Dialogue with Aros to indicate *flowers of antimony*. This process removes impurities and results in a higher quality product. The terms *shadow* and *blemish* are synonymous with corrosion. She is explaining that copper is brought into fusion with antimony as a first step before being tinged by the *elixir* in the second step.

> *This was graciously revealed to me by God, to know that the* **copper is first burnt with the sulphur [kibrīt; flowers of antimony], then with the body of magnesia [elixir; ferment]; and one blows it [keeps it in fusion] until the sulphurous parts escape with the shadow: [then] the copper becomes without shadow [corrosion-resistant].**

The passage above appears to suggest that *antimonial bronze* is brought into fusion with the *body of magnesia*. Magnesia (magnisía; μαγνησία) is commonly known as magnesium oxide. In an alchemical context however, *body of magnesia* is used as a cover-term for the *ferment, starter* or *leavening agent* such as *elixir, powder of projection* or synonymously, a small portion of ready-made *alchemical gold* from a prior batch; parallel to pre-fermentation *starter* customarily used in bread baking and beer brewing when starting a new batch.

Stephanos differs in his usage as he identifies both the *body of magnesia* (i.e. flowers of antimony) and *etesian stone* (Ios) both as *powders of projection*:

> *O inscription of the threefold triad and* completion of the universal seal, *body of magnesia* by which the whole mystery is brought about ...

> This is the etesian stone. *With these it is called by all names. It is the porphyry which is found in* the purple mineral, the purple-coloured substance ...

In the passage below preserved by Zosimus and attributed to Maria, *the body of magnesia* is employed in an alternate context to indicate *alchemical gold* as the composition of *antimony, Ios* and copper respectively:

> *The body of magnesia is the secret thing* which comes from [our] lead, from etesian stone, *and from copper.*

Body of magnesia may also be a double-entendre cover-term for the *powder of projection* and / or a reference to historical Corinth or Magnesia from Greek antiquity, where *body of magnesia* = product of Magnesia (Corinth), i.e. *Corinthian bronze*.

> *I am not speaking of ordinary lead, but of our black lead. Behold, how one prepares black lead [stibnite]: it is by cooking that one reaches [the purification of] black lead.*

> *Molybdochalkon is the etesian stone.*

CHAPTER 10 - POWDER OF PROJECTION

It is molybdochalkon which you must tint, by projecting on it **the motaria** *[ferment; raising] of yellow sandarac [powder of projection], so that the cooked gold should no longer exist [only] potentially, but actually.*

Molybdochalkon, used above as a cover-term, translates literally to *lead copper* (molyvdos chalkos; μολυβδος χαλκος), yet Maria clearly maintains that her lead is not ordinary lead, but rather "our black lead", meaning stibnite and *cooked* or *prepared* black lead meaning *flowers of antimony*. An alternate interpretation where *molybdochalkon* = *antimonial-bronze* is closer to the intended meaning.

⚨ = antimony + copper undergoing first fusion (burning)

☿ = antimonial bronze fused (*"fixt"*; burnt)

She is emphasizing the process of creating *antimonial bronze* as a preparatory stage prior to tingeing with the *powder of projection* or fermenting with *starter*, which served as a type of *ferment* or *leavening* agent. It must also be pointed out that the *starter* can also take the form of a small portion of ready-made alchemical gold (*molybdochalkon; etesian stone*) added into the antimony-copper fusion as alluded to in the final passage above. The term *projection* in *powder of projection* refers to the method by which the powder is introduced to the molten metal. The appropriate amount of *powder of projection* by weight was wrapped in beeswax and added to molten *antimonial bronze* in a crucible to achieve fusion.

According to Zosimus' interpretation, Maria used *body of magnesia* as a cover-term for alchemical gold and *etesian stone* as a cover-term synonymous with *molybdochalkon* to indicate *antimonial bronze* as a type of ferment. Stephanos uses the terms *body of magnesia* and *etesian stone* as specific *powders of projection*. In any case, they are describing the process of adding a type of ferment or leavening agent to *antimonial bronze*; the result being alchemical gold. Essentially, Maria was not transmuting copper to gold, but rather she exalted molten *antimonial bronze* to *Corinthian bronze* via the ferment of *powder of projection* or *etesian stone starter*. Alchemical silver was created via silver-based *powder of projection*. Pliny explained that *hepatizon*, or liver-purple bronze, was the second most valuable bronze after Corinthian, but that the secret of its manufacture had been lost. *Hepatizon* was the result of a purple-black patina achieved by surface treating silver or gold bronzes with ammonia or the vapors of an arsenic sulfide ore such as orpiment (As_2S_3) or realgar (α-As_4S_4). It is possible that the traditional alchemical color scheme was actually gradations of learning the Art of *Corinthian bronze* craft beginning with stibnite (Black) refinement to antimonial bronze, then crafting alchemical silver (White), followed by alchemical gold (Yellow) and finally royal-purple hepatizon bronze advanced patina techniques (Iosis – Tingeing/Purpling).

Maria most likely created her bronzes by means of artisanal crucible technique. Crucible work has historically been associated with smelting high-grade metal oxide ores. The crucibles used for this purpose in antiquity varied in size ranging from around 8-20 cm in diameter, 6-15 cm deep with a thickness between .5-2 cm and constructed of low-porosity material. In the southern Levant, crucible work was replaced by bowl furnaces much larger to facilitate mass production of commercial tin-bronze manufactured by Canaanite tribes in the region.

Throughout antiquity, gold was understood much the same way that bronze is today, meaning bronze by any number of alloys was still bronze. Likewise, gold to a certain degree of admixture, look, feel, corrosion-resistance, etc. was still considered gold. This sentiment was expressed in the 17th century by a leading figure in the scientific revolution, Royal Physician and Academician, Samuel Cottereau du Clos, who held that:

> *… the alloys that one makes of silver and copper with gold* change the karat of the gold without creating a new species.

That is to say, gold is generally considered gold regardless of the karat, and this must be true even if the karat is extremely small yet the product retains all the characteristics of a higher karat variety of the species. From this perspective, the notion that alchemists were in fact gold-makers of a specific type of gold appears to be justified, yet grossly misunderstood by later generations due primarily to the technological and cultural disconnects resulting from Diocletian's edict criminalizing *chrysopœia*. (See Appendix B for precise *chrysopœia* methodology)

> ... the changing of matter into something better has nothing incredible about it, since with us too, those who know matter take silver and tin, remove appearance, melt together and color, ennoble matter and produce gold, even the most beautiful.
> – Aeneas of Gaza, Theophrastus

European Alchemical Gold

> The truth of each thing, as Avicenna says in his Metaphysica, is nothing else than the property of its being which has been established in it. So that is called true gold which has properly the being of gold and attains to the established determinations of the nature of gold.
> – Thomas Aquinas, Quodlibeta VIII, q. 3

Thomas Aquinas essentially adopted the stance that if alchemical gold displays the general form and nature of gold, its essence, it could rightly be considered a species of gold: Aquinas once asserted, *"...the aim of the alchemist is to change imperfect metal into that which is perfect"*. This was a view held by alchemists of each tradition. During the 13th century, Thomas was sent to study at the University of Paris where he befriended Albertus Magnus, a German Dominican friar, bishop and then Chair of Theology at the Parisian College of St. James. Albertus Magnus was known for advocating an accord between science and religion. They each became accomplished alchemists yet held slightly different views regarding alchemical gold. Aquinas accepted alchemical gold as valid due to its attaining the nature of gold, while Albertus Magnus proved that alchemical gold failed *fusion fire assay* and was thus not identical to elemental gold. Holmyard quotes Magnus as follows:

> *I myself have tested alchemical gold and found that after six or seven ignitions it was converted into powder.*

This example serves to highlight two opposing views prevalent in Europe regarding alchemical gold. The question then is... Did initiated alchemist-adepts know that alchemical gold differed from elemental gold, or did they believe in the possibility of genuine transmutation? If transmutation was indeed possible, untold wealth lay in store for the adept or his patron. If they realized that transmutation yielded an artificial gold however, the subject of motive and technique demand further examination. As early as the 11th century CE, a Persian universal philosopher and physician began to openly question and debate the validity of gold-making. The theory of transmutation had been on shaky ground during the intense period of Islamic scientific scrutiny over the previous four centuries and it came to a head with Avicenna who, in an attempt to discredit the theory of transmutation, stated quite plainly that:

> ***Those of the chemical craft know well that** no change can be effected in the different species of substances, though they can produce the appearance of such change.* – Ibn Sīnā (Avicenna)

Avicenna's position sided with that of nature over art. Nature was God's creation and any creation by man was inferior to that of nature. He expressed this in the following manner:

> *Art is weaker than nature and does not overtake it, **however much it tries**.* – Ibn Sīnā

This modest and direct perspective is quite at odds with Alexandrian notions of exalting nature, thus elevating man to a position of being capable to manipulate and therefore in a manner of speaking, control or improve upon nature. From Avicenna's window of observation, humanity was subservient to the natural order and bound by its laws.

Alexandrian style copper-based alchemical gold had a few shortcomings; it was nowhere near the density of gold nor was it as malleable making it easily differentiated from genuine gold by European alchemists. During antiquity, it had served its purpose as an ornamental bronze highly prized to create adornments such as statuary, temple gates or pillars and pillar caps. To create an alchemical gold that could pass for gold currency however, European alchemists found that lead, mercury, silver or a combination thereof matched characteristics of mass and density much more closely than did Corinthian or *antimonial bronze*. It appears that European gold-makers were actively pursuing a closer approximation of genuine elemental gold to the degree that it could pass, at least superficially, for gold. Did alchemists believe that they had actually succeeded? For some, such as Thomas Aquinas, the answer is yes.

The standard accepted view is that alchemists observed nature and learnt from miners that metallic ores of varying composition could be found together with veins of metals. A certain fluid was observed that was identified in mines as the earth's natural solvent that facilitated the production of metals and that these metals were in organic geological evolution towards becoming gold in the furnace of the earth. This general theory was well stated in the late 15th century by renaissance Venetian scholar, poet and alchemist, Giovanni Aurelio Augurello:

> *The origin of the metals is the center of the earth... penetrated by the sun's rays and by other celestial rays which ripen and mature the assembled vapors that then pass upward and fill fractures in the crustal rocks.* Where the vapors are condensed and cannot move farther they solidify into those unripe metals *that fill veins in the earth's crust ... Finally with time Nature transforms these metals into gold,* silver, *copper, and so on ... – Giovanni Aurelio Augurello*

During the middle Ages, widespread notions circulated throughout Europe of the organic generation or growth of metals, particularly gold in the depths of the earth. The process of gold genesis and distribution is known today as *gold metallogeny*. Alchemists and miners exchanged theories of organic geology with one another, concluding that base metals underwent a slow process of transmutation or maturation into the noble metals. Alchemists believed that metals began their formation in a liquid solution called *gur* or *bur* based on observations of unctuous substances oozing out of fissures. According to their theory, these solutions solidified into ores and finally matured into gold.

Alchemists like Albertus Magnus observed and noted the precipitation of mineral matter from these solutions in their own jargon in an attempt to describe these geological processes, which at the time must have seemed rather mysterious. By comparing Albertus Magnus' interpretation of organic gold-formation with contemporary views, it becomes clear that he and others had observed gold's mobility based on the phenomena of ionic gold complexes and changes in stability. Lacking knowledge of atoms or ions, theirs was the best possible hypothesis for the period and actually shares some characteristics with modern *gold metallogeny* theories. A comparison between Albertus Magnus' gold deposit scheme and contemporary understanding reveals several parallels:

Albertus Magnus' View	**Contemporary Geological View**
In order to know the cause of all the things that are produced, we must understand that real metal is not formed except by the natural sublimation of moisture and Earth... *For in such a place, where earthy and watery materials are first mixed together, much that is*	**hydrothermal solutions** – *Mineral assemblages formed during hydrothermal alteration reflect the geochemical composition of ore-forming fluids. Gold is mainly transported in solution as gold chloride and sulfide complexes. Gold tends to be concentrated in the vapor phase of fluids at high temperatures and pressures. Au – As and Au – Sb associations are common in gold deposit. Native antimony and/or arsenic –*

CHAPTER 10 – POWDER OF PROJECTION

impure is mixed with the pure, but the impure is of no use in the formation of metal.

And from the hollow places containing such a mixture the force of the rising fume opens out pores, large or small, many or few, according to the nature of the [surrounding] stone or earth;

and in these [pores] the rising fume or vapour spreads out for a long time and is concentrated and reflected; and since it contains the more subtle part of the mixed material it hardens in those channels, and is mixed together as vapour in the pores, and is converted into metal *of the same kind as a vapour.*

native gold assemblages may precipitate *from hydrothermal fluids...*

metamorphic secretion – Most lode gold deposits are sourced from metamorphic rocks because it is thought that the majority of lode gold deposits are formed by dehydration of basalt during metamorphism. The gold is transported up faults by hydrothermal waters and deposited when the water cools too much to retain gold in solution.

oxidation and secondary enrichment - Laterite gold deposits are formed from pre-existing gold deposits during prolonged weathering of the bedrock. Gold is deposited within iron oxides in the weathered rock or regolith, and may be further enriched by reworking by erosion.

During late medieval Europe, international trade began to flourish due to the abundance of silk, spices and other exotic trade-goods arriving from India and the Far East. Gold and silver mining increased in an effort to meet commercial payment demands. It was due to these developments that Alchemists greatly benefitted from a growing body of firsthand knowledge of miners and smiths. By the time of the Renaissance, critical study of gold deposits led to increasingly advanced theories of *gold metallogeny* more in line with observed phenomena, yet these theories still retained Aristotelian and alchemical components. Secretion and emanation views prevailed based on observations of metamorphism, replacement processes and surface oxidation. Alchemists and miners recognized that *seeds of gold*, today known as *soluble ionic gold complexes*, were transported by hydrothermal solutions. Gold-carrying chemical complexes are at the core of modern *gold metallogeny* theories where thiosulfate, chloride and sulfate complexes eventually form metal sulfide deposits. In hydrothermal Carlin-type deposits such as those in Macedonia, gold occurs together with stibnite, cinnabar and other metal sulfides, known to alchemists as *atrament* or *vitriols*.

In light of the fact that metal deposits form due to very slow aggregation and/or oxide or sulfide dissociation, alchemical theory of organic geological growth and transformation of ores into gold was a valid observation. Their explanation was lacking, but only due to an immature theory of matter. Alchemical theory of organic geology was the best definition possible at the time, making the notion of creating the Philosophers' Stone and employing it in an attepmpt to produce gold, in a process that mirrors natural earth processes, rather reasonable. Bryant eloquently expresses this sentiment:

> *The whole process of "making gold" by Art is closely allied to that of Nature in the heart of the earth, only that the artistic process is exceedingly shorter and more efficacious.* – Delmar DeForest Bryant

Creating *powder of projection* involved reducing gold, dissolving it in antimony, joining it to a chloride solvent where it slowly aggregates and precipitates to the Stone, at which point it is understood to be a *ferment* or *seed of gold*. This closely parallels the natural process of *gold metallogeny* that occurs over geological time, yet an alchemist could simulate this process in a relatively quickened manner. Alchemical gold created via this *gold seed* or *ferment* was understandably viewed as a valid species of gold by alchemists such as Thomas Aquinas and others.

There is enough evidence however to indicate that a number of adept-alchemists such as Albertus Magnus and Petrus Bonus realized a profound difference between alchemical gold and elemental gold; a denial of which would discredit technical assaying abilities of skilled European alchemists and goldsmiths. It is quite possible that alchemical gold was sought after in Europe for the same utilitarian purposes as was done in Alexandria – luxury goods and adornment, the appearance of wealth or as a technology used in cathedral construction. Regardless whether alchemical gold was assumed a genuine gold product or artificial, it was safe that nobody would chip a piece away from a holy statue in the house of God, or destroy an item displayed in a palace of nobility simply to assay the gold content. The function of alchemical gold was likely to display the appearance of wealth in a decorative form to be admired at a distance:

> ***Now*** *alchemistic gold and silver cannot exhibit the same ultimate arrangement as natural gold and silver; consequently, they are not the same thing.* ***Hence,*** *if there be such a thing as alchemistic gold, it is specifically different from ordinary gold.*
> – Petrus Bonus, The New Pearl of Great Price

To gain a more comprehensive view of the subject, an opposing perspective must also be examined. Acknowledging that some alchemists such as Petrus Bonus were aware of the difference begs the question; what purpose would alchemical gold serve European alchemists, nobility and clergy that actively pursued it? The hypothetical answer that follows is not an accusation, but rather an unfettered pondering, a musing of possibilities and is not intended as conspiratorial allegations of intrigue or sedition. What follows is merely an unrestricted exploration of the other side of the debate supported by a history of human greed and the inevitable thirst for wealth and power consolidation historically exhibited by ruling classes.

Trade and economic expansion was bolstered by the invention of coinage. The economic breakthrough of a gold standard ultimately developed as an inevitable result of dividing gold into manageable pieces of equal weight and purity. Yet as long as gold coins have been in existence, so has recoinage and debasement schemes. Debasement is the practice of lowering commodity currency such as gold and silver coins gradually over time resulting in financial gain for the sovereign at the expense of the populace. The ruling body that controls debasement enriches itself while inflation affects the citizenry. If an alchemist could create a form of alchemical gold convincing enough to dupe both citizens and trade-partners, it promised a sure debasement method with great potential for generating and consolidating wealth. If the alchemical gold or silver used was of such a quality that it may perhaps go unnoticed and unchecked, it offered a relatively risk free venture for those involved in the deception; a practice with very ancient origins.

The phenomenon of replacing good currency with debased coins was first publicly parodied by Greek playwright Aristophanes in a comedic play known as *The Frogs*, presented at the Lenaia festival of Dionysus in 405 BCE. In Europe, Nicole Oresme discussed government debasement of coinage during the early 14th century. A short while later, Egyptian Islamic historian al-Maqrizi wrote of hoarding gold and simultaneously replacing it with debased copper coinage during the late medieval period. Finally, Nicolaus Copernicus followed forty years later by English financier for the Tudor dynasty Sir Thomas Gresham had each independently detailed this practice so clearly that it became an economic principle known as the *Copernicus-Gresham Law*, which states:

> When a government compulsorily overvalues one type of money [debased coinage] and undervalues another [gold or silver coinage], the undervalued money will leave the country or disappear from circulation into hoards [of the facilitators, those in the know], while the overvalued money will flood into circulation.

For those chiefly concerned with wealth and power building and consolidation, the technology of alchemical gold held great potential for a sophisticated coin debasement program, or as a form of payment for imports; essentially the potential for exporting debased currency abroad. This phenomenon had already been identified and addressed in Greece

CHAPTER 10 - POWDER OF PROJECTION

approximately two thousand years prior to Copernicus and Gresham and remained a valid concern throughout the history of many cultures.

Coincidentally alchemy arrived in Europe right around the same time that banking began in Florence, Venice, Milan and Western Europe. Magnificent hoards of gold, silver and riches were in the possession of the church, nobility and eventually merchant gold-depository banks. This resulted in very large stores of gold with only a fraction entering and leaving the reserve each day. The volatile combination of Bankers and large stores of unused wealth can be problematic even during the best of times, with bankers historically devising all manner of moneymaking ruses from such circumstances. Gold was held as collateral either privately, such as in a church or nobility keep, or in one of the many developing gold-depository banks. Adding alchemical gold to these accounts, thus bolstering equity in the form of additional stores of gold, would have offered the possibility to create an artificial facade of wealth or increased lending potential for the institution.

This notion is not as far-fetched or conspiratorial as it may first appear. On August 2, 2012, the Los Angeles Times newspaper released a story detailing the first known instance of a central bank having its gold audited, the U.S. Federal Reserve in this case, including drilling into the gold bars to extract samples for assaying. The treasury extracted over $110,000 USD worth of gold-bullion samples to be professionally assayed, destroying 1-1.5 grams per sample in the process. During antiquity, it is highly unlikely that any audit such as this would ever have posed a threat to nobility, clergy or bankers in possession of large stores of gold bullion. However, this strongly suggests the existence of a precedent and acknowledges valid concerns shared by the public and auditors over the possibility of debased gold bullion held in a central bank.

Throughout the history of European alchemy, the noble art of gold-making was often secretive, highly restricted and in some instances punishable by death for all but court or clergy sponsored alchemists. Those powers in control of currency valuation, circulation and guarding large gold-bullion stores have historically demonstrated the greatest interest in alchemical transmutation. Alchemical gold in Europe was far less about an unkempt alchemist in some sooty clandestine lab attempting to transmute lead into gold in an effort to ascertain the mysteries of matter, and more about the tremendous energy directed towards wealth and power accumulation by those establishments authorized to regulate and sanction the right to practice alchemy. European alchemists' efforts were largely patronized and financed by nobility or sanctioned by the church and their alchemical gold featured lead and mercury, thus more closely approximating gold-bullion. A small number of European alchemists were indeed adepts; they could create the *powder of projection* and from it, alchemical gold. Successful alchemical transmutations were often demonstrated by casting coinage.

Gold has been taken out of circulation as a currency for some time with many nations moving away from a gold standard. Historical inquiry may never expose the amount of wealth amassed owing to the use of alchemical gold coinage debasement or falsified collateral throughout history, or expose how much alchemical gold may still exist in the form of private hoards that remain unrevealed due to the threat of modern analysis. Irrespective whether they believed that their efforts produced elemental gold or a unique alchemical gold product, European alchemists of old were likely content and fortunate just to have their efforts legally sanctioned or patronized; concerning themselves more with mastery of their art than the power-plays and politics of their patrons.

What dost thou not compel mortals to do, accursed thirst for gold!

– *Virgil*, Aeneid

CHAPTER 10 - POWDER OF PROJECTION

ALCHEMICAL GEMS & STAINED GLASS

בינה מקום זה ואי תמצא מאין והחכמה יב	*Job 28:12* But where shall wisdom be found? And where is the place of understanding? ...
ותמורתה וזכוכית זהב יערכנה-לא יז פז-כלי	*Job 54:17-19* Gold and glass cannot equal it, nor can it be exchanged for jewels of fine gold.
חכמה ומשך יזכר לא וגביש ראמות יח מפנינים	No mention shall be made of coral or of crystal; the price of wisdom is above pearls.
טהור בכתם כוש-פטדת יערכנה-לא יט תסלה לא	The topaz of Ethiopia cannot equal it, nor can it be valued in pure gold.

As early as the 7th century BCE great wisdom was compared with gold, crystal and gemstones. The story of alchemical gems dates back to Egyptian, Phoenician, Babylonian and Assyrian glass technologies revealed as artifacts or recipes recorded on cuneiform clay tablets several thousand years ago. Alchemical gemstones were unique colored glass, crystal and crystalline objects of tremendous value created by gifted artisans and serve as a classic example of art imitating nature in alchemy. Maria appears to have been very skilled with glassmaking; she championed the use of glass vessels and according to Zosimus created them herself. The step from artisanal glassmaking to artificial gemstones is a small one and is the origin of alchemical gems and later the stained glass windows adorning Europe's great cathedrals.

Bronze and glass crafts share a common chemical technology – each art demands that the artisan be skilled at sourcing, identifying and / or producing relatively pure mineral and metal oxides artisanally. *Whitening the stone* is just a small example of metal oxide production, the product of which found applications in both bronze and glassmaking and therefore was a foundational technique likely mastered by an apprentice early on in one's training. The story of glass and artificial gemstones in antiquity is the story of mineral and metal oxides. When Maria instructed Aros to *make a glass of it*, she meant fuse the gold-antimony oxide and allow it to cool and solidify into a glass.

Like bronze-crafts, glassmaking was already an extremely ancient art by the time Maria inherited the tradition. Many writers in antiquity discussed various aspects of glassmaking:

> *The story is, that a ship, laden with nitre, being moored upon this spot, the merchants, while preparing their repast upon the sea-shore,* finding no stones at hand for supporting their cauldrons, employed for the purpose some lumps of nitre [natron] which they had taken from the vessel. Upon its being subjected to the action of the fire, in combination with the sand of the sea-shore, they beheld transparent streams flowing forth of a liquid hitherto unknown: this, it is said, was the origin of glass.
> – *Pliny the Elder,* Natural History, Book XXXVI Chapter 65

In a somewhat more sober account written around the time of Maria, Josephus details the location of glassmaking sands at the mouth of the Belus River:

> *This Ptolemais is a maritime city of Galilee, built in the great plain. ... The very small river Belus runs by it, at the distance of two furlongs; near which there is Menmon's monument, and hath near it a place no larger than a hundred cubits, which deserves admiration; for* the place is round and hollow, and affords such sand as glass is made of; which place, when it hath been emptied by the many ships there loaded, it is filled again by the winds, *which bring into it, as it were on purpose, that sand which lay remote, and was no more than bare common sand, while* this mine presently turns it into glassy sand. *And what is to me still more wonderful, that* glassy sand which is superfluous, *and is once removed out of the place, becomes bare common sand again. And this is the nature of the place we are speaking of.*
> – *Flavius Josephus,* The Wars of the Jews, Book II, section 188

CHAPTER 10 – POWDER OF PROJECTION

Despite Pliny's anecdote, the debate as to whether glassmaking originated with the Egyptians or another culture such as the Phoenicians, Babylonians or Assyrians remains unresolved. Solid glass beads that date from the 4th millennium BCE have been unearthed at archeological sites in Egypt, whereas chemical technologies arose in Sumeria prior to the advent of written records. Cuneiform clay tablets detail Babylonian perfumery and Assyrian glassmaking technologies from the 17th and 7th centuries BCE, the most interesting of which is a precise recipe for a decorative red glass made with gold and antimony called *red-coral* glass. These tablets reveal levels of critical observation and refined technique combined with telltale signs of spiritual or religious ritual, cover-names and levels of initiation typical of artisans, guilds and temple workshops of antiquity. Most importantly for this discussion, they provide archeological evidence for the types of chemical technologies that would later be described by the earliest Alexandrian alchemists as their Art – gold, silver and purple bronze making and the creation of alchemical glass gemstones with their trade secrets encrypted in cover-terminology.

Notable British archeologist, Assyriologist and cuneiformist R. Campbell Thompson excavated a number of Mesopotamian sites during the early 20th century unearthing some of the earliest glassmaking manuals, the oldest of which dates from circa 1,700 BCE. His find revealed an interrelationship between the priesthood, glassmaking as a temple craft, a guild of workers with specialized knowledge and the use of jargon and cover-terms antecedent to alchemy by nearly two millennia. All of these characteristics would eventually become associated with alchemical traditions:

> *Perhaps the most interesting text in all the cuneiform chemical tablets is* the glass-maker's document of the Seventeenth Century B.C. *which I have already mentioned,* written in a beautiful hand, and in a perfect state of preservation. It is first, I think, a duty to record his name as that

457

of the earliest glass-maker known, Liballit (?)-Marduk, the son of Ussur-an-Marduk, a priest of Marduk at Babylon.

Not the least interesting point is that the writer has anticipated the cryptic jargon of the alchemists of a later period, whereby he hopes to conceal his knowledge from those not in his Guild. He has made a point of using the rarest cuneiform signs, and of giving the most recondite values to such well-known signs as he uses.
– R. Campbell Thompson, A Survey of the Chemistry of Assyria in the Seventh Century B.C.

Ancient Egyptians accurately referred to glass as melted stones and their glass beads mimic gems so precisely that they can be difficult to distinguish from their authentic counterparts. During antiquity, the primary value of colored glass was as personal adornment, an example of which is the vulture collar laid on the chest of Tutankhamen's mummy. It is called the necklace of the sun and is crafted with gold, cornelian and glass beads, which suggests that glass held the same allure as precious metals and gems. Egyptian glass beads were exported to markets throughout the ancient world as far away as China. Apart from jewelry, the most accomplished Egyptian glasses from the 18th dynasty (1,549 to 1,297 BCE), were very nearly transparent indicating an already high degree of sophistication.

By the mid-7th century BCE, Assyria had become the powerhouse of the Near East and the eastern Mediterranean coastal region. The Assyrian homeland was located in Northern Iraq with a socio-cultural horizon of influence that included most of Syria and the eastern Mediterranean, Egypt and large stretches of Eastern Turkey and Western Iran. Babylon was Assyria's neighbor in southern Iraq and the two cultures shared much in common including relatively advanced chemical technologies. Babylonian texts reveal a sense of scientific objectivity and achievement in the fields of zoology, botany, medicine, mathematics and astronomy, yet each of these was infused with magic and mysticism to a certain degree. Babylonian perfume texts discuss crude plant extractions and distillation techniques that were adopted and improved upon by Islamic alchemists nearly two millennia later. Assyrian-Babylonian science and technology can be considered ancestors on alchemy's family tree as regards alchemical gemstones:

> *As applied chemists, there is little doubt that the Babylonians were highly skilled. Their apparatus was not elaborate but their objective approach to technology went a long way to make up for this lack.*
> — Martin Levy, Babylonian Chemistry

One of the most significant texts in the formative history of applied chemistry is a cuneiform tablet written by a glassmaker for the Assyrian King Ashurbanipal who reigned between 668 – 626 BCE. This and other glassmaking manuals were found along with Assyrian chemical dictionaries, which fortuitously allowed researchers to identify chemical ingredients. Some glassmaking texts were copies of originals kept at the Royal Library at Nineveh and the adjacent Temple of Nabu, yet now reside in the Kouyunjik Collection of the British Museum.

Of interest to this discussion of Alexandrian alchemy, are detailed instructions for constructing the two furnaces required for glassmaking, antecedents to the alchemists' athanor-style furnace. Furnace building demanded purity and was performed ceremoniously complete with offerings and magic ritual. The first furnace was a high-temperature *melting furnace* able to heat a *batch* for several hours or even days on end. The second type was a dual-purpose *fritting-annealing furnace*, known today as a type of *reverberatory oven* that operates at lower temperatures. As glass begins to cool and solidify, uneven cooling stresses the glass resulting in an inferior product. An *annealing oven* operates at a relatively low-temperature allowing for a more even cooling process; a practice still in use today.

In Alexandria, glassmaking evolved to a highly sophisticated art that incorporated a variety of intricate colorations, with the addition of gold resulting in deep red. Glass jewelry was fashioned into meticulous works of outstanding artisanship requiring the same attention to detail as any other form of jewelry making. Unfortunately, recipes for alchemical gems crafted in Maria's branch of *chrysopœia* remain to be found. An exploration of decorative glass used primarily for personal adornment or as artificial gems will provide insight into the origins of artificial gemstones and the technologies to create them that alchemists during the early Alexandrian period may have drawn upon.

Alchemical Gemstones in Antiquity

In Exodus 28:15-30, precious stones are chosen to adorn Aaron's gold *breastplate of judgment*. The first stone on the first row is called an *odem*, sometimes translated as ruby but more commonly as a *carbuncle* to indicate it was a red stone of a fiery glowing nature. In the book of Proverbs, wisdom and a virtuous woman suitable for marriage are both compared to rubies as a definitive measure of value. During antiquity, the ruby far overshadowed all other stones in value due to its blood-red color being associated with fertility, virility and resemblance to fire or a glowing ember. Cleopatra was fond of rubies as were many historical personages from antiquity. The Greeks called the ruby the *mother of all gems* and the Romans a *flower among stones*, more valuable than even diamonds. Diamonds and other colorless gems became fashionable only relatively recently due to advances in gem-cutting techniques and sophisticated marketing strategies.

In India, Sri Lanka and China the ruby corresponded to the sun and was believed to contain illuminating properties based on the observation that some rubies actually fluoresce quite noticeably. Aristotle's successor, Theophrastus, categorized gems by color and referred to any red gem by the term *carbuncle* to indicate the color of a glowing ember, where *karvo* (καρβο) originally meant burning charcoal. Pliny referred to the ruby by the term *carbuncle*, yet in a similar manner to Theophrastus, this term may have referred to any number of red gems. In antiquity, and this is especially true in ancient Egypt, color was often more important than the actual composition, therefore *carbuncle* served as a practical umbrella term.

By the Roman period in which Pliny wrote and around the period when Maria was believed to have been working, artificial gemstones were fraudulently sold as the genuine article. Pliny's writings reveal that glass and rock crystal, today known as colorless, transparent or translucent quartz, were keys to the counterfeiters' craft. He addressed this issue in detail in *The Natural History, Book XXXVII* chapters:

> [25] *In the first rank among these [fiery red gemstones] is carbunculus, so called from its resemblance to fire ...*
>
> [26] *Nothing is more difficult than to distinguish the several varieties of this stone... They are counterfeited, too, with great exactness in glass... and the weight of the glass counterfeit is always less. In some cases, too, they present small blisters within, which shine like silver.*
>
> [75] *... there are books in existence, the authors of which I forbear to name, which give instructions how to stain crystal in such a way as to imitate smaragdus [emeralds] and other transparent stones ... Indeed, there is no kind of fraud practised, by which larger profits are made.*

Theophrastus and Pliny were not the only chroniclers to generalize fiery red gems into a single category; Europeans typically referred to any red gemstone as a ruby until the early 19th century. The ruby has held a special allure in nearly every great civilization worldwide:

> The ruby is the world's most precious stone, and by far the rarest. Not even the diamond, with its unfathomable depths of light, nor the emerald, serene in its perfect greenness, is worth more, carat for carat, than the ruby.

> Yet for all its rarity, beauty, and passion, the ruby has always been a slightly sinister stone. It has none of the sublimity of the sapphire, the elegance of the emerald, the purity of the pearl, or the dazzle of the diamond. Of all the precious stones, the ruby is closest to a wild, living being: fiery, passionate, and dangerous.
>
> – Diane Morgan, Fire and Blood: Rubies in Myth, Magic and History

It is a relatively safe assumption that the most valuable type of alchemical gem Maria could have created would have been alchemical rubies. Unfortunately, no records have turned up for alchemical gems in her lineage, although Zosimus presents her as preferring the use of glass vessels and credits her with their design and manufacture. Just as Maria inherited an already ancient bronze-making technology, likewise artisanal glassmaking had begun in the region nearly two millennia prior to the period in which she operated. One of the most illuminating recipes for a red artisanal crystal product is preserved in an Assyrian glassmaking manual dating to the 7th century BCE. The recipe is for the production of [bah]-ri-e, interpreted by R. Campbell Thompson to be a type of *crystallin* called *Red Coral*. This formula allows for a critical look at ancient proto-alchemical gemstone materials and manufacture to understand their properties and purpose better.

[bah]-ri-e (Assyrian Red-Coral Crystallin)

zukû	clear glass previously prepared	60 shekels = about	99.260 %
[tus]kû	[zi]nc-oxide	16 carats = about	0.440 %
aba[ru]	antimo[ny]	10 carats = about	0.270 %
...mil'u	...of saltpetre	(estimated)= about	0.016 %
KÙ.[GI]	of go[ld]	½ carat = about	0.014 %
			100.000 %

The term *zukû* above was interpreted by Thompson as *clear glaze*, yet is likely what is referred to in modern terms as *frit*, meaning *fried* in Italian. The term *frit* can mean many things today, but during antiquity, it meant bringing the clean and prepared *glass former* to a molten state and allowed to cool before being ground to a powder. This process parallels Basil Valentine's method of preparing *glass of antimony* where *flowers of antimony* are brought to a molten state and allowed to cool before being ground to a fine powder. This is yet another example of an alchemical operative technique applied equally to both metallurgy and glassmaking. During the *fritting* stage, it is important that the heating environment be smoke-free. Techniques for creating *frit* in antiquity involved hammering and grinding or by bringing the glass to a red-hot state followed by quenching it immediately into cold water to cause fracturing due to thermal shock. In a modern sense, *frit* is a semi-finished crystalline product obtained by calcining in the *fritting-annealing furnace* below melting point. The aim of *fritting* was to make the *glass former* react with the *fluxes,* transforming the silica into alkaline silicates, which melt at greatly reduced temperatures and to remove carbon dioxide that forms through the decomposition of the carbonates. *Zukû* in the context of this recipe likely refers to a high-quality *frit* that has been reduced to a powder indicating that this recipe is actually a coloration step meant to take place in the *melting furnace* during the melt stage.

In the recipe, zinc oxide and antimony serve as fluxes. The form of antimony generally revealed in artisanal glass analysis is antimony pentoxide, yet this may occur because heating *flowers of antimony* causes it to absorb excess oxygen thus converting it to the pentoxide form. Adding a flux significantly reduces the melt temperature of pure silica. Today glassmakers create lead-free crystal known as *crystallin* by replacing lead oxide with zinc oxide. For this reason, the Assyrian recipe above appears to be an ancient crude lead-free *crystallin* product based on its composition, which suggests that the desired outcome was not a glass but rather an Assyrian alchemical gemstone or colorant. The name *red coral,* along with its gold content, suggests an artificial carbuncle.

When zinc oxide is used in glass, it imparts a unique combination of properties. It affects thermal expansion, adds greater brilliance and luster and acts as a stabilizer protecting against deformation caused by stress. Bubbles in the *melt* affect the product but this was often addressed by vigorous stirring during the *fritting* stage or by adding antimony. Antimony decolorizes by interaction with any existing metal oxide impurities. Gold served as the coloring agent to impart a ruby-red coloration. Since gold occurs in the matrix as a nanoparticle rather than an oxide, antimony would not affect the ruby-red color that gold imparts. Antimony is also an opacifier, meaning it makes the glass less translucent, yet this may have been counter-acted by the addition of saltpeter. Interestingly, zinc oxide was also used in antiquity, although rarely, to provide a red tint to glass.

The Assyrian red coral glass recipe provides a basis for understanding glass manufacture as it applies to decorative or colored glasses in antiquity. Artificial gemstones made of glass or rock crystal require the following components:

Glass Former

> Between Acê [Acre] and Tyre is a sandy beach, which produces the sand used in making glass. Now the sand, it is said, is not fused here, but is carried to Sidon and there melted and cast. Some say that the Sidonians, among others, have the glass-sand that is adapted to fusing, though others say that any sand anywhere can be used. I heard at Alexandria from the glass-workers that there was in Ægypt a kind of vitreous earth [perhaps flint] without which many-coloured and costly designs could not be executed, ...also, it is said that many discoveries are made both for producing the colours and for facility in manufacture, as, for example, in the case of glass-ware, where one can buy a glass beaker or drinking-cup for a copper. – Strabo, Geography Book XVI 2:25

The area described by Strabo in the passage above can be pinpointed as the Belus, modern Na'aman River (נחל נעמן) in northwestern Israel, where it enters the sea near Acre. St. Isidore of Seville described this area in his

CHAPTER 10 - POWDER OF PROJECTION

Etymologiæ as the birthplace of glassmaking. Scientists have analyzed the sands from this region and have discovered that it is composed of just the right balance of silica and other trace elements that qualify it as excellent starting material for glassmaking. Being a river delta that is very sluggish, it is likely that fresh sand is deposited by the tides and then washed by the freshwater of the river as the tides recede. Ancient accounts emphasize the need to wash any sand used for glassmaking though it appears here that nature at least partially performs the task.

The *glass former* is the primary ingredient and only the finest quartz sand or mineral containing as much silica as possible would have been used to create alchemical gems. It forms the glass network whereas the hollow spaces in the network are filled in by *fluxes and stabilizers*. Particle size is also important in order to achieve a homogenous *melt* and consistent fusion, the optimum particle size being between 0.1-0.3 mm in diameter. If sand were the primary starting material, it would have been thoroughly washed to remove any fine surface material impurities. Hypothetically, Maria would have chosen from among the following materials:

- a known quality glassmaking sand with a high silica content
- quartz pebbles
- rock crystal
- flint

Valuable genuine gemstones are hard crystals. A primary difference between a gem and glass is that a gem is a mineral with a regular and repeating crystal pattern, whereas a glass is essentially a frozen matrix with an amorphous or random atomic structure. Alchemists would have sought to begin with very fine quartz, such as *rock crystal* mentioned by Pliny, in order to mimic the properties of authentic gems. This would be combined with suitable *fluxes and stabilizers* to achieve these aims, such as antimony and zinc oxide in the Assyrian recipe.

During the Roman period in which Maria lived, a new technology in the form of *emery* was imported via trade routes to the markets of Alexandria. *Emery* is a very hard mineral consisting mainly of corundum (aluminium oxide), the primary mineral in rubies and sapphires, which found use as a glass-polisher. Emery provided a convenient medium by which glass could be easily shaped into faceted stones thus mimicking a crystal structure and provided a new technology for artisans to polish alchemical gemstones as well as all genuine gems with the exception of diamonds.

If Maria had chosen *white flint* as her *glass former*, this would have resulted in a type of lead crystal due to the lead content that occurs naturally in flint's quartz matrix. Lead oxide in the form of *litharge* was sometimes added to a batch along with the *glass former* and other *fluxes and stabilizers*. *Litharge* is the product of roasting another eye makeup, galena, which was preferred by the ancient Egyptians over stibnite favored in Jewish and Islamic traditions. Lead's function in glassmaking is to fuse the glass structure weakened by the fluxes, which makes the glass more resistant to surface corrosion. The Assyrian recipe employed zinc oxide and antimony rather than lead or *litharge*.

Fluxes and Stabilizers

Fluxes lower the melting point of silica from between 1,600-1,725 °C to around 1,000-1,200 °C, which falls within the range of ancient furnaces. *Natron* is a natural mineral deposit made up of various sodium salts (carbonate (Na_2CO_3), bicarbonate ($NaHCO_3$). It was mainly extracted from Beheira in Libya and Wadi Natrun in Egypt where large deposits form due to evaporation of water from the Nile and served as the main flux used in ancient glassmaking in Alexandria during the Roman period. Elsewhere, glassmakers used vegetable ash (potash-lime; K_2O) sourced from burning saltworts such as *Kali turgida* or *Salsola soda*. Saltworts grow in coastal areas and in saline soils and contain up to 30% sodium carbonate. Salts derived from these plants were known collectively in Arabic as *sal al-qalīy* (القلي), which entered Latin as *alkali*. Islamic and Indian alchemists made great use of *sal-alkali*.

The forms of soda ash discussed above typically contain 5-10% calcium oxide, which serves as a *stabilizer*. This type of glass is known as *soda-lime glass*. The *fluxes and stabilizers* in the Assyrian recipe were zinc oxide and antimony and although the recipe does not indicate what form of antimony, it was likely *flowers of antimony*. Antimony was employed in Egyptian glassmaking since antiquity. While it is uncertain which fluxes and stabilizers Maria's school may have used, antimony would certainly have been included due to its occurrence in the *powder of projection*.

Colorants

Gold introduced into molten glass results in ruby-red coloration while lower concentrations yield a less intense red known as *cranberry*. Metallic copper will also result in rich red coloration yet tends towards opaque, meaning that a clarifier would need to have been present in the matrix as well. Selenium, a byproduct of copper smelting, in combination with cadmium sulfide will also result in a rich ruby-red coloration. Maria would have included either her *powder of projection* or a portion of *gold-antimonial bronze* into her *melt*, either of which could potentially color glass the desired ruby-red to create alchemical rubies.

While it is tempting to posit a simple explanation for Maria's hypothetical alchemical rubies and possibly diamonds, emeralds and sapphires, the truth is that glassmaking and coloration is a refined art that requires tremendous experiential expertise even in light of modern technological advances. Many ancient colored glasses from antiquity have been professionally analyzed and their compositions ascertained. A large portion of these contain antimony present in some of the most ancient samples right up to medieval and early-modern European stained glasses. Alchemical gems and stained glass share a common ancestral technology in the form of mineral and metal oxides that are introduced into the glass matrix. Furthermore, this very same mineral and metal oxide technology is also prevalent in bronze smelting. These oxides link ancient colored glasses to medieval Islamic artificial gemstones and the magnificent stained-glass windows of Gothic cathedrals. Gold-ruby glass was first written about in 1689 by German pharmacist and alchemist Johannes Kunckel, yet this glass dates back to the Roman Empire. The history of stained glass may shed light on the importation of alchemical glass and gemstone technology into Europe.

Crafting Alchemical Gems

The relationship between alchemists and glassmaking artisans is exemplified in the magnificent explosion that was applied medieval Islamic alchemy. Islamic alchemists identified and categorized the various mineral-metal oxides and their resulting colors when applied to glassmaking. Al-Bīrūnī, Persian scholar and polymath writing in the 11th century, addressed the importance of practical experience and empirical observation in glassmaking in the following passage:

> *They have many methods and disclosures on the composition of basic glass and the quantities of the colouring materials; but none of these can be considered right except by observing the work of the distinguished artisans and by actually getting involved in the work and practicing it by doing experiments on the compositions. Glass, enamel and ceramics are close to each other and they have common techniques in pigments and in the methods of colouring.*

The glass industry that began in Syria in antiquity survived and continued at Ar-Raqqah, Aleppo, Damascus and elsewhere during the period of Islamic empire. An example of a medieval Islamic alchemical text definitively linking stained-glass with alchemical gemstones was the *Kitab al-Durra al-Maknuna* (The Book of the Hidden Pearl) authored by Islamic alchemist Jābir ibn Hayyān. The text dates to around the late 8th or early 9th century and contains forty-six recipes for stained glass. The work also includes the process for cutting glass into artificial gemstones. The recipes occurring in Jābir's work unequivocally present Islamic alchemical gemstones in the light of glassmaking technologies as evidenced by the recipes below:

Recipe 46, fol. 7b – Making ruby (yaqut ahmar) without equal
1. Take one hundred dirhams of cornelian stone ('aqiq), two hundred dirhams of rock crystal (billaur) and twenty-five dirhams of magnesia.
2. Heat each one alone and throw it in sour vinegar. Pound and cook with sour vinegar to which al-qali (alkali) has been added. Cook very well for half a day until it becomes dry and roasted.
3. Throw it in cold water and wash it in water and salt until its water and jawhar become clear.
4. Put it in a luted pot (qidr) with one hundred dirhams of natrun, twenty-five dirhams of al-qali (alkali) salt, forty dirhams of Armenian borax and ten dirhams of coarse salt (milh jarish).
5. Light up fire on it in the furnace two days and two nights or one day and one night. If it melts take it out when it cools and throw on it calamine (iqlimia).
6. Take out the melted ingot (nuqra) and pulverize it with sixty dirhams of red lead (isrinj), five dirhams of cinnabar (zanjufr), two dirhams of realgar (red zarnikh) pulverised in vinegar, five dirhams of magnesia, five dirhams of copper scales (rosakht), and ten dirhams of pulverised blood stone which is sadhan.
7. Mix and put in a luted pot (qidr) and place in a furnace. Blow on it continuously until it melts and becomes mature.
8. The sign of its maturity is that you put out one carat (qirat) of it on a clear surface until it cools. If you see it clear red with plenty of water that is the water of ruby (yaqut) then it has matured. If it has turbidity then blow on it until it matures.

Recipe 4, fol. 19a, colouring rock crystal (billaur) red
1. Four mithqals dragon's blood, one mithqal mastic, one mithqal rosin, one mithqal pitch, one mithqal balsam of Mecca (resin of commiphora gileadensis - duhn al-balasan).
2. Pulverize the drugs and knead with pitch and balsam of Mecca.
3. Heat the gemstones and bury them in the mixture.
4. Leave them until they cool down.

Throughout the period of Islamic alchemy beginning in the 7th century, the Venetian port of Murano was a major trading port with substantial Asian and Islamic influences. Glassmaking entered Italy through this port and nearby Venice became the glassmaking capital of Europe. Glassmaking artisans arrived from Constantinople in two waves of immigration, first following the sacking of the city in 1204 by the Fourth Crusade and again when the city fell to the Ottomans in 1453. The entire glassmaking industry was moved from Venice to Murano during the 13th century and remains there today. Murano glassmakers refined glassmaking in many ways including production of various crystal and *crystallin* glasses, beads and fantastic imitation gemstones.

Much of the technical jargon used in Murano glassmaking is either original or distinctly alchemical. They were the first to use the term *frit* (Italian: *fritta*) during the 15th century but it became obsolete due to the development of high-temperature furnace technology. The Murano term *corpo*, which literally means *body*, is an opacifier added to transparent glass to make it white. *Anima*, meaning *soul*, is the colorant added to molten glass that is responsible for the fantastic array of colors that Murano artisans are able to achieve.

Gothic cathedral construction began around the same time alchemy was introduced to European translators during the 12th century. A very distinct characteristic of gothic cathedrals is the fantastic quality of light attributable to their magnificent stained-glass windows. It is interesting to speculate whether the *powder of projection* ever played a role in the creation of these windows, but certainly alchemical glassmaking and gemstone artisanship donated key technology in the form of mineral and metal oxides and colloidal colorants. One of the earliest texts on medieval glass manufacture is the *De Divers Artibus* written by Benedictine monk, Theophilus Presbyter. The chapters on creating red, green and blue stained glass are omitted perhaps in an effort to protect the secret of their manufacture. Later these colors were achieved by the addition of ionic copper or copper oxides.

John Dee, a Welsh polymath, alchemist and consultant to Queen Elizabeth I writing during the 16th century, published a number of alchemical texts that, if decoded correctly, demonstrate that he was an adept who had achieved the Philosophers' Stone. Dee outlines his recipe for confecting the Stone in an alchemical treatise now in the care of the London Museum known as *MS Harleian 6485*.

> ... *Lastly note* philosophers had two ways, a wet one, which I made use of, and a dry one.
>
> There is one mystery more in Nature ... *I shall briefly declare it*, as also about the universal Stone of the Philosophers, how it resteth merely upon the white Spirit of Vitriol [butter of antimony], *and how that all three principles are found only in this Spirit and how you are to proceed in and to bring each into its certain state and order.*

The chemical identity of his *white spirit of vitriol* has erroneously been interpreted as sulfuric acid. In alchemy, the term *vitriol* was used synonymously with *sulfide-mineral*, also known as *atrament*. Dee's *vitriol* or sulfide-ore is stibnite and the *white spirit* or *distillate* is *butter of antimony*. Dee's elixir relied on *butter of antimony* for its creation, which demonstrates that he achieved his *elixir* via the archetypal recipe. He elaborates on the importance of *butter of antimony* and its unique properties:

> *I at last found the manner and true way of dissolving all bodies, separating and conjoining them, finding the composition of* their secret of secrets, that is to say Lac Virginis or Acetum acerrimum, and Raymund's Calcining Water [all being synonymous], wherewith I dissolved all bodies at pleasure **and perfected the gross work.**

In another work attributed to John Dee entitled *The Rosie Crucian Secrets*, a recipe for creating artificial carbuncles combines the *powder of projection* with standard glassmaking methodology. The recipe below represents a simple and straightforward method for incorporating the *powder of projection* into molten glass as a colorant resulting in a red alchemical gemstone in accordance with established glassmaking technique. His alchemical jargon closely parallels Murano glassmaking; the colorant represents the *dual-soul* principle joined to the *body* of glass.

The Operation of the Magistery of Carbuncles

... We now come to *the materials, which are two and are to be joined together; the first giveth the form, the other receiveth it. That which giveth the form is the Spirit and Soul of Sol or Gold, joined together in the Red Elixir and is the agent and, as it were, the man. That which receiveth the form is the hardest salt of the earth contained in glass,* and *is the power of heaven impregnating the earth; the patient is the power of the earth retaining the impression of the heaven.*

... And so *the Elixir is prepared for projection upon glass;* but *for the preparation of glass there is no more required but that it be made of the same matter that Venice [Murano] glass is made of [pure silica],* the composition of which, if ye know not ...

1. take as much Venice glass as you please and weigh it exactly, **upon which to project your Elixir. When you have done so,** put your glass in the crucible to melt, **and when it is well molten ...**
2. then take your corporeal red Elixir dissolved as before (or if you will, undissolved) as much as sufficeth to tinge the molten glass, **and put it tied up in a paper in the glass,** stirring it a little with a rod, and there let it stand the space of an hour.
3. Then take out the crucible and pour the matter into an ingot, **and it will be malleable but as hard as glass and stone-like to the sight;** and *you may either cut it like stone or work it with a hammer. This Carbuncle stone or metal hath the property of a Carbuncle in shining and glistring above all natural Carbuncles*

Dee's recipe references Venetian glassmaking in context, name and technique and correlates it with alchemical jargon and gemstone production. The process employs the *powder of projection* as a unique colorant applied to the standard method of glass coloration. Another method of dissolving and creating gemstones was alluded to by Johann de Monte-Snyder in his 17th century treatise *Universal and Particular Processes*:

> ... there are also other calcinations of gems according to the ways of some adepts as with pure and clear ♄ of ☉, or ☽ [sulfur; i.e. calx of gold or silver] with addition of a fixed Sal Commune, but these are long and difficult processes and the before described way is the best because given by experience and proved to be real by adepts.

Another mysterious alchemical text addressing the technique of alchemical gem manufacture takes advantage of the dissolving and coagulating properties of *butter of antimony*. The following recipes appear in a an anonymously authored text known as *Tractatus de Lapide, Manna benedicto* included as part of a collection of texts in *Aurifontina Chymica* published in London in 1680. Sir Isaac Newton possessed a copy in his personal library, now known as *Keynes MS 33*. Other copies are held at the British Library and Oxford's Bodleian library.

A careful reading of the manuscript suggests that it was written by a highly skilled pharmaco-alchemist, likely of iatrochemistry in the Paracelsian tradition such as Joseph Duchesne or Sir Theodore Turquet de Mayerne. This text is important to the discussion of alchemical gemstones because it emphasizes the use of *butter of antimony*, indicated by the cover-terms *clear water* or *white water* used synonymously in the text. The cover-terms that indicate gold-based red and silver-based white *powders of projection* are *red and white mercuries* respectively.

To Make Diamonds

1. Take the whitest Flint Stone you can get, beat off the outside, and dissolve the rest, as much as thou wilt, in the white Water:
2. when it is dissolved to clear Water, not to Pap, put it into a little Vial, stop it close, and set it in warm Ashes, and in twelve days it will congeal to a hard gray Stone:
3. then increase the Fire, that the Glass may be red hot, then let it cool;
4. take it out, and it will be like a flint; but polish it, and thou never sawst such a sparkling Diamond, nor so hard: but it will be better if thou dissolve little Diamonds.
5. All Stones that you dissolve in the white Water, the same colour they were of, the same will they be of;

[To Make Rubies and Carbuncles]

... but for Rubies and Carbuncles, and all red Stones, they are made of the red Mercury [gold-based powder of projection], and of [rock] Crystal;

1. and for a Carbuncle, you must add to ten parts of [rock] Crystal, dissolved in the white Mercury, one part of the red Stone brought to the highest,
2. and so as before congeal it with Fire, and being polished it shineth in the dark beyond all whatever.

To Make Stones [Pearls]

1. Having made Mercury of the Philosophers, and out of it the two Mercuries white and red, if thou wilt of small Pearls make great and Oriental ones, do thus:
2. Take white Seed Pearls, and dissolve them in the white Water, which will instantly of it self dissolve them:
3. when it is like Pap, that thou mayst work them with thy hands, make it into pearls; and have a round mold of pure Silver [at hand],
4. put thy Pap into this mould, but first anoint thy mould with the white Stone, which is an Oyl:

5. when they have layn three or four days, open it, and lay the Pearls in the Sun, but not too hot, and they will grow hard, and more orient than any Natural Ones.

The recipes above rely on *butter of antimony* as a chemical technology to dissolve flint, quartz or rock crystal and pearls. Techniques for creating alchemical gemstones via *butter of antimony* can be considered as originating from Maria's branch of alchemy only if a suitable argument can be made that alchemists in her tradition were indeed capable of creating *butter of antimony* during the early Alexandrian period.

Nevertheless, the final word on alchemical gems must honorifically go to one of Europe's most enigmatic and conspicuous gem-makers, the Comte de St. Germain, who was most certainly learned in Murano glassmaking. Manley P. Hall, in his introduction to *The Most Holy Trinosophia*, explains:

> *His skill as a chemist was so profound that* he could remove flaws from diamonds and emeralds, which feat he actually performed at the request of Louis XV in 1757. Stones of comparatively little value were thus transformed into gems of the first water *after remaining for a short time in his possession. He frequently performed this last experiment, if the statements of his friends can be relied upon. There is also a popular story to the effect that he placed gems worth thousands of dollars on the place cards at the banquets he gave.*

Baron Gleichen recounted his first meeting with St. Germain being extremely impressed by the quantity and quality of extraordinary sized diamonds, opals and a white sapphire. Madame de Pompadour had the following to say:

> *... he presented to the Marquise the most precious and priceless gems. For this singular man passed for being fabulously rich and he distributed diamonds and jewels with astonishing liberality.*

Legends abound regarding his technical ability to correct and enlarge flawed gemstones or to melt smaller gems into a single larger one. In written correspondence between St. Germain and Count von Lamberg, he revealed the source of his knowledge:

> I am indebted for my knowledge of melting jewels to my second journey to India.

The skill and technique that St. Germain repeatedly demonstrated to Europe's nobility in the form of crafted or improved jewels is known by today's trade-jargon as *fracture-filled* or *clarity-enhanced stones* and has ancient origins with the first ornamental Egyptian glass makers of the 18th dynasty at Akhetaten. Since then humanity has been fascinated by not only natural gems but also enhanced, imitation or synthetic variants, some of which can only be identified in a lab by a certified gemologist.

- ***natural stones*** – rare and costly gemstones of beautiful quality
- ***enhanced stones*** – natural stones treated to enhance desirable characteristics for a higher quality appearance
- ***imitation stones*** – have the same color and basic look as natural gemstones
- ***synthetic stones*** – have the same scientific characteristics – crystalline structure, refractive index, chemical composition – as natural gemstones

GIA Laboratory in Bangkok in collaboration with Mahithon Thongdeesuk is one of the world's leading authorities on colored gemstones and enhancement technologies of the type that St. Germain practiced. Today rubies and star-sapphires are routinely *fracture-filled* or *clarity enhanced* using the ancient alchemical substances *litharge* (PbO), *minium* (Pb_3O_4) or heat-treated *white lead* ($2PbCO_3 \cdot Pb(OH)_2$) resulting in lead-glass surfacing. From this understanding it is possible to hypothesize that St. Germain perhaps began by calcining the eye-makeup favored by the Egyptians – *galena* (PbS) – to *litharge* by the exact same *whitening the stone* technique of Maria Hebrea, followed by continued calcination to achieve *minium*, also known as *red lead*:

> Whitening the Stone $2\ PbS + 3\ O_2 \rightarrow 2\ (PbO + SO_2)$
> Reddening the Stone $6\ PbO + O_2 \rightarrow 2\ Pb_3O_4$

Alternately, he could have used the common white paint pigment known as *Venetian Ceruse*, also called *litharge* or *white lead* to achieve *minium / red lead*:

$$3\ Pb_2CO_3(OH)_2 + O_2 \rightarrow 2\ Pb_3O_4 + 3\ CO_2 + 3\ H_2O$$

Either white or red variations of lead oxide would have provided Comte St. Germain with the technology for gemstone enhancement that added to his alchemical fame and mystique. This alchemical technology ultimately yielded lead crystal created by adding lead oxide to molten glass. A simple way to comprehend these technologies is that the preferred Jewish / Islamic biblical eye makeup – *stibnite* – was fundamental to creating alchemical gold, whereas the preferred Egyptian eye makeup – *galena* – had practical applications concerning alchemical gems. The preparation of either followed the exact same foundational operative technique of open calcination – known originally from Maria Hebrea as *whitening the stone*.

Alchemists were able to effect change in material matrices to produce alchemical gold and gemstones by various methods. These chemical technologies did not originate with Alexandrian alchemists, nor did their written accounts and for this reason, Alexandrian alchemy must be considered as a continuation and subsequent evolution of existing and very ancient chemical technologies. What appears to have made Maria's tradition of alchemy so revolutionary was the potential for chemical reactions resulting from her artisanal use of sal ammoniac and possibly *butter of antimony* to create the *powder of projection*. By contemporary understanding, alchemists were working in a branch of *nanotechnology* that is known today as *nanocomposites*. Nanoscale particles of gold, silver and antimony inherent in *powders of projection* were added to a composite matrix thereby changing or improving its properties, which resulted in valuable new artisanal products. The belief that a substance could undergo transmutation derives from Heraclitus who held fire, humanity's first chemical reaction and object of veneration, to be the primordial element, and that transformation is the only actuality in nature. Alchemists mastered the pyro-technologies of bronze and glassmaking, observed and marveled at transformations brought about by them and called it transmutation, yet it was transmutation by fire.

Nothing is created and nothing is destroyed, all is transformed.
– Antoine Lavoisier

Chapter 11 – Aurum Potabile

Even in our degenerate age these wonders are still possible, even now the medicine is prepared which is worth twenty tons of gold, nay more, for it has virtue to bestow that which all the gold of the world cannot buy, viz., health. Blessed is the physician that knows our soothing medicine ...

Helvetius

THE ELIXIR OF LIFE

The universal medicine **which cures all human and metallic diseases** *is concealed in gold and its magnet* [antimony], *the Chalibs of Sendivogius.*
— *Johannes de Monte Snyder*

The notion that the Philosophers' Stone was a type of medicine can be traced back to Stephanos who used the term *pharmakon* (fármakon; φάρμακον), meaning *medicine* in reference to the liquid form of his White Stone. Applied industrially, the *elixir* as *powder of projection* was a *medicine* that cured impure metals and brought them to the perfection of alchemical gold. The European term for the Stone applied therapeutically as a *medicine* reputed to confer perfect physical and psycho-spiritual health was *aurum potabile*, meaning gold that is safe to drink. The story of *aurum potabile* is ultimately the story of humanity's relationship with death. The Philosophers' Stone applied as the *panacea* or *elixir of life* demonstrates that the prevailing attitude towards death was that it was something to be avoided if possible. Alchemists and their patrons optimistically placed their hopes on the *elixir* as a possible cure for humanity's greatest malady, inevitable mortality.

The term *aurum potabile* is somewhat misleading however, because the gold that the alchemists were alluding to was not elemental metallic gold but rather the Philosophers' Stone powdered and suspended in a liquid. It was alleged that the innate perfection of the Philosophers' Stone possessed an intrinsic potential to balance bodily fluids with the body's innate vital heat perfectly. The balance thus restored was thought to strengthen the inherent natural ability to repel disease, rejuvenate and preserve youthful countenance and unlock the doors to profoundly deeper human intellectual and spiritual capacities:

> *The fact, however, must not be ignored, that* gold which is found as yellow or saffron-colored becomes by decoction red ... **Hence the alchemists** *wishing to make gold* strive for the elixir to redden. This they call medicine. It is *the object of their desire that it embody four characteristics, namely that of color, penetration, imperishability in fire, and consolidation,* and they call this redness of the sun (rubeum solis). ... **Hence** Hermes says that this is the root on which all such philosophers are sustained, that the red of the sun is the medicine ...
> – *Albertus Magnus,* De Mineralibus IV:7

Rejuvenation, Longevity and the Elixir of Life

Humanity has always displayed an incredible fascination for the idea of personal immortality or extreme longevity since the Sumerian *Epic of King Gilgamesh* of Uruk, and interrelated stories retold in all Mesopotamian cultures. After many dangerous adventures, Gilgamesh's efforts earned him an audience with Ziusudra / Utnapishtim, (Sumerian / Akkadian version of biblical Noah) one of the few humans to have ever been granted immortality by the gods. Gilgamesh was refused the secret of the *Tree of Life* despite his royal and divine bloodline, yet acquired a special plant reputed to confer longevity, which he subsequently lost. Ever since Gilgamesh, humanity has continued the quest to recover extreme longevity and occult or magic substances that might enable it.

Revoked immortality appears in the narrative of Adam and Eve. Early Biblical authors broached the idea of a *Tree of Life* in Genesis, which became central to European notions of extreme longevity or immortality:

11 - AURUM POTABILE

מִן אֱלֹהִים יְהוָה וַיַּצְמַח ט
לְמַרְאֶה נֶחְמָד עֵץ כָּל הָאֲדָמָה
בְּתוֹךְ הַחַיִּים וְעֵץ--לְמַאֲכָל וְטוֹב
וָרַע טוֹב הַדַּעַת וְעֵץ הַגָּן

Genesis 2:9 And out of the ground the LORD God made to spring up every tree that is pleasant to the sight and good for food. The tree of life was in the midst of the garden, and the tree of the knowledge of good and evil.

הֵן אֱלֹהִים יְהוָה וַיֹּאמֶר כב
לָדַעַת מִמֶּנּוּ כְּאַחַד הָיָה הָאָדָם
יָדוֹ יִשְׁלַח פֶּן וְעַתָּה וָרָע טוֹב
וָחַי וְאָכַל הַחַיִּים מֵעֵץ גַּם וְלָקַח
לְעֹלָם

Genesis 3:22 And the Lord God said, "The man has now become like one of us, knowing good and evil. He must not be allowed to reach out his hand and take also from the tree of life and eat, and live forever."

Roman philosopher and Christian theologian St. Augustine of Hippo, writing during the early 5th century, offered an interpretation of Adam and Eve's revoked immortality that became the accepted Christian model for centuries thereafter:

> *[Their] bodies were not indeed growing old and senile,* **so as to be brought in the end to an inevitable death.** *This condition was granted them by the wonderful grace of God, and was derived from the tree of life which was in paradise.*

According to Augustine, Adam and Eve were stripped of their immortality upon being expelled from the Garden of Eden. Many European alchemists regarded Biblical narrative as literal historical account and adopted interpretations such as Augustine's as part of their conceptual frame of reference. They reasoned that Adam and Eve had indeed eaten the forbidden fruit of the *Tree of Knowledge*; thus, man acquired the knowledge of good and evil and was therefore able to comprehend sin. Comprehension of sin seemed self-evident and true implying that the remainder of the story must also be factual. Many of Europe's intellectual and alchemical elite looked to the lives of forty Biblical characters as veritable documentation of extreme longevity. Lifespans ranged from Joshua and Joseph at 110 years of age to eight of the eldest reputed to have lived for more than 900 years, the oldest being Methuselah at 969 years old.

This cannot be construed as immortality as such, yet Biblical literalists maintained that a human lifespan was potentially several hundred years and that perhaps the Patriarchs were in possession of a lost ancient and sacred technology responsible for such incredible longevity even without having eaten of the *Tree of Life*. A trend developed in European alchemy that identified the Philosophers' Stone as the *panacea*, the *elixir of life* and *food of angels* that had been hidden from the majority of humanity. A widespread belief in esoteric circles was that this occult secret had been carefully protected for thousands of years in philosophical and religious mystery schools originating in Babylon, Persia, Egypt, and the Levant. Admittance into the mystery schools challenged the greatest thinkers in antiquity. Pythagoras and Plato spent decades in preparation waiting for admittance to study in the Egyptian mystery tradition. The final mystery could be learnt only through an intense period of graduating instruction through many levels before the highest knowledge would be revealed. For alchemists, this meant access to the secret of the Philosophers' Stone and comprehensive knowledge of its potential transformative power.

The application of the Philosophers' Stone as an ingestible therapeutic or curative medicine is not explored in any detail by Alexandrian alchemists and opportunities abound for further research into this curious omission. The situation is much different concerning Islamic and European alchemical traditions. Aside from the golden calf episode in Exodus in which Moses had the children of Israel drink a solution containing powdered gold, there exists a bewildering scarcity of primary sources for the therapeutic application of the Philosophers' Stone until the first clear accounts appear in the Jābirian Corpus of Islamic alchemical texts dating from the 8th century CE. The *Kitāb al-Khawaṣṣ al-Kabīr* (Great Book of Chemical Properties) was written as a first-person narrative by Jābir ibn

Hayyān and was considered an authentic Jābirian text by researcher Ahmad Yousef al-Hassan Gabarin and others. Versions of the manuscript are held at the Alexandria and British Libraries. The *Jābirian Corpus* had a profound influence on early European alchemy and it is possible that Jābir's accounts of miraculous deathbed cures fueled European speculations that the Philosophers' Stone might in fact be the *panacea* or *elixir of life*. The question remains as to whether Jābir's *al-iksir* was the Philosophers' Stone, or could it have been some other medical powder.

Adept John Dastin, in a 14th century letter written in Latin summarizing his thoughts on the Philosophers' Stone, used Jābir's voice to explain that there is only one true *al-iksir*. Quoting Jābir, he wrote:

> ... as Jabir says, "Our art does not consist of a multitude of things. There is one stone, one medicine, to which nothing extraneous is introduced, nor added; nor is it reduced, unless superfluities are to be removed."

Jābir confessed that his mentor, Ja'far al-Sadiq, was reluctant to have knowledge of the Philosophers' Stone revealed so openly in the *Kitāb al-Khawaṣṣ al-Kabīr*:

> By God, my master disapproved of my having written this book, saying: By God, O Jābir, if I did not know that nobody will have access to it without meriting it ... I would have ordered you to destroy this book. Do you know what you have divulged to the public?

Despite Ja'far's reticence, Jābir appears to have justified publishing the work when he wrote:

> My master often used to say: Proceed as you wish, O Jābir, and reveal the sciences as you please – as long as only those who are truly worthy of it have access to it.

In the first section of the *Kitāb al-Khawaṣṣ al-Kabīr*, Jābir states clearly in a manner that would be echoed by Paracelsus centuries later that he would only discuss case studies that he had witnessed firsthand:

> *In this book we mention the properties of what we have seen after experiments and tests regardless of what we have heard or read. And thus we mentioned what proved to be right and we refused what proved to be wrong and we also compared what we discovered to what people mentioned.*

Jābir glosses the confection and character of the *al-iksir*:

> *...the operation is practiced via concentrated things which contain a great quantity of spiritual force, subtle and light.*

Dr. Gerald Gruman in *A History of Ideas About the Prolongation of Life* reveals that one of the earliest mentions of the Philosophers' Stone applied therapeutically as an ingested substance can be found in a case study from the *Kitāb al-Khawaṣṣ al-Kabīr* where Jābir was called to treat a favorite slave-girl on her deathbed:

> *I saw her almost dead, her strength very much depleted. But I had a little of the elixir with me and of this I made her drink the amount of 2 grains in 3 oz. of pure oxymel [spiced honey-vinegar tonic; julāb]. By God and my Master, I had to cover my face before the maiden, for in less than half an hour her perfection was restored even to a higher degree than she had formerly possessed.*

Gruman recounts two additional case studies of terminal patients treated by Jābir, another slave-girl who accidentally ingested a lethal dose of arsenic and a man suffering from a venomous snakebite, each spared certain death by the *al-iksir*. Jābir cured the slave-girl with a grain of the *al-iksir* suspended in honey-water and the snakebite victim with two grains suspended in cold water. Both completely recovered because of these treatments, just two of over a thousand cures according to Jābir's testimony.

The marriage between alchemy and medicinal mineral / metal powders is a complex story of intense experimental chemistry and cross-cultural intercourse between the eastern fringes of the Islamic Empire and India. This topic is covered in more detail in a separate work entitled *Gold Elixirs – A Cross-cultural History of Therapeutic Gold* by the author. Islamic alchemists Jābir (Geber), al-Razi (Rhazes) and Ibn-Sina (Avicenna) did much to develop new organo-metallic and metallo-therapeutic medicinal formulations. They were educated men concerned with the health and overall well-being of their patients and patrons and preserved their work in voluminous writings. This fact awarded them much fame as each held a prominent position in society. It also places them in the category of medico-alchemists that seem far more related to their Indian and Chinese cousins of the Eastern tradition in practice, than to the artisans and philosophers of Alexandria. They were addressed by the title *Ḥakīm* (حكيم) meaning to appoint, choose or judge, a title that distinguishes one as philosopher and physician typically associated with Sufism or mystic pre-Islamic Pahlavi (Persian) illuminism, *Hikmat-i ishrāq* (حكمت اشراق).

In his book *How Greek Science Passed to the Arabs*, De Lacy O'Leary presents these *Ḥakīm* as the Islamic Empire's version of renaissance men:

> *Arabic science flourished most in the atmosphere of courts. Scientists usually depended upon wealthy and powerful patrons.* **They appealed little to the average man, and this chiefly because scientific and especially** *philosophical speculation was regarded as tending towards free-thinking in religion, and so "philosophers" were classed as a species of heretics.* **Ultimately the** *philosophers themselves partially acquiesced in this judgment, and adopted the view that the inspired Qur'an was well adapted for the spiritual life of the unlettered and simple, but the illuminated (ishrāqiyya) saw beneath the written word and grasped an inner truth* **which it was not expedient to disclose to the simple.**

Ibn Sina (Avicenna) was most famous for his 10th century masterwork, *Al-Qanun fi'l at-Tibb* (The Canon of Medicine), which Encyclopedia Britannica describes as *"... the most famous book in the history of medicine, in East or West"*. His *Canon of Medicine* was translated into nearly every language in the civilized world including 28 Latin editions and served as the standard medical textbook until the 18th century. In section 683 of the *Canon of Medicine*, he applied the metaphor of an oil-lamp to convey his views on the aging process based on an innate interplay between heat and moisture, which had an enduring influence on European senescence theories thereafter:

> *A lamp is influenced by two types of moisture – oil and water. It remains lit with the help of one but is put off by the other.* **The case with the innate heat of the human body** *which thrives on the innate moisture and is overwhelmed by the extraneous moistures* *is, thus, similar to that of a lamp.* **The** *innate heat gradually decreases* **as the "extraneous moisture" goes on increasing from the growing weakness of digestion so that in the end it damps down the lamp as does the water. Thus** *when the innate moisture dries up the innate heat also comes to an end and physical death ensues.*

Roger Bacon was a 13th century English philosopher and Franciscan friar well versed in the study of natural philosophy, mathematics and logic. He studied at Oxford and excelled quickly, which earned him a fine reputation as a result. He lectured at Oxford on Aristotle and later at the University of Paris before becoming a friar in the Franciscan Order, but eventually his interests and views came to be considered mystical and heretical. This caused him to suffer a storm of indignation by the Franciscans over his "suspected novelties" resulting in a fourteen-year period of confinement in Paris towards the end of his life.

He was an incredibly talented practical alchemist. His unique approach to antimonial alchemy was highly influential to Basil Valentine's work. Bacon's recipe for confecting the Philosophers' Stone, which he termed

Lapidus Stibii, occurs in *De Oleo Antimonii Tractatus* and is an original, elegant and practical example of early chemistry. He was also outspoken concerning longevity and published his views in *The Cure of Old Age and Preservation of Youth* in two volumes in which he largely accepted Avicenna's lamp theory, yet combined it somewhat with Christian views on sin as playing an important role in shortened lifespans. He considered the soul immortal and reasoned that because of this, humanity had the potential to live to the age of the biblical patriarchs if it were not for increased sin and materialism since man's fall. For Bacon, this unfortunate human condition could be remedied via alchemy:

> *At the beginning of the world there was a great prolongation of life*, **but now it has been shortened unduly.**
>
> **For** *sins weaken the powers of the soul, so that it is incompetent for the natural control of the body; and therefore the powers of the body are weakened and life is shortened.*
>
> *The possibility of the prolongation of life is confirmed by the consideration that the soul naturally is immortal and capable of not dying.* **So after the fall, a man might live a thousand years; and since that time the length of life has been gradually shortened. Therefor it follows that** *this shortening is accidental and may be remedied wholly or in part.*

As a practical alchemist, he held to the notion that great secrets had been lost to antiquity and that alchemy could help one rediscover the *admirable virtue*, later known as the *quintessence*, found in all matter:

> *... the knowledge of those properties,* **that are in certain things,** *which the ancients have kept secret.*
>
> **I have also found this, that** *there is an admirable virtue placed in plants, animals and stones; which is partly hidden from the men of this age ...*

Bacon placed great stock in his alchemical substances, specifically his *Oil of Antimony* and *Stone of Antimony*, the latter being his unique and novel version of the Philosophers' Stone:

> *Thus you now have two separate things: Both the Spirit of Wine full of force and wonder in the arts of the human body: And then the blessed red, noble, heavenly Oleum Antimonii, to translate all diseases of the imperfect metals to the Perfection of gold. And the power of the Spiritual Wine reaches very far and to great heights. For when it is rightly used according to the Art of Medicine: I tell you, you have a heavenly medicine to prevent and to cure all kinds of diseases and ailments of the human body.*

In renaissance Europe, Paracelsus changed the landscape of alchemy by emphasizing the medicinal application of alchemical products over alchemical gold making. He was outspoken on the topic of potential physical longevity and psycho-spiritual awakening resulting from use of the Philosophers' Stone. The following appears in the introduction of his treatise *A Book Concerning Long Life*:

> *... no physician ought to wonder that life can be prolonged, by how much more can a living one be kept from decay?*

Paracelsus held longevity to be due to a three-fold plan of diet, physical fitness and most importantly in his system, medicines. He identified the life-principle to be *"a burning and living fire"* that was an everlasting universal property. He adopted the model of the *eternal flame* from antiquity as a metaphor in which the body is fuel that is consumed by the *living fire*. According to Paracelsus, death was a result of the fuel burning out resulting in a dimming of *living fire*, but that the *living fire* could be rekindled if a special type of fuel could be found. For Paracelsus every organic or inorganic substance concealed an active principle hidden within it called *quintessence* paralleling Bacon's *admirable virtue* believed to be innate to all plants, animals and stones. The *quintessence* could be isolated and applied much like *active ingredients* are today, and that unique expressions of the *quintessence* varied from substance to substance. The *quintessence* of cinnamon for example expressed itself differently than did the *quintessence* of gold. Paracelsus suggested that the most potent expression or manifestation of *quintessence* was found in the *Tincture of the Philosophers*, his personal jargon for the archetypal Philosophers' Stone.

Like *tincture*, the words *aurum potabile* and *quintessence* were used synonymously with the Philosophers' Stone in the European alchemical tradition:

> This Stone cures all **Leprous people, Plague, and all** Diseases which may reign upon Earth, or befal Mankind; this is the true Aurum potabile, and the true Quintessence which the Ancients sought; **this is what thing whereof the whole Troop of Philosophers speak so wondrously, using all possible skill to conceal its Name and Operation, as aforesaid.**
> – Johann Isaac Hollandus, A Work of Saturn, Its Use in Physick

During the early modern period, one of the most outspoken characters to champion a holistic approach to *rejuvenation* and *life-extension* by incorporating special medicines was early 17th century philosopher and scientist Sir Francis Bacon. Bacon accepted the *lamp theory* in which heat, or *native spirit* as Bacon termed it was responsible for vitality but that it also exhausted the body paralleling the manner in which fire consumes its fuel. He believed that this process was inevitable but could be delayed by various complex methods. In *The Advancement of Learning*, Bacon echoes Paracelsus and specifically Hippocrates by advocating a compound approach to health that:

> ... by ambages of diets, bathings, anointings, medicines, motions, and the like, prolong life, or restore some degree of youth or vivacity ...

Bacon's Hippocratic approach to health and well-being challenged the notion of a universal remedy, yet he understood the value of certain medicines and advocated regular purging as part of his regimen. This is important because one of the foremost properties of the immensely popular antimonial medicines of Europe was their purgative effects. Varying potencies and formulations caused effects including regurgitation, diarrhea or elevated body-temperature accompanied by mild to heavy perspiration based on the medicine's degree of refinement. The thermogenic phenomenon was construed as an indication of vital or innate heat being directly fueled by the medicine and taken as veritable proof of efficacy. The emetic properties signified purification or a purging of evils.

Although there were many critics, who believed that to prolong life would result in nothing more than an extended period of misery and suffering, Bacon introduced a new idealistic motive to the debate. For Bacon the true objective for extreme longevity was to attain an amalgamation of physical, intellectual, spiritual and social r/evolution. David Boyd Haycock, in his article *"A Thing Ridiculous"? Chemical Medicines and the Prolongation of Human Life in Seventeenth Century England,* explains:

> One of his major ambitions *for perfecting his project, therefore,* was the establishment of permanent, well-funded learned societies. *In his posthumously published New Atlantis, Bacon wrote of Salomon's House, an institution devoted to the collective, long-term advancement of learning.* The great enterprises pursued there included "the prolongation of life, the restitution of youth in some degree, [and] the retardation of age." In the decades after Bacon's death Salomon's House became the model for numerous scientific societies, culminating in 1660 with the foundation of the Royal Society of London ...
>
> *Samuel Hartlib and his circle — some of whom would be involved in establishing the Society —* advanced the Baconian project for the prolongation of life, in particular through the search for new, chemical

medicines, the philosopher's stone, the elixir of life and even, perhaps, a universal medicine.

Some of the finest minds known to science such as Sir Isaac Newton, Robert Boyle and John Locke had invested themselves in this pursuit by exploring every avenue including alchemy to achieve the aims of the Hartlib Circle. The reward for finding an *elixir of life* with revitalizing and life-extending properties combined with potential for intellectual and spiritual awakening was that humanity would then possess the means to live long enough to achieve the utopian goals disseminated throughout Europe by the Rosicrucians, Freemasons and other fraternal or royal *learned societies*.

Perhaps the most memorable emissary and model of unusual longevity due to ingesting the *elixir* is embodied in the figure of the Comte de St. Germain. His unfading youthful appearance and vitality, combined with an exceptional intellect and wit were the stuff of legends. He did not appear to age normally and reasons for this are explained by Manly P. Hall in his introduction to *The Most Holy Trinosophia*:

> *According to Madame de Pompadour, he claimed to possess the secret of eternal youth...* **Had it not been for his striking personality and apparently supernatural powers, the Comte would undoubtedly have been considered insane, but** *his transcending genius was so evident that he was merely termed eccentric. ... [He] whom Frederick the Great referred to as "the man who does not die".*
>
> *It was diet, he declared, combined with his marvellous elixir, which constituted the true secret of longevity,* **and although invited to the most sumptuous repasts he resolutely refused to eat any food but such as had been specially prepared for him and according to his recipes.**

The Comte de St. Germain was highly respected for his alchemy, *elixir* and efforts to institute fraternal societies in line with Baconian ideals of personal enlightenment and utopian social evolution. The recipe for his *elixir* remains a mystery because his multi-layered encryption is so complex that it has yet to be deciphered.

As early as the 13th century, mediæval alchemical authors such as Michael Scot, Roger Bacon and Arnaldus de Villa Nova left records of *alchemical aurum potabile* encrypted in alchemical texts along with accounts of its properties and method of confection. Renaissance period alchemists such as Paracelsus and others preserved and expanded the subject of *alchemical aurum potabile*. A contextual understanding of the term *aurum potabile* is necessary to comprehend the Philosophers' Stone as an ingested alchemical therapeutic substance.

Gold leaf is inert and can be safely eaten yet because it passes directly through the digestive system, it is not bioavailable and therefore inactive. The term *aurum potabile* in a general sense means *drinkable gold* yet used here refers to gold in a bioactive form. By reducing gold to a very small particle size (nanoparticle) or by chemically converting it to a salt or ash (ionic), it becomes bioavailable and active within the body and in the brain. In the original alchemical sense of the term, *aurum potabile* designated the Philosophers' Stone in powdered form or suspended in liquid and ingested as a therapeutic substance. However the term *aurum* in this context a misnomer, due to the fact that it was not simply gold, but rather a combination of gold and antimony as *alchemical elixir* that the term *aurum* in *aurum potabile* originally referred to; it being the first of three classes or categories of *aurum potable*:

> *Out of this spiritual Essence [al-ḳoḥl; butter of antimony], and out of this spiritual Matter [flowers of antimony], out of which first of all Gold was made into a Body [the Elixir], and became corporal, out of it is made a more true and compleat Aurum potabile than out of Gold it self, which must first of all be made spiritual, before a potable Gold can be prepared out of it.* – Basil Valentine, Of the Spirit of Gold

Two other types of medicine also termed *aurum potabile* became well known in Europe. The second type of *aurum potabile* known as *Mercurius Vitæ* is synonymous with the *Hermetic White Stone* of Cleopatra, Zosimus, Synesius and Stephanos' and Hollandus' *Stone of Saturn*. The chemical identity for this product is antimony-oxychloride, but is known to alchemists as *algarot powder*, named after Vittorio Algarotti who commodified the substance and distributed it throughout Europe. The third type of *aurum potabile* is best designated *apothecary potable gold* and is any proprietary gold-containing solution sold publicly by physicians and pharmacists that generally included other medicinal oils, herbs or even animal parts such as stag-horn or liver of viper. A variety of these *potable gold* formulations created according to a bewildering number of recipes became popular with European nobility and the general populace.

Alchemical Aurum Potabile – Tincture of the Philosophers
Great renaissance humanist philosopher, writer and reformer Giovanni Pico della Mirandola played a role in shaping attitudes towards *aurum potabile* in Europe. The passages below reveal two pertinent points regarding his views towards *aurum potabile*: 1) he understood *potable gold* to be "*conjunct essence*", that is to say the Philosophers' Stone, and 2) he was convinced enough of its therapeutic potential to publish his views for public scrutiny:

> *They say that the use of potable gold leads to the alleviation of illnesses, the preservation of health, and the lengthening of life, as far as nature allows; this I shall consider, not as settled, but as a matter for argument. In confirmation of this, I assert that it is a matter of controversy among those who are skilled in debate.* Perhaps those who drink up gold agree that it is not really gold that is drunk, but its pure and subtle portion, called the conjunct essence.
>
> I have supported the cause of those who would allow potable gold because I think something can be collected that can be very helpful for

> *our bodies and those of others, especially as everyone agrees that our surgeon Antonius in years gone by restored a marrried woman of Cornelius' house who was dying of consumption to health in a few days, free from her wasting disease, solely by the use of potable gold.*
> – Pico della Mirandola Jr., De Auro, Chapter 4

The *Tincture of the Philosophers* was Paracelsus' personal jargon for the archetypal Philosophers' Stone created via the combination of a fused and powdered *gold-antimony glass* reacted with *butter of antimony*. The term *tincture*, symbolized as ℞, became so synonymous with the Philosophers' Stone and *aurum potabile* that it was still in use during the late-18th century by Sigismund Bacstrom, among others. Paracelsus believed that the *tincture* was in fact a secret sacramental *elixir of life* known to ancient Egyptians and the key to incredible longevity and youthful appearance as indicated by the following passage:

> *Some of the first and primitive philosophers of Egypt have lived by means of this Tincture for a hundred and fifty years. The life of many, too, has been extended and prolonged to several centuries,* **as is most clearly shown in different histories, though it seems scarcely credible to anyone. For** *its power is so remarkable that it extends the life of the body beyond what is possible to its congenital nature, and keeps it so firmly in that condition* **that it lives on in safety from all infirmities. And although, indeed,** *the body at length comes to old age, nevertheless, it still appears as though it were established in its primal youth.*
> – Paracelsus, Archidoxis, De Tinctura Physicorum, Chapter IV

Paracelsus clearly asserts that the *Tincture* is the *Universal Medicine* and a most important key to perfect health and well-being. For Paracelsus the mechanism of action for life-extension and rejuvenation rested upon the *tincture's* ability to literally burn-up or expel disease in what might be construed as an ancient antibiotic:

> *So, then,* the Tincture of the Philosophers is a Universal Medicine, and consumes all diseases, **by whatsoever name they are called, just like an invisible fire. The dose is very small, but its effect is most powerful.** ...
> This is the Catholicum [universal principle] of the Philosophers, by which all these philosophers have attained long life for resisting diseases, **and they have attained this end entirely and most effectually, and so, according to their judgment,** they named it The Tincture of the Philosophers. ...
>
> This, therefore, is the most excellent foundation of a true physician, the regeneration of the nature, and the restoration of youth. **After this, the new essence itself drives out all that is opposed to it. To effect this regeneration,** the powers and virtues of the Tincture of the Philosophers were miraculously discovered, and up to this time have been used in secret and kept concealed **by true Spagyrists [medico-alchemists].**
>   – *Paracelsus,* Archidoxis, De Tinctura Physicorum, Chapters IV-VII

Testimonies such as were asserted by Jābir ibn Hayyān, Paracelsus and other alchemical authorities fanned the flames of legends that characterized *alchemical aurum potabile* as a miraculous wonder drug. Only an adept knew the secret of its preparation and this occult treasure was protected at all costs and had been throughout history. The stakes were high; if the testimonies were correct, pursuit of this arcane knowledge was among the most worthy human endeavors one could engage in. Success meant rejuvenation and the extreme longevity needed to fulfill a greater life's mission in the service of humanity. For many, confecting *aurum potabile* was nothing less than divine labor and required a profound level of spiritual purity and preparation as prerequisites:

> ***What say you, my Child,*** *is not this the true Aurum potabile, and the true Quintessence, and the thing which we seek? It is a spiritual thing, a Gift which God bestows* **upon his Friends, therefore, my Child,** *do not undertake this Divine Work, if you find your self in deadly Sins,* **or that**

> your intent be otherwise than to Gods Glory, and to perform those things which I taught you before.
> – Johann Isaac Hollandus, A Work of Saturn, Its Use in Physick

English botanist, herbalist and physician Nicholas Culpepper wrote *Treatise of Aurum Potabile* published in 1656, which addresses *alchemical aurum potabile* by inference, or as the subtitle suggests – *A Description of the Three-fold World containing The Knowledge necessary to the study of Hermetic Philosophy*. No clear recipe is discernible in the text although Culpepper's treatise presents orthodox alchemical views of the period. It is possible, if one is familiar with the archetypal Philosophers' Stone recipe to identify the figurative elements in the text, yet it would be extremely challenging to learn the recipe from this work without superior skill of extrapolation. What is clear is that the text was not addressing *apothecary potable gold*.

Unlike the direct transmission of Alexandrian Art (téchni; τεχνη) to the Islamic world via Morienus passing the alchemical torch to Khālid, Europeans had no other choice but to engage in a type of restorative or revivalist alchemy in an attempt to rediscover the secrets of confecting the Philosophers' Stone that had been lost in the interim. The accepted academic view is that they failed in their endeavors, but the failure is actually the failure to comprehend their work fully. European alchemists met with much success in that the Philosophers' Stone was indeed reproduced by adepts and employed therapeutically. European lineages and transmissions ensued resulting in an incredible outpouring of experimental chemistry that played no small role in laying the foundations for the Scientific and Chemistry Revolutions that followed. Up until around the 17th century, the term *aurum potabile* was generally synonymous with powdered *Tincture of the Philosophers* suspended in a liquid medium and applied therapeutically.

11 – AURUM POTABILE

Quintessence of Medicinal Gold – Algarot Powder

The substance known as *Algarot powder* was known from Zosimus' Fixed Mercury/Hermes, Synesius and Stephanos' Hermetic *White Stone*, Paracelsus' *Stone of the Philosophers,* Khunrath's *Great* or *Universal Stone* and Basil Valentine's recipe for the *arcanum of antimony*. Vittorio Algarotti developed a proprietary technique for creating the substance and called it *pulvis angelicus* (angel dust), marketing it as the *quintessence of medicinal gold*. Researcher Guerrero Jose Rodriguez published a comprehensive study of Vittorio Algarotti and revealed what might be called the first international pharmaceutical trade network. Algarotti's success revolved around an aggressive and sophisticated marketing strategy and distribution setup centered upon a single medicine.

Throughout his life, Algarotti attempted to keep his recipe a secret, even from his own distributors, in an effort to prevent others from copying or creating a generic version of his medicine. These efforts successfully protected his trade secret securing a network of wealthy patrons throughout Europe and North Africa. The resulting financial success earned him tremendous prestige as well as many detractors. Rodriguez explains that Algarotti's powder was controversially marketed as being derived from *a white mineral considered the root of gold*, and therefore implied to be a type of *potable gold*. Roger Bacon, Basel Valentine, Alexander von Suchten and Vittorio Algarotti each considered antimony to have the same inherent virtue as gold and that antimonial medicine was in effect a form of *potable gold*; a notion causing much confusion for the populace not versed in alchemical jargon.

Glauber and others soon developed various novel techniques for creating *algarot powder*. Its chemical composition was first revealed by Sabanejew in 1871. According to Sabanejew, hydrolysis of *butter of antimony* resulted in antimony-oxychloride ($SbOCl$) which, when treated with water for a second time resulted in *algarot powder* ($Sb_4O_5Cl_2$). The powder was studied in 1885 by LeChatelier who was most interested in reversibility of the reaction. Hydrolysis of *butter of antimony* can yield a number of different products based on water and antimony-trichloride ratios. Researchers F.C. Hentz Jr. and G.G. Long of North Carolina University carried out an in-depth study to determine which of the various possibilities most closely matched those of ancient recipes. Their findings were published in the Journal of Chemical Education in March 1975 corroborating those of Sabanejew. A sample of a traditional 17th century recipe for Algarot Powder occurs in Glaser's *Cours de Chymie*:

> **Emetic Powder or Algarot – Christophe Glaser, *Cours de Chymie* 1663**
> 1. Take about half of your *glacial oil of antimony [butter of antimony], put it in a bowl, in which there is a pint [568.26 ml] of warm water, you will see it immediately precipitate into powder white as snow;* the water held weak corrosive spirits, which held part of reguline [metallic] antimony in dissolution, having forced them to abandon this body.
> 2. The precipitation being completed, [we] must stir everything once again, then let the powder settle, & pour by inclination [decant] into a bottle the supernatant water [liquid layer], and *keep apart;* because the first lotion contains within itself all the spirits that were attached to saline Antimony. It has a pleasant acidity, which is why it is *called philosophical spirit of vitriol.*
> 3. Continue washing & watering down the powder, then *dry & keep.*

The dose of *the powder* is from two to six grains; *it is used for cleaning filth & viscosities from the stomach:* it purges upwards [by regurgitation] and downwards [by diarrhea]. It is also used to purge dropsy [edema; fluid imbalance], the mingling amidst other purgatives, which entertain its vomitive force, and make it do its full force at the bottom.

One uses the first lotion in juleps, and in beverages of fever patients [as a diaphoretic; sweat inducer], which makes it tangy and very pleasant.

The recipe results in two medicinal products – 1) emetic *algarot powder* (purgative by regurgitation and diarrhea), and 2) diaphoretic *philosophic spirit of vitriol* (sweat inducer) known as *the first lotion*. These medicines remained popular into the 19th century. The *julep* mentioned above is similar to *oxymel* used in Jābir's miraculous cure and derives from Arabic *julāb* (الجلاب), meaning syrup, Persian *gulāb* (گلاب) meaning rose-water syrup or *sekanjabīn* (سكنجبين) Arabicized from Persian *sirka-anjubin*, meaning *vinegar-honey*.

Apothecary Potable Gold

> *He who makes gold for the advantage of either the healthy or the sick brings great increase of wealth to mortals...*
> – Pico della Mirandola Jr., De Auro, Chapter 5

Around the turn of the 17th century apothecaries, pharmacists of their day, attempting to capitalize on the reputation of *alchemical aurum potable* began to develop gold-based medicines and dispense them to physicians and directly to patients. Apothecaries were not adept-alchemists as a rule and therefore were forced to develop their own formulations, which resulted in gold-based medicines fundamentally dissimilar to *alchemical aurum potable* in composition and efficacy, although boasting the same name. For purposes of differentiation, formulations such as these are termed *apothecary potable gold* in this treatment. Scathing criticism from Basil Valentine in his *Triumphal Chariot of Antimony* offers insight into the disdain some European medico-alchemists displayed towards apothecaries and physicians in this regard:

> *They inscribe upon a sheet of paper, under that magic word "Recipe," the names of certain medicines, whereupon the Apothecary's assistant takes his mortar, and pounds out of the wretched patient whatever health may still be left in him.*

> *Change these evil times, oh God! Cut down these trees, lest they grow up to the sky! Overthrow these overweening giants, lest they pile mountain upon mountain, and attempt to storm heaven!* Protect the conscientious few who quietly strive to discover the mysteries of Thy creation!

Apothecaries relied on basic gold-salts originally discovered by alchemists such as aqueous chlorauric acid ($HAuCl_4$) and gold trichloride crystals ($AuCl_3$) that result from dissolving gold in aqua-regia. These salts attract water and are therefore highly soluble in water, alcohol and ether, which makes them very apothecary friendly, easy to work with yet somewhat dangerous to ingest in all but the most minuscule or highly buffered doses. The story of *apothecary potable gold* and cordials is preserved mainly in the form of historical fragments. An incredible array of *apothecary potable gold* formulations were created during the 16th and 17th centuries, some fraudulent and containing no gold at all, some containing gold as dissolved gold salts while other formulations featured gold in novel forms such as acetates and nanoparticles. The following is neither thorough nor comprehensive and serves only to highlight a few of the more notable formulations that are historically relevant, potentially therapeutic and include gold as a primary active ingredient. The story of *apothecary potable gold* reveals a transition from alchemy to early-modern chemistry.

Dr. Anthony's Potable Gold

One of the most unique and interesting *apothecary potable gold* formulations was developed by noted English apothecary and physician Dr. Francis Anthony and published in 1610 in *Midicinae Chimicae, et assertio potabilis Auri*. Like Vittorio Algarotti, his practice revolved around a single unique and highly secret formulation. His father was a goldsmith in the service of Queen Elizabeth, a factor that may have influenced Anthony's decision to focus on gold as the foundation for his therapeutic medicine. He studied chemistry at Cambridge where he received his M.A.

and M.D. credentials. Anthony was a qualified physician yet this fact did not prevent him from becoming embroiled in a standing dispute with the Royal College of Physicians at London for practicing medicine without having addressed himself to the College for their license. He published a modest yet intelligent defense of his work in Latin, which included theory and history of therapeutic gold along with a number of case studies as evidence of efficacy. This was met with a quick counter-attack by Dr. Matthew Gwinne of the College of Physicians at London and physician Dr. John Cotta of Northampton, to which Dr. Anthony published his rejoinder in both English and Latin. Another aspect of his defense was the later publication of his recipe, yet the question remained as to whether his technique for dissolving gold and other details were truly forthcoming or not.

Dr. Anthony's recipe, although not specifically alchemical, follows the basic alchemical *modus operandi* or template published by Bacon and Valentine for creating oils of antimony and stibnite via acetic acid and ethanol. Dr. Anthony used gold in place of antimony and combined tin oxide with concentrated acetic acid to produce his menstruum:

Dr. Anthony's Potable Gold Recipe – from Biographia Britannica 1744

This account of Dr. Francis Anthony's method of making his aurum potabile was transcribed from his own manuscript, which was once in the possession of a Chemist well known to the author of this life from whom he had it. *The secret was long in Dr. Anthony's family,* and very beneficial to them ... The reader hath in this note, *the whole of Dr. Anthony's receipt [recipe] without the smallest alteration or omission* [pedagogical breakdown by the author].

The Philosophic Vinegar [Acetic Acid]

1. Take 6 gallons of the strongest red wine vinegar, and set as many Stills at work at a time as your Balneum will hold. Throw away the first pint that comes over,

2. wash and wipe the still, and then put in that which was distilled [redistill], putting away always the first pint for 5 times, so out of a gallon you shall have 3 pints, and out of a whole 6 gallons 10 quarts of spirit of vinegar [acetic acid],
3. which keep in glass bottles well corked with a leather over it.

The Philosophic Salt [Tin Oxide]

1. Take an iron pan, like a dripping pan, and having made it red hot,
2. put into it as much as you will of block tin, and stir it continually, until it turns to a kind of ashes or calx [tin (II) oxide; SnO], and keep the fire up to a good height all the time, which may be half a day or fifteen hours, some of these ashes will look red [metastable variant], which is a sign the operation is well performed.
3. These ashes thus obtained keep in a glass close covered.

The Process for Dissolving Gold [Menstruum Production]

1. Take 4 ounces of these ashes, and of the spirit of vinegar 3 pints, put them in a glass like an urinal, and let the ashes be put in first. Lute the vessel, set it in a hot bath for 10 days, then take it out and set it to cool, shaking every 2 hours, and in three days all the dregs will fall to the bottom.
2. Let that which is clear be drawn off into a glass bason by 2 or 3 woolen threads [wick distillation], then distill it; to this distilled water put 4 ounces of fresh ashes, put also a quart of spirit of vinegar on the first ashes; lute the glass as before, set it in a hot bath, and let it digest 10 days, filter this and distil as before;
3. thirdly, put on that ashes a pint of spirit of vinegar, set it in the hot bath 10 days filter and distil it, after the third infusion throw away the ashes.
4. Take this distilled water, pour on it fresh ashes, keeping the weight and order of fusions, filterings, and distillations 7 times, then the spirit will be well impregnated with the salt and you have the menstruum sought [tin oxide saturated acetic acid].

11 – AURUM POTABILE

The Process for Dissolving Gold [Gold Reduction]

5. Take an ounce of pure gold in the ingot, file it into small dust, put it into a crucible with as much white salt as will near fill the pot, and let it stand in a moderate heat 4 hours [gold and salt fusion, salt removes any trace silver as well as reduce gold particle size],
6. then take it out and grind it on a Painter's stone, return it thence to the crucible,
7. calcine and grind it 4 or 5 times until it looks red and blue [reduced to nanoparticles], and then it is fit for use.

The Process for Dissolving Gold [Gold Calcination]

8. Put it next into a glass bason, pour upon it scalding hot water, stir and decant it, till the water when settled has no taste of salt, which will take 2 or 3 days.
9. By this operation you will have 16 or 17 grains of a very fine white calx, which will swim on the top of the water, and may easily be blown over into another bason, and the water being evaporated by a gentle heat, it will remain a white powder.
10. By repeating the calcination and grinding, the whole ounce of gold may be reduced to such a calx.

The Process for Dissolving Gold [Creating the Tincture]

11. Take an ounce of this calx, put it in a urinal, pour upon it half a pint of the menstruum, lute it close and set it in a hot bath six days, shaking it often every day, let it cool 3 days and then pour it gently off.
12. Take this liquor, put it in a glass still, and with a gentle fire, evaporate it, till it becomes the consistency of honey, then remove the fire, take out the contents, put them in a glass bason, and with the bottom of another round glass, grind them to powder.
13. Put this powder into a urinal containing about a pint [volume], and add somewhat more than a half a pint of rectified spirit of wine [ethanol]; set it in a cold place for 10 days, shaking it often for the first 7 days, but afterwards let it stand without shaking and the tincture will appear of a fine red.

14. By putting a bare half pint of rectified spirit of wine on the dregs, a second tincture may be drawn, and if this be very high colored, you may draw a third.
15. Put all these colored liquors together, distil them, and there will be left behind a clammy substance of the consistence of honey, one ounce of which put into a quart of pure canary wine [Malvasia], is my Aurum Potabile.

The recipe above employs a solution of tin-ash dissolved in concentrated acetic acid originally sourced from a reduction of red wine vinegar. This menstruum is reacted with salt-fused and mechanically ground gold nanoparticles (red in color but blue as they aggregate) and concentrated to an oily reduction, which is further ground and dried to a powder. Dr. Anthony's recipe suggests that he was creating a matrix of gold nanoparticles and aqueous tin acetate salts in solution, although the published methodology is questionable.

He then creates a tincture by dissolving the matrix in ethanol and evaporating to an oily concentrate. His *potable gold* was then combined with *Canary wine* made famous by Shakespeare, who mentions it in *Henry IV, Twelfth Night* and *The Merry Wives of Windsor*:

> You have drunk too much Canaries, and that's a Marvelous searching wine. – Henry IV, by William Shakespeare

Canary wine is a fortified sweet white Malvasia with a somewhat yellow-amber tint and alcohol content of around 17%. It is still produced in the Canary Islands off the northwestern coast of Africa. The Malvasia grape is grown on the slopes on Lanzarote yet is thought to have originated in Ancient Greece. During Dr. Anthony's time, it was imported to England where it became immensely popular with royalty and the upper class. Dr. Anthony picked a very fitting medium for creating his *potable gold* and this may have added somewhat to its value-added prestige and appeal. During the Elizabethan period, *potable gold*, although not necessarily Dr. Anthony's and canary wine both occur in Shakespeare's play *Henry IV*:

> *I spake unto this crown as having sense, And thus upbraided it: the care on thee depending Hath fed upon the body of my father; Therefore, thou best of gold art worst of gold: Other, less fine in carat, is more precious, Preserving life in medicine potable.*
>
> — Henry IV, *by William Shakespeare*

Angelo Sala's Auro Potabili

Italian physician and chemist Angelo Sala published his *Processus de auro potabili* (Process of Drinkable Gold) in 1630. His methods reveal a focus on dissolving purified gold in aqua-regia to achieve gold salts, a process which served as the foundation for the advance of potable gold and colloidal gold to follow:

1. First gold is purified through antimony
2. The gold-leaf [is] reduced the same
3. It must be dissolved in aqua regia of [Basil] Valentine, or by one which is described by [Oswald] Croll

Johann Rudolph Glauber's Golden Carbuncle

Glauber shifted between the two worlds of applied chemistry and alchemy, which allowed him to earn a livelihood from his alchemical and chemical products and recipes. Glauber addressed recipes based on gold salts such as Sala's above with contempt in his *De auri tincture sive auro potabili vero* published in Amsterdam in 1646. In this work, Glauber extols *alchemical aurum potabile* as being the genuine article and derides *apothecary potable gold*. He provides a recipe for *alchemical aurum potabile* but this does not necessarily preclude Glauber's publishing recipes elsewhere for *apothecary potable gold* innovatively achieved by dissolving gold in *strong spirits of salt*. *Strong spirits of salt* is Glauber's jargon for a unique solvent achieved by distilling concentrated zinc chloride solution and combining it with hydrochloric acid. For administering these gold salts, he advised:

> ... therefore a drop of it is to be mixed, with a spoonfull of Beer, Wine, or warm Broth, before it be administred to weaken the spirit of salt; but if any desire to have it sweeter, instead of Wine, Beer, or Broth, it may be mixed with melted Sugar, or syrup of Roses.

Glauber wrote prolifically on chemical processes and is most known for his discovery of sodium sulfate (Na_2SO_4), which he referred to as *Sal Mirabilis* (Wonderful Salt); a salt known today as *Glauber's salt*. He innovatively applied his new *Wonderful Salt* to an *apothecary potable gold* product that he coined the *Carbuncle of Gold of the Ancients*. This is a strange name for an innovative and completely original *potable gold* product; "*ancient*" as a descriptive being far from applicable. Despite the appellation, his organo-metallic *Carbuncle* was a unique addition to the many apothecary style *potable gold* products during the 17th century.

> **The manner of conjoining Gold contrary to its nature, with any burning and Volatile Vegetable Sulphur, and of amending the other Metals, all done by the help of my Sal Mirabilis.**
>
> – Glauber, Centurys, Book 2: 74

1. Take one Quinta [quintal = jargon for full unit; hundred weight or quanta = plural of quantum; a quantity or portion] of small weight of Gold, more or less, reduce it;
2. add thereunto six, eight or ten parts [by volume] of Sal Mirabilis [sodium sulfate; Na_2SO_4], which matters you must melt in a Crucible with an accurate and strong fusion:
3. When they flow, throw in some pieces of Coals into the Salt and Gold as they are melting in the Pot, that the Sal Mirabilis may dissolve the Gold and Coals in the melting. Which usually is done in half an hour or thereabouts.

The matter being poured out will shew you whether or no you have well operated for all the Gold, as likewise the Sal Mirabilis and Coals will be dissolved and changed into a red Stone, *that bites the Tongue as if it were Fire.*

This Fire and red Stone, is the golden Carbuncle of the Ancients, *for it shines in the dark like a burning Coal, and produceth such wonderfull effects in Medicine and in Alchemy, which we have no mind at present to reveal. ...*

The way of making a most excellent Medicine out of the Carbuncle of Gold. – Glauber, Centurys, Book 2: 75

1. This Carbuncle is to be beaten into Powder, *and* the best Spirit of Wine [ethanol] is to be poured thereupon, which *may extract the Tincture.*
2. *This tinged Liquor is to be poured off into another Glass, and more fresh Spirit is to be again poured upon the matter, that it may again extract in the heat more Tincture; these Labours you must* repeat so often till all the Tincture is extracted, *and the Spirit will be no more coloured.*
3. The Spirit being drawn off by distillations in a Bath leaves behind a most red Tincture *in the bottom,* in the form of a Liquor named COS, for here are present, Colour, Odour, Savour or Taste; the Colour and Odour from the Gold and Sulphur; the Savour from the Salt. ...

Thus hast thou friendly Reader a Medicine of great moment and of great efficacy, *in which the most pure parts of the Gold and of the Vine are conjoined, nor can they be other than a most profitable Medicament for men and metals.*

Glauber may have been attempting to create a pun with the acronym C.O.S.T. for his product, but he also may have been using wordplay to suggest *cos* or *cosa* (Latin: grindstone or Italian; thing, matter, piece, object). He achieves gold reduction by fusing it with a salt much like Dr. Anthony's method, a process that may have its origins in ancient Egypt.

Hoffman's Tinctura Solis

In 1722, German physician F. Hoffmann described a potable "gold tincture" called *Tinctura Solis* made by causing a chemical reaction between cinnamon essential oil, ethanol and chloroauric acid digested with gentle heat. He achieved a dark reddish-brown resinous mass, which was then converted to a tincture with ethanol:

> *Hoffmann's recipe*
>
> It's time to examine that Solar tincture, used to heal, which is prepared based on gold with cinnamon oil, as follows:
>
> 1. thicken, the way you wish, a fully saturated solution of pure gold.
> 2. Then, a dram of pure cinnamon oil is dissolved in three drams of spiritu rectificadisimo (ethanol).
> 3. A portion of the gold solution is mixed with three parts of this other solution (the pure cinnamon oil and spiritu rectificadisimo) in a small pan placed on warm sand.
> 4. Thus, both of them bind together, resulting in a resin-like mass, black like pitch, which has to be dissolved in spiritu rectificadisimo, from which a dark brown essence, tasty, though bitter and subastringent, results.
>
> This [remedy] can be applied when the forces must be strengthened.

Potable Gold of Paris

> *Founded in Paris in 1666 under Louis XIV. Its [Académie Royale des Sciences] mission was to advance natural philosophy and mathematics and to apply the laws of nature to practical reforms. The academy quickly became a preeminent arbiter of scientific thought, a role it sustained until 1793, when it was dissolved, to be reincarnated in 1795 as part of the Institut de France. From the start, this company of savants was a monument to royal patronage and to the ideals of the Scientific Revolution.*
>
> – Wilbur Applebaum, Encyclopedia of the Scientific Revolution

11 - AURUM POTABILE

Potable gold of Paris was considered one of the finest *apothecary potable gold* preparations of the 17th century. The recipe was widely known throughout the 17th century, the following from *The New Dispensatory*, London published in 1753.

1. Dissolve, with a moderate heat, *half a dram of fine gold, in two ounces of aqua regia;*
2. and add to the solution *one ounce of the essential oil of rosemary.*
3. Shake them together, *and then suffer them to rest :*

... the acid loses its gold color; and the oil, which arises to the surface, becomes richly impregnated therewith.

4. Separate the oil *by decantation, and add to it four or five ounces of rectified spirit of wine :*
5. digest this mixture for a month, *and it will acquire a purplish colour.*

Gold trichloride gives up its chlorine ions when dissolved in rosemary essential oil and remains in colloidal form.

> *It would be a shame if we omitted to place rosemary on the list of strange and wonder-working herbs,* for, indeed, the virtues of rosemary were formerly very great, although now they appear to have fallen into abeyance. ... *It has taken away our charms and our philters, it has put to flight our familiar fairies, and dispersed most, if not all, our hobgoblins; it has removed the ancient landmarks and dealt a deathblow to the old superstitions.* ...but its most remarkable property consisted in making old folks young again. Precious, precious rosemary; could you but accomplish that now, gold of Ophir would be your price!
> – Andrew T. Sibbald, Curiosities of Ancient Medicine

The above passage was perhaps not addressing *potable gold of Paris* specifically, but it is apropos to the role essential oil of rosemary played in gold technology's shift from alchemy to chemistry. Based on today's understanding of the constituents of rosemary oil, it was a very interesting substance by which to make *potable gold of Paris*. Pliny, Dioscorides and Galen all wrote of rosemary. Rosemary has been found in Egyptian tombs and has been used therapeutically throughout history. Paracelsus held that rosemary could strengthen the entire body and that it was especially therapeutic for the brain, liver and heart; essentially acting as a nerve and cognition tonic. Oil of rosemary in former times was distilled only from the flowering tops, whereas modern commercial oil of rosemary is sourced mainly from the stems and leaves and is likely somewhat different from that originally used by Duclos.

The Académie Royale des Sciences was headed in the beginning by Dutch mathematician Christiaan Huygens and Bolognese astronomer Jean-Dominique Cassini, both incredibly gifted scientists backed by powerful families or patrons. They appointed four of France's most talented polymaths including Huguenot alchemist and doctor to King Louis XV, Samuel Cottereau Duclos. Duclos championed chemical medicines and was a staunch advocate of *potable gold*. There was a great intellectual exchange between England and the Académie during this formative period and Robert Boyle's was among the works most thoroughly reviewed by the Académie. The recipe and insight *potable gold of Paris* engendered for manipulating gold was intriguing to other chemists working in this area and specifically those of the Hartlib circle in England.

Even the most *sceptical chymist*, Robert Boyle, admitted that medicinal efficacy might be found in a specific type of *potable gold* or as he calls it, *tincture of gold*. He published his skepticism of gold-based medicines in *Usefulness of Natural Philosophy*, II: I, yet ultimately ended up publishing his own recipe for *potable gold*. A letter written by Boyle that provides unique insight into his change of heart was published in *Biographia Britannica* in 1744. Boyle wrote:

> *Tho'* I *have been long prejudiced against the pretended aurum potabile, and other boasted preparations of gold, for most of which I have still no*

great esteem; yet I saw such extraordinary and surprising effects from the tincture of gold I speak of, upon persons of great note, **with whom I was particularly acquainted, both before they fell desperately sick, and** *after their strange recovery, that I could not but change my opinion for a very favourable one, as to some preparations of gold.* **But tho' this simple medicine can only be made in small quantities, and that too not without a great deal of pains and time, I can speak thus circumstantially of it, because by the kindness of the artists, and the pains I had bestowed upon the working upon the same subject they use for their menstruums,** *I so far knew and partly saw the preparation of it, as to apply what has been said* **to the present occasion. ... [Boyle then details three case studies of miraculous cures] ...** *I could relate some other odd effects of this remedy;* **but the present may suffice to alleviate a prejudice against medicines, made of so fixed and supposed unalterable metal as gold.**

English chemist William Lewis contributed the recipe for *aurum potabile of the faculty of Paris*, in *An Experimental History of the Materia Medica*, published in 1761. His chapter on gold medicines demonstrates just how far gold technology had advanced during the century that followed Duclos' *potable gold of Paris*. He provided no further details on how the recipe came into his possession, the final form of the medicine or its dosage.

> AURUM Vel Sol Pharm. Paris – GOLD ... *easily dissolving in a mixture of the nitrous and marine acids, called* aqua regis, *into a yellow liquor which stains the skin purple.*
>
> *Essential oils, shaken with this solution, imbibe the gold from the acid, and carrying it up to the surface, keep it there for a time dissolved;* **but gradually throw it off again, on standing for some hours, in form of bright yellow films, to the sides of the glass.** *The ether or* spiritus vini æthereus *takes up the gold more readily and completely, and keeps it permanently dissolved. Rectified spirit of wine mingles uniformly with the acid solution; but on standing for some days, the gold separates from the mixture,* **and rises in films to the surface. A piece of tin, placed**

in the solution largely diluted with water, changes it red or purple, and throws down a precipitate of the same colour. By the appearances resulting from these additions, very minute portions of gold, dissolved in acid liquors, may with certainty be discovered. …

Potable Gold of Mademoiselle Grimaldi

French chemist and pharmacist Nicholas Lemery referred to *Potable Gold of Paris* as *Potable Gold of Mademoiselle Grimaldi* in his landmark *Course of Chemistry*. His editor and commentator, M. Baron, provided a recipe for the 1757 edition shown here. In the 1766 *Dictionary of Chemistry*, French chemist Pierre-Joseph Macquer explained that the product contained finely divided particles of gold (gold nanoparticles) "floating in the oily fluid". Lemery's process is identical to Potable Gold of Paris:

> **Lemery's Recipe**
> 1. Takes 1/16 [troy] ounces of the purest gold [1.94 grams], dissolve it in two ounces [56.83 ml] of aqua regia;
> 2. add on this solution, whose color shall be beautiful yellow, one ounce of rosemary essential oil;
> 3. mix together the two liquids; let them to stand quite. You will soon see the oil, colored in a beautiful yellow, over floating the aqua regia which should have lost all its color;
> 4. separate the two liquids by using a funnel, leaving all the aqua regia to flow out through the extremity [of the funnel], which you will close with your finger just before the oil is ready to leave it;
> 5. then, place the oil in a matrass and mix it with five times its weight of rectified wine spirit [ethanol];
> 6. close the matrass and place it on a sand bath for a month.
>
> After that, it will have a purple color and a graceful taste, though a little bitter and astringent. This will be the gold tincture, or the Mademoiselle Grimaldi's potable gold.

Essential oil extracted from *Rosmarinus officinalis* contains an incredibly dense package of more than 38 phytochemicals. Today many secondary plant metabolites present in essential and fragrance oils – terpenes, terpenoids, alkaloids and phenolic compounds – have been identified and studied in-depth. These compounds serve primary functions such as chemical defense system to ward off bacteria, fungi and to attract insect or animal pollinators. Rosemary oil exhibits known antibacterial and antifungal properties. Iatrochemists of the 16th and 17th century became masters at isolating these *quintessences* from plants and essential oil of rosemary is just one fitting example of the legacy of plant alchemy. Essential oil of rosemary employed to create *potable gold of Paris* is just one such example. As a combination of gold nanoparticles with a matrix of organic phytochemicals, *potable gold of Paris* qualifies as an *organometallic* compound.

Hoffman's *Tinctura Solis* and Lemery's *Potable Gold of Grimaldi* were replicated in a modern laboratory in 2014 by Mayoral et al. and the results published in the Gold Bulletin. Analysis by transmission electron microscopy (TEM) revealed that 99% of the gold content in the aqua regia was taken up by the rosemary essential oil. *Potable Gold of Grimaldi* contained stabilized gold nanoparticles with particle sizes from 1.89 to 18.80 nanometers (nm), 90% of which were between 2 and 8 nm with an average particle size of 5.54 nm at a concentration of 220 ppm. Hoffman's *Tinctura Solis* contained gold nanoparticles in small clusters of a few atoms up to particles 20 nm in size. Modern commercial colloidal gold typically contains gold at 10 to 20 ppm with an average particle size of approximately 27 nm. In both cases, the oils acted as both a reducing agent and a capping agent, which prevented the small crystalline gold particles from combining to form larger ones (aggregation).

Hoffman's *Tinctura Solis* and *Potable Gold of Paris-Grimaldi* were 18th century therapeutic organo-gold formulations with nano-gold in bioavailable and bioactive form comparable if not superior to modern colloidal gold products. *Potabile gold of Paris* may serve as the precise bifurcation point from which *aurum potable* opened the doors to a burgeoning scientific understanding of *colloidal gold* and ultimately today's *capped, conjugated* and *functionalized gold nanoparticles*. The shift from alchemy to gold salts and colloids resulted in the altering of traditional definitions for *elixir, tincture* and *aurum potable* thereafter.

Elixir – originally from Greek x̱irós (ξηρός; dry) or x̱írion (ξήριον; medicinal powder), transliterated to Arabic *al-iksir* (الإكسير) from which it entered mediæval Latin as *elixir*, all originally synonymous with the Philosophers' Stone. During the 16th century, the term came to indicate a medicinal tonic in a general sense. Today the word *elixir* is generally understood to mean a hydro-alcoholic solution containing at least one bioactive compound and may contain other inactive substances such as colorants, preservatives, and buffering agents, flavor enhancers, etc. The Merriam-Webster Dictionary defines elixir as follows:

1. *a (1): a substance held capable of changing base metals into gold*
   (2): a substance held capable of prolonging life indefinitely
   b (1): cure-all (2): a medicinal concoction
2. *: a sweetened liquid usually containing alcohol that is used in medication either for its medicinal ingredients or as a flavoring.*
3. *: the essential principle*

Tincture – originally from Greek v̱afí (βαφή; dye, pigment), which entered mediæval Europe as *baphe* in the compound word *baphe-met*, Latin *baphometh* or *baffometi*. During the early Alexandrian period the *elixir* or *tincture* was known in alchemical trade-jargon as *ios*, synonymous for *venom, poison, medicine, arrow, medicinal honey* or *purple* with the verb form *iosis* being jargon for *tingeing, dying* or *coloring* (with the royal purple vestment or robe). Paracelsus popularized the term *tincture* by using *Tincture of the Philosophers* synonymous with the Philosophers' Stone as a medicine. The alchemical symbol ℞ remained representative of the product of a Philosophers' Stone recipe into the late 18th century. *Tincture* eventually evolved to mean an alcohol extract similar to the standard definition of *tincture* in chemistry as a solution that employs alcohol as the solvent. Dictionary definitions of tincture are as follows:

1. a: archaic: a substance that colors, dyes, or stains
   b: color, tint
2. a: a characteristic quality : cast
   b: a slight admixture : trace (a tincture of doubt)
3. obsolete: an active principle or extract
4. : a heraldic metal, color, or fur
5. : a solution of a medicinal substance in an alcoholic solvent

Aurum potabile – originally meant the Philosophers' Stone in a liquid suspension to be ingested. During the 16th and 17th centuries, the term *aurum potabile* encompassed some antimony medicines on the basis that alchemists like Bacon, Valentine and Algarotti held the notion that antimony and gold ultimately derive from the same source. Eventually the term *aurum potabile* came to denote a liquid tonic or cordial containing gold in the form of soluble gold salts, acetates or nanoparticles in colloidal suspension. A dictionary definition of *aurum potabile* reads:

: a formerly used cordial or medicine consisting of some volatile oil in which minute particles of gold were suspended

The story of *aurum potabile* reveals alchemical origins of these terms in context, their evolution and modern use as commonplace and scientific expressions. Only just recently are scientists shedding light on what alchemists have always held to be valid – the therapeutic value of gold. *Alchemical aurum potabile* had opened the door to advances in gold research that seemed to shift the spotlight away from alchemy and in the direction of industrial applications.

In 1857, Michael Faraday used phosphorus to reduce gold chloride to achieve the first European sample of pure gold nanoparticles in solution. Since then great advances in gold nano-technology have led to the ability to control particle-size, stability and to combine gold nanoparticles with organic materials. This has resulted in thousands of papers and patents published about gold nanoparticles and their functionalization for various applications. One unique discovery was the fact that the traditional Indian Ayurvedic gold medicine Swarna Bhasma is composed of stabilized dry gold nanoparticles. Furthermore, the earliest recorded use of therapeutic gold nanoparticles occurs in the *Sushruta Samhita* (सुश्रुतसंहिता) and the *Caraka Saṃhitā* (चरक संहिता), dating from the 3rd to 4th century BCE or slightly earlier as an antitoxin and to improve intelligence, although these works fail to mention the method of particle-size reduction. What has been heralded as new gold nano-technology today is actually based on products and processes derived from ancient elixir alchemy.

SOPHIC BIOCHEMICAL DYNAMICS

One of the greatest challenges to an accurate understanding of the therapeutic potential of *alchemical aurum potabile* specifically, and gold therapeutics in general, is the current view of what qualifies as a medicine. Today's mainstream approach to medicine focuses on cure rather than prevention or maintenance and regards nutrition, health-maintenance or enhancement as alternative or outside institutionalized healthcare systems. Modern physicians and other healthcare professionals typically swear to some form of the *Hippocratic Oath*, yet completely ignore, deny or fail to integrate Hippocrates' first injunction that addresses diet as the primary approach to wellness and disease prevention:

> *I will apply dietetic measures for the benefit of the sick according to my ability and judgment; I will keep them from harm and injustice.*
> – *1st injunction addressing health in the Hippocratic Oath*

Throughout antiquity, anything that affected the body's ability to regulate temperature or to retain or expel fluid was considered a medicine and for this reason, foods and medicine were tantamount. Spicy ingredients were able to induce sweating or a runny nose and therefore were identified as diaphoretic. Salty substances helped the body avoid dehydration and retain fluids, while certain foods eased bowel movements and so forth. As a result, foods and medicines were loosely classified as either *diaphoretic* or *emetic* to varying degrees. A mysterious third class of medicines operated in a different manner as explained by Applebaum in the *Encyclopedia of the Scientific Revolution*:

> There had always been drugs that did not seem to operate on the body in the manifest ways of raising or lowering body heat, or through retaining or expelling fluid, **and the great medical authority** Galen (second century) had referred to these as working in some occult way on "the whole substance" of the body. **The introduction of new drugs during the Renaissance, particularly from the New World but also from the Far East, to say nothing of the efforts of the** Paracelsians and other iatrochemists to introduce mineral, as opposed to botanical, drugs, led to a vast increase in the number of drugs that were acknowledged to operate by occult means.

These medicines generate an overall sense of well-being paralleling the rejuvenating or tonifying medicines of Ayurveda, Unani-Tibb and Traditional Chinese Tonic Medicine; many preparations of which are mineral and metal-based or in combination with botanicals. Other examples might include medicines originating from homeopathy, traditional herbalism or even shamanic traditions that rely on the therapeutic use of non-harmful and non-addictive plant-based entheogens that would have been considered as operating by occult means in former times.

For anyone researching the Philosophers' Stone as a medicine, the overriding impression is two-fold:

1. There must have been a perceptible physical effect on the person ingesting the elixir to elicit such powerful testimonials as occur in alchemical writings, and;
2. It somehow seems to have triggered a noticeable psychoactive event or non-ordinary state of perception or awareness to a certain degree.

Today institutionalized medicine tends to consider a medicine in terms of curative potential only, one medicine for one ailment, and completely disregards organic macro-micro nutrition, phytochemicals, mineral and metal trace nutrients and non-essential accumulated elements as valid adjuncts to health and well-being. Yet these are precisely the substances that fascinated alchemist-physicians and renewed interest in their therapeutic potential is making resurgence today. In addition to antibiotic properties of some plant oils, other substances created by alchemists are known to possess many useful pharmacological properties. This has proven especially true in the *elixir alchemy* of Persia and India, which boasts a living tradition into the 21st century and fully incorporate *organometallics* and *metallotherapeutics*. The notion of an effective mineral or metal-based alchemical elixir should come as no surprise to health conscious individuals who routinely supplement their diets with iron, zinc, copper, magnesium, potassium, calcium, etc. Minerals, metals and salts of the type alchemists specialized in, subtle in effect though extremely vital to health and well-being, certainly have earned a place in today's wellness approaches.

The Philosophers' Stone in the form of *alchemical aurum potabile* has yet to be the subject of modern research to ascertain its therapeutic significance. Until *alchemical aurum potabile* is reproduced for this purpose and studied *in vivo* in randomized double-blind placebo-controlled trials and the results peer-reviewed, its potential remains a mystery. Alchemists historically relied on anecdotal accounts or resorted to risky self-bioassay to gain deeper insight into *alchemical aurum potabile's* effects. Recently, therapeutic gold salts and nanoparticles have been increasingly researched and experimented with resulting in a greater comprehensive understanding of gold's bioavailability, translocation and bioactivity than ever before. Indian scientists are leading the research into traditional gold-based elixirs. Unfortunately *alchemical aurum potabile's* effects *in vivo* remain a mystery, even more so the reported influence it seems to exhibit on consciousness and cognition. The following section will highlight research into gold's biological role and therapeutic applications. Finally, an example of a traditional 17th century recipe for *alchemical aurum potabile* and reports of its effects offer a point of departure for future research into its legendary health and cognitive potential.

Therapeutic Gold

> *Of all Elixirs, Gold is supreme and the most important for us... gold can keep the body indestructible... Drinkable gold will cure all illnesses, it renews and restores. – Paracelsus*

Gold's therapeutic value as preserved in the writings of Pliny the Elder was mostly as a topical treatment for skin disorders. Pliny provides recipes for creating medicinal gold ashes and liniment in *Natural History* 33: 25 in a section entitled *Eight Remedies Derived from Gold*. His recipe for gold ashes involves crucible work in a style similar to early Alexandrian alchemical technique and later cementation techniques for separating traces of silver from gold:

> *Gold, too, is melted with twice its weight of salt, and three times its weight of misy [roasted chalcopyrite]; after which it is again melted with two parts of salt and one of the stone called "schistos" [alum or red ochre]. Employed in this manner, it withdraws the natural acridity from the substances torrefied [roasted] with it in the crucible, while at the same time* it remains pure and incorrupt; the residue forming an ash which is preserved in an earthen vessel, and is *applied with water ...*

BOOK 3 – APPLICATIONS & OBSERVATIONS

> *Gold, boiled in honey with melanthium [fennel flower] and applied as a liniment* to the navel, acts as a gentle purgative upon the bowels.

Paracelsus was one of the pioneers in the use of therapeutic gold applied to treat nervous disorders, epilepsy and specifically melancholy because according to him, *aurum potable* "made one's heart happy". The notion that gold acts therapeutically as a *nervine* dates back at least to the period of Islamic alchemy as evidenced by a passage occurring in *An Experimental History of the Materia Medica,* published in 1761:

> *GOLD was introduced into medicine by the Arabians, and held to be one of the greatest cordials and comforters of the nerves.* As it apparently can have no medicinal effect in its gross state, not being dissoluble by any fluid that can exist: in the bodies of animals; the *chemists have attempted to subtilize and resolve it, and to extract what they called an anima or sulphur from it*. But as no means have been discovered of separating the component parts of this metal, their tinctures and aurum potabiles *either contained none of the gold [due to conversion to ionic form], or were no other than diluted solutions of its whole substance [in nanoparticle / colloidal form]. That* the aurum potobile *of the faculty of Paris, reckoned one of the best of the preparations of this kind, (made by shaking some oil of rosemary with a solution of gold in aqua regis, and afterwards digesting the oil for a month in rectified spirit of wine) retains none of the gold [in bulk form],* is obvious from the characters of this metal above laid down. ...

The same article addresses experimental therapeutic properties of *purple of Cassius* powder and *gold-fulminate*, now known as gold-hydrazide; *purple of Cassius* listed as a diaphoretic and *gold-fulminate* an emetic.

> *The purple precipitate, made by adding pure tin to the solution [purple of Cassius], is said to be diaphoretic: a precipitate made by alkalies [gold fulminate] is strongly purgative and emetic.* This last precipitate washed from the adhering saline matter by repeated affusions of water, purges more moderately, though rarely without gripes, and sometimes

operates by sweat: ...regarded only as a matter of curiosity, on account of its property of exploding violently **when heated or strongly rubbed.**

The reference to *Arabians* and gold medicine above is drawing on the Islamic tradition of prophetic medicine, meaning traditional remedies proposed by the prophet Muhammad (ﷺ) and recorded in collections dating as early as the 9th century. The following is from the *'Al-ṭibb al-Nabawī* (الطب النبوي) or *Prophetic Medicine* compiled by Egyptian writer, religious scholar and theologian Jalaluddin Al-Suyuti in the 15th century. Cyril Elgood published his translation of the text with commentary in the *Cambridge Journal of Medical History* in 1962, from which the following passage is sourced. The excerpt below highlights gold as an ingested and topically applied medicine yet does not detail what form it is to be taken orally:

Dhahab (ذهب) – Gold

Gold is evenly balanced. It contains a delicate heat. It enters into carminatives and strengthens the heart. **It is good for the mouth and to hold it in the mouth decreases pain in the throat. Used as a cautery it does not blister and heals rapidly.** *The Prophet has forbidden the use of gold or silver vessels, but has permitted their use as medicines.*

In Europe, therapeutic gold medicines were known by many names such as *Pulvis Auri, Tinctura Auri, Aurum potabile, Aurum potabile verum, Tinctura Solis, Tinctura aurea,* etc. Glauber was among the first to treat syphilis with gold salts. Glauber's views on gold were clear:

Aurum ... Medicina Catholica in senibus et juvenibus.
(Gold ... Universal Medicine for the elderly and youth.)
— Glauber, *De auri tinctura sive auro potabili vero* 1651

This work was likely known to Doctor André-Jean Chrestien who fervently pursued therapeutic gold technology in France despite strong opposition from mercurialists. Chrestien experimented with gold leaf, gold-chloride and trichloride salts around 1810 and found them very reactive and caustic in large doses. In an effort to buffer the extreme effects of gold salts, he developed gold sodium chloride (sodium-tetrachloraurate; $NaAuCl_4$), which he called *Aurum muriaticum natronatum*. His pioneering work was published in 1821 by himself and Dr. J.G. Niel in *Recherches et observations sur les effets des preparations d'or du docteur Chrestien*. Chrestien's work was confirmed by another Parisian, M. Legrand, and the term *auralist* entered the medical vocabulary to describe researchers specializing in therapeutic gold applications. *Auralists* such as Chrestien, Niel, Legrand and approximately eighty French physicians and surgeons in total employed various gold-treatments and reported success in treating sexually transmitted diseases and tuberculosis of the neck; conditions extremely difficult to treat at the time. Many gold formulations were developed based on Chrestien's new gold sodium chloride, particularly by German physicians who were reputed to have perfected the recipe as suggested by Dunglison in his 1839 publication *New Remedies: the method of preparing and administering them*:

> This preparation [gold sodium chloride] has a beautiful yellow colour, and appears under the form of four-sided prisms. It attracts moisture from the air, but to a less degree than the chloride of gold with excess of sulphuric acid.
>
> The Aurum Muriaticum Natronatum of the Germans is milder than the preceding preparations, and is more frequently administered, especially in Germany, than any preparation of gold. It is used both internally and externally.

Prussian (German) Salzsaures Goldnatrum
1. Take of gold 6 parts: Dissolve in a sufficient quantity of Muriatic acid, adding as much nitric acid is required to dissolve the gold.
2. Then mix ten parts of dry muriate of soda [NaCl];
3. and after evaporating the solution over a slow fire reduce it to a yellow powder.

What remains unspecified in the recipe for *Salzsaures Goldnatrum* is the precise method to *reduce it to a yellow powder*. At first glance, the product appears to be sodium tetrachloroaurate ($NaAuCl_4$). The German variety is milder, which begs the question how a milder version of sodium tetrachloroaurate may have been achieved. When this gold salt is thermally decomposed, the result is tiny particles of gold suspended in table salt; this being the more likely chemical identity of *Salzsaures Goldnatrum*. The unwritten key appears to have been a careful thermal decomposition procedure. A number of preparations were developed based on gold sodium chloride (auri et sodii chloride):

> **Pulvis auri et sodii chloride** *(Powder of gold sodium chloride)*
>   *– ground with orris root – Legrand*
> **Liquor auri et sodii chloride** *(Solution of gold sodium chloride)*
>   *– dissolved in distilled water – Grötzner*
> **Pilulae auri et sodii chloride** *(Pills of gold sodium chloride)*
>   *– with northern wolfsbane and marshmallow – Grötzner*
> **Pilulae auri et sodii chloride** *(Pills of gold sodium chloride)*
>   *– with nightshade root-starch and gum acacia – Chrestien*
> **Pastilli auri et sodii chloride** *(Lozenges of gold sodium chloride)*
>   *– with sugar and gum acacia – Chrestien*
> **Unguentum auri et sodii chloride** *(Ointment of gold sodium chloride)*
>   *– mixed into lard – Grötzner and Riecke*

The next stage of research into gold's bioactive properties was performed by German physician, developer of Homeopathy and once considered one of the greatest heroes of modern medicine, Samuel Hahnemann, who at tremendous personal risk attempted to reveal the effects of various medicinal gold preparations on the body by adhering to a rigorous protocol of extreme self-experimentation. This was the age of heroic medicine but Hahnemann was heroic in the sense that he and his colleagues would stop at nothing to characterize the medicines used in their practice.

Hahnemann and ten other medical professionals carefully performed systematic human trials, known as *prüfung*, upon themselves in a prelude to modern experimental pharmacology and clinical trials. This daring team ingested toxic levels of finely powdered gold leaf, gold-trichloride, gold-chloride, sodium gold-chloride, gold-fulminate and gold sulfate to elicit and record toxic responses. He did so based on foundational principles underlying homeopathy that *like cures like* (similia similibus curentur) and *less is more*, which is to say that if high doses of gold medicines could cause symptoms of disease in the healthy experimenters, it was then reasoned that homeopathic doses could cure those same symptoms in sick people. He published his findings in 1825 and again in 1835.

It was a risky prospect but the reward was that these brave pioneers revealed a wealth of firsthand evidence for the *pharmacodynamics* of gold and gold-salts *in-vivo*. A collection of Hahnemann's gold-toxicity human-trial results was recorded by Dr. Timothy F. Allen during the 1870's in his multi-volume *Encyclopedia of Pure Materia Medica*. Research during the 19th century into gold's effects on the human body was documented by Fellow of the Royal Geographic Society and physician Dr. James Compton Burnett in his detailed work *Gold as a Remedy in Disease*, published in London in 1879.

Most of the medical applications for gold as a curative remedy have been replaced by more efficient and modern medicines. Apart from gold as a pharmaceutical however, gold continues to show great promise as an adjunct to health and well-being due to its low toxicity and unique properties of reducing stress, anxiety and depression while enhancing cognition and creativity when taken in the proper form and at the correct dosage. During the 18th century, the emphasis was on characterizing the various forms of gold that might be applied therapeutically and their reproducible formulations. Throughout the 19th century, various gold-salts were the focus of research into therapeutic gold, which led to gold-drug compounds in the 20th century still in use today. In *Gold as a Remedy in Disease*, Burnett cites Legrand's account of the *pharmacodynamics* of gold, particularly in salt form:

> *To return to Legrand's account of the Medicinal Properties of Gold and of its mode of action on the economy, he says in substance that Gold is an excitant; seeming to act principally upon the* arterial, venous, and lymphatic *system. This excitation is always mild at first when the dose is small. A weakened stomach is strengthened, the appetite is at times incredibly increased, and the digestive functions become regulated. The patients feel an* indescribable sense of well-being, they feel themselves lighter *(as they express it), so that we may say that Gold possesses* hilariant properties. *The* intellectual faculties are more active. *It has been known to* produce frequent erotic salacity going on to painful priapism. M. Legrand, however, states that he has not used it as an aphrodisiac, but it has been used as such with success.

Potter's Materia Medica of 1906 illustrates that early in the 20th century the *pharmacodynamics* of gold-salts were still being discussed but with the addition of immediately addressed dosage concerns:

> **Physiological Action** *[the action of the salts of gold on the human organism]* – The *salts of gold, administered in small medicinal doses, increase the appetite and the digestive power and stimulate the functional activity of the secreting organs, especially the skin and the kidneys. They also stimulate the generative apparatus,* causing diaphoresis, and diuresis, exciting the menstrual flow in women and the sexual appetite in men. The *observations of several competent physicians have established the power of these salts to excite the vascular and muscular systems* and to produce fever, to increase the urine and the sweat, to cause salivation without stomatitis, a sense of heat in the stomach, headache and diarrhea, to *promote menstruation, excite the genitalia, and profoundly affect the nervous system.*

During the 19th century the focus was on the *pharmacodynamics*, which can be better understood as the "what does it do?" question as juxtaposed with the "how does it do it?" of *pharmacokinetics* that preoccupies modern research today. In 2002, the spotlight was once again cast upon the therapeutic potential of gold by Douglas G. Richards et al. of the Meridian Institute who reopened the debate with their informative article *Gold and its Relationship to Neurological / Glandular Conditions* published in the *International Journal of Neuroscience*. The article reviews the history of therapeutic gold, addresses the biological role of naturally occurring gold in the body, gold's effect on neuroglandular systems, dosage considerations, the increasingly valuable role of gold nanoparticles and finishes with suggestions for further research.

One of the most significant advances in therapeutic gold research is the discovery that the gold content is largely responsible for efficacy of gold-salts and gold-drug compounds, thus gold-salts serve as prodrugs for releasing nano-gold *in-vivo*. Whereas western researchers studied gold primarily as a technology for imaging and drug-delivery, Indian researchers did much to reveal the story of medicinal gold nanoparticles used in traditional Persian Unani-Tibb and Indian Ayurvedic systems of medicine. Gold's bioavailability, translocation within the body and its effects are largely based on particle size, stability and dosage. In a number of cases, gold nanoparticles were demonstrated to be more potent and effective than prescription gold-drug compounds. Relevant contemporary research corroborates 19th century claims for gold's effects:

- *1977 Alexiou, Grimanis, Grimani, Papavangelou, Koumantakis and Papadatous – possibility that gold is specifically involved in reproductive glandular activity*
- *1977 Hagenfeldt, Landgren, Plantin and Diczfaluzy – cyclic variations of gold in the body do exist*

- *1984 Kauf, Weisner, Niese and Plenert* – gold occurs in the hair of newborn infants **like the trace elements zinc and copper**
- *1984 Skandhan and Abraham* – **the richest source of** biological gold is found in semen
- *1996 Simon P. Fricker* – in gene expression, antimicrobial, anticancer properties and in rheumatoid arthritis
- *1996 Abraham* – **although originally believed inert,** gold nanoparticles can have significant biological effects
- *1997 Abraham and Himmel* – no clinical or laboratory evidence of toxicity resulted from gold nanoparticle treatment **indicating a possible non-toxic alternative to gold-drug compounds**
- *1997 Szentkuti* – gold nanoparticle diffusion rates depend on size charge and surface coating
- *1998 Abraham, McReynolds and Dill* – gold's potential as a nootropic (smart drug) and nervine (therapeutic against nervous-system disorders)
- *2000 Bajaj and Vohora* – anticataleptic, antianxiety, antidepressant, immunostimulant and analgesic actions with a wide margin of safety and without any discernible negative side-effects
- *2001 Hillyer and Allbrecht* – uptake of gold nanoparticles occurred in the small intestine **and that gold nanoparticles** less than 58 nm in size **reach various organs through the blood**
- *2001 Hillyer and Allbrecht* – **study indicates that** gold nanoparticles less than 58 nm in size would be orally effective acting as a depot for slow sustained release of Au(I) ions **required for therapeutic action**
- *2002 Shah and Vohora* – antioxidant and restorative effects
- *2002 Danscher* – in-vivo liberation of gold ions from gold implants
- *2002 A. Mitra et al.* – free radical scavenging action of swarna bhasma
- *2004 Hoet et al.* – smaller nanoparticles (<58nm) cross the mucus layer faster than larger ones
- *2006 Brown, Bushell and Whitehouse* – gold nanoparticles suppressed the development of three different forms of arthritis **and more potent than the leading prescription gold-drug compound**
- *2006 Brown, Bushell and Whitehouse* – standard colloidal gold preparation may work as a substitute for Swarna Bhasma

- 2006 DeDecker (Harvard Medical School) – *identified* gold's bio-mechanism of action in treating auto-immune diseases *such as* Juvenile Diabetes, Lupus and Rheumatoid Arthritis
- 2007 Lai et al. – *reconfirm Szentkuti's findings that* nanoparticle diffusion rates depend on size charge and surface coating
- 2008 Sherika Mahepal et al. – *in* antitumor activity
- 2008 Sonavane, Tomoda and Makino – *tissue distribution of gold nanoparticles is size-dependent* with smaller particle size showing the most widespread organ distribution
- 2010 Danscher and Larson – *gold placed intracerebrally liberates gold ions to suppress inflammation* in the brain and central nervous system
- 2010 Souza et al. – *demonstrated that* gold nanoparticles migrate to the thalamus and hypothalamus, hippocampus and cerebral cortex areas of the brain
- 2010 D. S. Agrawal – *potential treatment and prevention of Alzheimer's disease*

Therapeutic gold is available today as liquid *colloidal gold* at an average particle size of 27 nanometers, Ayurvedic *swarna bhasma* at an average particle size of 56 – 57 nanometers, dry *colloidal gold* tablets or Unani-Tibb's *kushta tila kalan*. Ancient alchemists would be pleased to know that *aurum potabile* is alive and well and that eastern and western researchers have collectively established gold's therapeutic potential beyond any reasonable doubt.

Alchemical Aurum Potabile

> ... *obtained in a saline form*, whether multiplied or not, it can only be used for the healing of human illnesses, preservation of health, and growth of plants. *Soluble in any alcoholic liquid, its solution takes the name of Aurum Potabile ...* – Fulcanelli, Dwellings of the Philosophers

During the early part of the 20th century, Richard and Isabella Ingalese developed a late in life fascintation with alchemical research. They devoted themselves to reproducibility of the Philosophers' Stone based on Paracelsus' method of combining oils of metals such as *oil of antimony* and *oil of gold* or silver. Richard Ingalese was an attorney at law by profession living in Southern California. Frater Albertus details the couple's account in an introduction to *They Made the Philosophers' Stone*:

> In 1911, Richard Ingalese then in his fifty-sixth year and his wife in her forty-eighth year were, in his words, "determined to put our conception of the teachings of Paracelsus to laboratory tests and commenced our experiments". It took them nine long years of continuous hard work before their labors finally met with success and they achieved their goal, namely the Philosopher's Stone. In the first six of those years they experienced many failures and heartbreaking disappointments.

The Ingalese story is important to the study of *alchemical aurum potabile* because they are the only researchers in modern history to detail not only their own ingestion of *alchemical aurum potabile*, they created a research group of participants willing to perform informal human trials upon themselves and record the empirical evidence. Their groundbreaking account demonstrates that the Philosophers' Stone was reproduced in the early part of the 20th century and ingested therapeutically. Yet their story can be considered valid only if it can be determined that they actually had achieved the Philosophers' Stone.

In his public lecture and subsequent 1928 essay, Richard Ingalese revealed his general approach yet remained silent regarding the details. He described achieving the white stage then explains that only after three succeeding years of research they had achieved the quintessential red Philosophers' Stone, albeit in a crude state. The Ingalese couple's inspiration and method of operation merits further scrutiny. They were avowed followers of Paracelsian alchemy as indicated in the following passage:

> *Forty years ago I first read the "Hermetic and Alchemical Writings of Paracelsus". Of all the books I have seen oil the subject – and I have seen many – there are no others which contain so much knowledge as those two volumes. Dr. Waite's collection and translation is the best.*

They also leave very little doubt that their primary interest was in laboratory operative alchemy although a careful reading of their account reveals that Richard was well versed in not only speculative alchemy, but had also been exposed to a range of other occult subjects:

> *We speak of two kinds of Occultists, the practical and the theoretical; so there are two kinds of Alchemists,* the *laboratory Alchemist and the library Alchemist. The latter claim that all Alchemy is symbolical. ...*

> *But for untold thousands of years all Alchemists, both library and laboratory, asserted that the science was a material one; and history shows that it gave birth to both chemistry and physics.*

This unfortunate division represents a relatively modern and somewhat artificial perspective. Throughout most of alchemy's history, speculative and operative approaches were polarities of a single magnet and were as inseparable as night and day. Some felt compelled to choose between the two, thus violating the underlying alchemical principle of uniting the two polarities. The passages above do confirm that the Ingalese couple considered themselves laboratory alchemists although their writings suggest a more unified perspective:

> *To be a successful student of laboratory Alchemy, one first must acquire the philosophy of the subject, and then live that philosophy until it transmutes one's nature and makes it conform to the ideals of the Occultist. This is not a very easy thing to do, for such ideals are higher than those of other cults and creeds, by reason of the very nature of the subject and, the power it confers when success crowns effort.*

11 – AURUM POTABILE

The first-person account of rediscovery is best summarized in their own words:

> *After we established our laboratory and commenced our experiments, it did not take us long to find that* we had enlisted, not only in a difficult study, but in a very expensive one; *and that our income would be overtaxed to meet the requirements. It was therefore agreed that I should return to the practice of law to supplement our resources and that Mrs. Ingalese should pursue the experiments. There have been women Alchemists in the past who have assisted their respective husbands in the work, but I believe* Mrs. Ingalese was the first woman [apart from Alexandrian matriarchs] to take the initiative in the art; and to her goes all the credit of the pioneer for the four long years of solitary effort and for the final discovery of how to make the stone. ...

> *The essential theory of the Alchemists is that* all metals have oils and these oils are the spirits, or virtues, of the metals. That was the first principle which confronted us, *and, necessarily, it was true or false.* ...

> In 1917 we succeeded in making the White Stone of the Philosophers. It looked like soft, white marble, and its effect upon the body was startling. *We dared not try it on ourselves at first. But there was a third member of our family, a beautiful Angora cat of which we were very fond. We took a vote to see which of the three should test out that Stone, and the cat, neglecting to vote, was elected.* It survived the first dose, and we repeated it on the two following days, with the cat becoming more frisky than usual. After that we tried it ourselves, *each taking a dose at the same moment so we would excarnate together if it should prove fatal. But it proved beneficial and energized our bodies.* ...

> *Shortly after that event, the wife of a prominent local physician died; and the doctor, knowing of our experiments and that the books*

> *claimed that such a Stone, if used within a reasonable time, would raise the dead, asked us to experiment on the body of his wife.* Half an hour had elapsed since her death and her body was growing cold. A dose of

531

the dissolved White Stone was put into the mouth of the corpse without perceptible result. Fifteen minutes afterward a second dose was administered and the heart commenced to pulsate weakly. Fifteen minutes later a third dose was given and soon the woman opened her eyes. In the course of a few weeks, the patient became convalescent, after which she lived seven years.

Encouraged by this success, we redoubled our efforts to make the Red Stone of the Philosophers, which is the one most mentioned in Alchemical writings. This effort was continuous from 1917 to 1920, when our quest was rewarded. True, the product was crude, but it answered to every test of a newly-made Stone ...

This fascinating account of early 20th century alchemists, married and living a professional life in an urban environment and successfully reproducing the stone, demonstrates the timeless and enduring grip alchemy can exert on aspirants. According to their accounts, they worked with *oils of metals* to produce the stone and specifically mention *oil of gold*. Paracelsus' description of an *oil of metal* indicates an aqueous solution containing a soluble metal-salt. The chemical identity of *Oil of gold* in this case is easily identified as chloroauric acid ($HAuCl_4$) or an aqueous solution of gold trichloride (Au_2Cl_6). The chemical identity for *oil of antimony* is antimony-trichloride ($SbCl_3$) as indicated by 17th century chemist, Christopher Glaser in his *Cours de Chymie*, termed *glacial oil of antimony* by Lemery, Glaser and others.

The Philosophers' Stone can indeed be created by a combination of aqueous *oil of gold* and *oil of antimony*, an alternate interpretation of *lion's blood* (aqueous gold salt) and *eagle's* gluten (butter of antimony) of Paracelsus, according to the following equations:

$$4\ HAuCl_4 + 4\ SbCl_3 + 10\ H_2O = 24\ HCl + 5\ O_2 + 4\ AuSbCl$$
$$2\ Au_2Cl_6 + 4\ SbCl_3 + 10\ H_2O = 20\ HCl + 5\ O_2 + 4\ AuSbCl$$

While this is not conclusive proof that the Ingalese couple had achieved the Philosophers' Stone, it argues strongly for the plausibility of their claims. They quickly drew a group of followers and formed *The Renewal Club*, including elderly, who ingested *alchemical aurum potabile* and recorded its effects. Therefore, their accounts serve as the only published reports of informal human trials of *alchemical aurum potabile* in recent history. Richard admits that the couple administered their medications to some without disclosing the identity of the medicine. This is termed *informed consent* in modern parlance and violating this protocol is now considered unethical. In a strange twist, this violation is actually invoked by Richard to strengthen his claim of efficacy.

> *The books, or manuscripts, claim that the Red Stone of the Philosophers will cure any illness, and that after one has taken it for five years one cannot contract any disease. We desired to test the truth of that statement and tried the Stone on many "incurables."* **The number of cases cured was remarkable, but we found it not infallible.** *Aside from personal benefit, the one reason we entered upon the great quest was to know the truth about medical Alchemy...*
>
> **The** *Alchemists who wrote on the subject usually did so within a period of a few years after obtaining the Stone.* **The marvelous work done by it, for themselves and others, stimulated their enthusiasm and warped their judgment.** *A careful observation over a greater number of years and a larger number of cases would have made them more accurate. These good men had no intention to deceive, but they spoke, or wrote, too soon.*
>
> **Mrs. Ingalese and I both know that if the Stone is** *administered to a young, or a matured, person in normal health, it will prevent old age;* **that** *if given to a healthy but aged person, it stops further physical deterioration* **and starts him backward toward youth.** *...*
>
> **We know that** *the Stone restores virility in men at any age, and normal desire in both sexes. If a woman has recently passed her change of life,*

it restores all normal functions of the sex organs. **But, if she has long passed that period, then, childbearing is out of the question.** ...

The Stone is an aid in acute diseases, but cannot be relied upon alone to cure since its action is too slow. **In chronic cases,** *where there are no complications and fair vitality, its action is certain in any disease; where there are complications and low vitality, other aids are advisable.* ... **If one disregards all laws of hygiene and misuses mind and body, then one must take the consequences of one's thoughts and acts; for there is no vicarious atonement either in medicine or in morals.** ...

I have been asked often if it were not the mind, or faith, of the patient which produced the marvelous results in the cases we have observed. I answer, "NO," because some of them did not know what they were taking, and others did not believe in its power **but took it as a "forlorn hope."**

This is our testimony in behalf of Alchemy and the Alchemists, which each person may accept, or reject, according to his conviction, **until such time as our bodies, now sixty-five and seventy-three years of age, respectively, by their youth and vigor, will compel acceptance of our statements.**

The Ingalese study reveals what might be expected from ingesting therapeutic gold – an overall sense of well-being, stimulation of the sexual glands and increased libido, antidepressant and antianxiety activity and so forth. That the Philosophers' Stone indeed affects the body in unique ways has been at the very core of the medico-alchemical tradition since the period of mediæval Islamic alchemy. *Alchemical aurum potabile* appeared to fade into history's background in the face of the scientific and emerging chemical revolutions that followed. Yet hints of the Stone's efficacy, and even of its manufacture remained in Materia Medica literature late into the 18th century:

> ... *others have melted the gold with twice its weight or more of martial regulus of antimony, and exposed the powdered mixture, in a glass*

vessel, to a moderate heat, till the powder became purple. That these kinds of preparations have very considerable medicinal virtues, is not to be questioned ...
— An Experimental History of the Materia Medica, 1761

Dosage and Administration

Upon achieving the Philosophers' Stone, the next question is how to convert it to *alchemical aurum potabile* and administer it according to appropriate dosage considerations. The history of *alchemical aurum potabile* brings to light the fact that there are no universal answers to either of these questions. Jābir ibn Hayyān's pioneering medical use of *al-iksir* was administered as 2 grains in 3 ounces of pure *oxymel*. Without a copy of the original Arabic manuscript, it becomes difficult to know how the translator interpreted and / or converted ancient Arabic units of measure. According to Cowan's *Arab English Dictionary*, an Arabic ounce is known as an *ûqiya* (أوقية), which can represent a weight anywhere between 26 grams and 320 grams depending upon the region. Jābir's *ûqiya* was likely the same unit as used in Alexandria, which was 37 grams. His *grain* is just as difficult to ascertain. In antiquity, *grain* as a unit of measure equating the mass of a seed applied to both barley and wheat, both being staples during the Islamic Empire. A *grain* based on wheat was approximately 50 milligrams whereas a *grain* based on barley was 65 milligrams. This represents an extremely high dosage of *al-iksir* even assuming the smallest units of measure where 2 grains is a 100-milligram dose of *al-iksir* administered in 3 *ûqiya* (111 ml) of pure *oxymel*. Pliny the Elder provided an early recipe for *oxymel* in *Natural History* 14: 21 as follows:

> **Oxymeli** – *Vinegar even has been mixed with honey; nothing, in fact, has been left untried by man. To this mixture the name of oxymeli has been given; it is compounded of ten [Roman] pounds of honey [3.289 kg], five semi-sextarii [aka hemina or cotyla] of old vinegar [1.365 l], one pound of sea-salt [328.9 g], and five sextarii of rain-water [2.73 l]. This*

> is boiled gently till the mixture has bubbled in the pot *some ten times, after which it is drawn off, and kept till it is old ...*

The identity of Jābir's *oxymel* remains undetermined and, based on later recipes, likely did not include salt. The translator may have justifiably interpreted Persian *gulāb* (گلاب), Arabic *julāb* (الجلاب) or *sekanjabīn* (سكنجبين) to mean *oxymel*. Syrups such as these were employed by Greek and Islamic physicians as a medium by which to administer powdered elixirs. Recipes for each have survived intact at least into the 13th century when they were recorded in the Arabic recipe book *Kitab al tabikh fi-l-Maghrib wa-l-Andalus fi ʿasr al-Muwahhidin* (The Book of Cooking in Maghreb and Andalus in the era of Almohads) by an unknown author. Note that an Andalusian *ûqiya* is 39 grams:

> **Syrup of Julep [rosewater syrup]**
> Take five ratls [1 ratl = 468g / 1lb] of aromatic rosewater, and two and a half of sugar, cook all this until it takes the consistency of syrup.
>
> Drink two ûqiyas [1 ûqiya = 39g / 7tsp] of this with three of hot water. Its benefits: in phlegmatic fever; it fortifies the stomach and the liver, profits at the onset of dropsy [swelling from water, edema], purifies and lightens the body, and in this it is most extraordinary, God willing.
>
> **Syrup of Simple Sikanjabîn [vinegar syrup]**
> Take a ratl [1 ratl = 468g / 1lb] of strong vinegar and mix it with two ratls [1 ratl = 468g / 1lb] of sugar, and cook all this until it takes the form of a syrup.
>
> Drink an ûqiya [1 ûqiya = 39g / 7tsp] of this with three of hot water when fasting. It is beneficial for fevers of jaundice, and calms jaundice and cuts the thirst.
>
> Since sikanjabîn syrup is beneficial in phlegmatic fevers: make it with six ûqiyas [1 ûqiya = 39g / 7tsp] of sour vinegar for a ratl [1 ratl = 468g / 1lb] of honey and it is admirable.

This fascinating recipe book lists 27 syrups in total and provides unique insight into the how Muslims attributed therapeutic qualities to everyday foods. For the anonymous author, food and beverage were medicinal. The key to administration likely lies with water solubility of the gold in the Philosophers' Stone. The water content of vinegar served to bring ionic gold into solution and the natural sugars in the honey then served to reduce it to colloidal form and cap it to prevent aggregation.

A number of recipes for administering the Philosophers' Stone in wine are preserved in European alchemical texts:

> *Take a Drachm of the Stone [3.41 grams], seeth it in a pottle of Wine in a Glass,* **the space of two or three Pater-nosters, that** *the Stone may melt, the Wine will be as red as Bloud, therewith wash the Sores morning and evening, ...in a short time, as in ten or twelve days the Sores will be whole; and* give him every day the quantity of a Wheat-Corn [50 mg], in warm wine till he be well.
>
> – *Johann Isaac Hollandus*, A Work of Saturn, Its Use in External Diseases

Among the most comprehensive European recipes for *alchemical aurum potabile*, the enigmatic and highly informative *Tractatus de Lapide, Manna benedicto* (Treatise of the Stone, Blessed Manna) introduced in the previous chapter stands out as being unique in its level of detail, suggesting authorship by a genuine medico-alchemy adept. It provides clear and concise instructions for converting the Philosophers' Stone to *alchemical aurum potabile* via a fail-proof method that ensures correct dosage. The author issues a stern warning concerning *aurum potabile*, stressing that it is a strong medicine responsible for the death of a number of alchemists who were unaware of the correct dosage and / or method of ingestion. The method provided below offers what appears to be a pleasant and effective way to administer *aurum potabile* at what the mysterious unknown author of the *Tractatus de Lapide* considers appropriate dosage for safe ingestion and therapeutic efficacy:

BOOK 3 - APPLICATIONS & OBSERVATIONS

In the use of this medicine, many great philosophers themselves, after they obtained this wonderful blessing, desiring to have perfect health, have been so bold as to take a certain quantity of it, some no more than a quarter of a grain, some less, some more, but all that did so with it, instead of health, took death itself... if this be not known, more hurt than good may be received by it. For the method of health, it is thus:

1. Take the quantity of four grains, I do not mean the grains of wheat, or Barley grains or corns, but four grains [jewelers or pearl grain; French grain] of gold weight [200 – 212.46 milligrams], and dissolve them in a pint [approximately 568 ml] of white or Rhenish wine [Gouais Blanc, Riesling, Gewürztraminer, Silvaner, etc.] ...put it into a great clean Glass, and instantly it will colour all the Wine almost as red as itself was, which is the highest red in the world: let it stand so, close covered from dust, four days.

2. ... then add to this a pint more by degrees, until it be not so red, stirring it with a clean stick of wood, not of metal, nor glass, and so continue the pouring on of fresh wine, until it be just of the colour of gold, which is a shining yellow. Beware there be no redness in it; for so long as there is any redness in it, it is not sufficiently dilated, but will fire the body, and exhaust the Spirits ...

3. ... neither is it sufficiently brought to yellow, until the wine have round about the sides a ring like hair, of a whitish film, which will show itself plain when well dissolved, if it stand but four hours quiet. As soon as you see that whitish film, then let it run through a clean linen cloth, or paper [filter], so the white film [insoluble antimony products] will stay behind and look like a pearl on the paper: and all the rest will be yellow like Gold. This is the token of truth, that you cannot wrong yourself by this Liquor; and without this token, it will be either too weak, or so strong that it will fire the Body. Know this to be a rare Secret.

4. Of this golden water let the patient – whatever sickness he may have – take a tablespoonful every morning, and it will drive the sickness, whatever it may be, away with pleasant perspiration. For it does not purge, nor does it cause vomiting, nor does it call forth

perspiration so strong and so much that it causes tiredness. On the contrary, it is rather invigorating, I tell the sick. If the sickness has lasted for several years, or if it is a chronic disease, it will go away in approximately 12 days, but otherwise in 24 hours, or in two, at most three days. This is the way in which it must be used for all internal diseases.

The recipe above suggests that powdered Philosophers' Stone as a salt of gold and antimony is converted to *alchemical aurum potabile* in the presence of acid and ethanol in wine. The white wine is tinged red, indicating that the salt is reduced to either an ionic solution or nanoparticles in colloidal suspension. The wine is then diluted as a measure to prevent over-ingestion of the elixir. The insoluble dregs are then separated via filtration. The result is *alchemical aurum potabile*.

Fulcanelli stated that multiplied or not, the Elixir is obtained in saline form and is alcohol soluble. The Elixir's identity as a salt was clearly stated long before Fulcanelli by French iatrochemist and royal physician Joseph Duchesne (Josephus Quercitanus) M.D.:

> *This universal cure, or* panacea, *thus artificially created, is called by various names. Some call it 'quintessence', for it is the fifth essence, not made out of a mixture of the four elements. Others call it 'elixir', for it is an incomparable medicine for preserving life and repelling disease. Others call it* aurum-potabile *or potable gold, not because it must always be made from gold... but par excellence, because gold excels all things. Others call it the philosopher's stone ... because of its eternal durability, or because it partakes of the nature of salt – salt which is the life of nature, the solid foundation of all its virtues.*

Once considered inert and completely inactive, gold reduced to nanoparticulate displays entirely different properties in comparison with its unreduced counterpart. Gold nanoparticles possess the potential to bond strongly to living molecules like proteins, peptides, antibodies, oligonucleotides, and pathogens such as bacteria, viruses, etc. Gold and affects the neuroglandular system at extremely minute amounts, which explains why most alchemical literature cautions strongly against over-ingestion and advises miniscule doses. All foods and medicines are composed of chemicals and this is the case with *alchemical aurum potabile*. Any chemical known can be toxic if over-ingested (eat, drink or absorb) too quickly, and this includes substances thought to be inert such as water. Paracelsus clearly understood dosage issues when he proclaimed:

> *All substances are poisons; there is not which is not a poison. The correct dose differentiates a poison from a remedy.*

In the passage below, powdered *elixir* is additionally prescribed for external use. Gold and silver exhibit oligodynamic effect, which is to say that when their ions come into direct contact with microorganisms they become antibacterial, antifungal and even antiviral to a lesser degree. This is also the case with copper compounds such as those found in brass doorknobs, which disinfect themselves in just hours. A drinking vessel created from alchemical gold or silver would likely be very sanitary indeed. By topically applying powdered Philosophers' Stone to an external sore, the *elixir* exerts oligodynamic effect.

> *But for all external diseases, such as ulcers, carbuncles, fistulas, noli me tangerea (touch—me—nots), etc., the spot itself must be oiled and spread with or smeared with the oil or the Stone which must NOT first be melted in wine. This must be done for 9 or 10 days, and whatever it may be, it will cure all internal and external diseases.*

The subjective effects of ingesting *alchemical aurum potabile* are also addressed in the *Tractatus de Lapide*:

> *Taking the aforementioned medicine early in the morning for 9 consecutive days, will render a man so light that he could fly [metaphorically speaking], and his body is so merry that it is almost*

impossible to believe it except from one who has himself experienced it. It has this wonderful property, it gives perfect health till God calls the soul, and it gives perfect knowledge if one knows its right use. But it is precisely this aspect that has been known to very few of those who made it, because it is a divine and so to speak angelic medicine.

The above passage is vital to understanding the overall therapeutic effects of *alchemical aurum potabile*. The author of the *Tractatus de Lapide* begins by suggesting that the medicine works according to a cumulative process. Today nano-gold is considered a non-essential accumulative trace element. Research conducted in 1998 by Abraham et al. and reported in *Effect of colloidal metallic gold on cognitive functions: A pilot study* demonstrated effects on cognition measured by significant increases in the trial subjects' I.Q. test scores once gold had reached a certain accumulation in the body. That same year, researchers Bajaj and Vohora showed gold nanoparticles to exhibit an analgesic (pain-relief) effect in *Analgesic activity of gold preparations used in Ayurveda & Unani-Tibb*.

The *light* and *merry* sensation reported in the passage above alludes to gold's antidepressant and stress-reducing properties reported in 2000 by researchers Bajaj and Vohora in *Anti-cataleptic, Anti-anxiety and Anti-depressant activity of gold preparations used in Indian Systems of Medicine*. In another study of the effects of gold nanoparticles performed a year later, researchers Bajaj et al. published *Augmentation of non-specific immunity in mice by gold preparations used in traditional systems of medicine*, which suggests that gold nanoparticles can affect the immune system. Most of the Indian studies were in-vivo trials on rat models. While this may be inconclusive concerning effects in humans, the results are encouraging, shed some light on the reported effects of *alchemical aurum potabile* in alchemical literature and offer hope for discovering a mechanism of action. Immunomodulatory activity, an increase in I.Q. scores, stress-reducing anti-anxiety properties and pain-relief all suggest that there might be more to the Philosophers' Stone and *alchemical aurum potabile* than previously considered.

> *I have cured several cases of Melancholy [with gold] ... promptly and permanently, and they were those of such who went about with the serious intention of committing suicide.* – Samuel Hahnemann

The Angelical Stone – Food of Angels

> *I was about to hand the cup back to him after moistening my lips in the liquor, when the old man said: "Drink it all; it will be thy only nourishment during thy journeys." I obeyed and felt a divine fire course through all the fibers of my body. I was stronger, braver; even my intellectual powers seemed doubled.* – Comte de St. Germain

Traditional accounts of the effects of *aurum potabile* suggest thermogenesis and non-ordinary states of consciousness after repeated ingestion. Further research may reveal that *aurum potabile* affects the limbic brain in such a way as to trigger a psychoactive event, resulting in a heightened state of awareness, inspiration or even possibly a type of religious or mystic insight elicited by changes in brain chemistry. Alchemical texts certainly suggest such effects:

> *You must know, before you do these things, you must take the Stone nine days, as I prescribed first [as aurum potabile], and it will make you have an angelical understanding; you will despise the World, and all in it; then you will know how to serve God, and understand the Scriptures.*

> *I have written that which has never been written before; think whether or not they are secrets and arcana's and whether you ought to show this or not to any man, but [only] to him that has the Stone.*
> – Tractatus de Lapide, Manna benedicto

The notion of a distinctly separate type of Philosophers' Stone, or as interpreted here a religious application or spiritual effect produced by it, occurs in the writings of German court alchemist Heinrich Khunrath.

Khunrath also equated these properties to the *Urim* and *Thummim* through which the High Priest of the Tabernacle entered into dialogue with the divine in order to seek counsel and revelation. During the 16th century, Simon Forman addressed these properties in his manuscript *Of appoticarie druges*:

> The angelical stone is true medison to mans bodie against all infirmities and makes a man live longe and by that stone he obteined wisdom and knowledge of things in dreams and otherwise.

Elias Ashmole credits al-Razi and Petrus Bonus with knowledge of these properties yet specifically cites a 16th century work written by Edwardus Generosus Angelicus Innominatus entitled *The Epitome of the Treasure of Health*, in which the Stone is described as *food of angels*. This text was known to Robert Boyle, Sir Isaac Newton and others who conceded that the Stone might actually offer the potential for entering into some form of spiritual dialogue. Ashmole believed the passage had originated from St. Dunstan when he wrote:

> *S. Dunstan calls it the* Food of Angels, and by others it is termed The Heavenly Vaticum [foretelling]; The Tree of Life ... There is a Gift of Prophesie hid in the Red-Stone.

One of the unusual effects gold appears to exert on those who ingest it is a pronounced influence on dream states. These states can manifest in range from incredibly lucid and colorful to horrific and nightmarish. This effect suggests mild psychoactivity and appears to have startled test subjects enough to record them in medical literature. The 1874 *Encyclopedia of Pure Materia Medica* lists *Religious excitement* first under mental effects brought about by ingesting finely powdered gold. A small sampling of gold's effects on dreams states from the same Encyclopedia are as follows:

- *Invincible sleep after dinner;* during this sleep his mind was thronged with ideas ...
- *Agreeable and sensible dreams,* but which cannot be remembered ...
- *Vivid dreams at night,* which cannot be remembered ...
- *Dreams with constant erection* every night ...

In Burnett's 1879 text *Gold as a Remedy in Disease,* he shares his observations of the psychoactive properties of gold in the following excerpts:

- *One sits moping in a comer, desirous of being left alone, while another is all vivacity, and has a lively word for everybody.*
- *In some the memory is rendered very acute,* while in others it becomes almost annihilated.
- *Not only* does Gold thus affect the brain, but it is a great disturber of the cranial circulation ...
- *Gold is an excitant ...* This excitation is always mild at first when the dose is small.
- *The* patients feel an indescribable sense of well-being, they feel themselves lighter (as they express it) ...
- *... Gold possesses hilariant properties.*
- *The* intellectual faculties are more active.

In 2010 researchers, Souza et al. published *Functionalized gold nanoparticles: a detailed in vivo multimodal microscopic brain distribution study,* which demonstrated that 19 hours after gold-nanoparticles entered the bloodstream they had migrated to the brain and nested primarily in the thalamus, hypothalamus, hippocampus and the cerebral cortex in animal studies. The gold nanoparticles passed the blood-brain barrier and accumulated in specific and highly defined areas of the brain in such a way that indicated active uptake rather than barrier damage.

One of the most critical relays in the brain is the thalamus, which works by collecting all of our sensory data and routing it to the higher parts of the brain allowing us to experience a sense of reality, an ongoing story or dialogue between self and landscape. This sensory theater is communicated from one part of the brain to another with the help of the thalamus resulting in our overall cognition, conscious awareness and sense of reality in general. Various different neurotransmitters are important mediators for interpreting the drama of sensory input. Receptors regulate concentrations of neurotransmitters with the result being our emotional and sensual (psychosomatic) experience that profoundly influences and shapes our dream state, waking reality and world-view. The hippocampus is the gateway to memory and spatial navigation. This is not an attempt to reduce the magic and mystery of the human condition to brain chemistry, but rather to highlight the fact that the brain can be and often is manipulated, which affects each person's experience of reality in ways that range from extremely subtle to quite dramatic.

It is therefore plausible that alchemical accounts dating to antiquity that touch on what might be described as a non-ordinary state of consciousness or heightened awareness were actually describing genuine phenomenology, albeit in the idealistic and often religiously charged language of the time.

> *I no longer wonder, as once I did, that* the true Sage, though he owns the Stone, does not care to prolong his life; for he daily sees heaven before his eyes, *as you see your face in a glass. When God gives you what you desire,* you will believe me, and not make yourself known to the world. *– Michael Sendivogius,* The New Chemical Light

The interplay between spiritual practices, ritual ingestion of *alchemical aurum potabile* as a sacrament in this case, and the profound experiences they elicit both affect and are affected by brain chemistry in a dynamic and integrated way, that is to say they are interdependent and symbiotic. One primary effect is how the stress response may be affected or dampened and reduced, which often results in contemplative or tranquil awareness or even insightful inspiration. These states are dependent upon neurotransmitter reward response, which manifests as a sense of happiness and well-being, enthusiasm, delight or even euphoria. It is perhaps more the language employed by alchemists and their incredible claims that strain credibility. However, the scientific basis on which *aurum potabile* may affect the body by interfering with stress response, and / or affect cognition or even cause a mild psychoactive episode due to activity in the limbic brain should be reconsidered and further studied.

Achieving the Philosophers' Stone, allowed the aspirant to externalize his inner journey, the struggle, his mindset and world-view and merge with the process and product finally fully realized before him. Important to the totality of the adventure through the labyrinth of speculative and operative alchemy was the actual ingestion of the substance once achieved. This sacred act intimately united the seeker with the goal of his search and thus transformed him into the very object of his pursuit. The adept became aware that he had been wealthy in mind and immortal in spirit all along; that the awareness and insight gained by undertaking such a journey of discovery had culminated in a uniquely alchemical self-redemption.

Ingesting *alchemical aurum potabile* at this point was a symbolic climax, a sacrament or type of communion that transmutes the base physical, mental and spiritual tria-prima reborn as a new triune being. Here is the moment when the Philosophers' Stone is made real in all its forms. The aspirant becomes the adept made real by speculative and operative success – where to make real = to realize. The entire alchemical process coagulated all the previously dissolved, separated and purified aspects of the alchemical pursuit together again into a final climactic alchemical self-realization – the *All* has always been *One*, the adept becomes the

Universal Medicine incarnate with the power to tinge and transmute all he comes into contact with.

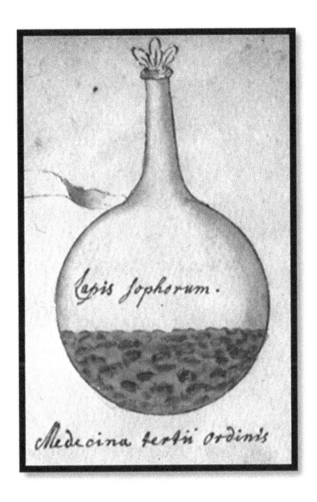

Chapter 12 – Universal Medicine

At that instant a loud and melodious voice exclaimed, "The work is perfect!"
Hearing this, the children of light hastened to join me,
the doors of immortality were opened to me,
and the cloud which covers the eyes of mortals, was dissipated.
I SAW and the spirits which preside over the elements knew me for their master.

Comte de St. Germain

According to Sir Isaac Newton, no great discovery is ever made without first making a bold guess. Today we might add that the guess should not only be bold, but also based on as much information as possible. This treatment of the origins and evolution of the Philosophers' Stone is just one of many such guesses as to its chemical composition. We have endeavored to present a model that takes into consideration the industrial, ideological and therapeutic properties traditionally associated with the Stone throughout its long history. The adjustable element appears to be the various salt / flux / menstruums that alchemists employed to unite gold with antimony philosophically. These menstruums evolved along with the Stone as alchemy transitioned from culture to culture, yet color-indicators and a ramped heating regimen in a single sealed glass vessel remained constant factors throughout. These three constants, color-regimen, the application of low yet ascending heat and a sealed glass digestion vessel demand that any model for confecting the Stone via Ars Magna technique must match these three criteria. The territory of alchemy and the Philosophers' Stone has traditionally been mapped out based upon cultural shifts, valuation of chemical technologies and their applications, all of which point to Alexandrian origins for recognizable Western alchemy. Yet the bold guesses and explorations in this work suggest that Alexandrian alchemy was simply a branching or divergence from more ancient bronze and glassmaking technologies prevalent in Mesopotamia and the ancient Near East. (See Appendix A for a timeline of developmental alchemy) Evolution of alchemical technology during the early Alexandrian period was based on the use of sal ammoniac as a sophic substance that allowed for a shift in technique from the purely pyrotechnical to an Art based on chemical reactions. This early-Alexandrian bifurcation as interpreted here was the origin of recognizable western alchemy, the primary products of which are the Philosophers' Stone and the archetypal recipes developed during the formative periods. To explore further back into history or climb out onto the branches of alchemy that emerged in Asia or Europe is perhaps to venture too far afield. This treatment remains an effort to expose the roots and trunk of alchemy and the archetypal Philosophers' Stone, and perchance in turn offer insight as to how and why the various side-branches developed thereafter.

Origins

Happy the man who has been able to learn the causes of things.

– Virgil

Material Composition – The material composition of the Philosophers' Stone is the matter from which it is composed, i.e. gold, antimony and flux. Antimony is the volatile component bursting with potentiality when viewed from an alchemist's perspective because of its tremendous inherent transformative qualities. Alchemists identified the prime material, stibnite, and recognized in it the many, multiplicity in-potentio, the *All* – sulfur (raw ore), oxides (flowers and glass), chlorides (butter), fusions and alloys (rebis, antimonial bronze, artificial gems) – and understood the *All* as innate potentiality embodied within this unique and volatile *one*. The products of antimony comprised the soul and spirit of the union, bursting with growth and regenerative principles yet the body of stibnite or antimony was fragile, weak and easily broken. Gold was the body of the king, incorruptible, enduring and extremely difficult to destroy, yet gold was viewed as having reached its final perfection and contained almost no growth or regenerative principles left to impart. By uniting the growth and regenerative principles inherent in antimony to the enduring perfection of the body of gold, the new elixir was understood as a combination of the best of both – the incredible growth and transformative soul and spirit of antimony come to life in the perfect enduring corporeal body of gold. These opposites thus united, potentiality had been brought to actuality by the alchemist and the Elixir became fit for one of its transformative applications.

No man enters life but through woman – mother – and the child of alchemy is no different in this regard. During the pre-Alexandrian period, the uniting principle took the form of common fluxing salts used in metallurgy and glassmaking, which were brought into fusion with gold and antimony. At some point around the 1st or 2nd century CE, alchemy's matriarch, Maria, experimented with sal-ammoniac and discovered that a unique union could be affected by the novel application of this *much-honored* sacred salt, by which her Art *could not be completed but by it*. Further experimentation revealed that sal ammoniac would vaporize upon heating and a new technique using low heat to create a reaction between sal ammoniac, antimony and gold in a sealed-environment led to a technological bifurcation initiated by Maria and thus she became the mother of alchemy. This new process came to be the archetypal recipe that yielded a novel and more sophisticated product – *Ios, tincture, elixir, and powder of projection* – all synonymously known later as the Philosophers' Stone. Generations of alchemists subsequently discovered that urine menstruums, quicksilver or acetate-ethanol oils and tinctures could also affect a union between antimony and gold, which gave rise to the various side-branches of alchemy that emerged thereafter. The material composition of the archetypal Philosophers' Stone is antimony and gold united, whereas the Stone in a more general or generic sense is antimony and gold united via any reagent able to dissolve and coagulate.

Formal Classification – One of the criteria that must necessarily be satisfied for a substance to qualify as the Philosophers' Stone is its form. The substance can only be recognized as being the Philosophers' Stone in virtue of having the proper form according to traditional descriptive indicators, i.e. heavy vermilion powder, ruby-colored waxy substance, deep-red translucent crystal, saffron powder, golden liquid, etc. The Philosophers' Stone has been described in many different ways throughout the history of alchemy. This treatment has endeavored to demonstrate the reasoning behind the various different forms the finished Philosophers' Stone can assume. It can be either white or red in color, crystal when fused or powdered when ground, waxy in a molten state and of varying colors or consistencies when fused with additional gold or treated again with the fluxing agent. Dissolved into a liquid medium as *alchemical aurum potabile*, it can take on a thick consistency such as is the case with Greek oxymel and Islamic syrups or either a red or golden thin liquid based on the European wine used and the level of dilution.

Efficient Cause – If the art of alchemy were simply the geologic formation of minerals and metals, the Philosophers' Stone would form naturally. One will never see the Philosophers' Stone, the *universal medicine*, *aurum potabile* or the *powder of projection* coming about because of natural forces working on matter over time. To possess the Philosophers' Stone it must first be confected, which is to say that skilled artisans must set to work on the material components and fashion them into the very form of the Philosophers' Stone. Each step of the process elicits a change in the original components of the resulting Philosophers' Stone product. This treatise has attempted to uncover the means by which alchemists developed techniques necessary to bring about such changes, and how these techniques became more sophisticated and varied, sometimes lost only to be rediscovered over time. The human ingenuity revealed in this study demonstrates that there is more than one way to confect the Philosophers' Stone as is evidenced by so many different approaches recorded throughout alchemical history but each based on a unifying principle.

Final Purpose – Aristotle felt that determining the final purpose – *telos* – was more important than the others were, as did Thomas Aquinas who gave the *final cause* priority in his hierarchy. This is likely because the *final purpose* is the root-cause that moves the artisan to collect the materials and fashion them into the product. Yet, this raises other questions such as how did each alchemist know how to proceed? How did they identify the prime material and other components? What criteria informed these artisans how to accomplish the technical processes to create the Stone? To understand the Philosophers' Stone requires informed knowledge of its application, its function, the purpose or goal it serves. This is what motivated adept-alchemists from the outset, and what material, formal and efficient causes ultimately led to. They worked according to a blueprint, template or design. In the order of time, the application or *final purpose* for creating the Philosophers' Stone is realized last, though according to Aristotle and Aquinas in the order of conception it is paramount. It comes before the choice of materials and methodology. All of these subsidiary causal ingredients are to be understood as engendering an object, the Philosophers' Stone, which serves the purpose of realizing an eventual objective. This ultimate plan and purpose is the *final cause*, which is to say the very reason for the Stone itself.

Berthelot quotes Palagus as saying that the final purpose of the Art is either creation of the Tincture itself and/or to produce a certain quality (likely *Corinthian bronze*; alchemical gold):

BOOK 3 – APPLICATIONS & OBSERVATIONS

The tinctorial art has been invented in order to make a certain tincture and to produce a certain quality. Also, this is the end of the art.

The *final purpose* is what has been historically enigmatic, which is partly responsible for mainstream incredulity surrounding the Philosophers' Stone. Gold-making in Alexandria meant the production of a highly valuable commodity in the form of a unique alloy that displayed the characteristic features of gold, i.e. non-corrosive, lustrous golden color and malleable; making it perfectly suited to temple ornamentation adornment and for the appearance of immense wealth. It was so valuable that according to biblical and historical sources it was of equal or greater value than genuine elemental gold. *Alchemical gold* was harder, lighter in weight yet indistinguishable from gold in appearance, thus making it perfectly suited for statuary, adornment and small luxury items such as goblets and plates made even more valuable by its oligodynamic properties. *Powders of projection* allowed for greater quality controls with regard to alloying technology. However, this was just the first of a trio of final purposes that the stone served. Something was lost in translation along the course of history due to a technological disconnect. A misunderstanding ensued which held that the *final purpose* of the Stone was in a very literal sense to transmute base metals into genuine gold. This interpretation was void of cultural context, which resulted in broad skepticism displayed later in history. Even today, the evidence offered to prove that alchemy was a misguided failure is often based on this misunderstanding. Each day in the modern world, the alchemical process of transmuting lead into gold is played out in the form of base-materials being routinely fashioned into extremely valuable commodities according to well-kept trade secrets. The true story of alchemical gold as a final purpose accorded to this very process, in which simple copper and antimony were "fermented" with the elixir and thus transmuted, which is to say changed or manufactured into highly valuable *gold-antimonial bronze* that came to be known as alchemical gold.

12 – UNIVERSAL MEDICINE

As *universal medicine*, the Philosophers' Stone not only purified base metals into alchemical gold by mere contact or projection, it also served as a dynamic physical model of early Jewish, Neoplatonic, Hermetic and Gnostic philosophical and mystical currents that dominated early-Alexandria and Byzantium. As such, confecting the Stone served as a philosophical rite of passage, an initiation into deeper mysteries of the nature of reality, growth, evolutionary change and transformation, that is to say a medicine capable of curing the universal illness of mankind – psychospiritual disconnection, the feeling of being divided. By observing a perfect working physical model representative of philosophical or religious edifices, one gained unique insight and veritable proof of universal truth underlying the respective ideology. During the 6th century, Stephanos highlighted this application of the Philosophers' Stone by repeatedly implying that in-depth knowledge of the Stone and its confection was an initiation into the mysteries of his unified philosophic-religious system. He used the term *pharmakon*, meaning medicine perhaps to suggest not only remedying physical ailments but also those of a psychospiritual nature. The Philosophers' Stone's role as a *monument to the mysteries* remained consistent throughout later alchemical traditions. Christians, Sufis and Renaissance Humanists would also adopt this application of the Philosophers' Stone. Occultists of the 19th century and Carl Jung in the 20th chose this aspect over all others as the *final purpose* of the Philosophers' Stone. Some have interpreted the entire body of alchemy in this manner forsaking all others to the degree that it became the prevailing elucidation of alchemical models, symbols and imagery into the 21st century.

Alchemy merged with medicine and chemical pharmacy during the Islamic period as is evidenced by Jābir's celebrated pioneering accounts of employing the *al-iksir* as a treatment in one thousand cases. From that point onward, the therapeutic application of the elixir emerged as the third *final purpose*. Paracelsus prioritized this application and *alchemical aurum potabile* became a principal driving force behind the quest for the Philosophers' Stone until around the 17th century. As a motivating factor, the therapeutic application of the Philosophers' Stone continued to inspire aspirants and adepts such as Bacstrom in the 18th century, and Cockren, the Ingalese couple, Fulcanelli and others into the 20th century. Just as the Philosophers' Stone is triune in its composition, its *final purpose* and prime motivating factor is also triune in nature.

> Surely to alchemy this right is due, that it may be compared to the husbandman whereof Aesop makes the fable; that, when he died, told his sons that he had left unto them gold buried in his vineyard; and they digged all over the ground, and gold they found none; but by reason of their stirring and digging the mould about the roots of their vines, they had a great vintage the year following: so assuredly the search and stir to make gold hath brought to light a great number of good and fruitful inventions and experiments. – Francis Bacon

Furrowing the earth of alchemy fertilized the soils of science and technology to follow, yet in so doing the Philosophers' Stone became buried and forgotten about by all but a few. Today's mainstream has replaced gold-making with technology, elixirs with pharmaceuticals and philosophical models with science and religious orthodoxy. Philosophers addressed theories of matter whereas alchemists were fascinated with change and transformation occurring in the natural world. Alchemists ultimately questioned how man might achieve an analog of these in substances and self via the alchemical process. Chemistry in its modern form emerged at a time when the Philosophers' Stone ceased to be a prime motivating factor behind humanity's quest for understanding, and impulse to model, processes of transformation. Yet alchemy and the Philosophers' Stone remain alive in the form of appreciable aspects in the field of abstraction, representations in consciousness that cannot be rendered solely in material terms. The alchemical process is timeless and still has much to offer humanity as an artisan's path towards reconnecting with and directly experiencing change, transformation and harmony of natural law.

> *The scientist is possessed by the sense of universal causation. His religious feeling takes the form of a rapturous amazement at the harmony of natural law, which reveals an intelligence of such superiority that, compared with it, all the systematic thinking and acting of human beings is an utterly insignificant reflection.* – Albert Einstein

Abstraction

> *... now I know that one ends with abstraction, not starts with it.*
> *I learned that one has to adapt abstractions to reality*
> *and not the other way around.*
>
> *– Alexander Stepanov*

Anyone who ever experiences operative alchemy firsthand quickly realizes the tremendous difference between alchemical abstractions and practical lab-work. There is a remarkable variance between a template or recipe for confecting the Stone and the material reality, replete with small subtleties, in working with substances and processes. Abstractions can come before manual labor in the form of the inception of ideas, recipes, symbols and templates, or after the fact as poetic or iconic renderings of one's manual labor. Operative alchemical artisanship however rests squarely in the middle. This is so because the work begins with an idea or instructions in the form of recipes that include symbols and imagery coupled with imagination; the combination of which comprises the initial abstraction. This is followed by the practical operative work, which generally leads to the realization that there are countless unwritten nuances and small details that can be reconciled only through firsthand experience. Upon achieving mastery over alchemical processes, what follows is a natural human inclination towards individual abstract expressions of the work in the form of poetry, imagery, literature or even music.

Alchemy served as a window to a deeper reality where alchemical processes of dissolution, reduction and purification resulted in new growth, generation and transformation not only in the mineral and vegetable arenas of the natural world but also in individual and social contexts as outgrowths of the natural order. Insight, inspiration and observation of processes aroused emotions and impulses that imbued alchemy with significance and staying power among generations of alchemists throughout history. Alchemy offered the possibility of a lab-bench realization of deeper processes of creation and the natural order that lent additional weight and credibility to philosophical and religious ideologies. It must be understood however that without the operative aspects of alchemy, its abstractions become reduced to reiterations of ideological themes, which is to say simply another abstract wisdom model. The beauty of the alchemical process is the way in which operative and speculative aspects of alchemy rely on each other, identify and complement one another in such a profound manner.

Alchemy, considered from the physical standpoint, was the attempt to demonstrate experimentally on the material plane the validity of a certain philosophical view of the cosmos. – H. S. Redgrove

Alchemical work allowed aspirants and adepts to learn and master abstract thought and expression. The result was a deep understanding that everything is potentially symbolic of something else, or as Plotinus might say, "All is noble image". One avenue of approach towards entering into deeper gnosis of unified reality was through the doors of symbolism, metaphors, emblems and allegory. A realization ensues that what at first appeared haphazard and chaotic is in fact an ordered reality of fractal dimensions iterating at micro and macro levels wherever one looked. Even today, physics and chemistry rely on abstractions in the form of mathematical or chemical equations. It is often these very abstractions that are offered as proof or validity of a theory or chemical reaction, which is to say that the manner in which we use abstractions has not changed much – only the form of the abstractions themselves have been altered and refined.

One of the first lessons an aspirant must master concerning alchemical substances and processes is to interpret abstractions in their material context, usually by decrypting cover-names and symbols or today by learning mathematical formulas or the table of elements. Working with materials provided an abstract model useful for grasping greater natural, individual or even social processes taking place everywhere on a grander scale – alchemists saw the observable world as a reaction taking place in vessel Earth. Alchemical abstraction can better be understood by the illustration of a Persian carpet displaying intricate patterns and designs rather than being a jumble of nonsensical abstractions in a cacophony of swirling colors and patterns. From the conventional perspective, the patterns appear to be separate and disconnected images on the surface, yet further examination from every vantage reveals that each individual image on the front is interwoven and connected at the rear in purposeful and complex embroidery suggesting tremendous skill of the weaver. So too, it is the case that alchemical abstractions when viewed and understood from every possible perspective, reveal networks of interconnected meaning applicable to material, intellectual, psychospiritual, social and natural patterns intrinsic to any creative or transformative process.

These impressions, patterns and recipes, captured and compacted into abstract models, were based on and intended to represent a greater reality, albeit in a tremendously compressed form and accessible only when the aspirant became liberated from formal discursive reasoning. The abstraction is therefore, the tip of the iceberg so to speak and represents what might be considered by some an overwhelming looming complexity. A natural response to which is to reduce the subject matter alchemically to manageable pieces in order to purify it and understand it more completely, thus making it easier to work with. This natural process is used in all fields of discovery encountered in the modern world.

> This art and mystery should only be revealed in parables, which must be considered and weighed exactly, it is necessary also to be familiar with the books and the writings of the rest of the philosophers. Thus, in order to reach this Art, it is not necessary to make tremendous works, many efforts nor great expenses. ... the artists who know the end try to hide it, and secretly guard this craft ... – Basil Valentine, Azoth

Purification and Reduction

> To make progress in understanding all this,
> we probably need to begin with simplified models
> and ignore the critics' tirade that the real world is more complex.
> The real world is always more complex, which has the advantage that we
> shan't run out of work.
>
> – John A. Ball

Reduction follows abstraction as the initial operative stage of the alchemical process, and is closely associated with purification and a type of death and rebirth. The prime material, stibnite, offers a wonderful example for how this process is carried out manually. The first task was to identify and source the prime material, a substance imbued with the correct attributes and potentiality for the Grand Work. Identifying potentiality comes naturally to some and is an attribute intentionally developed by others but is ultimately a requisite of all alchemists. Stibnite is unspectacular as an ore and slightly less so as mśdmt / kaḥâl / koḥl eye cosmetic, yet it possessed great potential for purposes of alloying and glassmaking. It was easily overlooked by all but the informed; to the layperson, it was a useless lump of earth or powder whereas the artisan saw a valuable resource imbued with meaning and pregnant with potentiality. As the alchemical ouroboros with its tail in mouth, it was the circle containing the *All* in-potentio due to its innate venom. Following identification came a breaking down and grinding of the matter to a fineness limited only by the patience of the operator, yet fineness was not enough. The matter then needed to be purified by the fire, a sulfurous emanation indicating that noxious impurity was departing and the matter was undergoing transformation to a pristine new form ready to combine with other bodies once purified. The prime material's innate pure nature and potentiality were revealed by the process of being ground down, purified in the fire and further reduced before being fit for use. Alexandrian and later alchemists depicted the purification and reduction of gold in terms of killing the body of the king. Paracelsus demonstrated that this process is what happens each time we reduce food by grinding with our teeth, dissolving it with saliva and gastric juices and further digestion in the gastro-intestinal tract before it becomes pure enough to act as nourishment.

The above process serves as a perfect model for anyone embarking upon a journey of transformation whether it is personal or professional. Psychospiritual transformation usually begins with a death of the old self, reduction in the form of an ordeal, a fasting period or other extremely ego-challenging scenario followed by ritual cleansing in some manner prior to or as part of initiation. When applied to the study of a subject like alchemy, it allows for a breaking down of an overarching and sometimes overwhelming subject, reducing it to bite-sized pieces in order to digest more easily, and deduce all the various elements in order to arrive at how the small pieces fit together in the larger perspective. Modern science is still firmly entrenched in this reductionist stage of the process of discovery. In short, reduction is simply a removal of that which will hinder further progress, yet the inherent danger is to become lost in the fine details, lose sight of the bigger picture and ignore the need for future holistic or systemic integration. Reduction is a death that takes place in order for a new configuration or rebirth to occur. The dissolution absolutely must transpire before the process of reconfiguration can begin. The alchemical process of reduction as a crucial purifying step is applicable to physical, biological, psychospiritual, social and natural transformative functions. A natural result of the chaos of reduction is that this multiplicity eventually yields an inherent binary nature over time.

> The nigredo is the precondition for transformation. And what is it? It's shit! It's detritus... it's flotsam... it's debris... it's being HIV positive... it's being deep into your fourth marriage and sinking fast... it's bankruptcy... it's, you know, serum hepatitis... it's the inevitable dark night of the soul that comes upon us, and these dark nights of the soul come upon all of us. Nobody gets through the world without a little dung raining down on them, believe me. I mean you may evade it for decades, but then there'll be a knock on the door. You know, it's said that the millstones of fate grind slowly but they grind exceedingly fine.
>
> So what do we do with that? Well the answer is ... we welcome it! This is what the alchemists awaited, the nigredo, the prima-materia, the dark matter, the chaos ... the chaos that is the precondition then for redemption. And God knows we've got lots of chaos right now, I mean we have wars, famine, revolution, millions of homeless people on the move, the nation state is dissolving, the relief agencies of the world can't keep up, the various secret societies, mafias and cabals that have always tried to tie us into chains ... they are all working overtime. We are in the nigredo condition. Hallelujah. This means ... this means that the kissing has to stop but the fun can begin ... the real fun.
>
> – Terrence McKenna

Opposites as Polarity

> We must no more ask whether the soul and body are one than ask whether the wax and the figure impressed on it are one.
>
> – Aristotle

Gold and antimony in combination with salt-flux comprise the tria-prima symbolized by the alchemical triangle. Continuing with reduced and purified antimony as a conceptual and physical model of the alchemical process, its dualistic nature reveals the paradox that both positive and negative poles are present in a single entity such as antimony, and indeed all things. Alchemists first discussed binary duality in terms of fixed and volatile, hot and cold or dry and moist. Purified and reduced antimony was divided into two portions and each portion applied in a diametrically different manner. The first portion was fused in a crucible with gold pyrotechnically, the second portion distilled with sal ammoniac via chemistry. Antimony is the volatile element united with gold as its fixed counterpart during the creation of latten / rebis, yet gold causes the rebis to assume the form of a cold and dry celestial-earth – the fixed element. This served as the model of the material body of things. Inversely, antimony is extremely volatized when combined with sal ammoniac, the result of which is corrosive eudica / butter of antimony that readily assumes the form of a hot and wet fiery-fluid – the volatile element. This served as the model of energetic growth and regenerative potential innate to all things. Apart or in combination, they formed polar matches to each other symbolized by the alchemical square. In these two initial preparatory processes, the conceptualized binary nature of antimony is revealed and opposites are demarcated. These two complimentary substances were encrypted by abstract cover-names such as *red man ↔ white wife, red gum ↔ white gum, latten ↔ eudica, sulfur ↔ mercury, lion's blood ↔ eagle's gluten, rebis ↔ butter* of antimony and many others.

The alchemical process in its simplest form, *solve et coagula*, is profoundly dualistic and thematic of human experience. Perhaps one reason why the human mind tends towards duality is the brain's bicameral structure in combination with our dualistic physiology – left-right brain, binocular vision, stereo hearing, bipedalism and so forth.

Experiencing complexity or multiplicity from a binary or polar frame of reference is perfectly natural. Everything experienced in the realm of psychological time – thought, feeling, perception, action – is an incredible expression of being that is dualistic in nature as is human experience of transformation and change. What one experiences through the senses is delivered by waves whether it is sound, color, vibration, heat and so forth. Consciousness too reveals crests and troughs in the vicissitudes of ordinary life situations. Dualistic experience is to descend into the abyss and to return triumphally from the darkness into the light. Arriving at a binary perspective helps counter the chaos of reductionism by providing a means of sorting through an overwhelming amount of information in order to make sense of things; the natural process of categorization. We rely on this psychological function as a crucial step in how we learn, work and progress through complexity as part of growing and developing a functional sense of reality. In a broader sense, experience polarizes reality in terms of self and landscape in the same manner in which a magnet polarizes itself into north and south embodied in a single entity.

With an appreciation of the dualities in alchemy, the alchemist comes to recognize a divided nature within and an apparent rift between self and landscape. Gerhard Dorn referred to this artificial rift as "the binary of multiplicity and confusion". Such a profound experience of perceived innate duality was the bedrock for formative philosophies and religions the world over and remains a conceptual frame of reference for many today. In material and energetic senses, alchemists ultimately attained a deep understanding of the binary nature of reality in the realm of time and upon doing so, sought to experience the reunification of perceived duality directly – operative and speculative, above and below. They did so by assuming the role of mediator whose function it is to bring about this union of opposites – or alternately, to simply realize the artificiality of separation altogether.

... the inside and the outside, the subjective and the objective, the self and the other go together. In other words there is a harmony, an unbreakable harmony and when I'm using the word harmony I don't necessarily mean something sweet, I mean absolute concordant relationship between what goes on inside your skin and what goes on outside your skin. It isn't that what goes on outside is so powerful that it pushes around and controls what goes on inside. Equally so, it isn't that what goes on inside is so strong that it often succeeds in pushing around what goes on outside. It is very simply that the two processes, the two behaviors are one. What you do is what the universe does, and what the universe does is also what you do. Not you in the sense of your superficial ego, which is a small very tiny area of your conscious sensitivity, but you in the sense of your total psychophysical organism; conscious as well as unconscious. – Alan Watts

Synthesis as Reunification

The whole is more than the sum of its parts.

– Aristotle

The alchemical process developed over time concurrent with confecting the Philosophers' Stone. The ouroboros, the serpent with its venom was identified as the prime material containing within it potentiality; it was reduced to fineness, purified by the fire and divided into dual portions. The tria-prima emerged as sulfur, mercury and flux. Simple duality was then converted to a compound duality where the alchemist fused the celestial-earth *latten* and distilled fiery-liquid *eudica*, arriving at the components of the tetrasomia. The unification of *lion's blood* with *eagle's gluten* revealed the illusory nature of a division of opposites, reunification of which brings the *All* of the preceding multiplicity into unity, duality rectified by dialectic synthesis in a triune body. *Solve et coagula*, the *All* becomes the *one* – circle, triangle and square united in the greater circle of the Philosophers' Stone. The Tincture was then applied towards perfecting base substances such as metal and glass; icons of durable and friable inorganic matter, or to perfect the greatest icon of organic matter – humanity.

If the reader has been capable of following the conceptual model of the alchemical process as depicted above, in antiquity this understanding would signify a type of initiation into the mysteries of the alchemical process and confection of the Philosophers' Stone. The implementation of holistic observational comprehension results in insight, causing the mind of the alchemist to become conditioned to ever-deeper levels of intuitive understanding and abstract thought. The finished Stone served as a monument or physical model commemorating inspiration, technical aptitude and most importantly a deeper experience of the processes of change and transformation manifest in all creation. Thus, multiplicity is experienced in a single model via the process of dissolution and reconfiguration. The alchemical process of *causation* → *abstraction* → *reduction and purification* → *polar opposites* → *dialectic synthesis* demonstrated in a very real physical sense the act of creation. The artisan began with a single raw substance and worked artisanally through multiplicity to arrive at not only a fine product that can be further applied, but also to arrive at a new gnosis, a new experience of reality. Even in its simplest form – *solve et coagula* – the alchemical process reveals that the most laborious stage is the preparation, after which having been performed correctly the entire process that follows is self-growing and self-organizing, autopoeisis as an expression of natural law and its revealed superior intelligence – a sacred confluence of *all* alchemical substances and process into *one* final expression.

The sum total of biological, social, psychological and spiritual domains collectively make up the holistic mind as a natural flow of expression or manifestation emerging from that which is prior – the unmanifest as potentiality. Extroversive unitary gnosis arrived at via the alchemical process engenders an inner unifying state where the alchemist senses *all* things and events, thoughts and feelings, as *one* undivided reality.

Alchemical manipulations of substances and processes and foundational idealistic motivations play an important role in the journey to adept-hood – from the very practical and mundane to the incredibly esoteric and mystical – ending with gnosis of the *All* as being embodied within each and every being, an awareness of being beyond theoretical or conceptual models – beyond discursive reasoning and mind itself. There is the very real sense of a sacred reality being played out as reunification of consciousness with its expression. These powerful experiences brought about through the alchemical process can seem paradoxical, very difficult to linguistically express and thus yield to abstract expressions more readily than concrete ones – the result being a genuine sense of transformation for the alchemist. Thus the alchemist arrives at a new comprehension of the natural order, innate makeup and humanity's role within it and it is here where alchemy becomes a mirror by which the *One* recognizes itself in the *All* as its expression of being; the unmanifest source of all manifestation of which man is a microcosm.

However, this begs the question… So what? Who cares? What does an undivided awareness of being have to do with anything in the practical sense? Can we commodify the new awareness, bottle it, label and sell it … and should we? How does humanity profit in any way from gnosis arrived at via the alchemical process? Can the Philosophers' Stone still be understood as a model of individual or social transformation today? How is the alchemical process relevant to our complex modern world? Actually, it has everything to do with the direction our species takes and perhaps even its very survival. We can continue to hope that the experiment of humanity in Hermetic vessel Earth will ultimately become optimized, elegant and as efficient as a chemical reaction, free from waste and unrecyclable byproducts.

Just as confection of the Philosophers' Stone is based upon a seemingly self-synthesizing process of evolution towards novelty and progressive complexity only to return to singularity, so it is with the natural order and our life's journey in relationship to it when viewed from the alchemical perspective. It may just be possible to look at the reductionist and specialized direction humanity has taken recently and see that we are just now beginning to put the pieces together, that is to say we've passed through the dissolve stage and signs that the inevitable and badly needed reconfiguration are just beginning to appear.

All of the impressions upon which alchemical models are constructed are real in the sense of apprehending natural law as an undivided whole – nature is brutish as well as intelligent, organic clothed as mechanistic, ordered yet also chaotic, built upon partnerships disguised as domination where all components in accord with its processes reveal perfect dynamic equilibrium and harmony – in other words, a union of opposites. Duality is fundamental to existence in the sense that every thing or event is a function of relationship, but this dualistic interaction, this apparent opposition is always essentially a dialectic synthesis. The profound and wonderful experience of reconciling multiplicity into an integral whole is primary and quite essential to a deeper understanding of alchemy and its role as a multifaceted yet unified wisdom tradition, replete with doctrinal elements and practical works. The alchemical process reveals a unitary continuum that begins with a distinct identification and purification of the prime material – self – and transitions to a sense of interconnectedness between the landscape and human expression as outgrowths of the natural order; a view of the whole of reality as an expression of a single evolving process. It was this sense of reunification that the word *religion* originally suggested, where *re + liga* or *ligare* = *connect, tie* or *bind once again* – to reconnect.

The alchemical process has always lent itself to as many interpretations as there are interpreters and it is not wholly incorrect to say that it has often served as a sort of ever-changing transcendental Rorschach blot to which each generation added another splotch of ink. This treatise is just such an interpretation but it is one in which every aspect of alchemy and every alchemist serves a purpose, each approach is and has been useful, the ensuing development of which has been systemic and organic rather than hierarchical, with time-honored lessons being continually recycled. The secrets of alchemy and the Philosophers' Stone were hidden, concealed, protected for nearly two millennia, and available only to those who could rediscover the keys to unlock the sophisticated abstractions that secured the Art from the uninitiated.

There will be those who feel that this book should have never been written, that the revelation is too plain, vulgar or oversimplified, that some secrets are best kept hidden or that the interpretations are incorrect altogether. In this journey through alchemical things, events and ideas, nothing presented in this treatise is new or original since all alchemical secrets have always been revealed in-perpetuity, exposed to view in their abstract forms, hidden in plain sight – a fact that relegates this work to being little more than an echo of collective voices from the past. If the *bold guesses* in this treatise are correct, then the subject will yield more treasures. If however a more accurate explanation backed by reproducible chemistry proves this interpretation incomplete or incorrect, alchemical research moves forward still – a win-win in any case for the subject of alchemy.

Traditionally it is forbidden to reveal alchemy so openly. That being said, the profound lesson preserved in the motifs and myths of humanity throughout history is that life is only ever truly initiated after committing the one forbidden act.

> *Now today, we are living in an age, which is quite peculiar, because in the world of science there are no longer any secrets, because the method of science requires that all scientists be in communication with each other. And therefore that every scientist, as soon as he has discovered something or got a good idea, he rushes into print. And it's important for him to do so because some other scientist somewhere else in the world might be thinking about something on the same lines and*

would be stimulated in his work by this man's speculations even if not by discoveries. ... In former ages, that [maintaining secrecy] might have been managed, because there were many secrets once upon a time and people were not admitted to these secrets unless they were in some way tested and found capable of handling them without running amok. We live in such a dangerous age because all the secrets are out in the open and anybody can run amok with them, and that's just the situation we have to face and that is just the situation we have to handle. It is too late to stop it because that would be, as they say, "locking the door after the horse has bolted". – Alan Watts

APPENDIX A

Timeline of Developmental Alchemy

- **c. 200,000 BP**: Sometime during the Middle-Pleistocene, man encountered the first ternary alchemical reaction in the form of fire.
- **c. 26,000 BP**: Around 26,000 years ago, this new chemical technology gave rise to humanity's first great alchemical transmutation; with the aid of fire man learned that he could magically transmute soft wet clay into dry hard stone and thus developed earthenware pottery and the use of kilns and ovens.
- **c. 9,000 to 8,000 BCE**: The earliest evidence of native copper exploited by humans in Sumeria and southern Anatolia, modern Turkey, at the Neolithic settlement of Çayönü Tepesi, situated at the foot of the Taurus Mountains near the upper Tigris River. During this period, copper metallurgy reached technological breakthroughs in basic cold and heat hammering, annealing, melting and casting. The word for copper in Sumerian is *urudu*, which was also the name of the Euphrates River (Copper River). The word for *stibnite makeup* in Sumerian was *šembi*, whereas *gùnu* or *gùn* was the verb form meaning *to apply šembi* to the eyes or face.
- **7,000 to 1,500 BCE**: Copper thoroughly dominated metallurgy.
- **7,000 BCE**: Evidence of the earliest use of lead in the Near East.
- **6,000 BCE or earlier**: Evidence of the earliest gold technology in Turkey and the Middle East.
- **c. 5,000 BCE or earlier**: Crucible smelting developed in Iran, Northern Mesopotamia, Central Europe, the Balkans and Greece. *Silver-antimonial bronzes* from the Bell-Beaker culture of the Carpathian region were some of the recurrent prehistoric bronzes and were likely precursors to *gold-antimonial bronze* and *Corinthian bronze* of

APPENDIX A

Greek traditions. Antimony-bearing *arsenical bronze* craft appeared on the Iranian plateau. The earliest evidence for the use of silver in the Middle East and iron in Egypt dates to this period.

- **c. 4,500 BCE**: Copper smelting and foundry sites began to crop up in Jordan and around the Dead Sea ushering in complex crucible technologies.
- **c. 4,200 BCE**: Egyptians began mining malachite copper ore by exploiting Midianite and Amalekite labor at Timna in the Sinai. Near the high place of offering at Timna, the earliest furnaces for reducing malachite to copper were established.
- **c. 4,000 BCE**: Refined copper smelting and solid glass beads were manufactured in Egypt. The area from the Dead Sea in the East to Gaza in the west and south into the Sinai as far as Timna became the homeland of proto-industrial bronze metallurgy. Early copper finds in Mesopotamia at Tepe Gawra and slightly later in Iran at Tepe Yahya date from this period.
- **c. 3,750 BCE**: Prestige and ceremonial *antimonial bronze* artifacts from the *Nahal Mishmar horde*, possibly linked to the Ghassulian Temple of Ein Gedi contain up to 12% antimony. Ghassulian antimony was sourced from Ghebi in the southern Caucasus Mountains, intentionally imported and specifically employed to create the 416 cupreous religious artifacts for use in Ghassulian rituals. The earliest evidence for the use of carbon in metallurgy in Mesopotamia and Egypt dates to this period.
- **c. 3,500 BCE**: Artisanal crucible smelting gave way to the more industrial bowl-furnace. Earliest examples of the deliberate tin-bronze production appeared near Ur of Chaldea at al 'Ubaid.
- **3,000 BCE**: Evidence of Babylonian gilding technologies found at the Royal Tombs at Ur date from this period.
- **c. 2,500 BCE**: The oldest trace of occupation at the Hathor Temple at Serabít el-Khâdim includes a statue date back to the reign of Sneferu from this period.

TIMELINE OF DEVELOPMENTAL ALCHEMY

- **c. 1,700 BCE**: Earliest Assyrian glassmaking cuneiform tablets from this period reveal an interrelationship between the priesthood, glassmaking as a temple craft, a guild of workers with specialized knowledge and the use of jargon and cover-terms.
- **1,833 to 1156 BCE**: Hathor Temple at Serabít el-Khâdim was active with mining and metallurgy. Each Pharaoh enlarged the complex, beginning with Amenemhat III and concluding with inscriptions from the Ramses VI reign. Biblical Moses is hypothesized to have practiced *antimonial bronze* technology sometime during this period.
- **c. 1,500 BCE**: Arsenical and *antimonial bronze* as it pertains to alchemy diverged or branched off from utilitarian tin-bronze.
- **1,549 to 1,297 BCE**: The most accomplished Egyptian glasses from the 18th dynasty were very nearly transparent indicating an already high degree of sophistication.
- **1,450 to 1,200 BCE**: Canaanite literature from Ugarit, modern Syria, preserved a poem that provides insight into ritualized gold and silver making with references to the Canaanite (Kenite) god of metallurgists and their supreme deity El-Elyon.
- **668 to 626 BCE**: Glassmaking manuals existed alongside Assyrian chemical dictionaries. Some glassmaking texts were copies of originals kept at the Royal Library at Nineveh and the adjacent Temple of Nabu. Included in these was a recipe for the production of *[bah]-ri-e*, interpreted by R. Campbell Thompson to be a type of decorative *crystallin* glass or artificial gem called *Red Coral*. Use of gold and antimony together first recorded in this recipe.
- **7th century BCE**: Hesiod mentions *aurichalcum* in a Homeric hymn, interpreted here as an early term synonymous with antimonial or *Corinthian bronze*. By this time, antimonial bronze technology in Europe, around the Mediterranean, Anatolia, Mesopotamia, the Iranian plateau and the Levant was already over 2,000 years old.
- **5th century BCE**: Herodotus wrote Histories in which he recorded an account of the Libyan salt trade in connection with the established

trans-Saharan caravan route linked by a series of oases. The first two stops were famed for salt of Amun and Augila, known historically and today as sal ammoniac. Biblical Ezra describes a *"fine bright bronze as precious as gold"* during this period.

- **early 5th to late 4th centuries BCE**: Ostanes, Persian magus who accompanied Xerxes on his invasion of Greece, reputedly introduced the Magic Arts to the Greeks. His tradition was typically credited as being studied by traveling Greek scholars such as Pythagoras, Empedocles, Democritus and Plato among others.
- **400 BCE**: Wenamun recorded his travels, populated the Siwa Oasis and established the Temple of Amun.
- **c. 460 to c. 370 BCE**: Democritus, Greek polymath and philosopher, traveled to Egypt and extensively throughout ancient Near East and Persia, and according to Cicero and Strabo as far as India to the east and Ethiopia to the south. He is typically portrayed as having studied Chaldean Magic in the tradition of Ostanes. He is credited as formulating the first atomic theory of the universe and according to Celsus and Hippocrates as being the teacher of Bolus of Mendes.
- **356 to 323 BCE**: Aristotle, student of Plato for nineteen years, left Athens and tutored Alexander the Great. He is credited as the first genuine scientist in history. His five-element theory, four-cause theory, and metaphysics of abstraction and change as it relates to substance, potentiality and actuality were influential to all alchemical traditions.
- **332 BCE**: Alexander the Great liberates Egypt, consulted the oracle at the Temple of Amun at Siwa and established Alexandria and the Ptolemaic dynasty.
- **3rd century BCE**: Bolus of Mendes worked in Ptolemaic Egypt and wrote of magic, natural medical remedies and astronomical phenomena. He is typically portrayed as having studied with Democritus.

- **20 BCE**: Herod completely rebuilt and enlarged the *Second Temple of Jerusalem* and outfitted it with massive pillars that featured pillar-caps and gates of *Corinthian bronze* manufactured and shipped from Alexandria.
- **1st century CE**: Josephus described *Corinthian bronze* crafted in Alexandria at Herod's command and shipped to Jerusalem.
- **1st to 3rd centuries CE**: Maria Hebrea's Judeo-Egyptian school of *chrysopœia* flourished during this period. Zosimus attributed the discovery of numerous alchemical equipment, the production of *divine water* and the secret of gold-making to her. Most of her original works are non-extant. Other alchemists in this tradition such as Comarius, Moses, Cleopatra and Isis among others (all alchemical pseudonyms) were active during this period. The Tincture of the Philosophers, the archetypal Ars Brevis and Ars Magna recipes to confect the Philosophers' Stone originated with this school. The Philosophers' Stone during this period was valued primarily as a trade secret and essential ingredient to creating alchemical gold and silver, *Corinthian Bronze*.
- **2nd century CE**: Pseudo-Democritus' Persian-Egyptian style gilding school flourished. He authored *Physical and Mystical Matters* that preserved recipes for the manufacture of imitation gold and silver by surface-treatment and alloying technologies.
- **2nd to 3rd centuries CE**: The *Hermetica*, an intellectually eclectic collection of Greco-Egyptian texts also known as the *Corpus Hermeticum*, was compiled and addressed divinity, mind, nature, alchemy, astrology and spiritual transcendence. The *Emerald Tablet of Hermes* and the archetypal White Stone of Hermes / Philosophers' Mercury may have either influenced or originated with this movement.
- **269 CE**: Zenobia proclaimed herself Queen of Egypt. She was defeated and taken hostage by Emperor Aurelian five years later in 274 CE.

APPENDIX A

- **292 to 298 CE**: Diocletian secured control of Alexandria and issued an edict to destroy all works and books that addressed the *Chimeía* of gold and silver *Corinthian bronze* making, and violently suppressed the Art.

- **c. 300 CE**: The *Papyrus Graecus Holmiensis*, also known as the *Stockholm Papyrus*, preserved craft recipes for dying, coloring gemstones, cleaning (purifying) pearls, and the manufacture of imitation gold and silver. Democritus was mentioned by name in this text, suggesting a style typified by the Persian-Egyptian gilding tradition of Pseudo-Democritus. The *Leyden Papyrus X* likely written by the same scribe as the *Graceus Holmiensis*, preserved recipes for precious metal extraction, counterfeit precious metals, gems, and the manufacture of artificial *Tyrian purple* dye. In addition, it detailed the manufacture of textiles, and gold and silver inks. The end of the text included short extracts from the *Materia Medica* of Dioscorides. The notion of Greek medicine having anything to do with early Alexandrian alchemy derives from this movement typified by the Pseudo-Democritan school of alchemy.

- **c. 300 CE**: Zosimus of Panopolis, mystic and alchemy anthologist, attempted to rescue, catalogue and preserve alchemical texts banned by Diocletian. He was the first to clearly present alchemy as reflecting Hermetic and Gnostic spirtualty. Much of what is known of pre-Diocletian alchemy derives from what remains of Zosimus' writings and fragments.

- **389 to 391 CE**: Theodosian Decrees established a practical ban on non-Christians, forbade visitation to non-Christian temples, non-Christian holidays were abolished. This marked the beginning of strict and violent Christian subjugation.

- **391 CE**: Synesius initiated Dioscorus, Alexandrian priest of the Temple of Serapis, by tutoring him on cover-names that encrypted the primary material, finally revealed by Synesius as *"Roman antimony"*, for creating *divine water* and *philosophers' mercury (Hermes)*. He became Hypatia's disciple two years later. The *True Book of Synesius* and the *Epilogue According to Hermes* are attributed to him. These texts closely parallel the methods of operation preserved in the Cleopatra fragments and Zosimus' writings *On the Evaporation of Divine Water that Fixed Mercury* and *Authentic Memoirs on Divine Water*. The Philosophers' Stone was valued during this period primarily as a cosmological model of Hermetic, Gnostic and Neoplatonic philosophical currents.

- **415 CE**: Hypatia's tragic murder marks the symbolic end of the Golden Age of classical antiquity typified by a cultural and intellectual climate that fostered free and open scientific inquiry and philosophical expression.
- **491 CE**: Anastasius was forced to sign a written declaration of orthodoxy prior to his crowning, Christianization of the Roman Empire completed.
- **564 CE**: Olympiodorus, the last Neoplatonic teacher of the Alexandrian School, delivered a series of lectures during May-June in Alexandria on astrology. An alchemical text is attributed to him and his teachings and approach to unified philosophy and mysticism influenced Stephanos of Alexandria.
- **617 CE**: Stephanos of Alexandria travelled to Constantinople to cast a horoscope for the Emperor, where he remained and lectured on Plato and Aristotle, the Quadrivium (arithmetic, geometry, music, and astronomy) along with astrology, medicine and alchemy at the court of Heraclius. He was tutor to Morienus, and became known as Istafan or Adfar of Alexandria in the Islamic tradition of alchemy. A series of alchemical lectures titled *Chrysopœia* (gold-making) and a highly alchemical *Letter to Theodoros* are attributed to him. The Philosophers' Stone was valued during this period primarily as an expedient means of initiation into Stephanos' unified wisdom tradition and as a material monument or memorial of psychospiritual

APPENDIX A

redemption. His approach to alchemy influenced Islamic and European alchemical traditions.

- **632 CE**: Death of Muhammad (ﷺ) and the beginning of the Islamic Empire.

- **c. 683 to 704 CE**: Morienus al-Rumi tutored Khālid ibn Yazīd in alchemy in Damascus, thus transmitting late-Alexandrian / Byzantine alchemy to the Islamic Empire. The transmission was recorded as an eyewitness account by Khālid's non-Arabic Muslim retainer, Ghalib. Morienus' brand of alchemy was highly practical yet also belied deep spiritual foundations inherited from Stephanos. The book of the *Composition of Al-Kīmyā'* is attributed to Morienus and his personal instructions to Khālid. This text was the foundational Islamic alchemical text and was among the first alchemical texts translated to Latin in Europe. Morienus' methodology highly influenced the sulfur-mercury theory of confecting the Philosophers' Stone medieval Islamic alchemy. His particular recipe to confect the Philosophers' Stone served as the archetypal template or blueprint for Islamic and foundational European traditions of alchemy. The Philosophers' Stone as inherited by Islamic alchemy as typified by Jābir ibn Hayyān (Geber), Muhammad ibn Zakariyā Rāzī (Rhazes) and Abū 'Alī al-Ḥusayn ibn 'Abd Allāh ibn Al-Hasan ibn Ali ibn Sīnā (Avicenna) was valued primarily as an ingestible therapeutic or curative medicine known as the al-Iksir (elixir).

- **10th to 13th centuries**: The transmission of Alexandrian / Byzantine alchemy to the Islamic Empire marked the end of developmental Alexandrian alchemy. European alchemy developed following the Crusades, particularly after the fall of Constantinople in 1204 when scholars began to have access to Greek manuscripts containing important theological, philosophical and scientific learning previously unknown to the West. The second wave of alchemical influence resulted from Islamic alchemy imported through Spain and Sicily with Islamic immigration into Europe. Early European alchemy was largely a rediscovery or restoration effort that matured into highly experimental proto-chemistry. The Philosophers' Stone became valued during the period of European alchemy primarily as riddle or test that secured access to a fellowship of alchemical insiders and served as an object of quest, the pursuit of which catalysed what might be described as experimental chemistry. It also became symbolic or representative of renaissance to early-modern humanist movements and occult sciences.

APPENDIX B

Dialogue of Maria and Aros
(Latin Translation with Commentary)

It is unknown whether the translators who initially translated from Arabic to Latin did so linguistically or technologically. Were they practicing alchemists familiar with the substances and process described in the text? This concern also applies for the initial translation hypothetically from Hebrew to Greek, or historically from Greek to Arabic. Certain chemical misunderstandings compounded with the difficulty of accurate expression that always accompanies transliteration must be taken into consideration. A beautiful or poetic translation, as engaging as it might be, does disservice to the subject of practical alchemy if linguistic expression is at expense of chemical technologies. It is for this reason that I have attempted to reinterpret the Latin version of the Dialogue of Maria and Aros with emphasis on substances and processes.

Likewise Hermes [said]: Have you heard of the Stone mixed of saffron and white, and [that its] pulverization evinces separation, which whitens the gem's core (κέντρο γεμ; kéntro gem)?

Item Hermes: A Capite lapidem mixtum croceum album, & comminuite evinæ sepa quæ albificat Hendragem?

Then answered Aros: It is as you say, oh Lady but in a long time.
Deinde respondens dixit ei Aros: Ita us dicis est, o domina in tempore longo.

APPENDIX B

Maria answered: Hermes mentioned in all the books of his that Philosophers whiten the Stone in one hour of the day.

Respondit Maria: Hermes dixit in omnibus libris suis, quod Philosophia albificant lapidem in hora diei.

Then answered Aros: Oh how excellent that is!

Dixit ei Aros: O quam illud est excellens.

Said Maria: It is most excellent to the person unaware of it.

Inquit Maria: Excellentissiumum est hoc apud illum qui ignorant.

Aros said: Oh Prophetess, if in the presence of all people [humanity] there are 4 Elements [Tetrasomia; earth, air, water, fire], he [Hermes] said they could be completed, and complexed, and their fumes coagulated, and retained in one day, so long as they follow [preparation].

Dixit Aros: O Prophetissa, si sunt apud homines Omnia 4 elementa, dixit compleri possent, & complexionari, & coagulari eorum fumi, ac retineri die uno, donec impleant consequens.

Maria said: Oh Aros, by God, if your understanding was not firm, you would not hear these words from me, so long as the Lord filled my heart, by grace of divine will.

Dixit Maria: O Aros, per Deum si non essent sensus tui firmi, non audires a me verba haec, donec compleret dominus cor meum, gratia voluntatis diuinae.

Nevertheless: Here is the Recipe –alum of Spain – [i.e.] white gum and red gum – which is Philosophers' sulfur, their Sol [☉; gold] and greater Tincture – and marry gum with gum by a true union. Asked Maria: Do you understand Aros? Certainly Lady [replied Aros].

Veruntamen: Recipe alumen de Hispania, gummi album, & gummi rubeum, quod est kibric Philosophorum, & eorum Sol & tinctura maior, & matrimonisica gummi cum gummi vero matrimonio. Inquit Maria: Intellexisti Aros: utique domina:

Maria said: Make it [the compound] flow like water; vitrify this water [convert it to glass] for a day, the fixed body of two components. Liquefy by the secret of nature in philosophic vessel [crucible]. I don't understand, Oh Lady [replied Aros].

Dixit Maria: facilla sicut aquam currentem, & vitrifica hanc aquam diem laboratam ex duobus zubech super corpus fixum, & liquefac illa per secretum naturaru in vase Philosophiæ. Intellexisti no, o domina.

At any rate, said Maria: Watch the fume carefully that none of it escape, with a gentle fire such as is the measure of the heat of the sun in June or July. Stay near the vessel and carefully observe, how black to red begins in less than three hours of the day, and the fume penetrates the body, the spirit binds [it] and will become like milky wax, melting and penetrating. And that is the secret.

Utique dixit Maria: Custodi sumum, & cave ne fugiat aliquid de illo, & isto mensura igni levi, sicut mensura caliditatis Solis in mense Iunij vel Iulij, & morare prope vas, & intuere cura, quomodo nigrescit & rubescit & albescet, in minus quam in tribus horis diei, & fumus penetrabit corpus, & spiritus constringetur, & erunt sicut lac incerans & liquefiens & penetrans: & illud est secretum.

APPENDIX B

Aros said: Do not say [I can't believe] that it will always be thus.

Dixit Aros: Non dico, quod erit hoc semper.

Maria replied: Aros, there is a more marvelous thing, of which did not exist among the ancients, nor available to them as anything medicinal. And that is: Take the white herb, bright and much honored, occurring on small mounds [volcanic or mineral deposits], grind it fresh in the hour [of its collection], and this is the true body that does not flee from the fire.

Dixit ei Maria: Aros, & mirabilius est, de isto quod non suit apud antiquos, nec accessit ad eum per medicationem. & illud est: Accipe herbam albam, claram, honoratam, crescentem super monticulis, & tere ipsam recentem sicut est in sua hora, & illa est corpus verum non fugiens ab igne.

And Aros said: Is not this the true Stone?

Et dixit Aros: Nunquid ille est lapis veritatis?

And Maria said: Yes truly. But others do not know of this regimen with its quickness.

Et dixit Maria: Utique. Verumtamen nesciunt homines hoc regimen cum velocitate sua.

Aros said: And then what?

Dixit Aros: & postea quid?

Maria said: Make a glass of this sulfur and mercury, and they are the two fumes comprised of two Lights, and project therein the completion [fulfilling ferment] of the Tincture and spirits, according to the true weight, ...

Dixit Maria: Vitrifica super illud Kibrich & Zubech (alias Zibeic) & ipsa sunt duo fumi complectetes duo Lumina, & proijce super illud complementum tincturarum & spirituum & pondera veritatis,

... then pulverize it, and put it into the fire, and you will see marvels brought about by them [the compound].

& tere totum, & pone ad ignem, & videbis de ipsis mirabilia.

... The entire regimen depends on the temperature of fire. Oh how wonderful, how it [the Stone] is moved from colour to color in less than an hour of the day, until it reaches the goal of red and white [white and red according to Bacstrom's interpretation from German]. Extinguish the fire, and let it cool, and open it [the vessel],

Totum regimen consistit in temperie ignis. O quam mirum, quomodo mouebitur de colore in colorem in minus quam in hora diei, quousque veniet ad metam rubedinis & alboria: & deice Ignem, & dimitte confrigidari & aperi,

... and you will find a bright pearly body in the color of wild poppy mixed with white [red and white mentioned previously], and it is waxy, melting and penetrating, and its golden color falls upon one thousand two hundred [can be projected upon or tinge 1,200 parts]. That is the hidden secret. Then Aros kneeled and prostrated.

& inuenies corpus margaritale clarum, esse in colore papaveris syluestris, mixtum albore, & illud est incerans, liquesiens & penetrans, & cadit eius aureus super in millia & ducenta. Illud est secretum occultum. Tum procidit Aros in faciem suam.

APPENDIX B

Maria said to him: Raise your head Aros, because I will abbreviate it for you, take that bright body strewn on mounds (small hills), which is not subject to putrefaction or movement. Here is the Recipe – and grind it with gum Elsaron, and unite the two powders, for gum Elsaron is a seizing body, and grind it all.

Dixitque illi Maria: Leva caput tuum, Aros: quia abbreviabo super te rem, ut illud corpus proiectum super monticulos clarum, quod non capitur putrefaction vel motu. Recipe & tere ipsum cum gummi Elsaron, & cum duobus fumis, quia corpus coprehendisicens est gumm Elsaron, & tere totum.

... Approach therefore [draw near; pay attention] – it all melts. If you project [the husband] upon this wife, it [the liquid] will be as distilling water [water dripping], and when it cools, it coagulates [freezes] and becomes a unified body, and make projection of it [into molten copper], and you will behold wonders.

Apropinqua igitur, quia totum liquescit. Si proieceris super ipsam uxorem, erit ficut aqua distillans, & quando percutiet ipsum aer, congelabitur, & erit corpus unum, & proijce de ipso, & videbis mira:

... Oh Aros that is the hidden secret of the Art [schools; guilds]. And know that the two aforementioned fumes [prepared alchemical reagents] are the roots of this Art, they are the white Sulfur and fused powder, but the fixed body is from the Heart of Saturn which preserves the tincture, and the campus of wisdom or of the Art [schools; guilds].

O Aros, illud est secretum occultum scholiae. Et scias quod praedicti duo fumi sunt radices huius artis, & sunt Kibris album & calx humida, sed corpus fixum est de corde Saturni comprehenificans tincturam, & campos sapientiæ, sive scholia.

... And the Philosophers called it by many and all names [this fixed body], taken from mounds [hills; hillocks] this bright white body. These are the reagents of this Art, some [of which are] prepared and some found on small hills [hillocks]. And know Aros, although sages have not mentioned it by name in any thesis-dissertation, except that the Art cannot be completed unless by it. And in this Art there are nothing but wonders.

& Philosophi nominaverum illa multis & omnibus nomibus, & acceptum de monticulus est corpus clarum, & album. & ista sunt medicina huius artis, pars comparatur, & pars invenitur su per monticulos: & scias Aros, quod sapientes non nominaverunt illud campos sive scholie, nisi quia scholia non complebitur nisi per illud, & in hac scholia non sunt, nisi mirabilia.

Initiation: and in this [work] are 4 stones, and their regimen is true as I have said. And this is primary, the compulsory study [of] Ade and Zethet [Azoth], by allegory [as found in] Hermes' books and Teachings.

Intrant namq; & in illa 4 lapides, & suum regime verum est sicut dixi: & illud est primum, scoyare Ade & Zethet, per illud allegoriza vi Hermes in libris suis Scoyas,

Philosophers have always indicated a long regimen, and have pretended that the work consisted of various things that [in fact] were unnecessary to perform the work. And they [pretend to] take a year to perform the magistery; but only for hiding it from the ignorant, until they can comprehend it, because the Art is not completed except with gold; which is a great and divine secret. And those who hear our secret cannot verify it because of their ignorance. Do you understand Aros?

& semper longi ficaverunt Philosophi suum regimen, & simulaverunt opus pro quodlibet, quid non oportet facere illud opus, & faciunt magistrum in uno anno, & hoc non nisi occultatione ignorantis populi, donec firmatum

APPENDIX B

sit in cordibus corum & in sensib. quia ars non complebitur praeterqua in auro, qui est secretum Dei magnum: & qui audiunt de secretis nostris, non verificant ea propter corum ignorantiam. Intellexisti Aros.

Aros said: Certainly. But tell me about that vessel, without which the work cannot be completed.

Aros dixit: utique, sed narra mihi de isto vase sine quo non complebitur opus.

Maria said: It is the vessel of Hermes, although Stoics [Philosophers] have concealed it, and it is not a vessel used for magic or divination [νεκρομαντία], but is the measure of your fire.

Dixit Maria: Istud est vas Hermetis, quod Stoici occultaverunt, & non est vas nigromanticum, sed est mensura ignis tui.

To which Aros said: Oh Lady you have obeyed the society of Scholars: Oh Prophetess, have you found the secret of the Philosophers, who maintain in their books that one can make the Art from one body [i.e. a single substance]?

Cui dixit Aros: O domina obedisti in societate Scoyari: O Prophetissa an invenisti in secretis Philosophor. qui posuerunt in libris suis quod aliquis possit sacere artem de uno corpore?

And Maria said: Certainly. Hermes did not teach that because the root of our Art [school; guild] is the unbreakable body, venom and a toxin that kills all bodies and pulverizes [powders] them, and its odor solidifies quicksilver.

Et dixit Maria: utique, quod Hermes non docuit, quia radix scholiae est corpus indolabile, insanabile & est toxicum mortificans Omnia corpora, & pulverizat ea, & coagulat Mercurium odore suo.

And she said: I swear to you by eternal God, that when that venom is dissolved until it becomes a subtle [divine] water, no matter how the solution is done, it solidifies Mercury [quicksilver] into Luna [silver] with the power of truth, and falls into the Throne of Jupiter [tin], and forms [silvers] itself into Luna [silver].

Et dixit: Ego iuro tibi per Deum aeternum, quod illud toxicum quando soluitur donec fit aqua subtilis, non curo qua solution siat: coagulat Mercurium in Lunam cum robore veritatis, & incidit in solium Iovis, & lunificat ipsum in Lunam.

... There is a science of all bodies [substances], but the Stoics because of life's shortness and the work's duration have concealed this, and they found tingeing [coloring; potent] elements, and they increased them, and all Philosophers teach those things rather than the vessel of Hermes, because it [the vessel of Hermes] is divine, and, by the Lord's wisdom, hidden from the Gentiles. And those who do not know it are ignorant of the true regimen due to their ignorance of the vessel of Hermes.

Et in omnibus corporibus est scientia, sed Stoici propter eorum vitæ brevitatem & operis loginquitatem hoc occultaverunt, & haec elementa tingentoria invenerunt, & ipsi increverunt ea, & omnes Philosophi docent illa præter vas Hermetis, quia illud est divinum, & de sapientiae domini Gentibus occultatum: & illi qui illud ignorant, nesciunt regimen veritatis, propter vasis Hermetis ignorantiam.

APPENDIX C

Scorpion Formula for Gold-making

The Scorpion Formula appears in the *Codex Parsinus Græcus* and is a pedagogical template-outline preserved by Zosimus for creating alchemical gold, silver and purple-patina bronze via antimony, copper, silver or gold and the Tincture. These four ingredients actually follow the canonical color pattern found in the writings of Zosimus – Black (copper oxide), White ('silvered' copper), Yellow (alchemical gold) and Red / Purple (patinated bronze). The color scheme was important to any Greek school of thought and was not necessarily unique to alchemy. It can become confusing when these terms are encountered as substances, again in the confection of the Philosophers' Stone and once again in the process of creating alchemical gold, silver and colored patinas. However, these are not mutually exclusive and actually suggest a sense of coherence rather than division. Calcining metals or ingredient-preparation produced these color variations. Confecting the Philosophers' Stone at the next level of complexity produced these colorations, as did the highest level of complexity – artisanally creating alchemical gold, silver and patinated bronzes. The color-regimen can be better understood as fractal reiterations occurring at all levels throughout the Art of Alexandrian *chrysopœia*.

The *Formula of the Scorpion* when understood in its entirety can be easily interpreted as being three overlapping recipes, or alternately a single 3-dimensional recipe. The overlap occurs with levels of complexity. For example, if the artisan intended to create alchemical silver, step 6 indicates completion of this stage by the addition of silver. The next level

APPENDIX C

of complexity is to create alchemical gold in which case step 6 indicates a second fusion of antimonial bronze as a purifying step with step 9 indicating completion of the process for creating alchemical gold. The highest level of complexity in this or any form of bronze-craft is the patina stage, in which case the bronze will be surface-treated to effect a change in coloration. Various shades of patina arise from the corrosion of any surface silver or copper present. At this level of complexity, step 13 or undefined 14 is the final step.

If this overlap theory is correct, it suggests a tremendous level of forethought and complexity on the part of Alexandrian *chrysopœians*. Alloying metals can be problematic even today, whereas these *chrysopœians* appear to have minimized any chance of error and maximized exemplary output by using purified and reduced metal-oxide *powders of projection*. The creation of the Philosophers' Stone or Tincture at first might appear to be an overly elaborate, redundant and unnecessary procedure from the perspective of modern alloying, yet it is an example of a chemical technology that offered a strong level of reproducibility and consistent quality during antiquity. The Scorpion Formula/e is a fantastic iconic artifact of authentic Alexandrian gold making in its original form.

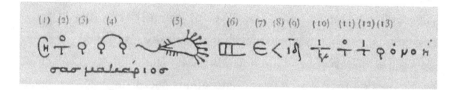

σασ μαμάρ [μαρμαίρω] ⸂ασ; sas̱ mamár [marmaíro̱] ios̱
(your mother [glittering] tincture)

1. σημεῖον or σημείωσα⸂παῖ; si̱meíon or si̱meío̱sai paí
   (take note – or – begin here child[ren])

The symbol is a stylized ouroboros with the head and tail connected by the letter H, possibly a connection to Hermes if it was written as Hρμής instead of the modern Ερμής. From a Judeo-Egyptian perspective, it might refer to the Ugaritic Canaanite word for gold – ḫrṣ (Biblical Hebrew; hārûs הרס) or Egyptian Horus. Combining Horus (gold) + -boros (resource) results in the Ouroboros or serpent biting its tail so prevalent in alchemical imagery and literature and is symbolic of the gold-based red Tincture of the Philosophers. If this interpretation is correct, the first step in alchemical gold, silver and purple manufacture was to create the Tincture. This was known later in European alchemy as the *Tincture of the Philosophers* and the *Lesser* or *Specific Stone*.

2. τὸ πᾶν; to pan
   (the *All*; The white stone; the composition)

   Cleopatra, Synesius and Stephanos each referred to the Hermetic White Stone by the cover-term *the All – comprehensive magnesia*. The sequence suggests that step 2 is a preparatory stage and refers to an essential ingredient – purified antimony in some form. The *All* as antimony can be understood in general or specific terms. This step might simply refer in general terms to calcined antimony as *flowers of antimony*. In the specific sense used by Cleopatra, Synesius and Stephanos, it clearly indicated the Hermetic White Stone, known to European alchemists as the *Stone of the Philosophers* and the *Great or Universal Stone*. Steps one and two may refer to either stone sequentially or to a single stone. The meaning is unclear. The recipe can be dated no earlier than Zosimus, yet in that period, there is evidence for the existence of these two variants of the Philosophers' Stone. It is possible that they were used together, each with a specific role to play.

APPENDIX C

3. χαλκοῦ ἰός; chalkoú iós
   (copper-calx – or – tinged copper)

   If the first numbers in the sequence represent an ingredient list, *chalkoú iós* indicates the copper ingredient in the form of cuprous oxide or cupric oxide via roasting;

   $$4\,Cu + O_2 \rightarrow 2\,Cu_2O \;-\;or\;-\; 2\,Cu + O_2 \rightarrow 2\,CuO$$

   ...or as is more likely, roasting natural copper carbonate in the form of malachite or azurite mined in the Sinai at Serabít el-Khâdim and Timna by a calcination process known as *blackening the stone*. These minerals decompose at 290 °C, giving off carbon dioxide to yield jet-black cupric oxide powder:

   $$CuCO_3 \rightarrow CuO + CO_2$$

   If however the beginning of the sequence represents sequential steps, step three could refer to tingeing copper via the product in step 2, resulting in antimonial bronze of the following step.

   Blackening is a *chrysopœian* term that refers to blackened-copper calx – a stage known as *melanosis* in Greek and *nigredo* in Latin.

4. μολυβδοχαλκος καίε▨καμμένο; molyvdochalkos kaíei kamméno
   (antimonial-bronze alloyed calxes)

   This stage is a clear indication that the list actually represents sequential steps for creating alchemical silver, gold and their patinas. The Greek term *kaíei kamméno* literally means *burning the burnt*, which can be better understood as *alloying* (burning) *the calcined* (burnt) antimony and copper powders. Here the antimony and copper ingredients of the previous steps were combined in a crucible and alloyed to create the first of two fusions of antimonial bronze. Lead-copper (*molyvdochalkos*) here refers to unrefined antimonial bronze – antimony being the chemical identity of philosophers' lead. This process likely included charcoal and a fluxing salt.

5. αργυρωχαλκος καίει▢καμμένο κα▢πεπηγμένος; argyrochalkos kaíei kamméno kai pepigménos
   (silvered-copper [antimonial bronze] twice-molten [and now] fixed)

This symbol represents Scorpio, the final astrological sign. It is meant to indicate the final stage of preparing the base metal for transmutation – i.e. to receive its venom. The three lines radiating on one side represent the three potential alchemical products – silver, gold or purple. The four radiating lines on the other side represent the tetrasomia of the *All*. The eight claws may perhaps indicate the first eight steps of this process – an emblematic encryption of *chrysopœia* in condensed form.

In this step, the copper is fully fixed (alloyed) with antimony. The term *silver* is an adjective likely meant to indicate that the alloy has reached a higher commercial value, i.e. trade-jargon to differentiate a more refined double-fused form of antimonial bronze (silver-copper) from the single-fused product (lead-copper) of the previous step. The final word implies fixation and links step 5 to step 6 sequentially.

Whitening is a *chrysopœian* term that refers to *silver-copper* – a stage known as *leukosis* in Greek and *albedo* in Latin.

6. πεπηγμένος; pepigménos
   (fixed, solidified; lit. "coagulated")

This step is merely a statement that the antimonial bronze has been brought to "fixation" via two fusions and thus it has been purified or exalted.

APPENDIX C

7. ἐμέρ☐τος (επίτ☐μος); emeritos (epítimos)
   (important or impressive but without power)

   The symbol for step 7 indicates that there are 3 options available to the artisan at this stage – alchemical silver, gold or purple.

   This term is uncertain and hints of Latin, which would not be unusual during the Roman period of Alexandria. If it is a Greek transliteration of the Latin *emeritos* (vetran, retire) or *emeritus* (earned, gained, deserved, veteran, retired). The Greek equivalent is *epítimos*, meaning honorary, possibly implying that although the metal has been exalted, it is still impotent in that it remains to be tinged or patinated, which is to say transmuted into alchemical gold or hepatizon.

8. δραγμαὶ (δραχμή) – or – δράκμαϊ; dragmaí (drachmí) – or – drák maï
   (Drachma – or – dragon of May; i.e. Tincture)

   The symbol for step 8 indicates that there are two remaining options available to the artisan at this stage – 1) *alchemical gold,* or 2) *hepatizon*.

   The Greek appears to suggest drachma[i], indicating a specific weight of an unnamed substance, in which case it might be interpreted as *"...now go get your scale and weights at this point"*. Either Greek *drachmí* is horribly misspelled or it is an example of typical alchemical encryption via wordplay to indicate the Dragon (serpent) of May. This is in keeping with the alchemists' admission that theirs was *"An art, purporting to relate to the transmutation of metals, and described in a terminology at once Physical and Mystical"*.

   In this scenario, the serpent is the Tincture that began in Aries, the sign of antimony, with May representing the following sign of Taurus, the Tau symbol being replete with symbolic meaning and imagery. The insinuation is to the serpent or scorpion envenomating the bull of antimonial bronze. In either case, this appears to be the step in which a measured amount of something, likely Tincture, is included into the prepared and fixed base metal composition – refined antimonial bronze known as silvered-copper (argyrochalkos; αργυρωχαλκος).

9. δεκατεσσάρων; dekatessáron
   (fourteen; 14)

This is may refer to the quantity of the substance referred to in step 8 above, yet without exact ratios it becomes impossible to determine the final product. Since there are 13 steps in this process, the number 14 may also indicate that this is the end of the process for transmuting copper to alchemical gold via the Tincture. In this scenario, number 14 would indicate, *"If you are after alchemical gold, this is the 14th step. It is finished. Happy is he who understands this. If you want to create a patina, continue to the following steps"*.

Another interesting possibility linking the number 14 to Egyptian culture was presented by Maspero:

> *The Nile must be considered high enough to submerge the land adequately before it is set free. The ancient Egyptians measured its height by cubits of twenty-one and a quarter inches.* At fourteen cubits, they pronounced it an excellent Nile; *below thirteen, or above fifteen, it was accounted insufficient or excessive, and in either case meant famine, and perhaps pestilence at hand.*
>
> ... More or less authentic traditions assert that the prelude to the opening of the canals, in the time of the Pharaohs, was the solemn casting to the waters of a young girl decked as for her bridal—the "Bride of the Nile." – Maspero, A History of Egypt, Vol. I

APPENDIX C

In this scenario, the number 14 was eagerly awaited by all Eygyptians as the number indicating perfection in which an offering or sacrifice was cast into the waters. The number 14 may have been a cultural slang term popular with Egyptians indicating *perfection* in a popular sense that made its way into *Corinthian bronze* making as artisanal trade-jargon.

This step in the process of alchemical gold making according to the Scorpion Formula coincides with the step in which the *Tincture* or synonymously the *Powder of Projection* is cast into perfectly prepared antimonial bronze as a ferment as the final step in creating alchemical gold – also known as the Yellowing stage of the work.

Yellowing is a *chrysopœian* term that refers to alchemical gold – a stage known as *xanthosis* in Greek and *citrinitas* in Latin.

10. τῖτανος χαλκὸς τὸ πᾶν ὄστρακον; titanos chalkós tó pán óstrakon (copper corrosion [is the] test [of] the composition)

The word *titanos* means *calx, lime* or *plaster* in primitive Greek, which in combination with copper suggests surface copper corrosion. The Greek word *óstrakon* can mean *shell* or *shellfish* implying a casing, yet also means *"a test"*. In context of later stages of creating alchemical gold, one of the tests performed on the newly transmuted metal was to see if it was without shadow or blemish, meaning testing for corrosion resistance.

Another valid interpretation of this passage and symbol might perhaps suggest corroding the copper via a shell or plaster in order to create a patina such as the liver-purple colored patina known as *hepatizon* deemed the most valuable *Corinthian bronze* by Pliny. In either case, a corrosion test is indicated for this step.

SCORPION FORMULA FOR GOLD-MAKING

11. **το παν όστρακον; to pan óstrakon**
    (the composition tests [positive])

    This step is a confirmation that the previous process was successful, which is to say that whether it was a test of corrosion-resistance or corrosion-capability, in either case the product tests positive.

12. **τίτανος; titanos**
    (corrosion [patina])

    This step confirms that steps 9-11 are the steps known in Alexandrian *chrysopœia* as *Iosis*. The Greek word *Iosis* is loaded with symbolic meaning, such as envenomated, medicated, purpled or tinged. The symbolism suggests covering Royalty with the Royal-purple Robe – the King's purple vestment. The royalty in this case is the newly adorned alchemical gold or silver covered or tinged in the form of a liver-purple patina, a product of immense value in antiquity – via a finishing process known as *Iosis*, meaning *purpling* or *reddening*, *tingeing* or *envenomating*. Color possibilities begin with pink hues and extend to deep purples leading to blackish liver color depending on the composition of the bronze and the corrosive being used. There is strong evidence to suggest that Maria relied on arsenic sulphide ores such as orpiment and realgar to affect her patinas.

    Purpling or Reddening is a *chrysopœian* term that refers to creating a black, purple, red or pinkish patinas – a stage known as *Iosis* in Greek and *rubedo* in Latin.

13. **χαλκο; chalko**
    ([this] from copper)

    This is not really a step as much as it is a declaration, which might be understood as the following. It is intriguing to consider that this may have been what Maria shrieked:

APPENDIX C

"[And all of this] from copper!"

ὁ νοήσας [νόησης] μακάριος μακαριος; o nóisis makarios
(the being who understands is blessed / blissful)

"Blessed is he who understands."

ALCHEMICAL IMAGERY

Page number:

11. *Purification of Gold by Stibnite* © Lauren Khu 2009
23. *Stibnite and Flowers of Antimony* © Lauren Khu 2009
34. *Sal Ammoniac – Eagle Salt* © Lauren Khu 2009
46. *The Rebis – Gold-Antimony Glass* © Lauren Khu 2009
57. *Griffin's Egg* © Lauren Khu 2009

Images that follow are copyright-free and in the public domain:

63. *Philosophical Lamp Furnace* (3 wicks) – From LeFebre's *Compleat Body of Chemistry*, 1670
67. *Alchemist's Laboratory* from Heinrich Khunrath's *Amphitheatrum sapientiae aeternae*, 1595 by Peter van der Doort
112. *Moses and the Brass Serpent* – from Illustrators of *Bible Pictures and What They Teach Us* by Charles Foster, 1897 – Moses is standing up and pointing to a great snake, or serpent. The serpent is made of brass/bronze. It is fastened to the top of a pole in the shape of a Tav or Tau symbol. Moses is telling the people to look at it. The brazen serpent, interpreted here as antimonial bronze, healed the Israelites and strengthened the developing nation.
164. *Maria Hebrea* – from Michael Maier's *Symbola Aureae Mensae Duodecium Nationum*, 1617 – She points to the two fumes of two lights interpreted here as gold and antimony, separating from a single source and recombining in the context of above and below. The fumes are joined via the herb found on small hills identified as sal ammoniac.

171. ***Chrysopœia of Cleopatra*** fragments dating from 1st to 3rd centuries CE – from the Venice *Codex Marcianus Græcus* MS 299 (7th to 9th centuries) fol. 188v – Original fragments of sketch drawings for a template that encrypts the processes of distillation and evaporation to confect the Philosophers' Stone.

178. ***Distillation apparatus of Cleopatra*** fragments preserved by Zosimus – from the *Codex Parisinus Græcus* BNF 2327 (15th century) – Sketch drawings for a template that encrypts the process of distillation and evaporation to confect the Philosophers' Stone.

182. ***Distillation apparatus of Zosimos*** – from the *Codex Parisinus Græcus* BNF 2327, fol. 81 – Sketch drawings for a template that encrypts the process of distillation and evaporation to confect the Philosophers' Stone. Equipment from images 3, 4 and 5 match descriptions of equipment used in Zosiumus' *On the Evaporation of the Divine Water that fixes Mercury*.

189. ***Distillation apparatus of Synesius*** – from the *Codex Parisinus Græcus* BNF 2327, fol. 33 and 2325 (13th century) – With receiver and nearly identical to the images of stills occurring in the Chrysopœia of Cleopatra and writings of Zosimus.

213. ***Hermes the Egyptian*** – from Michael Maier's *Symbola Aureae Mensae Duodecium Nationum*, 1617 – The astrolabe represents the heavens, the landscape the earth: *"That which is above is as that which is below,"* etc. Sol and Luna, the Sulfur and Mercury principles interpreted here as sal ammoniac and antimony are brought into combination by the agency of fire, the product of which is the *secret fire* of the philosophers, which is to say the *divine water that burns*.

239. ***The Three Primes*** – from Michael Maier's *Tripus Aureus*, 1618 – The Tria-prima is represented as a crowned lion (gold), serpent (antimony) and eagle/bird of Hermes (butter of antimony).

255. ***Man as Microcosm*** – from Robert Fludd's *Utriusque Cosmi Historia*, 1617 to 1621 – Illustration (adaptation) of man as microcosm within the universal macrocosm.

296. ***Morienus and Khālid*** – from Michael Maier's *Symbola Aureae Mensae Duodecim Nationum*, 1617 – Morienus (left) explaining to Khālid (right) that *"... whosoever shall seeke any other thinge than this storn for this Magistery shall be likened unto a Man that endeavoreth to clime a Ladder without steppes, which thing he being unable to doe, he falleth to the Earth on his face..."*.

315. ***Lion Devouring an Eagle/Bird of Hermes*** – from Basil Valentine's *Azoth ou le moyen de faire l'oder Cache des Philosophes* (Paris), 1659 – Fourth woodcut depicting the Philosophers' Stone as sol and luna being united in the form of a serpentine Lion Devouring an Eagle/Bird of Hermes, in the spirit of Paracelsus' union of *lion's blood and eagle's gluten*.

370. ***Winged and Wingless Serpent Ouroboros*** – from the *Book of Abraham the Jew* published in Germany, 1774. This is the first engraving that indicates *whitening the stone* by removing the wings of the serpent (stibnite) resulting in *flowers of antimony* suggested by the flower pointing upward to the wingless serpent and downward to the earth. The tree in the background is perhaps the tree of knowledge.

373. ***Green Lion Devouring the Sun*** – from the St. Gallen colored version of the *Rosarium Philosophorum*, 17th century – The 18th image features a colored drawing of a Green Lion Devouring the Sun. The associated text reads: *"Of Our Mercury which is the Green Lion Devouring the Sun"* and *"I am the true green and Golden Lion without cares, in me all the secrets of the Philosophers are hidden."* Our Mercury is interpreted here to mean refined antimony, whereas *the true green and Golden lion* refers to *gold-antimony glass* as suggested by Morienus and Sir Isaac Newton.

385. ***Il Clavis*** – from Basil Valentine's 12 Keys as engraved by Matthaeus Merian (1593–1650), and published in the collection "Musaeum hermeticum", Francofurti: Apud Hermannum à Sande, 1678 – *The Second Key* is an image depicting sal ammoniac (right figure) with a

603

eagle/bird of Hermes on his sword being distilled with flowers of antimony (left figure) depicted by a crowned wingless serpent on his sword. The result is Mercurius Duplicatus – Living Doubled Sophic Mercury interpreted here as butter of antimony.

391. **Distillation apparatus of Synesius** – from the Codex Parisinus Græcus BNF 2327, fol. 33 and 2325 (13th century) – With receiver and nearly identical to the images of stills preserved in the Chrysopœia of Cleopatra and writings of Zosimus.

474. **Scorpion Formula/e for Gold-Making** – from the Codex Marcianus Græcus folio 193, fragment (cropped) preserved by Zosimus c. 300 CE – Operative template encrypts the chrysopœtic process for creating alchemical silver, gold and patinated bronzes.

543. **Lapis Sophorum, Medicina Tertii Ordinis** – from the Cabala Mineralis illustrated manuscript anonymously authored during the 17th century – An image of the finished Philosophers' Stone in a vessel, with Latin inscriptions that read "Stone of the Philosophers, Medicine of the 3rd Order". The recipe encrypted in this illustrated manuscript closely follows the work described by Morienus.

575. **Synesius' Ouroboros** – from the Codex Parisinus Græcus BNF 2327, fol. 279 – This and another ouroboros located earlier in the manuscript are similar, but not identical. Each features three spiky scales on the head (tria-prima) and four feet (tetrasomia). This version has more detailed scales and features only two concentric rings (gold and red) whereas the other has three rings. The ouroboros represents antimony and the finished elixir.

578. **Scorpion Formula/e for Gold-Making** – from the Codex Marcianus Græcus folio 193, fragment (complete) preserved by Zosimus c. 300 CE – Operative template encrypts the chrysopœtic process for creating alchemical silver, gold and patinated bronzes. For a detailed interpretation and explanation of this process, see Appendix B.

586. **Armadillo girdled lizard** – Image of live specimen Ouroborus cataphractus endemic to desert areas of southern Africa. It possesses

an uncommon predator-defense adaptation – when frightened, it takes its tail in its mouth and rolls into a ball. It is hypothesized that this African lizard may have served as the archetypal model for the various 4-legged Ouroboros images preserved in alchemical texts. Incidentally, it has a yellow abdomen just as depicted in alchemical imagery.

AFTERWORD

The research, discoveries, and writing of this book took place under quite unique circumstances, circumstances most would deem as inconducive to writing and researching any historical work. Situated on a tiny coconut island in southern Thailand, Erik was without libraries, learning institutions, research grants or academic collaboration. Moreover, he was faced with unreliable electricity and Internet connection, yet the most notable hurdle was his lack of a formal academic education. Although this was initially thought of as a significant disadvantage, it became increasingly evident that it was not only an actual advantage, but also perhaps a necessity in uncovering the mysteries behind the Philosopher's Stone.

So how did Erik uncover the Stone's riddles? Although most authorities on Alchemy are either professional researchers or scholarly academics with numerous degrees and publications to their names, they tend to be rather specialized. Throughout Erik's work, it became increasingly apparent that solving the riddle of the Philosophers Stone required a generalist, a non-specialist and without an allegiance to any one school of thought. In other words, to decipher the stone, one should share the qualities of a genuine alchemist. Alchemists required a number of talents such as sourcing of raw materials, producing basic reagents, metallurgical skills, furnace and lab-equipment design and implementation, the literary and artistic ability to write and publish alchemical texts, Latin and other language skills, socio-political networking and the shrewdness to aquire a patron (Erik helped me write that sentence).

Daring and stubborn, their rebellious nature in opposing the status quo and undertaking forbidden endeavors required a person of special character and worldly views. Erik is a self-taught polymath, a generalist, a jack-of-all-trades, a non-academic with no peers and no authority to adhere to, with complete freedom to follow any red thread with only truth as intention.

There is an unexpected lesson to be learnt from Erik's inspiring journey: a perceived shortcoming to most can be an unexpected advantage to those who recognize it. When it comes to great and unexpected feats, underdogs may prevail if they are able to see the world differently from their peers. Erik's work demonstrates just this since, despite his lack of a higher education, he was able to accomplish what many educated critics believed improbable. This is a lesson to any struggling misfit, battling the status quo as the scrappy underdog: A disadvantage can be your strongest ally for success if you can see the world differently from those around you.

Victor Hörnfeldt

be obtained